DEMYSTIFYING SUSTAINABLE FARMING USING SMART TECHNOLOGIES

DEMYSTIFYING SUSTAINABLE FARMING USING SMART TECHNOLOGIES

Gunjan Chhabra, PhD
Keshav Kaushik, PhD
Priya Chugh, PhD
Susheela Dahiya, PhD
Aryan Chaudhary
Editors

AAP | APPLE ACADEMIC PRESS

First edition published 2026

Apple Academic Press Inc.
1265 Goldenrod Circle, NE,
Palm Bay, FL 32905 USA

760 Laurentian Drive, Unit 19,
Burlington, ON L7N 0A4, Canada

CRC Press
2385 NW Executive Center Drive,
Suite 320, Boca Raton FL 33431

4 Park Square, Milton Park,
Abingdon, Oxon, OX14 4RN, UK

Library and Archives Canada Cataloguing in Publication

CIP data on file with Canada Library and Archives

Library of Congress Cataloging-in-Publication Data

CIP data on file with US Library of Congress

ISBN: 978-1-77964-092-5 (hbk)
ISBN: 978-1-77964-093-2 (ebk)
DOI: 10.1201/9781779640932

Dedication

To my family, thank you for encouraging me in all my pursuits and inspiring me during this journey. I am especially grateful to my parents and wife, who supported me emotionally. I always knew that you believed in me and wanted the best for me.

—Dr Gunjan Chhabra

This book is dedicated to my beloved parents—Sh. Vijay Kaushik, Smt. Saroj Kaushik, wife, Priyanka, daughter, Kashvi, and son, Harshiv. May God always bless us, Har Har Mahadev!!

—Keshav Kaushik

This book is dedicated to my children, Daksh and Ranvijay, who believe in me and continue to inspire me to do my best.

—Susheela Dahiya

This book is dedicated to my family.

—Aryan Chaudhary

About the Editors

Gunjan Chhabra, PhD
Associate Professor, Department of Computer Science and Engineering, School of Science and Technology, Swami Rama Himalayan University, Dehradun, Uttarakhand, India

Gunjan Chhabra, PhD, is an accomplished academician, researcher, and innovator with over a decade of extensive experience in the field of computer science and engineering. Currently working as an Associate Professor at the School of Science and Technology, at , Swami Rama Himalayan University (SRHU), Dehradun, India, he has consistently demonstrated excellence in teaching, research, and academic leadership. His academic journey is marked by a PhD in Computer Science, complemented by a strong foundation in software systems, intelligent technologies, and emerging areas like artificial intelligence (AI), Internet of Things (IoT), and human-computer interaction (HCI). Dr. Chhabra's research interests span interdisciplinary domains, including healthcare technologies, biofield analysis, smart systems, and ethical AI. He has published numerous research articles in reputed conferences and journals (SCI, Scopus), authored multiple books, and served as editor for several international publications. His contributions are recognized globally, with invitations to serve as a reviewer, editorial board member, and guest editor for prestigious journals and conferences. A passionate innovator, Dr. Chhabra has been granted eight patents, including recent innovations focused on emotion-biofield analysis and IoT-based healthcare applications. He actively mentors undergraduate, postgraduate, and doctoral students, guiding them on real-world projects that have led to patents, publications, and product development. His commitment to student growth and academic excellence is evident through his role as a research coordinator and his involvement in curriculum design, academic auditing, and outreach activities. Beyond academia, Dr. Chhabra is involved in incubation and consultancy projects, aiming to build bridges between research, industry, and societal needs. His vision includes establishing state-of-the-art labs, fostering global collaborations, and leveraging AI and wellness technologies for impactful solutions. Known for his discipline, integrity, and forward-thinking mindset, Dr. Chhabra continues to contribute meaningfully to both national and international academic communities.

Keshav Kaushik, PhD
Center for Cyber Security and Cryptology, Sharda School of Computing Science & Engineering, Sharda University, Greater Noida, Uttar Pradesh, India

Keshav Kaushik, PhD, is an accomplished academician and expert in cybersecurity and digital forensics. He serves as an Associate Professor in the Center for Cyber Security and Cryptology, Sharda School of Computing Science & Engineering, Sharda University, Greater Noida, India. As a key member of the Cybersecurity Centre of Excellence and Vice Chairperson of the Meerut ACM Professional Chapter, he has significantly advanced the field through teaching, research, and leadership. Recognized among the world's top 2% scientists by Stanford University and Elsevier in 2024, his scholarly portfolio includes over 150 publications, with 25 in SCI/SCIE/Scopus-indexed journals, and over 50 in Scopus-indexed conferences. He has authored 40 books, contributed 25 book chapters, and holds one granted patent, six published patents, and five copyrights. Dr. Kaushik is a Certified Ethical Hacker (CEH v11), ISO/IEC 27001 Lead Auditor, and IBM Cybersecurity Analyst. A Bentham Ambassador and Guest Editor for the IEEE Journal of Biomedical and Health Informatics, he has delivered over 50 talks globally and earned accolades, including recognition from The Indian Space Research Organisation (ISRO), the *All India Council for Technical Education (AICTE),* and Uttarakhand Police. A two-time GATE qualifier, he has a rich academic and professional trajectory marked by impactful mentorship and innovative research.

Priya Chugh, PhD
Assistant Professor, School of Agriculture and Technology, Maya Devi University, Selaqui, Dehradun, Uttarakhand, India

Priya Chugh, PhD, is an accomplished academician and researcher with expertise in plant physiology and biochemistry. She earned her PhD from Punjab Agricultural University, Ludhiana, India, and has over five years of experience in both teaching and research. Her academic contributions include more than 20 research papers and nine book chapters, reflecting her commitment to advancing scientific knowledge. With a strong interest in interdisciplinary research, she actively explores innovative approaches that foster creativity and generate impactful insights. Dr. Priya has participated in numerous national and international conferences, sharing her expertise with the global academic community. Beyond research, she has played a pivotal role in institutional development, contributing to curriculum design, organizational setup, and student mentoring. Through these efforts, she has significantly contributed to the growth and quality of higher education.

Susheela Dahiya, PhD
Professor, Department of Computer Science and Engineering, Graphic Era Hill University, Dehradun, Uttarakhand, India

Susheela Dahiya, PhD, is a Professor in the Department of Computer Science and Engineering at Graphic Era Hill University, Dehradun, Uttarakhand, India. She earned her M.Tech. in Computer Science and Engineering in 2008 and completed her Ph.D. in 2015 from the Indian Institute of Technology Roorkee, Uttarakhand. She has qualified both GATE and NET in Computer Science.

With over eleven years of combined experience in academia, research, and industry, Dr. Dahiya has made notable contributions to higher education. In addition to her research activities, she is actively involved in curriculum development, student mentoring, and accreditation-related processes that support institutional growth and quality enhancement. Her research interests include image and video processing, the Internet of Things (IoT), cybersecurity, cloud computing, and deep learning–enabled computational techniques across interdisciplinary research areas. She has published extensively in reputed conferences as well as Scopus- and SCI-indexed journals and has also edited scholarly books.

Aryan Chaudhary
Chief Scientific Advisor at BioTech Sphere Research, India

Aryan Chaudhary is the Chief Scientific Advisor at BioTech Sphere Research, India. He continues to make groundbreaking contributions to the industry. Having served as the Research Head at Nijji HealthCare Pvt. Ltd, he has demonstrated his expertise in leveraging revolutionary technologies such as artificial intelligence, deep learning, IoT, cognitive technology, and blockchain to revolutionize the healthcare landscape. His relentless pursuit of excellence and innovation has earned him recognition as a thought leader in the industry. His dedication to advancing healthcare is evident through his vast body of work. He has authored several influential academic papers on public health and digital health, published in prestigious international journals. His research primarily focuses on integrating IoT and sensor technology for efficient data collection through one-time and ambulatory monitoring. As a testament to his expertise and leadership, he is not only a keynote speaker at numerous international and national conferences but also serves as a series editor of a CRC book series and is the editor of several books on biomedical science. His commitment to the advancement of scientific knowledge extends further, as he acts as a guest editor for special issues in renowned journals. Recognized for his significant contributions, he has received prestigious accolades, including the "Most Inspiring Young Leader in Healthtech Space 2022" by Business Connect and the title of the best project leader at Global Education and Corporate Leadership. Moreover, he holds senior memberships in various international science associations, reflecting his influence and impact in the field. Adding to his accomplishments, Aryan Chaudhary is currently serving as a guest editor for a special issue in the highly regarded journal, *EAI Endorsed Transactions on AI and Robotics,* and he has joined the editorial board of *Biomedical Science and Clinical Research* (BSCR). Additionally, he is a respected professional member of the Association for Computing Machinery (ACM). He has been recently elected as Chair for an ACM professional chapter.

Contents

Contributors

Annu
Department of Post Harvest Process and Food Engineering, Govind Ballabh Pant University of Agriculture and Technology, Pantnagar, Uttarakhand, India

Arun Agarwal
Ramanujan College, Delhi University, Delhi, India

Ambika Aggarwal
School of Computer Science, UPES, Dehradun, Uttarakhand, India

M. M. Ahmad
Department of Agricultural Economics and Extension, BUK, Kano, Nigeria

Nitin Arora
Department of Computer Science, Thapar Institute of Engineering and Technology, Patiala, Punjab, India

M. Babawachiko
MCA- Google Apps, Suresh Gyan Vihar University, Jaipur, Rajasthan, India

Rakesh Kumar Bhadula
Department of Mathematics, Graphic Era Hill University, Dehradun, Uttarakhand, India

Akansha Bhandari
Graphic Era (deemed to be) University, Dehradun, Uttarakhand, India

Neha Bhardwaj
Department of Mathematics, Sharda School of Basic Sciences & Research, Sharda University, Greater Noida, Uttar Pradesh, India

Rajneesh Bhardwaj
School of Agriculture, Graphic Era Hill University, Dehradun, Uttarakhand, India

Leela Bhatt
Department of Genetics and Plant Breeding, G. B. Pant University of Agriculture and Technology, Pantnagar, Uttarakhand, India

Ashutosh Bhatt
Department of Computer Science Engineering, School of Science and Technology, Swami Rama Himalayan University, Dehradun, Uttarakhand, India

Pallavi Bhatt
Graphic Era Hill University Dehradun, Uttarakhand, India

Sayantika Bhattacharya
School of Agriculture, Graphic Era Hill University, Dehradun, Uttarakhand, India

Anchit Bijalwan
School of Computing and Innovative Technologies, British University Vietnam, Hanoi, Vietnam

Saurabh Chandra
School of Law, Bennett University, Greater Noida, Uttar Pradesh, India

Arun Kumar Chaudhary
Industrial and Production Engineering Department, G. B. Pant University of Agriculture and Technology, Pantnagar, Uttarakhand, India

Aryan Chaudhary
BioTech Sphere Research, Kolkata, West Bengal, India

Nikita Chauhan
Graphic Era Hill University, Dehradun, Uttarakhand, India

Shiv Vendra Singh Chauhan
Department of Agronomy, Rani Lakshmi Bai Central Agricultural University, Jhansi, Uttar Pradesh, India

Sachi Choudhary
School of Computer Science, UPES Dehradun, Uttarakhand, India

Priya Chugh
School of Agriculture and Technology, Maya Devi University, Dehradun, Uttarakhand, India

Priya Gond
Department of Post Harvest Process and Food Engineering, Govind Ballabh Pant University of Agriculture and Technology, Pantnagar, Uttarakhand, India

Deepak Gupta
Mechanical Engineering Department, Graphic Era Hill University, Dehradun, Uttarakhand, India;
Mechanical Engineering Department, Graphic Era (deemed to be) University, Dehradun, Uttarakhand, India

Supriya Gupta
Graphic Era Hill University Dehradun, Uttarakhand, India

Khushboo Jain
UPES, Dehradun, Uttarakhand, India

Richa Jaswal
Department of Soil Science, CSKHPKV, Palampur, Himachal Pradesh, India

Pallavi Joshi
Graphic Era Hill University, Dehradun, Uttarakhand, India

Christian Kaunert
Dublin City University, Dublin, Ireland & University of South Wales, Wales, UK

Deepak Kholiya
School of Agriculture, Graphic Era Hill University, Dehradun, Uttarakhand, India

Subodh Kumar
Ramanujan College, Delhi University, Delhi, India

Mamta Martolia
Department of Cyber Security, COER University, Roorkee, Uttarakhand, India

Bhavana Mathur
Department of Mechanical Engineering, Anand International College of Engineering, Jaipur, Rajasthan, India

Makkhan Lal Meena
Mechanical Engineering Department, Malaviya National Institute of Technology, Jaipur, Rajasthan, India

Amit Kumar Mishra
Department of CS&E, Graphic Era Hill University, Dehradun, Uttarakhand, India

Dipesh Kumar Mishra
Mechanical Engineering Department, Madan Mohan Malaviya University of Technology, Gorakhpur, Uttar Pradesh, India

Somya Misra
Graphic Era Hill University Dehradun, Uttarakhand, India

M. K. Nautiyal
Graphic Era Hill University, Dehradun, Uttarakhand, India

I. K. Nazifi
Information Technology Unit, Federal Inland Revenue Service, Nigeria

Vivek Kumar Pathak
Graphic Era Hill University Dehradun, Uttarakhand, India

Manpreet Singh Preet
Graphic Era Hill University Dehradun, Uttarakhand, India

Kamlesh Kumar Raghuvanshi
Ramanujan College, Delhi University, Delhi, India

Pankaj Rautela
G.B. Pant University of Agriculture and Technology, Pantnagar, Uttarakhand, India

Muskan Rawat
Graphic Era (deemed to be) University, Dehradun, Uttarakhand, India

Anupama Rawat
Graphic Era Hill University Dehradun, Uttarakhand, India

M. S. Sadiq
Department of Agricultural Economics and Extension, FUD, Dutse, Nigeria

Rohit Sahu
Department of Mechanical Engineering, GL Bajaj Institute of Technology and Management, Greater Noida, Uttar Pradesh, India

Madhvendra Saxena
Department of Mechanical Engineering, Roorkee Institute of Technology, Roorkee, Uttarakhand, India

Ishu Sharma
School of Engineering and Technology, CGC University, Mohali, Punjab, India

Jagdeep Sharma
School of Management and Commerce, RIMT University, Mandi Gobindgarh, Punjab, India & School of Management Studies, CGC University, Mohali, Punjab, India

Subhash C. Sharma
Electronics & Computer Discipline, Indian Institute of Technology, Roorkee, Uttarakhand, India

Prashant Shukla
School of Advanced Engineering, UPES, Dehradun, Uttarakhand, India

Vriksha Singh
School of Agriculture, Graphic Era Hill University, Dehradun, Uttarakhand, India

Warshi Singh
School of Business Management, CSJM University, Kanpur, Uttar Pradesh, India

Bhupinder Singh
Sharda University, Greater Noida, Uttar Pradesh, India

Rashmi Singh
Amity Institute of Applied Sciences, Amity University, Noida, Uttar Pradesh, India

I. P. Singh
Department of Agricultural Economics, SKRAU, Bikaner, Rajasthan, India

K. P. Singh
G.B. Pant University of Agriculture and Technology, Pantnagar, Uttarakhand, India

Shreejaya Sivadas
Department of Post Harvest Process and Food Engineering, Govind Ballabh Pant University of Agriculture and Technology, Pantnagar, Uttarakhand, India

Kalpana Srivastava
School of Computer Science, UPES, Dehradun, Uttarakhand, India

Savita Tamta
Department of Vegetable Science, College of Horticulture, VCSG Uttarakhand University of Horticulture and Forestry, Bharsar, Pauri Garhwal, Uttarakhand, India

Ratnesh Kumari Thakur
College of Basic Sciences and Humanities, PAU, Ludhiana, Punjab, India

Alka Verma
College of Agriculture, G. B. Pant University of Agriculture and Technology, Pantnagar, Uttarakhand, India

Komal Vig
Sharda University, Greater Noida, Uttar Pradesh, India

Priyank Vyas
Graphic Era (deemed to be) University, Dehradun, Uttarakhand, India

Abbreviations

AGB	above ground biomass
AI	artificial intelligence
BD	Big Data
BGB	below ground biomass
BTC	block transaction counter
CEA	controlled environment agriculture
CIoT	climate IoT
CNNs	convolutional neural networks
CPS	cyber-physical systems
DBH	diameter at breast height
DL	deep learning
DPMS	digital pest management system
DSS	decision support systems
eSAP	electronic solutions against agricultural pests
FAO	Food and Agriculture Organization
FFD	full-function device
FFPS	food and fiber production systems
GA	genetic algorithms
GHG	greenhouse gas
GIS	geographic information system
GMO	genetically modified organisms
GPRS	general packet radio service
GPS	global positioning system
HAIs	hospital-acquired infections
HIPMS	hospital integrated pest management system
HPC	high-performance computing
ICT	Information Communication Technologies
IIoT	industrial IoT
IoAT	Internet of agricultural things
IoT	Internet of Things
IPMRS	integrated pest management research system
IR	industrial revolution
LAMP	loop-mediated isothermal amplification
LP	linear programming
MFA	multifactor authentication
ML	machine learning
MLR	multiple linear regression
MQTT	Message Queuing Telemetry Transport
NDVI	normalized difference vegetation index
NFMI	near-field magnetic induction
NGS	next-generation sequencing
NIR	near infrared

P2P	peer-to-peer
PA	precision agriculture
PCR	polymerase chain reaction
QoS	quality of service
RF	radio frequency
RFD	reduced-function device
RS	remote sensing
SA	smart agriculture
SDG	sustainable development goal
SF	smart farming
SHPMS	smart hotel pest management system
SWIR	shortwave infrared
TIR	thermal infrared
UAV	unmanned aerial vehicle
VIS	visible light
VRA	variable rate application
VRT	variable rate technology
VTOL	vertical takeoff and landing
WSN	wireless sensor networks

Foreword

In an era of steady global population growth and increasing pressure on agricultural systems, the need for innovative and sustainable farming practices has never been more urgent. The book, *Demystifying Sustainable Farming Using Smart Technologies,* is a timely and crucial contribution to the field of agriculture, offering a comprehensive exploration of how technologies can revolutionize the way we grow our food.

This book delves into the transformative potential of artificial intelligence (AI) and machine learning (ML) in agriculture. By harnessing the power of these technologies, we can optimize crop yields, enhance resource efficiency, and make informed decisions that promote sustainability. The integration of AI and ML into farming practices is not merely a futuristic concept; it is a present reality that holds the promise of addressing some of the most pressing challenges in agriculture today.

Throughout this book, we journey through a myriad of topics that encompass the essence of smart agriculture. From precision farming and predictive analytics to automated machinery and IoT-enabled devices, each chapter unveils the practical applications of technology in agriculture. These innovations are not only making farming more efficient but are also paving the way for more resilient and adaptable agricultural systems.

This book is a beacon of hope and a call to action for researchers, practitioners, policy-makers, and anyone invested in the future of agriculture. It underscores the critical role of interdisciplinary collaboration and continuous innovation in ensuring that our agricultural practices can sustain and nourish future generations.

I invite you to delve into the pages of this book with curiosity and an open mind. May it inspire you to envision a future where technology and sustainability go hand in hand, fostering a thriving and resilient agricultural landscape.

Happy reading!

—Prof. (Dr.) Sanjay Jasola
Vice Chancellor
DBS Global University
Dehradun, India

Preface

Agriculture, the backbone of human civilization, is undergoing a transformative shift in the 21st century. As we grapple with the increasing demands of a growing global population, the challenges of climate change, and the necessity for sustainable practices, the integration of advanced technologies into farming has never been more critical. This book, *Demystifying Sustainable Farming Using Smart Technologies,* aims to bridge the gap between traditional agricultural practices and the innovations that are revolutionizing the field.

In the pages that follow, we delve into the myriad ways in which artificial intelligence (AI) and machine learning (ML) are reshaping the agricultural landscape. From predictive analytics and precision farming to automated irrigation systems and crop health monitoring, these technologies are not just enhancing productivity but also promoting sustainability. By leveraging the power of data and intelligent algorithms, we can optimize resource use, reduce waste, and ensure that farming practices are both economically viable and environmentally friendly.

Our exploration is not confined to current technologies alone. We also cast our gaze forward, examining the future possibilities and the challenges that lie ahead. The rapid pace of technological advancement brings with it both opportunities and uncertainties. How will smallholder farmers in developing regions access and benefit from these technologies? What ethical considerations must we address as AI becomes more integrated into food production systems? How do we ensure that technological advancements lead to inclusivity and equity in the agricultural sector?

As you embark on this journey through the pages of D*emystifying Sustainable Farming Using Smart Technologies,* we invite you to envision a future where agriculture is smarter, more efficient, and more sustainable. A future where technology and nature coexist harmoniously, ensuring that we can meet our present needs without compromising the ability of future generations to meet theirs.

We hope this book serves as a valuable resource and a catalyst for positive change in the agricultural sector. Together, let us embrace the potential of smart technologies to build a resilient and sustainable agricultural future.

Preface

CHAPTER 1

Introduction to Agriculture 5.0

VRIKSHA SINGH[1*] and PRASHANT SHUKLA[2]

[1]*School of Agriculture, Graphic Era Hill University, Dehradun, Uttarakhand, India*

[2]*School of Advanced Engineering, UPES, Dehradun, Uttarakhand, India*

*Corresponding author

ABSTRACT

Digital agriculture, often termed Agriculture Technology 5.0, represents the next generation of tools and innovations aimed at enhancing agricultural productivity. Technologies, such as 5G, precision irrigation, advanced fertilization schedules, crop rotation, and tailored tillage methods are transforming conventional farming practices. Key components of this revolution include data-driven risk prediction algorithms, proximal and remote sensing tools, high-resolution satellite imagery, smart farm management software, and IoT-enabled devices. These tools provide farmers with real-time insights, enabling optimized resource use, improved decision-making, and enhanced crop yields. Modern agriculture has evolved beyond basic planting and harvesting. Recent advancements include resilient seed breeding, efficient postharvest processing, and integrated labor management through agricultural software. Traditionally, field monitoring required physical presence, but now remote technologies, especially agricultural satellites, offer efficient alternatives for real-time field assessment. Rapid technological growth is reshaping farming systems by improving input management, operational efficiency, and overall productivity. The integration of satellite-based monitoring, artificial intelligence, and connectivity tools signifies a paradigm shift toward smart, data-centric agriculture. This transition not only supports sustainable farming but also ensures better resource utilization and resilience to climatic challenges, making digital agriculture a cornerstone of future food security.

1.1 INTRODUCTION

The agriculture industry has experienced huge changes lately due to technological advancements. From traditional approaches to precision farming, all levels of farming have become more productive, sustainable, and efficient. The next phase is known as Agriculture 5.0 which takes in the most advanced techs such as robots, artificial intelligence and big data analytics into every aspect of farming. In this chapter, we investigate the concept of

Demystifying Sustainable Farming Using Smart Technologies.
Gunjan Chhabra, Keshav Kaushik, Priya Chugh, Susheela Dahiya, and Aryan Chaudhary (Eds.)

DOI: 10.1201/9781779640932-1

Agriculture 5.0; its components, applications, and implications for future food production are also examined [1].

1.2 UNDERSTANDING AGRICULTURE 5.0

The combination of digital, biological, and physical innovation in agricultural systems is referred to as "agriculture 5.0." It means a shift from traditional farming practices that are more and more automated, information-oriented, and interconnected. Agriculture 5.0 wants to address some of the major challenges that face agriculture, such as resource scarcity, population increase, global warming, and the need for sustainable food production (Figure 1.1).

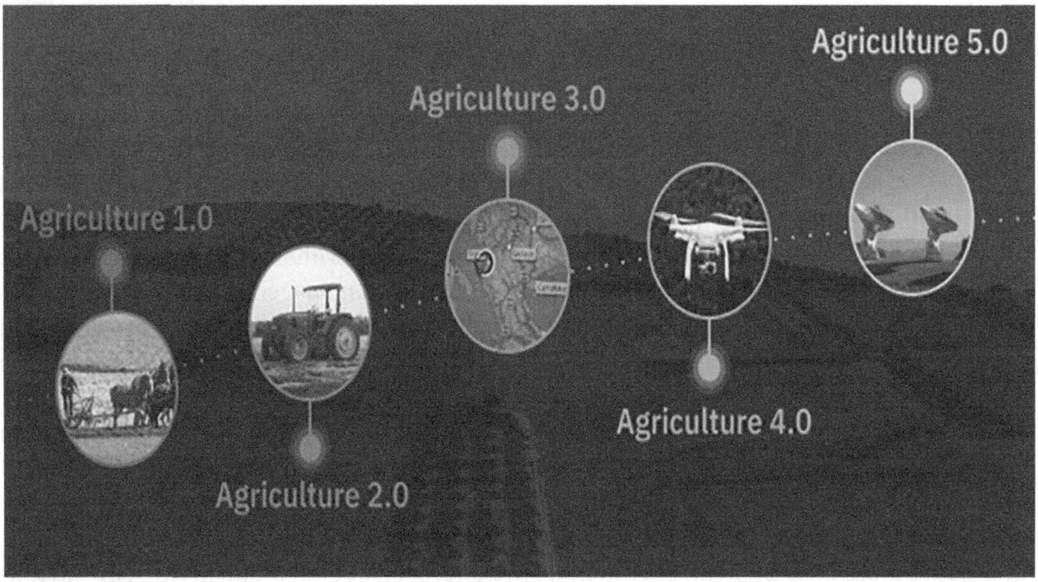

FIGURE 1.1 Stages of agricultural innovation.

1.3 PRINCIPLES OF AGRICULTURE 5.0

Agriculture 5.0 is an all-encompassing system of farming that incorporates sophisticated technology as well as ecological principles. There are several key concepts:

Sustainability: In Agriculture 5.0, it means adopting methods that do not cause harm to the environment, save natural resources, and promote biodiversity.

Resilience: This is realized through the adoption of data-driven decision-making approaches, embracing regenerative agriculture, and diversifying farm produce to enable adaptation to climate change and other disturbances.

Efficiency: Resource utilization optimization is achieved by using automation, AI, robotics for the reduction of waste, as well as increased productivity in Agriculture 5.0.

Connectivity: The networked technologies that are blockchain technology and operational technology make it possible for data to be exchanged and communication among all sectors of agriculture.

1.4 COMPONENTS OF AGRICULTURE 5.0

Artificial intelligence: AI is crucial to Agriculture 5.0 because it allows for the development of decision support systems, autonomous machinery, and predictive analytics. It involves machine learning algorithms that can predict agricultural diseases, improve irrigation plans, as well as schedule crops.

Robots and automation: Robots change how farming is done by doing everything from planting to harvesting, weeding, and pest control. This makes work easier, hence more accurate in carrying out tasks using sensors and actuators with the help of autonomous drones and robots, which saves on their employees' wages while mitigating environmental concerns with regard to their negative effects [2].

Internet of Things (IoT): Sensors, drones, and smart agricultural equipment are interconnected via IoT devices in order to allow for real-time data collection on soil moisture levels, temperature changes, or crop health conditions. This data are then transmitted through cloud-based technologies, where it can be analyzed for farmers' use in real time.

Blockchain technology: Blockchain technology complements traceability and transparency in the agricultural delivery chain. Blockchain protects meals product integrity, strengthens meals protection processes, and boosts customer–producer trust by documenting transactions and records in an immutable, decentralized ledger. With growing issues approximately meals safety, sustainability, and deliver chain integrity, blockchain technology is rising as a powerful device for growing agricultural transparency and traceability. Blockchain technology enables stakeholders to trace food objects from farm to desk with previously unheard-of precision with the aid of creating immutable, decentralized statistics of facts transfers [3].

These statistics consist of info concerning the agricultural practices employed, the usage of insecticides and fertilizers, and certifications for herbal or honest-alternate production. This degree of transparency encourages openness and responsibility, which promotes sustainable agriculture techniques at the same time as additionally improving consumer–manufacturer self-warranty. Furthermore, blockchain decreases inefficiencies and susceptibility to fraud and counterfeiting via enabling better supply chain control.

Vertical farming: In vertical farming, both efficiency and innovation are equally vital. New technologies maximize photosynthesis, resulting in better crop boom charges and dietary costs. One instance is LED lighting fixtures structures adjusted to certain wavelengths. Furthermore, automation is vital in vertical farms, where robots cope with environmental management, planting, and harvesting.

This technological integration allows for the 12 month-spherical production of fresh, domestically grown food, transforming metropolitan areas into agricultural hubs.

1.5 APPLICATIONS OF AGRICULTURE 5.0

1.5.1 *PRECISION FARMING*

Agriculture 5.0 permits precision farming practices, inclusive of variable-fee pesticide and fertilizer application. These technologies rely on the gift of information about crop and soil fitness. This focused technique increases output while reducing enterprise waste and environmental damage. Precision agriculture is a leap forward in agricultural methods that uses technology to achieve gold standard resource distribution and improve decision-making strategies. Farmers may additionally advantage gift time imaginative and prescient on crop repute, characteristic climate, and soil fitness via combining sensors, drones, GPS, and data analytics. With the potential to precisely administer inputs together with water, fertilizer, and pesticides, this granular stage of records reduces waste even as growing yields [4]. The use of drones for aerial monitoring and surveillance is a key issue in precision agriculture. Drones, the use of multispectral imaging sensors, and high-resolution cameras may also capture designated images of farms and come across insect infestations or stressed regions. This permits farmers to precisely target treatments, reducing the need for extensive chemical packages while minimizing expenses and environmental effects.

Farmers might also now make data-driven choices with previously unheard-of accuracy, thanks to advances in synthetic intelligence and machine getting to know. Predictive algorithms can estimate crop yields, ailment outbreaks, and the most efficient planting seasons with the aid of analyzing big portions of agricultural data. This proactive method lets farmers decrease dangers while growing yield throughout the growing season [5].

1.5.2 *SUPPLY CHAIN TRACEABILITY*

The blockchain era keeps track of the beginning place, manufacturing process, and distribution of food merchandise, growing transparency and traceability in the agricultural supply chain. Consumers may additionally get this information through cellular applications or QR codes, permitting them to make informed decisions about the meals they eat.

1.6 SATELLITE REMOTE SENSING TECHNOLOGY IN AGRICULTURE

The advent of satellites in the field has led to a great increase in the amount and frequency of agricultural data that is collected. Satellite sensors are used to observe, measure, and record electromagnetic radiation which is reflected or emitted by the Earth and its environment for further analysis and data extraction on a given parameter. The use of these technologies, which include available and archived satellite imagery, can enable monitoring of crop development throughout the season, even in large and inaccessible areas. This information has several applications, including evaluating how effective farming activities have been.

EOSDA Crop Monitoring—It is a one-stop platform for landowners and farmers. Satellite imaging data are evaluated by the EOSDA Crop Monitoring one-stop platform so that it may be accessed easily along with vegetation indices as well as other essential field indicators meant to assist land owners and farmers in making informed decisions (Figure 1.2).

FIGURE 1.2 Remote sensing technology in agriculture.

1.7 MOBILE TECHNOLOGY IN AGRICULTURE

Mobile apps targeting agriculture are experiencing exponential growth in their market share as well as functionalities. With the mobile app EOSDA Crop Monitoring, you can chart any field's vegetation into a map from anywhere using your mobile phone without having to carry around a heavy laptop. You can organize and manage scouting missions, keep track of the weather in the field, and get updates on your modern-day situation all from the ease of your mobile tool. The capacity of mobile applications to function without an internet connection is only one of many benefits of employing them in agriculture (Figure 1.3) [6].

FIGURE 1.3 Mobile technology in agriculture.

1.8 CHALLANGES

The subsequent degree of the rural revolution, called "Agriculture 5.0," is characterized via the usage of modern-day technologies, inclusive of massive data analytics, robotics, the IoT, and AI.

While Agriculture 5.0 has significant potential, its vast implementation confronts several demanding situations:

1. **Access and affordability:** Small-scale farmers in underdeveloped nations might not have access to the devices and resources required to undertake Agriculture 5.0 practices.
2. **Ethical and regulatory concerns:** To foster accountable innovation, questions regarding the security, ownership, and equitable distribution of genetic resources and data ought to be addressed.
3. **Digital divide:** Inequalities within the agricultural economy can be exacerbated by way of disparities in get right of entry to virtual infrastructure and technological competencies.
4. **Technological adoption gap:** Not every farmer has the necessary infrastructure, budget, or understanding to properly take delivery of and use modern-day technology. The distinction between technologically savvy farming operations and those that are not creates a digital divide.
5. **High preliminary prices:** Agriculture 5.0 generation installation from time to time necessitates huge initial expenses in sensors, software, hardware, and education. It can be hard for small-scale farmers, specifically, to pay these upfront fees, stopping them from modernizing their groups.
6. **Data management and privacy concerns:** Data evaluation and collection is an important issue of Agriculture 5.0. It may be challenging to address the large volumes of facts generated by means of satellites, drones, IoT sensors, and other gadgets. In addition, a number of humans are involved in approximately shielding sensitive agricultural facts' safety and privacy, in particular in light of the risk of cyberattacks.
7. **Infrastructure limitations:** Many farms are located in rural regions, which commonly lack the infrastructure and dependable internet connectivity required to apply Agriculture 5.0 technology. Farmers may additionally warfare to absolutely make the most and get right of entry to these improvements in the absence of a steady connection.
8. **Skills and training:** Using modern-day agricultural technology requires unique know-how and teaching. Farmers and agricultural laborers ought to learn to use new equipment, renovation practices, and software interfaces. Agriculture 5.0 era adoption can be confined within the lack of suitable schooling and assistance.

9. **Environmental impact:** Agriculture 5.0 technologies are meant to enhance sustainability and efficiency; however, there are worries approximately how they may impact the surroundings. For instance, if not treated properly, growing reliance on generation and automation might also bring about increased power intake or disruption of natural ecosystems.
10. **Regulatory challenges:** Regulations typically struggle to keep up with speedy technological improvements. The adoption of Agriculture 5.0 technologies can be hampered by ambiguous or old policies, which can also stymie innovation in the areas of statistics ownership, legal responsibility, and highbrow belongings protection.
11. **Social and ethical implications:** The great adoption of Agriculture 5.0 era may have poor social and ethical effects, which include adjustments in rural groups, job losses, and issues of ownership and management over agricultural data. Sustainable agricultural expansion necessitates balancing capability advantages with broader societal problems.

    To make certain that Agriculture 5.0 supplies its promised advantages in an equitable and sustainable way, stakeholders—consisting of governments, agricultural businesses, generation providers, researchers, and farmers themselves—ought to collaborate to triumph over those challenges [7].

1.9 ADVANCEMENTS AND IMPROVEMENTS IN AGRICULTURE ERA

Agricultural generation has advanced considerably from the times of simply planting, being concerned with, and harvesting crops. Numerous tremendous improvements in agricultural manufacturing planning and processing, as well as the improvement of extra resilient seeds, have passed off currently. Technology improvements in agriculture can even enhance logistics and field-to-desk transport.

The discipline of agricultural software and generation is also advancing fast, allowing for the control of diverse food supply chain additives in addition to in-discipline labor. Monitoring the fields used to contain physically traveling them, but nowadays it is viable to do it truly. Agricultural satellite TV for PC technology is ideal for this reason as a new way to look at the sector.

Every step of the selection-making technique is knowledgeable through pertinent records while using EOSDA Crop Monitoring, from comparing historical and meteorological area facts to organizing agricultural operations. Even in conditions while someone has to be present in the subject, this precision agriculture platform is helpful. It can be used to designate discipline scouts to analyze a doubtful area and return with their findings. Using a telephone, pictures can be promptly added to the process report.

1.10 ADVERSE EFFECTS OF FARMING TECHNOLOGY

Although technological trends have absolutely helped present-day farming, especially in terms of better yields, lower costs, and more efficiency, there is another facet to the tale that is in particular related to large-scale, extensively practiced agriculture. The damage it does to the surroundings is the maximum considerable.

The following are the main issues with agricultural technology which are negatively affecting the atmosphere:

- contamination of the water and land due to extensive pesticide use.
- decrease in biodiversity due to local species being replaced by using agricultural crops.
- emission of greenhouse gases because of forest elimination for agricultural purposes and expanded equipment use. Not every downside of the era in agriculture is related to its effects on the environment. Some problems with manufacturers' and their employees' use of agricultural era consist of the following:
  - Farmers are unable to use the advanced agricultural era available nowadays because they lack the necessary education and actual-world experience, making it tough for them to operate equipment and software properly.
  - Chemical insecticides and fertilizers may be harmful to the health of farmers and different agricultural laborers who paintings in the fields.
  - Maintaining machinery can be highly priced.

We can brush aside agricultural advancements due to the fact they are critical to offering food for the arena increasing populace. However, we will lessen these terrible effects with the aid of using and developing environmentally friendly techniques in addition to precision agricultural technologies. These technologies can mitigate, if no longer definitely put off, the harmful outcomes of conventional agricultural strategies, which enables us to solve a number of environmental problems. This might assist industrial agricultural manufacturers acquire two desires without delay: improving their competitiveness and boosting global well-being within the long time [7].

1.11 IMPLICATIONS FOR THE FUTURE

The future of Agriculture 5.0 seems very promising for the reason that this subsequent degree of the rural revolution has the power to definitely rework how food is produced, dispensed, and consumed. Here are a few crucial future possibilities:

Increased efficiency and productivity: Agriculture 5.0 technologies offer particular monitoring and control of crops, animals, and sources, increasing productivity and performance. Using current sensors, drones, and satellite TV for PC imaging, farmers can manage inputs like water, fertilizer, and insecticides with the aid of making data-driven decisions in real time.

Resource conservation: Precision farming practices, supported with the aid of Agriculture 5.0 technology, minimize waste and useful resource use. Smart irrigation structures, as an example, store water and feature a lower environmental effect with the aid of turning in water exactly where and while it is needed. Similarly, converting the application rates of fertilizers and insecticides reduces chemical runoff and soil erosion.

Sustainable practices: Agriculture 5.0 employs automation, gadget mastering, and records analytics to help farmers adopt sustainable agricultural techniques. Precision farming techniques help to cope with worries such as soil degradation and weather alternation via decreasing waste, limiting environmental effects, and encouraging resource conservation.

Enhanced resilience: Climate variability and excessive weather conditions provide sizable problems for agriculture. Predictive modeling and smart irrigation systems are two examples of Agriculture 5.0 era that could boost agricultural machine resilience by way of permitting proactive chance control and adaptive techniques to counteract the outcomes of disruptions linked to climate change.

Enhanced food safety and quality: Traceability and first-class agricultural goods are assured from farm to fork way with Agriculture 5.0 generation. The blockchain era, with IoT sensors, provides transparent supply chains that beautify food safety protocols via promptly detecting and controlling any contaminants. This allows customers to monitor the starting place and direction of their meals.

Data-driven decision-making: To optimize their operations, farmers may also now access real-time data, actionable insights, and choice-help gear to Agriculture 5.0. Predictive analytics, danger evaluation, and scenario making plans are made feasible by means of superior analytics and modeling, which assist farmers make better choices that raise output, profitability, and resilience in the face of uncertainty.

Empowerment of farmers: Agriculture 5.0 offers choice support equipment, marketplace insights, and actionable information to farmers. By using virtual structures and cell packages to get entry to real-time marketplace expenses, weather forecasts, and agronomic hints, farmers may additionally growth productivity, decrease risks, and increase profitability.

Development of rural areas: The use of Agriculture 5.0 era has promise for promoting advancement and financial growth in rural regions. Because these technologies increase agricultural manufacturing and diversify revenue streams, they have the ability to improve livelihoods, draw investment, and generate employment. They can therefore aid in the fight in opposition to poverty and promote socioeconomic development.

Collaboration and creativity: Agriculture 5.0 encourages cooperation and innovation within the fields of agriculture, generation, and academia. Multidisciplinary studies and collaboration sell ongoing technological development on the way to deal with changing agricultural possibilities and demanding situations, growing a resilient and dynamic food gadget for the destiny.

Global food security: Agriculture 5.0 is becoming increasingly vast as the world's populace rises. Because these technologies increase agricultural output, decrease food waste, and promote sustainable practices, they enhance the availability, affordability, and accessibility of nutrient-dense food for expanding populations, particularly in regions confronting food poverty and malnutrition [8]. Overall, Agriculture 5.0's future potentialities are characterized by resilience, inventiveness, and sustainability. It will offer novel answers to the complex difficulties that the agriculture business faces in the 21st century.

Agriculture 5.0 has the capability to modify food production and create an extra sustainable and equitable destiny for agriculture and society as a whole, provided that stakeholders throughout the arena preserve to make investments, collaborate, and demonstrate their commitment to the undertaking [9].

1.12 WHAT KIND OF PRECISION AGRICULTURE TECHNOLOGY ARE CURRENTLY IN USE?

Farmers must stay current on the latest technological innovations and how they relate to the agricultural business in order to enhance production and remain relevant.

1.12.1 GPS TECHNOLOGY IN AGRICULTURE

Precision agriculture technologies reduce unnecessary spending on inputs like fuel, seeds, fertilizer, and insecticides while increasing production through the usage of GPS facts. When paired with a farm control software program, GPS no longer only affords area-primarily based agricultural information however will also be utilized to improve document-maintaining and automobile communiqué.

The following are just a few of the several programs of this generation in agriculture:

- maintaining an eye on and supervising fieldwork;
- amassing and evaluating area information; unique soil sampling;
- yield mapping;
- working agricultural equipment;
- maneuvering and controlling it; and
- ability to paint in low-visibility conditions, which includes constant rain or fog.

1.12.2 ROBOTIC TECHNOLOGY IN AGRICULTURE

Because robotic farm labor can do repetitive obligations while keeping accuracy, it appears to be a viable option for precision agricultural needs. These self-sufficient robots' potential to find out about their surroundings will allow ongoing field control to further accelerate agricultural productiveness and performance. The most popular and beneficial agricultural robot generation to be had these days includes telemetry-controlled autonomous gadgets (Figure 1.4) [10].

1.12.3 DATA COLLECTION AND STORAGE

Big statistics has been incorporated with agricultural technologies to enhance information flow, enable quicker and greater accurate analysis, and ultimately cause higher strategic planning and decision-making. Analyzing prior agricultural records lets us to predict potential outcomes, examine the risks, and choose the most excellent line of action. Advances in agricultural generation have made it feasible to gather and save a growing amount of farm-related records. Using agricultural generation, subject information may be used to

FIGURE 1.4 Robot technology in agriculture.

Source: Techslang. https: //www.techslang.com/agricultural-robots-are-we-ushering-the-age-of-robot-farmers/

get agricultural indicators which include soil pH, relative humidity, nutrient ranges, soil moisture, and historical climate records.

EOSDA Crop Monitoring gathers a plethora of records on agricultural practices (sowing, harvesting, and irrigation), plants productiveness, crop kinds, subject peak, and weather conditions, together with ancient temperature and precipitation readings.

You may also evaluate prior harvests, discipline volume, and potential harvests the use of this effectively available agricultural information that is shown on the web page (Figure 1.5). Furthermore, this know-how can help with crop planning, cooperative control, and detecting agricultural risks.

1.13 AGRICULTURAL TECHNOLOGY—HOW DOES IT AFFECT AGRICULTURE?

The information age has delivered the capability to use technology in farming than before. Agricultural firms rely on technology to enhance area control; this is just one way that farming has evolved due to the improvements in technology. With advanced agricultural software, fertilizer and pesticide usage can decrease.

This was also made possible through simplifying and eliminating processes from agricultural production. Through gene editing technology, farmers globally can breed seeds specifically tailored to the unique attributes of their local soil and environment. Climate-smart agriculture provides targeted crop-growing techniques addressing a number of food production concerns. Moreover, when fertilizers are applied correctly into the soil, new crops can be generated. Furthermore, all activities, including planting, harvesting, and distribution, are mechanized, leading to increased productivity and efficiency in agribusiness.

These days, information and communication technologies, nanotechnologies, and biotechnologies make it feasible to increase food yields in the most sustainable way possible—to the point that even drones are used to monitor agricultural areas [11].

1.14 CASE STUDIES

1.14.1 CASE STUDY 1: AEROFARMS—VERTICAL FARMING AND CONTROLLED ENVIRONMENT AGRICULTURE

AeroFarms is one of the leading businesses within the subject of Agriculture 5.0, it has been a key disruptor and innovator in agriculture due to its innovations in vertical farming. The employer is primarily based in Newark, New Jersey. AeroFarms has redefined indoor agriculture via using particularly superior aeroponics era with AI and information analytics to domesticate plants of herbs and layers of leafy vegetables stacked on pinnacle of every different in a vertical rely without soil and but little water [11].

1.14.1.1 KEY FEATURES

Vertical farming: AeroFarms maximizes region and maximizes resource performance by the use of vertical stacks of boom trays, interior enclosed chambers with managed environments.

Aeroponics: Growing vegetation in a mist environment freed from soil can save up to 95% of water in comparison to standard agricultural techniques.

On AI and records analytics: The organization employs AI algorithms to screen plant growth, optimize nutrient distribution, and adjust environmental conditions being managed in real-time via records supplied by means of sensors integrally integrated into its agricultural equipment. In order to lessen pollution and promote sustainable agriculture, it refrains from making use of pesticides, herbicides, or excessive water.

1.14.1.2 IMPACT

Increased crop yield: AeroFarms supplies better crop yields than traditional farming tactics because it provides ideal developing conditions that are tailor-made to the wishes of every plant.

Decreased resource footprint: Vertical farming complements environmental sustainability by drastically slicing the amount of water used, quantity of land required, and fee of transportation related to traditional agriculture.

Year-round production: AeroFarms' year-round production of sparkling, locally grown items are made feasible through their managed environment agriculture, which also lowers seasonal constraints and guarantees a constant supply for customers.

1.14.2 CASE STUDY 2: JOHN DEERE'S AUTOMATION IN FARMING AND PRECISION AGRICULTURE

John Deere, a distinguished agricultural equipment employer, has used precision agriculture and the automation era to enhance farm manufacturing and sustainability [12].

1.14.1.1 KEY FEATURES

Precision farming equipment: John Deere develops and manufactures precision agricultural equipment along with GPS-guided tractors, self-riding drones, and machinery equipped with sensors which can swiftly acquire and examine field information.

Data-driven decision-making: By combining field-particular information and predictive analytics with John Deere's virtual systems and agronomic insights, farmers may also beautify their planting, watering, fertilizing, and pest control techniques.

Automation and robotics: John Deere's robotic device and self-sustaining tractors make it possible to perform operations like planting, sowing, and harvesting more exactly, constantly, and economically, which reduces labor expenses and operating costs.

1.14.1.2 IMPACT

Enhanced efficiency: By allowing farmers to make records-driven alternatives and administer centered remedies, precision agriculture raises crop yields, streamlines farm operations, and lowers input waste.

Sustainable practices: John Deere generation promotes sustainable agriculture and soil health control by enhancing resource utilization and lowering environmental impact through precise input utility.

Farmer empowerment: By presenting choice support tools, actionable insights, and get right of entry to agronomic knowledge, John Deere's virtual structures facilitate data sharing amongst farmers and assist them in managing their farming operations.

John Deere, a prominent agricultural machinery company, has used precision agriculture and automation technology to boost farm production and sustainability.

1.14.3 CASE STUDY 3: UTILIZING AGRICULTURE 5.0 IN HILLY TERRAIN: SIKKIM'S ORGANIC FARMING INITIATIVE

Introduction: Using Agriculture 5.0 thoughts to tackle the specific issues supplied with the aid of its steep terrain, the little country of Sikkim in northeastern India has begun a top-notch path closer to natural farming. Through a mixture of sustainable practices, technology integration, and network engagement, Sikkim has emerged as an international pioneer in natural agriculture, demonstrating the progressive potential of Agriculture 5.0 in mountainous regions [13].

Challenges of hilly terrain: The scarcity of water, confined quantity of arable land, soil erosion, and susceptibility to the effects of weather exchange, in conjunction with landslides and unpredictable weather patterns, are some of the particular problems that hilly terrains pose to agriculture. Conventional farming practices typically fall short of those dreams, which results in lower yields and damage to the surroundings

1.14.1.1 IMPORTANT ELEMENTS OF ADOPTION IN AGRICULTURE 5.0

Precision farming and IoT integration: Sikkim has employed IoT-enabled devices, weather stations, and soil sensors along facet precision farming strategies to screen environmental factors and enhance useful resource management in high-altitude areas. If farmers have up-to-date information on soil moisture, nutrient ranges, and climate situations, they may make well-informed decisions approximately crop preference, irrigation, and fertilization.

Agroecology and sustainable practices: Sikkim's herbal farming challenge prioritizes agroecological requirements which encompass crop diversification, intercropping, and natural pest management as a manner to beautify soil fertility, biodiversity, and weather resilience. In Sikkim, farmers may additionally produce crops in difficult alpine environments in a sustainable way with the aid of encouraging ecological balance and minimizing their reliance on chemical inputs.

Community involvement and potential building: Sikkim's fulfillment with organic farming may be linked to the government's resource of farmer training projects, capability constructing programs, and understanding-sharing systems, similar to the lively participation of the local community. In the rural zone, Sikkim has encouraged innovation and entrepreneurship by manner of giving the specified resources, statistics, and expertise

1.14.1.2 EFFECTS AND ADVANTAGES

Environmental sustainability: By keeping biodiversity, soil, and water, Sikkim's herbal farming strategies assist to lessen the harmful consequences of traditional agriculture on the environment in steep areas. Sikkim's herbal farms keep ecological balance and natural resources, which allows with resistance to and variation from climate change.

Economic prosperity: The use of Agriculture 5.0 era and herbal farming practices has stepped forward Sikkim's agricultural sustainability, ensuing in extended farm earnings, advanced market gets proper of entry to, and more opportunities for rural living. Farmers in Sikkim have advanced their profits streams and reduced their dependency on outdoor property by way of capitalizing on professional markets for herbal produce.

Social empowerment: The herbal agricultural initiative in Sikkim has supplied neighborhood citizens with sources, market linkages, and training, consequently empowering ladies and smallholder farmers mainly. Sikkim's Agriculture 5.0 version has helped to beautify rural areas and decrease poverty through promoting inclusive growth and social fairness, in particular in hilly terrain

1.15 SUSTAINABLE PROSPECTS FOR AGRICULTURE TECHNOLOGY

The rapid progress of the generation has had a splendid influence on farming. The modern satellite era for agriculture is being advanced to decorate input management, simplify operations, and growth farm productivity. Improved yields and decreased renovation expenses can cause better agricultural profits. In light of this, employing satellites in agriculture represents a massive step closer to greater sustainable and powerful farming strategies able to assemble the sector's growing food needs.

1.16 CONCLUSION

Agriculture 5.0 represents a fundamental shift to a more resilient, sustainable, and efficient food manufacturing gadget. Farmers may additionally overcome conventional hurdles and usher in a brand-new era of site-specific agriculture through using the power of automated era, robotics, and operational technologies. To create a future wherein meals are plentiful, without problems accessible, and environmentally benign, we should prioritize cooperation, innovation, and inclusiveness as we clutch the possibilities provided by means of Agriculture 5.0.

KEYWORDS

- **agricultural satellite technology**
- **digital agriculture**
- **agricultural software**
- **agricultural outputs**
- **streamline process**
- **Agriculture 5.0**

REFERENCES

1. Smith, P., et al. (2020). Agriculture 5.0—the future of farming technology. *Agricultural Systems*, *186*, 103933.
2. Botta, A., Cavallone, P., Baglieri, L., Colucci, G., Tagliavini, L., & Quaglia, G. (2022). A review of robots, perception, and tasks in precision agriculture. *Applied Mechanics*, *3*(3), 830–854. https: //doi.org/10.3390/applmech3030049
3. Qin, Z., et al. (2021). A review of blockchain applications in agriculture: Current status, challenges, and future directions. *Computers and Electronics in Agriculture, 184*, 106027.
4. Bhutani, A., & Wadhwani, P. (2019, April). Precision farming market size by component, by technology, by application, industry analysis report, regional outlook, growth potential, competitive market share & forecast, 2019–2025. Global Market Insights.

5. Guebsi, R., Mami, S., & Chokmani, K. (2024). Drones in precision agriculture: A comprehensive review of applications, technologies, and challenges. *Drones, 8*, 686. https: //doi.org/10.3390/ drones8110686
6. Godfray, H. C. J., et al. (2020). The future of food. *Nature Reviews Earth & Environment, 1*(1), 21–23.
7. Lowder, S. K., Skoet, J., & Raney, T. (2016). The number, size, and distribution of farms, smallholder farms, and family farms worldwide. *World Development, 87*, 16–29.
8. Tilman, D., Balzer, C., Hill, J., & Befort, B. L. (2011). Global food demand and the sustainable intensification of agriculture. *Proceedings of the National Academy of Sciences, 108*(50), 20260–20264.
9. FAO. (2020). *The State of Food Security and Nutrition in the World 2020*. Food and Agriculture Organization of the United Nations.
10. Khaliq, T., et al. (2022). Role of artificial intelligence in agriculture: A comprehensive review and future directions. *Computers and Electronics in Agriculture, 194*, 106494.
11. Mohd, J., et al. (2022). Enhancing smart farming through the applications of Agriculture 4.0 technologies. *International Journal of Intelligent Networks, 3*, 150–164. https: //doi.org/10.1016/j.ijin.2022.09.004
12. Reij, C., Tappan, G., & Belemvire, A. (2005). Changing land management practices and vegetation on the Central Plateau of Burkina Faso (1968–2002). *Journal of Arid Environments, 63*(3), 642–659.
13. World Bank. (2021). *The State of Food and Agriculture 2021*. World Bank Group.

CHAPTER 2

Agriculture 5.0: Future and Challenges

LEELA BHATT[1*] and M. K. NAUTIYAL[2]

[1]*Department of Genetics and Plant Breeding, G. B. Pant University of Agriculture and Technology, Pantnagar, Uttarakhand, India*

[2]*Graphic Era Hill University, Dehradun, Uttarakhand, India*

*Corresponding author

ABSTRACT

Due to factors including population growth, the loss of agricultural farmlands, the depletion of renewable resources, and rise in erratic environmental circumstances, the availability of food has become a serious concern for almost every nation in the world. Global warming results from environmental stress caused by population growth and various sectors, especially the agriculture industry. Numerous human activities cause land degradation that impacts crop quality, while chemical pollution creates dead zones and risks marine life. Because of these problems, the agricultural industry is shifting toward "smart agriculture," which makes use of the Internet of Things (IoT) and artificial intelligence (AI) to increase productivity and optimize processes. IoT and AI incorporate several cutting-edge methods and modern advanced technologies, such as networking, sensing technology, big data, cloud computing, and end-user apps. Farmers who use smart technologies can increase agricultural yield while saving time. AI, IoT, and advanced technologies enable farmers to access vital information instantly, overcoming traditional limitations. Agriculture 5.0 will feature unmanned farming, robotics, and data-driven methods, ensuring sustainability and productivity in the face of labor shortages and rising food demand.

2.1 INTRODUCTION

Agriculture has advanced from its early stages to the present era, depicted in Figure 2.1, focusing on land cultivation and animal husbandry for sustenance.

Demystifying Sustainable Farming Using Smart Technologies.
Gunjan Chhabra, Keshav Kaushik, Priya Chugh, Susheela Dahiya, and Aryan Chaudhary (Eds.)

DOI: 10.1201/9781779640932-2

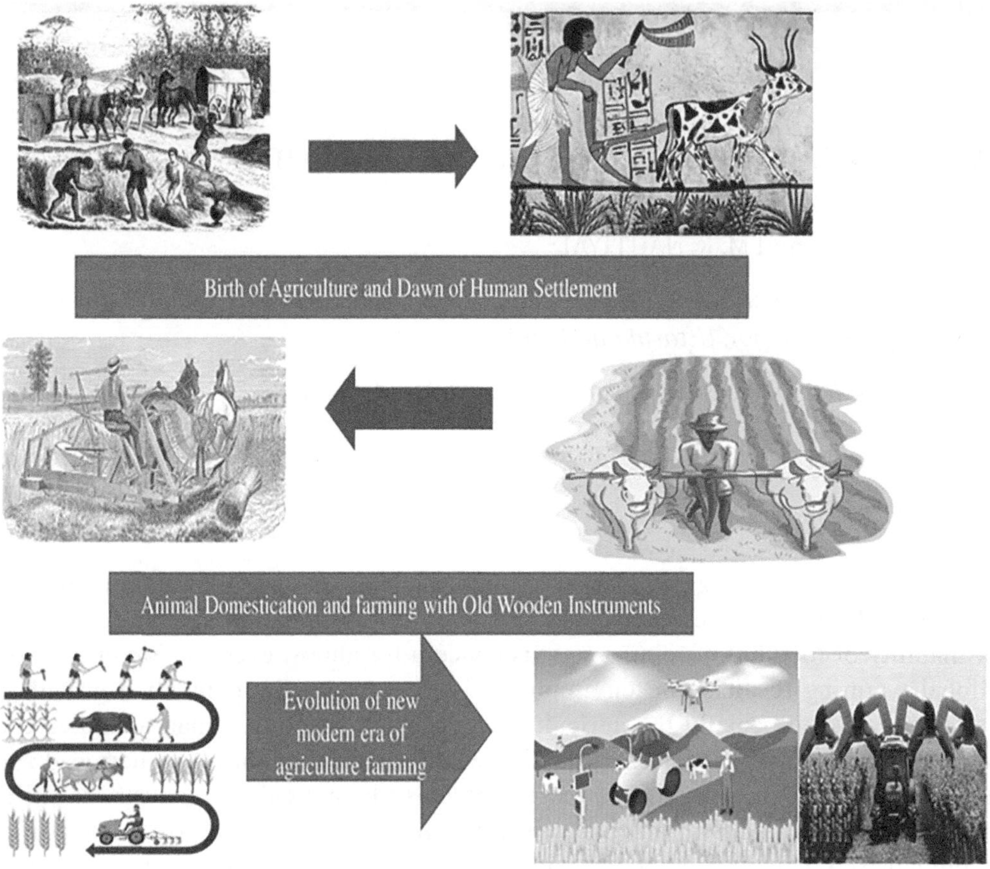

FIGURE 2.1 The evolution of agricultural farming.

The steam engine, developed in the 19th century during the Industrial Revolution, revolutionized agriculture by enhancing efficiency through mechanization. Subsequent advancements, including Agriculture 3.0 driven by computer technology, further optimized agricultural operations through automation and precision. The existing concerns from Agriculture 2.0 are completely fixed in Agriculture 3.0. The efficient delegation of responsibilities to farming equipment decreased the need for pesticides, increased the accuracy of irrigation, and others. Thanks to the application of cutting-edge of advance technologies like artificial intelligence (AI), large-scale data, Internet of Things (IoT), cloud-based computing, and remote sensing (RS), Agriculture 4.0 has become a reality [1].

As information technology in agriculture continued to advance, Agriculture 4.0—is referred as "Smart farming" or "Digital agriculture"—was born. AI, big data (BD), IoT, 3D printing, quantum information processing, both augmented as well as virtual reality, application of robotics and blockchain are the key Agriculture 4.0 technologies. Automatic control of agricultural farming resources and its mechanisms can be achieved through the construction of unmanned robotic systems through the use of reduced and inexpensive microelectronic component manufacturing technologies.

Real-time agriculture management decisions are made possible by high-throughput communication technology and cloud storage systems. Systems using AI assist in controlling vehicles, tractors, combines, and drones by recognizing reality. Agriculture 4.0 is a contemporary phase in the history of agriculture, with its instruments and practices being used in all industrialized nations' crops. Currently, agricultural technologies are developing in the direction of Agriculture 5.0, a new paradigm. But a lot of Agriculture 5.0's questions are up for debate, and its responses are only indicative. The application of bio-economics to the theory of Society 5.0 indicates its mass personification and individualization of the final product, resource optimization, the deployment of self-governing, automated systems that make decisions based on complex robotic applications, and innovative product differentiation are the sectors of agricultural growth that show the most promise. When there are no obstacles to communication in human–machine interaction—including when bioinformatic technologies are used—we may concentrate on improving the heuristics of creative agricultural decision-making. In agricultural science, research on Agriculture 4.0 and Agriculture 5.0 is extremely prevalent nowadays.

The content of practices and conditions of the present agriculture growth technical stage, additionally its effects on food security and food production, will be evaluated together with the principal theoretical and methodological approaches.

1.2 RECENT AGRICULTURE DEVELOPMENTS

The idea of "Industry 4.0," that categorizes the development of the industries into four stages, guided the evolution of the "Agriculture 4.0" concept. If the invention of vapor machines in the 1880s marked the beginning of Industry 1.0, the introduction of computer science and automation marked the beginning of Industry 2.0 with the advent of electricity. Improvements have been achieved in the introduction of digitization through Industry 4.0. The agricultural industry has experienced similar shifts as a result of AI at 5.0 [10]. The base of digital farming or smart agriculture (SA) is the transmission of data via remote and data storage for use in various agricultural data analyses to inform decision-making [11]. Research on the integration of agricultural technology and AI is currently being conducted all over the world. This research is primarily based on mechanized agricultural networks, unique agricultural sensors, large agricultural data sets, smart factories for plant production, biological knowledge, closed ecosystems, drones, equipment of intelligent research, precision agricultural technology, and so on [12, 13].

Eventually, Agriculture 4.0 will not rely on spraying insecticides, fertilizer, and water evenly across entire regions. After accounting for everything, farmers may focus on or employ the base sums needed and particular, very target-specific topics. The report on research goes on to suggest that, because of technological advancements like sensors, devices, machinery, and information development, farms and agro-enterprises need to be managed in a unique approach. In the future, aerial photos, robotics, temperature and humidity sensors, and GPS advancements will all be utilized in agriculture. Precision farming, automated structures, and these basic devices will make fields more profitable, usable, safe, and ecologically friendly [22] (Figure 2.2).

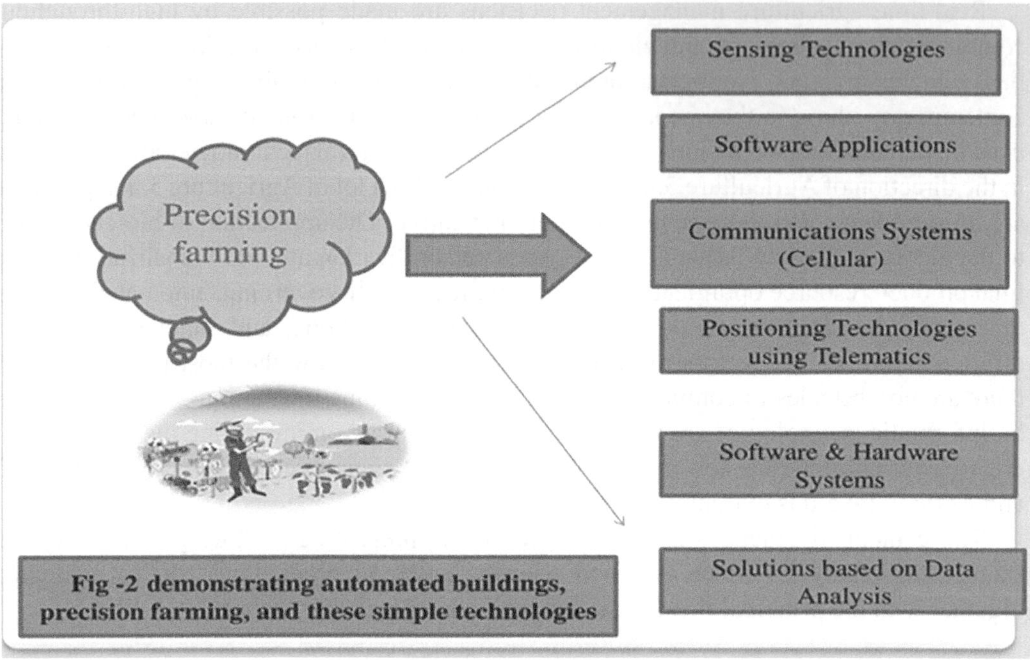

FIGURE 2.2 Demonstrating automated buildings, precision farming, and these simple technologies.

The development of agricultural crops and managing them are difficult tasks since environmental conditions, soil characteristics, and parameters related to water quality vary globally. Worldwide, there is a need for high agricultural productivity per unit hectare due to the growing population and constantly shifting environmental conditions [6, 7]. Climate or weather change is a product of stress caused by environmental factors brought on by growing populations and various sectors, particularly agriculture. Numerous human activities result in land degradation that has an effect on crop levels; chemical pollution endangers marine life and produces dead zones. Thus, providing workable answers to major issues, including infestations of disease and pests, inadequate biological control, improper application of water and removal, weed control, and yield forecasts may depend on the usage of smart applications in agriculture [5, 8, 9].

The production yield can be increased by agricultural crop protection efforts against climate or weather change, soil nutrients, land suitability, culture method, cultivation techniques, artificial nutrient applications, irrigation management, and so on. Agrochemical use can effectively raise agricultural output and quality, but in recent years, it has also brought along more environmental challenges. When it comes to chemical plant protection activities, overuse of pesticides and fertilizers or improper spray equipment may result in detrimental impacts on both human health and the environment. Smart cultivation, often known as farming, is the application of technological communication to automate and mechanize agricultural operations [2].

A new category of machine learning (ML) known as deep learning (DL) has emerged as a result of the last 10 years of increasing computing power and data in many scientific domains. DL allows for the implementation of more complex models that were previously unfeasible for current computers, but are now practical and functional [14, 15]. Applications of DL have transformed fields including speech recognition, object recognition, and other correspondingly hard tasks related to large-scale data processing [16, 17]. High levels of computer system performance have been attained in some applications, enabling them to successfully supplement human intervention in some situations, if not entirely replace it. In this modern era of "smart farming or digital farming," advancements in AI, DL, the IoT, blockchain, and further digital based electronic devices are now using in agricultural production for crops as well as animals additionally transitional industries [18, 19].

In an effort to achieve a sustainable agriculture sector, smart farming (SF) is designed to (1) encourage making decisions when managing a crop and maximize the range of sound effected decisions made for both each animal or agricultural land area accounting for time constraints and (2) smart farms should include renewable energy sources. Considering the first objective, it can be accomplished by making it easier to gather the necessary field data and to process it quickly and reliably so that applied agricultural techniques can be modified. ICT is used in conjunction with precision equipment, the IoT, robotics application, large-scale data, sensors, non-GPS manned vehicles, geo-location systems (GPS), 3D printing, and management information systems to perform these activities [20, 21].

2.2.1 SMART FARMING'S TRANSITION TO AGRICULTURE 4.0 AND 5.0

A significant change in agriculture is needed to deal with the global challenge of the 21st century. This revolution must be able to allow technology to boost productivity while maintaining or preserving environmental sustainability. The concept of "Agriculture 4.0" describes the fourth agricultural revolution, utilizing technological advancements to make agriculture more productive, intelligent, and environmentally conscientious [48, 49].

Agriculture 4.0 technologies must be needed if agriculture is to fully utilize or reap the potential of precision farming. Based on a broad coverage area and real-time data, the infrastructure for connected farming has to meet the demands of the agricultural industry and the objective of IoT adoption [48, 49]. Agriculture 4.0 aims to improve farming practices in addition to revolutionizing agri-food systems. The agriculture food supply chain is a series of actions that begin with the manufacturing and distribution of agricultural commodities to consumers [7].

Four pillars of Agriculture 4.0 are technological advancement, the employment of sophisticated equipment, environmental sustainability, and scaled-up data for oversight of productive farming. A vital component that goes across all architectural levels, the quality of the data collected is essential to the Agriculture 4.0 system's success [51]. The introduction of Agriculture 4.0 and the evolution, Agriculture 5.0, allow for a creation of novel innovations for the agricultural environment with less resource. According to the "Agriculture 5.0" theory, fields use precision farming techniques and machinery that allow them to make decisions on their own and run automatically. Therefore, Agriculture 5.0 implies the use of robotics when combined with a particular kind of AI.

Given the long useful life of agricultural equipment, care is taken to ensure that any technical improvement is backed by producers and other economic sectors and remains compatible with updated equipment [32]. Locating technology, such as remote drones, GPS, and unmanned aerial vehicle (UAV) operations, has possibilities to improve weather forecasting accuracy and other environmental aspects. The main issue regarding the idea of Agriculture 4.0 is the requirement for data interchange and channels of communication to connect the different systems into an integrated framework which takes into account all aspects used in agriculture. The ability of farmers to invest in and advance their agricultural techniques is a further vital aspect of Agriculture 5.0 adoptions [32, 50].

2.2.2 AGRICULTURE 5.0 INNOVATIVE SOLUTION FOR AN "ENERGY SMART FARM (SF)"

To maximize the agricultural sector's potential, SF is an integrated farming system that combines all production aspects, integrates knowledge and agricultural research, and makes use of technological advancements along with particular decision-making processes [26, 27]. As discussed below, SF is associated with three interrelated fields of technology: Management information systems: Computerized programs designed to collect, evaluate, store, and distribute data in the manner needed to carry out an agricultural enterprise's tasks and duties [28]. Precision agriculture is the practice of controlling temporal and spatial variation to enhance profitability while lowering input use and environmental effects. The use of GPS, GNSS (Geo Location Systems), aerial photographs taken by drones (UAVs), and the most recent creation of surreal pictures captured by the sentinel satellites are characteristics of DSS (based on decision support system) for to enhance produces while preserving assets for every aspect of management of farming. As a result, maps of the geographical variability of several quantifiable variables are produced, including soil topography, organic matter, nitrogen levels, moisture content, and crop yield [29, 30]. The application of robotics, AI, and automated control approaches to all stages of agricultural crop production and farm drones and FarmBots is known as agricultural automation and robotics [31]. SF, also known as precision farming, aims to maximize agricultural yields by using the best practices available, protect the environment by reducing chemical use and making appropriate use of farm equipment, maximize labor productivity while increasing producer profits, and ultimately, maximize the maximum amount of environmental protection and the manufacturing of healthy products [26, 30]. The idea of "Agriculture 5.0" represents the direction that the agriculture industry is going to go in the upcoming years. To boost the productivity of agricultural businesses, it involves focusing agriculture on cutting-edge or modern technology like AI, ML and IoT [18, 32].

On the other hand, in European countries, Agriculture 5.0 is currently seeing growth. Precision farming combined with agricultural engineering is known as "agricultural 5.0 digital farming." The general public started using precision farming as GPS signals became accessible. In addition to enabling site-specific monitoring and vehicle guiding, precision farming improves operational accuracy. With just one click in digital farming, one may access weather forecasts, agricultural production, and other associated data (Figure 2.3).

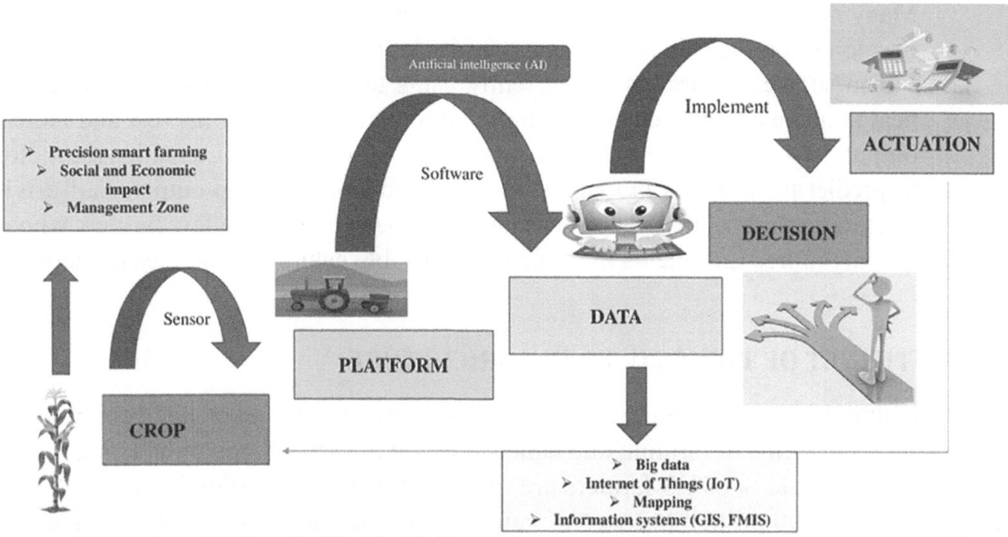

FIGURE 2.3 A cycle of information-based management for smart agriculture.
Source: Adapted from Ref. [59].

2.3 WHAT IS THE USE OF AI TECHNIQUES IN FARMING?

AI is defined by Professor McCarthy as " Developing intelligent machines and sophisticated software using engineering and scientific methods." The fundamental AI has long been supplied by controlled programs that demonstrate a basic form of "intelligence" under specific circumstances [3, 4].

AI stands as a pivotal domain in computer science research, distinguished by its extensive applications, rapid technological progress, and resilient performance in scenarios surpassing human or traditional computing capacities. Consequently, AI is swiftly becoming standard practice. Generally, crop management systems furnish a comprehensive platform for overseeing all aspects of farming operations, encompassing the entirety of crop management [23].

In agriculture, the challenge posed by insect pest infestation remains a prominent concern due to its substantial economic impact. Over the course of several decades, scholars have endeavored to mitigate this issue by exploring electronic or digital technologies capable of detecting pest activity and proposing effective remedial measures [24]. Crop diseases pose significant concerns for farmers. Recognizing unhealthy plants and applying appropriate remedies necessitates extensive expertise. Globally, computer-assisted methods detect diseases and recommend treatments. Agriculture involves crop storage, drying, grading, and AI-based food monitoring systems. Effective soil health and irrigation management are critical, impacting crop quality and yield efficiency [25].

1. Indian grape growers who have started using geographic location to identify crop diseases or famines, enabling them to more precisely and sooner manage infestations. Additionally, this results in less toxic pesticides being applied to the crop.

Many new farmers use soil mapping software to determine how much fertilizer is best to use on their fields. To spray herbicides more precisely, they use drones.

2. Indian sugarcane farmers are currently using technology to determine when it is best to harvest their crops, allowing them to better plan their harvest and increase output. Several Indian farmers have also started utilizing AI/ML-enabled technology to predict price trends in mandis, weather patterns, and crop output. Self-driving tractors and robots that plant seeds have also been tested by a few farmers who are seeking to free their fields from the unpredictable nature of shortages of labor.

2.4 INTERNET OF THINGS USED IN AGRICULTURE

The agricultural sector, the process of employing sensors and other devices to convert every aspect and action of farming into data is referred to as the "Internet of Things" (IoT). Between 10% and 15% of US farmers are reportedly using IoT technologies on 250,000 farms covering 1200 million hectares of land [33]. IoT technologies are the driving force behind Agriculture 4.0 [34], and they also contribute to the wealth of useful information that agriculture can offer [35]. The agriculture sector is anticipated to be greatly impacted by these technological advancements. By 2050, it is predicted that new methods enabled by the IoT might boost agricultural production by 70% [36].

Higher crop yields and lower costs are the two key benefits of using IoT. According to research from on-farm, for instance, an average farm employing IoT sees an increase in output of 1.75%, a decrease in energy expenses of 17–32 dollars per hectare, and an 8% decrease in water use for irrigation [34].

An exceptionally promising technological advancement aimed at revitalizing the agriculture sector and bridging industrial divides is the IoT. Academic and research institutions are actively creating products and solutions utilizing IoT to address various agricultural challenges. Policies for IoT-based agriculture were also explored at the national level. Ultimately, a list of unresolved issues and difficulties in the IoT agriculture space has been supplied to help researchers think about their alternatives moving forward [37].

Uses of IoT in the following categories have generally been determined to be frequent areas where IoT solutions are being deployed in Figure 2.4.

The agriculture sector can benefit from IoT technology in several aspects [58].

A few advantages are enumerated in the following section:

1. **Collective information:** The expansion of cooperative farming can be aided by IoT technology, especially in remote and rural areas. IoT can help with the development of apps which facilitate the exchange or sharing of information, collaboration of farm data management, and improved communication among agriculture experts and farmers.

2. **Safety and security:** Ranchers face difficulties that extend beyond the requirement to grow or need larger quantities of food production; these challenges include ensuring that food is dependable, secure, and readily available. Issues concern the quality of food being delivered for the market, including such issues as food

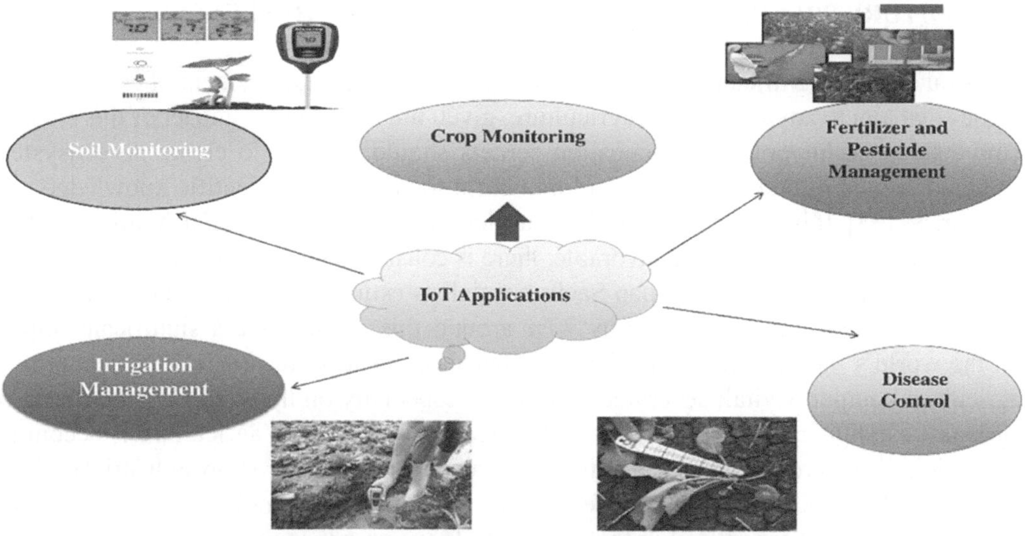

FIGURE 2.4 Benefits of IoT technology for the agricultural industry.

adulteration, counterfeiting, artificial plant augmentation, and large fruit sizes, have been reported on multiple occasions. These issues, which primarily influence medical conditions, could be harmful to economic expansion.

3. **Competition and challenging nature:** Agriculture is turning into an increasingly dynamic economic sector due to the increasing need for higher for greater agricultural production and the integration of novel cutting-edge technologies. It fortifies the framework in which the use of IoT technology for information sharing in the farming sector will provide further marketing, management, and monitoring possibilities.
4. **Growth and development**: Using IoT technology could offer new insights into the development of jobs. These strategy benefits farmers since it does away with a requirement for "intermediate men," frequently abused, and might interact with direct customers, resulting in more profits.
5. **Low costs and damage:** The primary benefit of IoT technology is the ability to recognize distant sensors and gadgets. IoT technology gives farmers greater decision-making power by informing them when to use herbicides and insecticides. As a result, crop damage and agriculture expenses may be reduced.
6. **Awareness:** The IoT device network can manage cost-effective services and control access to wireless sensor networks in the agricultural sector. Apps for smartphones offer data about products, services, costs, and market research. Government initiatives, policies, and regulations of various agricultural farm produce remain in place.
7. **Managing of resources:** IoT can enable real-time to sense and monitor agricultural investments in actual time to instantly protect different pieces of equipment from theft, update individual parts, and finish tasks on schedule.

1.5 FUTURE PROSPECTS OF AGRICULTURE 5.0 AND CHALLENGES

It is strategically significant for the government to assist the establishment of creative businesses that aim to advance digital agriculture, given the high capital impact of this change. Universities' technopoly and incubation centers should be able to create an ecosystem centered around digital agriculture and transform the corpus of scientific knowledge into enterprises [38]. It is uncertain if digitizing agriculture has been beneficial. While scientists concur that the effect would be favorable, there is considerable variation in the anticipated rate of efficiency growth. Goldman Sachs specialists estimate that new technology might boost agricultural output globally by 70% around this might have a significant impact on the world's food supply [44]. The importance of ensuring the social sustainability of agricultural output is vital; several authors focus especially on the severe social effects of agriculture's digitization [39]. Scientists frequently evaluate how agriculture is becoming more digitalized through industrialization. Researchers have studied how industrial agriculture evolved while considering the effects of the revolution in technology [45]. Industry 4.0 and Agriculture 4.0 are related in that they both focus on advancing fieldwork automated operations through the use of automation [46].

Robotics applications, Al, ML, and DL can detect crop diseases and estimate yields a probability of 75% chance, which improves crop quality and boosts farm profitability, according to researchers studying Agriculture 5.0 and the availability of food [40]. Sustainability of agriculture and increasing productivity and sustainability should go hand in hand with the shift in crop production from SA to Agriculture 5.0 [47]. Thus, it makes sense to think about Agriculture 5.0 in connection with Industry 4.0 and Society 5.0, as well as creating eco-friendly innovations technologies, building smart and ecologically sound cities, and securing industrial progress as a prerequisite for human welfare or mankind [41].

The modern world is distinguished by extreme differences between nations when it comes to food production and consumption. This defines their different levels of food safety, or the availability and supply of food required for an active and healthy life. The index of global food security, which is a composite statistic derived from 34 food security measures in 113 nations and produced according to *The Economist*'s methodology, is used to evaluate the degree of global food security worldwide [42].

Low-income nations like Nigeria and Pakistan have ineffective agricultural systems and a negative food net trade. These nations have fast population growth and low agricultural productivity. These countries are unlikely to increase food production through intensive techniques due to their poor potential for innovation, and even national food self-sufficiency is a concern. Due to their export-focused agriculture, Brazil and Indonesia may have an impact on the global food supply. When it comes to net trade in food, Brazil leads the world.

Brazil's agricultural sector has changed over the past 10 years from a traditional production system that used a few new technologies to a global leader in agriculture. The most productive farms have been able to adopt agricultural advances at a rapid pace [43].

Modern methods for addressing some problems in the farming and agricultural sectors have been merged with *artificial intelligence* of things based smarter applications [56].

Lack of structure, security, privacy, power, data merging, a wide range of IoT devices, and inconsistent business strategies are among some of the problems. There are several costs associated with using IoT in agriculture, including organizational and managerial costs [52]. Moreover, failure to take proper security precautions can lead to private information, like field data, as well as intellectual property, data loss, and privacy violations.

Farmers need to know about the best resource mechanisms for analyzing the quantity of data to be exchanged, the several gateways or IoT devices needed, and the amount of RAM needed for cloud storage to maximize revenue. Outdoor conditions can have an impact on sensors used in IoT devices, progressively decreasing their effectiveness and potentially leading to gearbox system failure. Protecting IoT sensors and structure installations is necessary to prevent expensive equipment from being damaged by various natural disasters, such as floods, heavy rain, cyclones, and so on [54]. Many IoT innovations are currently in the process of evolving, even though some are still being investigated. Selecting the best strategy is crucial since new technologies need to be thoroughly studied before they can be used. Before selecting a new technology, several aspects are considered, such as the farm's size and form, location, soil types, and surrounding conditions [53, 55, 57].

2.6 CONCLUSION

Boosting agricultural productivity is the primary means of securing worldwide food security in the 21st century. Technology must be introduced for agriculture to evolve to Agriculture 5.0. This would allow farmers to raise production while minimizing agriculture's adverse environmental impact and resolving several major political and social problems with food systems. A long-term solution to the postpandemic demands of the world, as well as to climate change, population increase that is taking place too rapidly, and environmental degradation, is Agriculture 5.0. Signifies the arrival of Agriculture 5.0, the next era of intelligent farming. This new period in the postpandemic world promises significant advancements in the agriculture field. The past two years of crop growth caused by the pandemic diving into two categories: (1) preserving food or nutritional sufficiency and (2) preparation for the phase of vertical production and new technologies exploitation. Consequently, the change from a conventional smart farm to an energy-smart farm is being made possible by Agriculture 5.0 when combined with smart equipment and green energy sources. The primary feature of Agriculture 5.0 is the requirement for RS combined with the adoption of renewable energy sources, including biogas and digestion by an anaerobic process. As a result, it may be achieved through the use of various smart devices that are low cost and environmentally benign, such as sensors, smart tractors, UAVs, and much more. These devices are based on the latest developments like AI, and BD analytics in the fields of ML. Persistent national inequality, however, makes it unlikely that the world will soon be able to produce more high quality, reasonably priced food. Thus, ensuring global food security is not a challenge that can be resolved by digital transformation technologies.

In a nutshell, SA farming will meet the needs of a large scale, flexible agricultural system which preserves the sustainability and environment, shows adaptability in the face of impending difficulties, and generates a reliable, long-term source of income within restricted factors. Furthermore, IoT and AI technology will help farmers and the agriculture sector satisfy the world's food needs by 2050.

KEYWORDS

- **artificial intelligence**
- **Internet of Things**
- **Agriculture 5.0**
- **data**
- **sensing**

REFERENCES

1. Zhai, Z., Martínez, J. F., Beltran, V., & Martínez, N. L. (2020). Decision support systems for agriculture 4.0: Survey and challenges. *Computers and Electronics in Agriculture*, *170*, 105256.
2. Pivoto, D., Waquil, P. D., Talamini, E., Finocchio, C. P. S., Dalla, V. F., Corte, G. de., & Vargas, M. (2018). Scientific development of smart farming technologies and their application in Brazil. *Information Processing in Agriculture*, *5*, 21–32.
3. Dutta, S., Rakshit, S., & Chatterjee, D. (2020). Use of artificial intelligence in Indian agriculture. *Food Science and Technology*, *1*, 65–72.
4. Pathan, M., Patel, N., Yagnik, H., & Shah, M. (2020). Artificial cognition for applications in smart agriculture: A comprehensive review. *Artificial Intelligence in Agriculture*, *4*, 81–95.
5. Bannerjee, G., Sarkar, U., Das, S., & Ghosh, I. (2018). Artificial intelligence in agriculture: A literature survey. *International Journal of Scientific Research in Computer Science Applications and Management Studies*, *7*, 1–6.
6. Boliko, M. C. (2019). FAO and the situation of food security and nutrition in the world. *Journal of Nutritional Science and Vitaminology*, *65*(Supplement), 4–8.
7. FAO. (2017). *The Future of Food and Agriculture—Trends and Challenges*. Rome.
8. Gowda, V. D., Prabhu, M. S., Ramesha, M., Kudari, J. M., & Samal, A. (2021). Smart agriculture and smart farming using iot technology. *Journal of Physics: Conference Series*, *2089*, 012038.
9. Srivastava, K. (2021). Farming automation. In *Artificial Intelligence and Iot-Based Technologies for Sustainable Farming and Smart Agriculture* (pp. 68–82). IGI Global.
10. Tăbăraşu, A-M., Anghelache, D., & Dumitru, D. (2020). Benefits Regarding the Implementation of Agriculture 4.0 in the Current Context, Annals of the University of Craiova—Agriculture, Montanology, Cadastre Series.
11. Vladut, V., et al. (2020). Agriculture 4.0—The Use of Smart Technologies for High-Performance Agriculture, Annals of the University of Craiova—Agriculture, Montanology, Cadastre Series.
12. Ling, Y. (2017). Application of artificial intelligence technology in agriculture. *Computer Knowledge Technology*, *29*, 181–183.
13. Chen, T. J., Zeng, J., Xie, C. J., et al. (2019). Intelligent system for identifying pests and diseases based on deep learning, China PlantProtect. *Guide*, 4,
14. Song, J., Rondao, D., & Aouf, N. (2022). Deep learning-based spacecraft relative navigation methods: A survey. *Acta Astronautica*, *191*, 22–40.
15. Majid, A. A. M., Alshaibi, A. J., Kostyuchenko, E., & Shelupanov, A. (2021). A review of artificial intelligence based malware detection using deep learning. *Materials Today: Proceedings*; *epub ahead of print*.

16. Fan, H. (2022). The digital asset value and currency supervision under deep learning and blockchain technology. *Journal of Computational and Applied Mathematics*, *407*, 114061.
17. Kumar, L. A., Renuka, D. K., Rose, S. L., & Wartana, I. M. (2022). Deep learning based assistive technology on audio visual speech recognition for hearing impaired. *International Journal of Cognitive Computing in Engineering*, *3*, 24–30.
18. Spanaki, K., Karafili, E., & Despoudi, S. (2021). AI applications of data sharing in agriculture 4.0: A framework for role-based data access control. *International Journal of Information Management*, *59*, 102350.
19. Doshi, M., & Varghese, A. (2022). Smart agriculture using renewable energy and AI-powered IoT. In *AI, Edge and IoT-based Smart Agriculture* (pp. 205–225). Academic Press: Cambridge, MA, USA.
20. Zhang, M., Wang, X., Feng, H., Huang, Q., Xiao, X., & Zhang, X. (2021). Wearable Internet of Things enabled precision livestock farming in smart farms: A review of technical solutions for precise perception, biocompatibility, and sustainability monitoring. *Journal of Cleaner Production*, *312*, 127712.
21. Mahmud, M. S., Zahid, A., Das, A. K., Muzammil, M., & Khan, M. U. (2021). A systematic literature review on deep learning applications for precision cattle farming. *Computers and Electronics in Agriculture*, *187*, 106313.
22. Manida, M., & Ganeshan, M. K. (2021). New agriculture technology in modern farming. In *Proceedings of the 2nd International Multidisciplinary Conference on Information Science, Management Research and Social Science*. Chennai, Tamil Nadu.
23. Chukkapalli, S. S. L., Mittal, S., Gupta, M., Mahmud, A., Joshi, A., Sandhu, R., & Joshi, K. (2020). Ontologies and artificial intelligence systems for the cooperative smart farming ecosystem. *IEEE Access*, *8*, 164045–164064.
24. Jerhamre, E., Johan, C., Carlberg, C., & Zoest, V. V. (2022). Exploring the susceptibility of smart farming: Identified opportunities and challenges. *Smart Agricultural Technology*, *2*, 100026.
25. Daponte, P., De Vito, L., Glielmo, L., Iannelli, L., Davide, L., Picariello, F., & Silano, G. (2019). A review on the use of drones for precision agriculture. In *IOP Conference Series: Earth and Environmental Science* (p. 275). IOP Publishing.
26. Fraser, A. (2021). You can't eat data'?: Moving beyond the misconfigured innovations of smart farming. *Journal of Rural Studies*, *91*, 200–207.
27. Cordeiro, M., Markert, C., Araújo, S. S., Campos, N. G., Gondim, R. S., da Silva, T. L. C., & da Rocha, A. R. (2022). Towards Smart Farming: Fog-enabled intelligent irrigation system using deep neural networks. *Future Generation Computer Systems*, *129*, 115–124.
28. Rodríguez, J. P., Montoya-Munoz, A. I., Rodriguez-Pabon, C., Hoyos, J., & Corrales, J. C. (2021). IoT-Agro: A smart farming system to Colombian coffee farms. *Computers and Electronics in Agriculture*, *190*, 106442.
29. Carrer, M. J., de Souza Filho, H. M., Vinholis, M. D. M. B., & Mozambani, C. I. (2022). Precision agriculture adoption and technical efficiency: An analysis of sugarcane farms in Brazil. *Technological Forecasting and Social Change*, *177*, 121510.
30. Bwambale, E., Abagale, F. K., & Anornu, G. K. (2022). Smart irrigation monitoring and control strategies for improving water use efficiency in precision agriculture: A review. *Agricultural Water Management*, *260*, 107324.
31. Shrivastava, A., Nayak, C. K., Dilip, R., Samal, S. R., Rout, S., & Ashfaque, S. M. (2021). Automatic robotic system design and development for vertical hydroponic farming using IoT and big data analysis. *Materials Today: Proceedings*; epub ahead of print.
32. Fraser, E. D., & Campbell, M. (2019). Agriculture 5.0: Reconciling production with planetary health. *One Earth*, *1*, 278–280.
33. Alpha Brown. What is IoT in agriculture? Farmers aren't quite sure despite $4bn US opportunity—report [Internet]. [cited 2019 Dec 28]; Available from: https: //agfundernews.com/iot-agriculture-farmers-arent-quite-sure-despite-4bn-usopportunity.html.
34. Gralla, P. Precision agriculture yields higher profits, lower risks [Internet]. [cited 2019 Dec 29]; Available online: https: //www.hpe. com/us/en/insights/articles/precision-agriculture-yields-higher-profits-lower-risks-1806.html.

35. Tzounis, A., Katsoulas, N., Bartzanas, T., & Kittas, C. (2017). Internet of Things in agriculture, recent advances and future challenges. *Biosystems Engineering, 164*, 31–48.
36. Sarni, W., Mariani, J., & Kaji, J. *From Dirt to Data: The Second Green Revolution and IoT*. Deloitte insights [Internet]. [cited 2019 Sep 18]; Available online: https: //www2.deloitte.com/insights/us/en/deloitte-review/issue-18/second-greenrevolution-and-internet-of-things.html#endnote-sup-9.
37. Jain, V. R., Kumar, A., & Ranjan, P. (2008). wildCENSE: GPS based animal tracking system. In *International Conference on Intelligent Sensors, Sensor Networks and Information Processing*. IEEE.
38. Ozdogan, B., Gacar, A., & Aktas, H. (2017). Digital agriculture practices in the context of agriculture 4.0. *Journal of Financial Economics, 4*, 184 http: //doi.org/10.17261/Pressacademia.2017.448
39. Rose, D., Wheeler, R., Lobley, M., Winter, M., & Chivers, C. A. (2017) Agriculture 4.0: Making it work for people, production, and the planet. *Land Use Policy*, 100, 104933 https: //doi.org/10.1016/j.landusepol.2020.104933
40. Murugesan, R., Sudarsanam, S. K., Malathi, G., Vijayakumar, V., Neelanarayanan, V., Venugopal, R., Rekha, D., Saha, S., Bajaj, R., Miral, A., & Malolan, V. (2019). Artificial intelligence and agriculture 5.0. *International Journal of Recent Technology and Engineering (IJRTE), 8*, 1870–1877.
41. Berawi, M. A. (2019). Managing nature 5.0 in industrial revolution 4.0 and Society 5.0. *Era International Journal of Technology (IJTech), 10*, 222–225 https: //doi.org/10.14716/ijtech.v10i2.3084
42. Global Food Security Index. (2020). https: //foodsecurityindex.eiu.com/Index
43. Rada, N., & Valdes, C. (2012). *Policy, Technology, and Efficiency of Brazilian Agriculture, ERR137*. U.S. Department of Agriculture, Economic Research Service
44. Revich, J., Koort, R., & Archambault, P. (2016). *Precision Farming: Cheating Malthus with Digital Agriculture*. Goldman Sachs.
45. Liu, Y., Ma, X., Shu, L., Hancke, G., & Abu-Mahfouz, A. (2021). From industry 4.0 to agriculture 4.0: Current status, enabling technologies, and research challenges. *IEEE Transactions on Industrial Informatics, 17*, 4322–4334 https: //doi.org/10.1109/TII.2020.3003910
46. Kovács, I., & Husti, I. (2018). The role of digitalization in the agricultural 4.0 – how to connect the industry 4.0 to agriculture? *Hungarian Agricultural Engineering, 33*, 38–42 https: //doi.org/10.17676/HAE.2018.33.38
47. Saiz-Rubio, V., & vira-Más, F. (2020). From smart farming towards agriculture 5.0: A review on crop data management. *Agronomy, 10*, 207 https: //doi.org/10.3390/agronomy10020207
48. Javaid, M., Haleem, A., Singh, R. P., & Suman, R. (2022). Enhancing smart farming through the applications of agriculture 4.0 technologies. *International Journal of Intelligent Networks, 3*, 150–164.
49. Jellason, N. P., Robinson, E. J., & Ogbaga, C. C. (2021). Agriculture 4.0: Is sub-Saharan Africa ready? *Applied Sciences, 11*, 5750
50. Siddharth, D., Saini, D. K., & Kumar, A. (2021). Precision agriculture with technologies for smart farming towards agriculture 5.0. In *Unmanned Aerial Vehicles for Internet of Things (IoT) Concepts, Techniques, and Applications* (pp. 247–276).
51. Gyamfi, E. K., ElSayed, Z., Kropczynski, J., Yakubu, M. A., & Elsayed, N. (2024). Agricultural 4.0 leveraging on technological solutions: Study for smart farming sector, preprint, arXiv: 2401.00814.
52. Cesco, S., Sambo, P., Borin, M., Basso, B., Orzes, G., & Mazzetto, F. (2023). Smart agriculture and digital twins: Applications and challenges in a vision of sustainability. *European Journal of Agronomy, 146*, 126809.
53. Ayaz, M., Ammad-Uddin, M., Sharif, Z., Mansour, A., & Aggoune, E.-H. M. (2019). Internet-ofthings (iot)-based smart agriculture: Toward making the fields talk. *IEEE Access, 7*, 129551–129583.
54. Cesco, S., Pii, Y., Borruso, L., Orzes, G., Lugli, P., Mazzetto, F., Genova, G., Signorini, M., Brunetto, G., & Terzano, R. (2021). A smart and sustainable future for viticulture is rooted in soil: How to face Cu toxicity. *Applied Sciences, 11*, 907.
55. Glaroudis, D., Iossifides, A., & Chatzimisios, P. (2020). Survey, comparison and research challenges of IoT application protocols for smart farming. *Computer Network, 168*, 107037.
56. Ray, P. P. (2017). Internet of things for smart agriculture: Technologies, practices and future direction. *Journal of Ambient Intelligence and Smart Environments, 9*, 395–420.

57. Terence, S., & Purushothaman, G. (2020). Systematic review of Internet of things in smart farming. *Transactions on Emerging Telecommunications Technologies*, *31*, e3958.
58. Priya, P., & Kaur, G. (2021). Smart sensors for smart agriculture. In *Artificial Intelligence and IoT-Based Technologies for Sustainable Farming and Smart Agriculture* (pp. 175–191). IGI Global.
59. Verónica, S. R., & Francisco, R. M. (2020). From smart farming towards Agriculture 5.0: A review on crop data management. *Agronomy, 10*, 207.

CHAPTER 3

The Rise of Agri-Drones: Transforming Food Supply Chains Through Robotics and Smart Devices in Agriculture

DEEPAK GUPTA[1,2], ALKA VERMA[3*], ARUN KUMAR CHAUDHARY[4], MAKKHAN LAL MEENA[5], DIPESH KUMAR MISHRA[6], and WARSHI SINGH[7]

[1]*Mechanical Engineering Department, Graphic Era Hill University, Dehradun, Uttarakhand, India*

[2]*Mechanical Engineering Department, Graphic Era (deemed to be) University, Dehradun, Uttarakhand, India*

[3]*College of Agriculture, G. B. Pant University of Agriculture and Technology, Pantnagar, Uttarakhand, India*

[4]*Industrial and Production Engineering Department, G. B. Pant University of Agriculture and Technology, Pantnagar, Uttarakhand, India*

[5]*Mechanical Engineering Department, Malaviya National Institute of Technology, Jaipur, Rajasthan, India*

[6]*Mechanical Engineering Department, Madan Mohan Malaviya University of Technology, Gorakhpur, Uttar Pradesh, India*

[7]*School of Business Management, CSJM University, Kanpur, Uttar Pradesh, India*

*Corresponding author

ABSTRACT

A prime example is the usage of unmanned aerial vehicles, or agri-drones, or in this novel subdomain such as agricultural robotics. After a general introduction to the field, the chapter will focus on the evolution of drones and how they overhaul food supply chains. A closing historical scope will provide an overview of drone technology in the agriculture domain from its emergence to the present. This will be followed by a description of the numerous drone types and classifications and their relation to various applications in precision agriculture, such as crop health monitoring, targeted spraying, and irrigation management. Next, the chapter will discuss if and how drones are used throughout the food supply chain and the way in which their use improves efficiency, cost-effectiveness, and sustainability. Finally, the chapter will focus on the challenges, limitations, and hindrances linked with drones and using drones mentioned regulatory environment and issues, safety, and privacy

Demystifying Sustainable Farming Using Smart Technologies.
Gunjan Chhabra, Keshav Kaushik, Priya Chugh, Susheela Dahiya, and Aryan Chaudhary (Eds.)

DOI: 10.1201/9781779640932-3

concerns technology and discussed future developments. To demonstrate drones' influence on agricultural productivity and sustainability more precisely, actual-world studies and achievement examples will also be presented. Thus, while drones have the potential to alter the safety of the food market, there is still much more research to perform.

3.1 INTRODUCTION

The current global agricultural soil is at the peak of an essential transition, invented by the confluence of robotics, smart devices, and data-centered decision-making. Agriculture 5.0 represents this age—a stepwise grid from traditional procedures to a much more exact, manageable, and productive way of food production. An agri-drone is the innovation behind the prediction. Agri-drones are unmanned aerial vehicles or UAVs precisely like those in the introductory note. However, no longer will we elevate our food—the drones, equipped with sensors and smart devices, will gather data used by scientists to get in real-time information on the crops' health, the colony and damage by pests, the soil, and other environmental determinants. The real-time data are consumed by robotics and any other type of automation, such as the system-level know-how described earlier, decreases the time from data gathering to decision-making. The consequence of this is that farmers can allocate scarce resources presently secure one and the rationale for pest control and irrigation, which in turn means greater output and lower spoilage.

Moreover, the implementation of agri-drones, robotics, and smart devices plays a critical role in the revolution of food supply chains due to improved productivity, efficiency, and sustainability [1, 2]. According to the identified studies, agri-drones with advanced sensors impact crop monitoring and soil analysis; in combination with pest management, they lead to better yield and minimal resourcing waste [3]. Moreover, smart contracts, blockchain technology, and the Internet of Things (IoT) in agriculture are south-out measures that allow for real-time monitoring of activities, secure transactions, and automation of the supply chain [4, 5]. These approaches and technologies developed do not only optimize but ensure traceability, transparency, and quality assurance throughout the agricultural supply chain. Lastly, Industry 4.0, along with artificial intelligence (AI), big data, and robotics, accelerates the implementation of digital transformation in the agri-food industry and ensures that climate change, population growth, and diseases are addressed efficiently.

3.1.1 OVERVIEW OF AGRICULTURAL DRONES

Agriculture drones, also referred to as unmanned aerial vehicles help to achieve various new dimensions in agricultural practices. These drones facilitate comprehensive real-time data collection, high-resolution onboard imaging, and aid in decision making in irrigation, fertilization, and pest management [6]. Additionally, drones play a major role in precision spraying, reducing chemical wastages, and optimizing the use of other resources [7, 8]. Moreover, these drones also help in the mapping and survey of an area, allowing for more accurate and systematic-based crop planning as well as allocation of resources [9]. UAVs also assist\ in sustainable agriculture through tasks such as spraying, irrigation scheduling, monitoring of the crop, soil analysis, and bird scarring project. Besides, drones are also integrated with sophisticated technologies and sensors to promote precision agriculture, data management, collection, and analysis and create a safe working environment. Generally, drones in agriculture promise efficiency, productivity, and sustainable practices.

3.1.2 IMPORTANCE OF DRONES IN TRANSFORMING FOOD SUPPLY CHAINS

Overall, drones significantly transform food supply chains by revolutionizing the key subsystems of the agri-food sector [10, 11]. In the agriculture systems, they enable effective monitoring of crops, estimation of biomass, and mapping of yield, which improves productivity in small-holder farming systems [12]. In addition, drones integrated with AI and blockchain technology allow for traceability, transparency, and management of inventories and food chains between producers and consumers. The potential use of drones in food delivery services is limited to the level of technological advancement and maturity, and the level of legislation that restricts drone use in different jurisdictions. In addition, drones with or without AI applications have eased sustainable agriculture through efficient production processes, prediction of diseases, and better utilization of resources (Figure 3.1). Thus, drones are essential in transforming how food chains are monitored, managed, and maintained.

FIGURE 3.1 Drones in transforming food supply chains.

Source: Adapted from https: //agritimes.co.in/investment-finance/india-s-kharif-crop-sowing-touches-13-13-million-hectares/drones-revolutionizing-indias-agriculture-landscape

This chapter focuses on the introduction of agri-drones, their importance in different levels of the food chain, the diverse use of flying robots, and the future. The chapter also details on the coming major uses of agri-drones, such as field surveillance, precision spraying, planting of seeds, and related work. The chapter further outlines the challenges such as policymaking, security, and ownership of information. The section seeks to state a long-term view of what drones have to offer and their impact on the agricultural field on 5.0 through a comprehensive review.

3.2 HISTORY AND EVOLUTION OF AGRICULTURAL DRONES

The rapidly growing field of agricultural drones is an intriguing convergence between technological developments in aeronautics and the age-old drive to enhance food production. Although the idea of unmanned aerial vehicles has a longer gestation, their targeted application in agriculture is relatively new. This chapter seeks to explore the brief history of the alignment, outlining the major emersions that have led to the emergence of agri-drones and their industry-altering influence.

3.2.1 EARLY DEVELOPMENTS IN AGRICULTURAL DRONE TECHNOLOGY

The first emergence: from crop dusting to precision mapping (early 2000s), the roots of agricultural drones stretch back to at least the early 2000s. Farming deployed UAVs for the first time at this time, even if crop dusting lines the UAV's practice back to human-filled aircraft in the 20th century. It was distinguished by the Swedish University of Agricultural Sciences: pioneering the creation of drones fit with cameras, this was the first time farmers may get comprehensive aerial views of their fields. They might make accurate field maps and recognize possible problems before they became large ones. Technological dominance: specialized drones and advancements (2010s onward). Precision farming has seen a massive rise in the adoption of agricultural drones in the 2010s; owing to the battery life, flight stability, and sensor-quality enhancements, agricultural drones employed have grown increasingly specialized. Multispectral imaging drones to detect diseases and nutrient deficits have become prevalent during this period. LiDAR technology has been developed to produce complex 3D models of the terrain. The most remarkable agri-drones impact might have been when steadier advancements were coupled with refined pesticide applications. Integration into Agriculture 5.0 and beyond Agriculture 5.0 is now being defined as precision farming since it is often powered by robotics and big data, utilizing smart technology in a targeted way [13].

Agricultural drones are thus integrated with other linked devices like automatic irrigation systems and farm management software, allowing a data-driven strategy that offers farmers real-time ways to optimize crop health and resource use. Furthermore, as technology advances at a significant rate, more applications may be anticipated. These could encompass automated gaming and planting, constant searching of cows, and on-demand crops analyzed via ADS-B incorporated drones. The agricultural drones narrative is far from complete. As technology evolves and regulations change, the flying machines have a lot more potential to improve food cultivation. Farming a more sustainable and efficient future [14].

3.2.2 MILESTONES IN THE EVOLUTION OF AGRI-DRONES

The success story of agri-drones is essentially a tale of innovation that never seems to end, because there is always room for improvement in agriculture. The following is a brief history of some significant milestones: early 1990s: the first agricultural application of drones to the work. First-generation UAVs could only take aerial photographs of crops and spray chemicals as the forerunner of manned aircraft method. 2006: The US Department of Agriculture took a giant step toward the future with the launch of the UAV for Agricultural Use Program [15]. The program actively promotes the use of drones in agricultural operations through research and development. 2013: The global agricultural UAV market exceeded the threshold of $200 million (Table 3.1).

TABLE 3.1 Milestones in the Evolution of Agri-Drones

Period	Key Development	Impact on Agriculture
Early 1900s	First use of aerial vehicles for crop dusting (proto-drones)	Pioneered the concept of aerial application in agriculture
1930s	Use of airplanes for seed spreading and pest control	Increased efficiency and scale of aerial applications
1960s	Advancements in RC technology and hobbyist drones	Laid the foundation for smaller, more maneuverable aerial platforms
2000s	Development of dedicated agricultural drones	Increased affordability and accessibility of drones for farmers
2010s	Integration of cameras, sensors, and GPS	Enabled data collection and precision agriculture practices
Present day	Advancements in AI, automation, and battery life	Improved autonomy, efficiency, and adoption rate of drones

This rapid growth indicates that drones are progressively regarded as a viable approach.

2015: The Asian nation's Ministry of Agriculture released detailed guidelines for the use of drones in agriculture. As a result of the increased governmental support for drones, their use skyrocketed in the Chinese agricultural sector.

2016: Overcoming bureaucratic difficulties. The U.S. Federal Aviation Administration adopted new regulations that govern commercial drone use. This makes it easier for agricultural producers to integrate drones into their daily operations. Today in 2024, in this time, agri-drones have been effectively integrated into Agriculture 5.0. It is the time of precision farming and data-driven decision-making [16]. Drones collaborate seamlessly with other systems and farm management software. They work hand in hand and create a highly connected system to optimize agricultural activities.

- **Automation:** However, the story of agri-drones is not over. On the contrary, the new technological front promises to be more thrilling, more automation. The full automation of planting, spraying, and harvesting operations with specially equipped drones.
- **Data analysis:** Real-time data collated by drones will be processed using advanced AI technology to provide information on crop health and resource management.
- **Robotics integration:** Drones will be integrated with ground-based similar cyborgs to form a connected farm management system. The summary of this path from the past to the future shows that drones will be a vital factor in the foreseeable future.

3.2.3 ADOPTION TRENDS AND GROWTH TRAJECTORY

The soaring Adoption and Trajectory of Agri-drones: What to expect in the Future? Agri-drone adoption has tremendously soared over the years and the future projections are pointing to an even higher trajectory of the use of the technology (Figure 3.2).

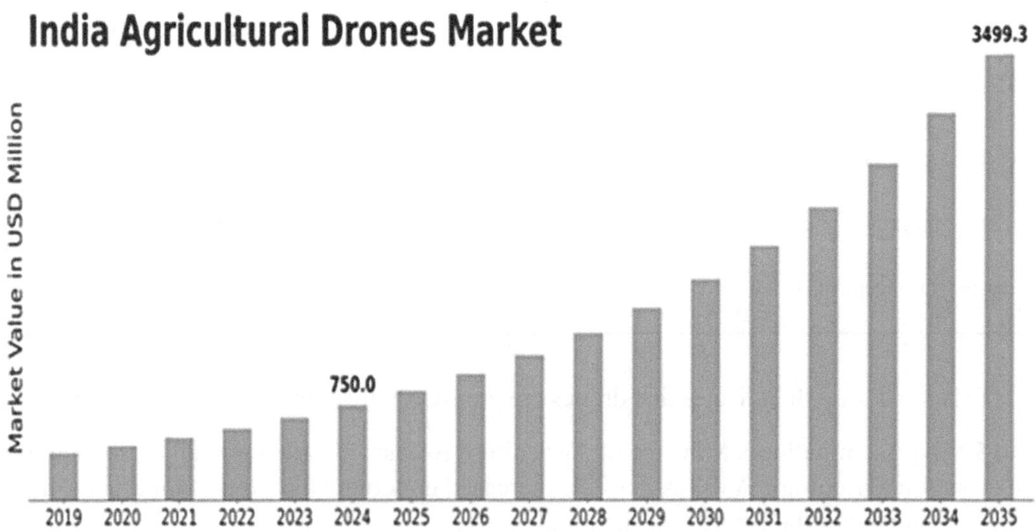

FIGURE 3.2 Growth trajectory of agri-drones.

Source: https: //www.marketresearchfuture.com/reports/india-agricultural-drones-market-44405

This subsection will delve into the various drivers that have fueled this adoption as well as give a view of what to expect in terms of the future of agri-drones in the agriculture scope.

3.2.3.1 THE MARKET PROJECTIONS

Market expansion choose credible template online. For instance, from Exactitude Consultancy, the projected valuation of the global agriculture drones and robots is pegged to hit an unprecedented US$12.3 billion by 2027 [17]. This will be a monumental rise from the estimated US$3.8 billion in 2020. Parameters that will drive this expected market expansion [18].

3.2.3.2 REGIONAL VARIATIONS

In the agri-drone market, North America is currently the leader, with Europe and Asia-Pacific following suit. However, emerging markets such as Africa and South America are expected to experience the fastest growth as the technology has the potential to significantly boost agricultural productivity.

3.2.4 BEYOND MARKET METRICS

Growing the market size is only part of the growth story for agri-drones. Other significant advancements should be:

- **Diversification of applications:** In addition to collecting data and spraying, the drones will be used for planting seeds, monitoring livestock, and even assessing the health of crops postharvest harvesting.
- **Technological:** The battery life, flight range, and the set of sensors will increase further, enhancing the drones' capabilities in agricultural applications.
- **Integration with automation**: In all likelihood, the drones will be part of the broader networks of interconnected devices and agricultural robots, creating fully automated farming systems.

3.2.5 CHALLENGES AND CONSIDERATIONS

- **Regulation:** While the outlook is positive, regulation of drone flying varies significantly by region. Regulators can do more by calibrating many aspects of the drone policy to foster adoption.
- **Security and privacy:** People can collect an enormous amount of information with drones. It is critical to store and retrieve it safely and not misuse the data.
- **Infrastructure and training:** Developing and operating a charging network and providing farmers with the necessary education and support to help them adopt drones in their workflows if drones are going to play an essential role in the future of your business, what impact do you anticipate they will have?

To conclude, the embrace of agri-drones is on a steep upward trajectory due to various drivers. Continued improvements in technology and the resolution of the regulatory bottlenecks are expected to push the role of agri-drones even more critical in the realization of a more sustainable, productive, and data-based future for agriculture.

3.3 TYPES AND CLASSIFICATION OF AGRICULTURAL DRONES

In the context of precision agriculture, unmanned aerial vehicles are undoubtedly a game-changer. While understanding the types of drones and the varied nature of these classifications is critical for successful utilization of this technology [19]. Therefore, this chapter will focus on the main types of classed that are identified in the agricultural drone sector.

3.3.1 FIELD DYNAMICS

Fixed-wing drones: A drone may have a fixed-wing drones or a UAV type that looks like a miniature airplane; it can fly for a few hours and is the best option for covering large fields in one go (Figure 3.3).

FIGURE 3.3 Fixed-wing drone.

Source: Adapted from https: //www.jetphotos.com/aircraft/Schempp-Hirth%20Ventus%202cT

At the same time, flat-wing drones require specific airports for taking off and landing; thus, they are less mobile than other drones.

Multirotor drones (vertical take-off and landing): Multirotor drones offer the comparative advantage of vertical takeoff and landing, making them ideal low-flight workhorses in the field. Quadcopters and hexacopters are several designs on the market (Figure 3.4).

FIGURE 3.4 Multirotor drone.

Source: Adapted from https: //iotechworld.com/agribot/

Their flight times are somewhat shorter and limit their use to field scouting, targeted spraying, and as-needed data collection in specific areas, such as the corner of a field.

Hybrid drones: Hybrid drones possess the potential to enable flight times matching fixed-wing drones and the freedom and flexibility of multirotor drones (Figure 3.5).

FIGURE 3.5 Hybrid drones.

Source: Adapted from http: //www.dronefromchina.com/new/long-endurance-hybrid-agriculture-drone.html

This feature makes them ideal for use in capturing vast sequences of field data with pinpoint accuracy. Complex operation, however, demands highly skilled operators.

3.3.2 FUNCTIONAL SPECIALIZATION

Spraying drones: Spraying drones are built with specially designed tanks and nozzles used to dispense various pesticides, herbicides, or fertilizers in a targeted manner (Figure 3.6).

FIGURE 3.6 Spraying drones.

Source: Adapted from https: //www.piffle.me/drone-sprayer/

Spray drones are more efficient in applying the concoction, resulting in less wastage, more effective use of resources, and less harm to the surrounding environment and its various ecosystems.

Monitoring drones: Monitoring drones act as the farmer's "eyes in the sky," with powerful high-resolution cameras and multispectral sensors attached (Figure 3.7).

FIGURE 3.7 Monitoring drones.

Source: Adapted from https: //insidefpv.com/blogs/blogs/agriculture-crop-spraying-top-10-best-drones-india

Monitoring drones conduct an intensive crop health assessment that enables farmers to identify diseases, deficiencies, and other potential problems that could impact quality. The early identification of crop issues in this way allows farmers to proactively address problems as they arise.

Mapping drones: Mapping drones use LiDAR technology to create detailed three-dimensional maps of the terrain (Figure 3.8).

FIGURE 3.8 Mapping drones.

Source: Adapted from https: //uavcoach.com/agricultural-drones/

The data acquired from this process can be used to optimize field planning, better manage irrigation, and drill down on soil composition, which will allow farmers and scientists to increase resource allocation and profits.

3.3.3 SELECTING THE OPTIMAL DRONE

The ideal agricultural drone selection hinges on a variety of factors. The factors that may suggest how an agricultural should choose which drone to use may include: Overall, multirotor drones are quite friendly to operators and affordable. On the other hand, fixed-wing drones require more experience and high capital while hybrid drones are a balance. With the expanding playing field of aerial technology, one can anticipate specialized agricultural drones to become even more advanced in the future [20]. Thus, the ever-growing drone playbook will provide precision choice and effectiveness, designing a data-enhanced, long-lasting future reversed for farming.

3.4 APPLICATIONS OF AGRICULTURAL DRONES

The agricultural landscape has entered an era of transformation through the utilization of unmanned aerial vehicles, popularly referred to as agricultural drones. Agricultural drones are versatile aerial platforms that have a wide range of applications which considerably improve operational efficiency, resource optimization, and data utilization across the agricultural domain. This chapter explores the major applications of agricultural drones, illustrating how they have influenced farm management in different sectors.

3.4.1 ENHANCED FIELD MONITORING AND CROP SCOUTING

In the past, field scouting has required manual examination, which is a time-consuming and tireful endeavor. Agricultural drones, which are equipped with high-resolution cameras and multispectral sensors, supply farmers with a more comprehensive and quicker alternative. As a result, farmers have access to a bird's-eye perspective of their fields in real-time that facilitates the following: so the following represents a series of essential benefits it offers [21].

First, growers can recognize discoloration, pest infestations, nutrient shortages, and other stress factors and promptly act to minimize damage and potential lost yields. Second, through careful assessment of patterns in crop development throughout huge fields, drones consume more time; farmers can optimize resource usage and alter their irrigation and fertilization routines to guarantee the most ideal growing circumstances on their farm [22]. Third, sophrolaelia drones with drones facilitate targeted identification of weed areas, which has wide implications for agricultural sustainability; farmers can identify the exact location of weed species among their crops because of the detailed pictures. On the whole, this helps minimize herbicide usage while preserving top-of-the-range agricultural practices.

3.4.2 REVOLUTIONIZING PRECISION SPRAYING

Chemical overuse is another challenge associated with spraying methods, including pesticides, herbicides, or fertilizers. Traditional methods are unlikely to be as detailed and controlled, and the application occurs over wide areas, while drones are equipped with a spraying mechanism designed to overcome the abovementioned challenge and enabling the following solutions. Drones are characterized by precise issue control [23]. Capable of spraying precise doses, the tool makes it possible to apply a chemical strictly in the area, reducing the required quantity to unprecedented proportions: to drops rather than liters. In addition, using a drone means that no other cropland will be shrunk, as it can protect it from chemical drops. Drones might help farmers in the following ways.

3.4.3 OPTIMIZING IRRIGATION MANAGEMENT

Water scarcity is becoming an increasingly pressing issue for modern agriculture. Scientists argue that drone technology in agriculture, fitted with thermal imaging sensors, helps farmers maintain well-thought water-irrigation strategies: a natural solution to water scarcity. First and foremost, drones can help:

- Highlighting potential water stress locations: Drones may identify variations in crop canopy temperature and reveal that some areas might be experiencing water stress. Such information aids in directing water resources to the areas that need them most or entirely avoiding irrigating some parts of the field.
- Monitoring soil moisture: By detecting soil moisture levels across the field, farmers may respond to the crops' individual water needs. From that perspective, it allows for more efficient water distribution, the reduced waste, and efficient water use.
- Mapping fields' topography: Drones can provide detailed terrain maps, which can help in the design of efficient water-irrigation systems. For instance, if the field has a slight elevation toward one corner, by pouring more water into that corner helps avoid runoff.

3.4.4 EMPOWERING DATA-DRIVEN PRECISION AGRICULTURE

Currently visiting agricultural drones collect the very foundations of precision agriculture—a data-driven approach to farm management:

- Early yield estimation: Drones provide farmers with actual crop health data, making it possible to estimate the yield shortly after the beginning of the growing season. This information helps to plan marketing and marketing strategies and optimize the application of the resources throughout the growing season.
- Field mapping statistics: Drones can develop field maps providing farmers with a birds-eye perspective of the field. This information helps to plan patterns in

planting and fertilization and optimization of the harvest pipeline—leading to a better managed farm.

- Easy integration into the workshop platform: Finally, data from drones can easily be incorporated into advanced farm management platforms. This data stream lets farmers monitor crop health, plan their rest of the resources, and remain in control, utilizing the holistic view of their businesses.

3.4.5 EXPANDING APPLICATIONS IN LIVESTOCK MANAGEMENT

"Drones are not just huge assets to crop fields; they also provide innovative solutions for innovative ways of managing livestock." Some of these solutions may include the following: "scope":

- Improved herd monitoring,
- Complexity: tracking sickness,
- Identifying lost animals;
- Simplicity: regularly scheduled drone checks;
- Tracking potential breaches before becoming a problem;
- Functionality: dynamic grazing check-ups;
- Tracking and checking animal health.

Some other things technological advancement of drones will make possible because the technology is rapidly evolving.

3.5 INTEGRATION OF DRONES IN FOOD SUPPLY CHAINS

Drones, or unmanned aerial vehicles, are poised to revolutionize the food supply chain. The result is not only the field, but also increases the scale, transparency, and quality of food in progress from farm to cookware. This chapter provides a new approach to understanding the toolkit-like use of drones in the food supply chain.

3.5.1 REVOLUTIONIZING FARM OPERATIONS

Today, agricultural drones are a primary element of farming. Farmers are empowered with real-time data and insights through sensors and cameras embedded in drones. This includes the following provisions:

- Enhanced field monitoring and crop scouting.
- Revolutionized precision spraying.

Drones facilitates targeted approaches in spraying of pesticides, herbicides, and fertilizers as compared to the traditional blanket methods of spraying that leads to wastage and pollution. Additionally, drones are used in irrigating through optimization of irrigation management by collecting data on soil moisture and health to ensure that irrigation is data-driven, leading to sustainable water utilization. Finally, drones are essential in collection and utilization of data to promote data-driven precision agriculture since their data integrates with the farm management software across the seasons of the production through entire season from seeding till harvesting.

3.5.2 *STREAMLINING POSTHARVEST ACTIVITIES*

Drones are changing the way postharvest management is done. The uses include: drones with scanning technology can be used in storage facilities to keep inventory accurate and minimize spoilage. Drones with thermal imaging can be used to monitor and evaluate the quality of stored produce. This allows for early identification of potential problems, such as temperature changes and spoilage. They can also be utilized for increasing security and surveillance around the perimeter of storage facilities.

3.5.3 *ENHANCING TRANSPORTATION AND LOGISTICS*

Food transportation and logistics: the involvement of drones in the farm extends beyond the farm. Generally, it can also be integrated in food transportation and logistics:

Delivery to isolated areas: Drones provide a novel alternative for delivering foodstuffs to isolated or calamity-stricken areas with poor or malfunctioning traditional transport facilities.

Warehouse stocking: Drones can enhance the proper stocking and stoking of food commodities in the warehouse.

Cold chain digitization: Food materials requiring cold storage must be monitored, and drones integrating temperature sensors can play a significant role in ensuring compliance with the cold chain storage condition.

3.5.4 *INNOVATION IN RETAIL AND DISTRIBUTION*

The following part of the food supply chain may also find drone integration useful:

Delivery to customers: Seems limited due to the ongoing evolution of regulations and safety concerns, but the ability to deliver groceries or food products to customers through drones has great potential for the future.

In-store inventory management, this aspect can be facilitated with the help of drones to provide real-time tracking and fill up of stock in stores, which contributes to the better customer experience.

Marketing and customer engagement—Drones might also be used to make entertaining marketing campaigns or to develop beautiful aerial views of the produce.

3.6 BENEFITS OF USING AGRICULTURAL DRONES

Agriculture is currently in the process of a profound transformation, which is actively promoted by the integration of unmanned aerial vehicles in the sector, also known as agricultural drones. Modern technological platforms provide a variety of unforgettable advantages in terms of quantifiable outcomes that translate into improved operational capability, increased resource efficiency, and realized data-driven decision-making capabilities at all stages of the agricultural production process. This chapter presents the main advantages of incorporating drones into modern agricultural practices.

3.6.1 *ENHANCED PRECISION AGRICULTURE AND RESOURCE OPTIMIZATION*

- Real-time field monitoring and targeted interventions: High-resolution cameras and multispectral sensors on agricultural drones provide farmers with information on the state of the field in real time. As a result, drone technology is used to identify crop stress, nutrient deficiency, and pest infestations ahead of schedule and allows farmers to intervene and allocate available resources of the field based on need. Prompt location of the problem affords the farmer an opportunity to reduce potential losses and make better use of fertilizers, pesticides, and water.
- Precision spraying and reduced environmental impact: Most importantly, the use of agricultural drones eliminates the need for traditional spraying, which results in chemical over-application and overspill on a broader area, including the crops intended for treatment. These drones are armed using a hollow-cone nozzle to reduce drift, allowing for treatment to occur with minimal spills. Essentially, precise spraying only applies necessary chemicals such as pesticides and herbicides to treat fertilizers or designated crop areas. Consequently, the use of agricultural drones almost eliminates chemical wastage and reduces environmental impact, reinforcing more sustainable agriculture.
- Data-driven irrigation management and water conservation: Water scarcity: At the current rate of consumption, water overuse will become a major issue for the agriculture of the present and the near future. Using thermal imaging sensors, agricultural drones allow farmers to create irrigation plans that are optimized. Sensor data regarding soil moisture and crop health spread throughout the field allows farmers to apply irrigation on an as-needed basis. A farmer is wasting less water by making irrigation decisions based on collected data, which leads to better water consumption and less water waste.

3.6.2 *EMPOWERING INFORMED DECISION-MAKING THROUGH DATA ACQUISITION*

- Early yield estimation and strategic planning: Drones that take crop health data can facilitate farm early estimates on the expected yields. The reflected gross margins by such estimates help in planning marketing and tactics on the appropriate use of resources during the growing period. Moreover, early hint on estimated yield enables the farmer to know the labor for harvesting and postharvest.

- Detailed field mapping and improved farm management: For example, drones can produce high-resolution, three-dimensional field maps which give a complete outlook of the farm because the quality of this information may be used to perform many tasks such as planting patterns, fertilization strategies, and harvesting operations. As a result, this attained information boosts an increased form of the farm's management and decision-making.
- Seamless integration with farm management software and real-time monitoring: The second option can also be achieved by means of drone-collected data seamlessly integrated with advanced farm management software. This stream of real-time information enables farmers to keep track of crop health and resource spending and plan their activities accordingly. In this way, an integrated technological solution can help farmers reach an unprecedented level of precision in managing their agricultural activities.

3.6.3 ADDITIONAL ADVANTAGES: SAFETY, LABOR OPTIMIZATION, AND SUSTAINABILITY

Minimize exposure of workers to chemicals: Use of drones for spraying fertilizers reduces the health risks of chemicals to presidents when sprayed on the crops. This eventually reduces the health levels of the workers in the agricultural sector. Field scouting will soon take out large labor costs: Drones will be used to scout fields on mobile measurement is expected to rise. This will cut down the labor used entirely in scouting and instead introduce a more quick, simple, and accurate approach to assessing fields. Drones can help reduce chemical usage and conserve water, thus minimizing agriculture pollution and waste of water; precision spraying will use less fertilizer in farming, and the water will be well applied to the fields.

As one may see, the advantages of agricultural drones listed above are not the only benefits. Since drone technology is rapidly developing, there will be even more new opportunities in this field. Such improvements as increased battery capacity, larger payloads, more automation processes, and more coherent regulations will unleash drones' full potential to create a more productive and environmentally friendly future for agriculture. Therefore, drones bring transformational change that not only accelerates the adoption of innovative processes and practices but also ensures that they are put to good use.

3.7 CHALLENGES AND LIMITATIONS

Despite offering a revolutionary approach to reshaping food supply chains, there are multiple technical, regulatory, economic, and social challenges associated with the widespread adoption of agri-drones. The current section aims to explore these limitations alongside the promising opportunities of using drone technology in agriculture. Specifically, the technical constraints include the following aspects:

- Battery technology does not allow for long-lasting missions because of the small range of drones. Therefore, it is reasonable to anticipate the need for battery swapping or recharging that can undermine the efficiency of operations.
- Similarly, drones are also environmental dependent since multiple precipitation events, high winds, and extremely low or high temperatures can suspend their operations.
- Payload capacity could also be defined as a technical constraint because drones have limited carrying capacity. Consequently, drones can carry a small amount of material, for example, seeds or pesticides, and should complete multiple trips when fulfilling larger spatial tasks.
- Lastly, data protection and management are relevant technical constraints because the big data collected by drones require a comprehensive approach to storing and securing information.
- Air traffic management integration, drones must efficiently and safely be integrated into the shared airspace with manned aircraft. Thus, this needs the development of broad-spectrum air traffic management systems to prevent any collisions and ensure maximum safety.
- Privacy concerns, drone surveillance is often a concern, especially for data collection over private property. The use of agri-drones necessarily needs regulations and protocols regarding data collection and data privacy.
- Licensing and training, for safe and responsible use of agri-drones, the licensing of drone operators needs to be standardized. The same has to be done with training programs to ensure the operators have the requisite knowledge to safely use their drones legally.
- Cost of implementation, the purchase and build-up prices for agri-drone technology and the accompanying infrastructure, such as charging stations or data processing systems, may be costly for small farms with less financial resources due.
- Digital divide, the lack of stable internet in many rural areas may make the monitoring and managing of drones impossible due to most agri-drone applications requiring real-time data sharing and processing.
- Job displacement, agri-drone technology is designed to substitute human labor in many tasks. Therefore, its implementation might lead to job losses, which can be mitigated by emotionally stabilization and reskilling options.
- The environmental impact of drone operation, the noise pollution caused by the drones, as well as their possible interference with wildlife habitats.
- Maintenance and repair infrastructure, the need to provide access to technicians who can repair and maintain drones, especially in remote areas.

In reflecting on these challenges and constraints, a more holistic comprehension of agri-drones and the way they can feature in the future food supply chains can be promoted. By overcoming these barriers through critical technological research, regulatory guidelines, and ethical implementation pathways, the potential of agri-drones to foster a more efficient, sustainable, and data-oriented agricultural industry can be released.

3.8 FUTURE TRENDS AND INNOVATIONS

The field of agri-drones is growing fast and the future opens up unprecedented possibilities to revolutionize food supply chains. This chapter gives insight into the four main trends that will shape the future of agricultural robotics and smart devices.

3.8.1 ENHANCED AUTONOMY AND INTEGRATION OF AI

Swarm technology: with the introduction of swarm technology, huge fleets of drones can work together in a coordinated manner. This will make it possible to solve major agricultural problems as quickly and effectively as never before. For instance, swarms of drones can monitor crops across an entire field within minutes or target pest control with surgical precision.

Advanced AI and machine learning (ML): Additionally, drones will feature integrated advanced AI and ML algorithms that will allow them to access data and make autonomous decisions. Exemplary applications consist of performing weed recognition and selecting weeds for effective fulmination or adjusting crop irrigation based on soil moisture levels. This way, the farming industry will be able to most efficiently utilize resources thereby reducing expenses and environmental impact.

3.8.2 PRECISION AGRICULTURE ON A GRANULAR LEVEL

Hyperspectral imaging: Hyperspectral cameras integrated into drones will provide agriculture with a novel layer of data collection. The data gathered by these cameras can record down to coloraria of plant health, exhibiting subtle changes that can indicate a nutrient deficiency or the start of a disease outbreak. The possibility of early detection that will allow for prophylactic measures has the potential of drastically reducing losses caused by undetected diseases [24].

Microbiome analysis: Microbiome analysis using drones for collecting sampling: however, drones capable of collecting samples can be used to take air samples, as well as soil ones, to test the present microbial community. This data can be used to design targeted bio-control measures and adopt measures to make soil healthier.

3.8.3 CONVERGES WITH ROBOTICS AND IOT

Collaboration among agri-bot and drones: Agricultural robots working autonomously on the ground would not be all that new once wind and solar startups got used to the idea in machine fields. More interesting is the day not too far off when agri-bots can work together with drones in formation, planting seeds and gathering crops in a well-coordinated robotic workforce.

Data sharing: Drones will be linked into overlapping IoT networks in order and will be able to swap up-to-the-second data with networks tied to farm equipment, weather stations, and top-of-the-line farm management software. Integrating the suite will allow for data-guided decision-making throughout the farm and support a more connected approach to farm experimentation.

3.8.4 SUSTAINABILITY AND RESOURCE EFFICIENCY

Biodegradable materials: Changing the paradigm of building drones from biodegradable materials will be environment friendly by mitigating the pollution caused by drones when out of use traditionally. This will help not only form a more favorable agricultural ecosystem but also enhance sustainability in drone manufacturing.

Precision spraying technology: Considering that drones will have highly sophisticated targeting system, the systems will ensure efficient use of water and other sprays. This will reduce the cost of water and the spraying materials used on farming.

3.8.5 THE EVOLVING LEGAL AND REGULATORY LANDSCAPE

Standardization and regulatory frameworks: Moreover, drone technology is evolving, and these trends will lead to the emergence of new standards and demands for regulatory bodies. In the case of drone usage, this will create a need to develop proper regulatory frameworks building on existing principles to integrate drones into airspace in a responsible, safe manner.

Data privacy and security: Additionally, drones collect vast amounts of data, raising concerns about security and privacy; the chapter could propose practices of secure data storage in the cloud, in-line with state-of-the-art trends and requirements, and ensuring they comply with regulatory needs while fostering stakeholders' trust.

In delving into these trends, the present chapter aspires to deliver a well-rounded and professionally oriented perspective on the promising future of agri-drone technology. In taking into account the extensive opportunities for improvement and the remarkable breakthroughs in the industry, the chapter takes on a well-informed and cautiously optimistic attitude to the future development of agri-drones in the framework of food supply chains.

3.9 CASE STUDIES AND SUCCESS STORIES

3.9.1 PRECISION SPRAYING

1. **Krishiviman:** These Indian drones saved millions of dollars by targeting spraying wheat fields of Haryana (Figure 3.9).

FIGURE 3.9 Krishiviman case study.

Source: https: //www.krishiviman.com/blog/revolutionizing-indian-agriculture-krishiviman-s-pioneering-drone-technology/

As a result, the company did not have to use dozens of barrels of pesticides per hectare, which made fruits healthier and reduced the entire environmental footprint.

2. **Banana cultivation:** Such drones as T40 by DJI Agriculture are very effective and energy- and cost-efficient for banana pests and illnesses control (Figure 3.10).

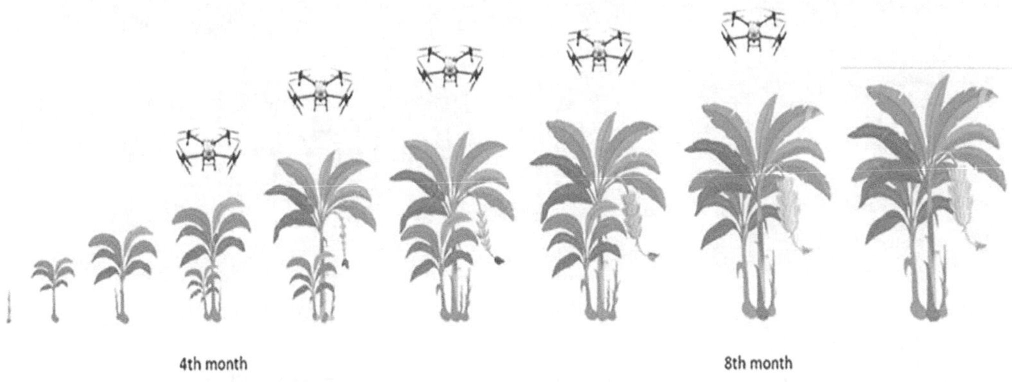

FIGURE 3.10 Banana plant disease care.

Source: https: //ag.dji.com/case-studies/ag-case-study-banana-t40

They do not miss one spot and are not forced to spray more than necessary, as a result, fruits are healthier.

3.9.2 CROP MONITORING AND DATA COLLECTION

1. **Vineyard analysis with drones:** In India, the state of Maharashtra, a drone with high-resolution cameras and sensors morphed into a tool to create vineyard maps which helped get precise measurements on soil health (Figure 3.11).

FIGURE 3.11 Study vineyard (1) monitored in dormant period and UAV platform (2) equipped with DJI Zenmuse P1 RGB camera.

Source: Reprinted from Ref. [27]. Copyright © 2023 by the authors. Licensee MDPI, Basel, Switzerland. https://creativecommons.org/licenses/by/4.0/

Thus, they refined their irrigation and fertilizing to get better grape quality and harvest.

2. **Unlocking sugarcane potential:** South African smallholder sugar farmers have also found a use for drones as they lack the personnel to monitor huge fields.

FIGURE 3.12 Agri-drone in sugarcane cultivation smallholder farmers.

Source: https: //ag.dji.com/case-studies/ag-case-en-sugarcane

Utilizing Dj's drone technology, these farmers are able to control many hectares in a matter of minutes to identify problem areas and optimize resource planning and deliver boosting productivity.

3.9.3 IMPROVED DECISION-MAKING

Weed control in pastures: Drones with sophisticated high-resolution imaging allowed farmers in Lithuania to target invasive weeds in their pastures with a high degree of accuracy. As a result, there was a 50% decrease in the amount of herbicides used, and the pastures' condition improved.

3.10 KEY TAKEAWAYS AND RECOMMENDATIONS

3.10.1 PROPER SELECTION OF A DRONE

The following selection criteria should be taken into consideration: farm size, purpose (spraying, imaging, etc.), and amount of budget.

- Data analysis and application: The farming plan for the data collected needs to be developed to process the information and apply it in decision-making.
- Regulatory background: It is critical to review the legal background for drones regarding permits, zones, and prohibited areas to operate safely and legally.

- Training activity: Finally, one should consider investment in pilot training and technical support for the drone and develop it.

3.10.2 *IMPACT ON AGRICULTURAL PRODUCTIVITY AND SUSTAINABILITY*

- Increased efficiency: Drones quickly and accurately cover vast field areas and may thereby reduce the need for farmer involvement on the land, thus saving time and labor costs.

Reduced waste: Drones apply resources, such as water, fertilizers, or spray chemicals only to the targeted areas that require them minimizing their waste and the environmental footprint and residual agri-waste can convert to byproducts [29, 30]. This can likewise include the reduction of excessive runoff and leached nutrient.

Improved crop health: Drones monitor crop and field health, provide early warnings in case of problems, and allow for timely farmer intervention. This can contribute to higher crop health, crop yield, and quality.

Data-driven decision-making: Drones collect a mass of complex data about fields and crops, allowing for effective farmer decision-making regarding resource allocation and overall farm management.

3.11 CONCLUSION

The potential of the agri-drones, as we have seen, is immense, and indeed the use of the agri-drones represents a watershed moment for the agricultural sector. It is not merely a replacement for traditional methods, but the dawn of a new age of precision agriculture. It is, in essence, a farming process in which decision-making is guided by data and the resources are efficiently allocated, and the results are demonstrably more efficient. This chapter outlined various uses and applications of the agri-drones, from the precision spraying and meticulous crop monitoring to the farm management quality optimization. Furthermore, the above examples mentioned various case studies that have demonstrated several merits of the new technologies, including the dramatic reduction in waste, often a clear crop health improvement, and fewer stages of the production, which led to the increased productivity. All of the above should not distract us from the fact that challenges remain. Namely, the regulatory framework should be adjusted to accommodate for the wide use of drones as soon as possible. Finally, the issue of social justice—the smallholders should not be deprived of the chance to exploit it same as the big conglomerates. The future of the agri-drones seems bright from where we stand. The continued evolution of the technology is clear—the AI and automation are the most probable developments onboard. Agri-drones are on the verge of becoming a true game-changer. With their help, there should be no reason for the food shortage to ever occur.

KEYWORDS

- **agricultural drones**
- **precision agriculture**
- **food supply chains**
- **robotics**
- **smart farming**

REFERENCES

1. Hassoun, A., Marvin, H. J. P., Bouzembrak, Y., Barba, F. J., Castagnini, J. M., Pallarés, N., Rabail, R., Aadil, R. M., Bangar, S. P., Bhat, R., Cropotova, J., Maqsood, S., & Regenstein, J. M. (2023). Digital transformation in the agri-food industry: Recent applications and the role of the COVID-19 pandemic. *Frontiers in Sustainable Food Systems*, *7*. https: //doi.org/10.3389/fsufs.2023.1217813.
2. Raza, Z., Haq, I. U., & Muneeb, M. (2023). Agri-4-all: A framework for blockchain based agricultural food supply chains in the era of fourth industrial revolution. *IEEE Access*, *11*, 29851–29867. https: //doi.org/10.1109/access.2023.3259962.
3. Kumar, S. P., Subeesh, A., Jyoti, B., & Mehta, C. R. (2023). Applications of drones in smart agriculture BT—Smart agriculture for developing nations: Status. In: Pakeerathan, K., (Ed.). *Perspectives and Challenges* (pp. 33–48). Springer Nature Singapore: Singapore. https: //doi.org/10.1007/978-981-19-8738-0_3.
4. Emmi, L., Fernandez, R., & Guerrero, J. M. (2022). Editorial: Robotics for smart farms. *Frontiers in Robotics and AI*, *9*, 1113440. https: //doi.org/10.3389/frobt.2022.1113440.
5. Adow, A. H., Shrivas, M. K., Mahdi, H. F., Zahra, M. M. A., Verma, D., Doohan, N. V., & Jalali, A. (2022). Analysis of agriculture and food supply chain through blockchain and IoT with light weight cluster head. *Computational Intelligence and Neuroscience*, *2022*, 1296993. https: //doi.org/10.1155/2022/1296993.
6. Eliana, N. (2023). Employing drones in agriculture: An exploration of various drone types and key advantage.
7. Rathod, P. D., & Shinde, G. U. (2023). Autonomous aerial system (UAV) for sustainable agriculture: A review. *International Journal of Environment and Climate Change*, *13*, 1343–1355. https: //doi.org/10.9734/ijecc/2023/v13i82080.
8. Bhushan, M., & Negi, A. (2023). Impact of UAVs in agriculture. In: Goel, N., & Yadav, R. K., (Eds.), *Handbook of Research on Machine Learning-Enabled IoT for Smart Applications Across Industries* (pp. 258–268). IGI Global: Hershey, PA, USA. https: //doi.org/10.4018/978-1-6684-8785-3.ch013.
9. Daniel Lawrence, I., Vijayakumar, R., & Agnishwar, J. (2023). Dynamic application of unmanned aerial vehicles for analyzing the growth of crops and weeds for precision agriculture. In: Gupta, R. K., Jain, A., Wang, J., Bharti, S. K., & Patel, S., (Eds.), *Artif. Intell. Tools Technol. Smart Farming Agric. Pract.* (pp. 115–132). IGI Global: Hershey, PA, USA. https: //doi.org/10.4018/978-1-6684-8516-3.ch007.
10. Panday, U. S., Pratihast, A. K., Aryal, J., & Kayastha, R. B. (2020). A review on drone-based data solutions for cereal crops. *Drones*, *4*, 41. https: //doi.org/10.3390/drones4030041.
11. Zawish, M., Ashraf, N., Ansari, R. I., Davy, S., Qureshi, H. K., Aslam, N., & Hassan, S. A. (2022). Toward on-device AI and blockchain for 6G-enabled agricultural supply chain management. *IEEE Internet Things M*, *5*, 160–166. https: //doi.org/10.1109/iotm.006.21000112.
12. Yaqot, M., & Menezes, B. C. (2021). Unmanned aerial vehicle (UAV) in precision agriculture: Business information technology towards farming as a service. In *2021 1st International Conference on Emerging Smart Technologies and Applications (eSmarTA)*. https: //doi.org/10.1109/esmarta52612.2021.9515736.
13. Veroustraete, F. (2015). The rise of the drones in agriculture. *EC Agridulture*, *2*.

14. Rejeb, A., Abdollahi, A., Rejeb, K., & Treiblmaier, H. (2022). Drones in agriculture: A review and bibliometric analysis. *Computers and Electronics in Agriculture, 198*, 107017. https: //doi.org/https: // doi.org/10.1016/j.compag.2022.107017.
15. Hafeez, A., Husain, M. A., Singh, S. P., Chauhan, A., Khan, M. T., Kumar, N., Chauhan, A., & Soni, S. K. (2023). Implementation of drone technology for farm monitoring & pesticide spraying: A review. *Information Processing in Agriculture, 10*, 192–203. https: //doi.org/https: //doi.org/10.1016/j. inpa.2022.02.002.
16. Mahadasa, R., Surarapu, P., Vadiyala, V. R., & Baddam, P. R. (2020). Utilization of agricultural drones in farming by harnessing the power of aerial intelligence. *Malaysian Journal of Medical and Biological Research*, *7*, 135–144.
17. Agriculture Drones and Robots Market Growth, Trends & Forecast. (n.d.). [cited 8 April 24]; Available from: https: //exactitudeconsultancy.com/reports/40911/agriculture-drones-and-robots-market/.
18. Agriculture Drones Market Size, Share, Forecast and Growth [Latest]. (n.d.). [cited 8 April 24]; Available from: https: //www.marketsandmarkets.com/Market-Reports/agriculture-drones-market-23709764.html.
19. Maes, W., & Steppe, K. (2018). Perspectives for remote sensing with unmanned aerial vehicles in precision agriculture. *Trends in Plant Science*, *24*, 152–164. https: //doi.org/10.1016/j.tplants.2018.11.007.
20. Daneshpajooh, E. (2021). Internship report design of a multi-purpose agreculture drone.
21. Delavarpour, N., Koparan, C., Nowatzki, J., Bajwa, S., & Sun, X. (2021). A technical study on UAV characteristics for precision agriculture applications and associated practical challenges. *Remote Sensing*, *13*, 1204. https: //doi.org/10.3390/rs13061204.
22. Kumawat, V., Dulawat, L., & Verma, S. (2020). Agri-drones for field surveillance. *Turkish Journal of Computer and Mathematics Education*, *11*, 1105–1111. https: //doi.org/10.61841/turcomat.v11i1.14409.
23. Abbas, A., Zhang, Z., Zheng, H., Alami, M. M., Alrefaei, A. F., Abbas, Q., Naqvi, S. A., Rao, M. J., Mosa, W. F. A., Abbas, Q., Hussain, A., Hassan, M. Z., & Zhou, L. (2023). Drones in plant disease assessment, efficient monitoring, and detection: A way forward to smart agriculture. *Agronomy*, *13*, 1524. https: //doi. org/10.3390/agronomy13061524.
24. Bajpeyi, M. M., Kumar, A., Aman, A. S., Mishra, P., Pandey, V., & Mishra, S. (2023). Use of drone technology in precision farming. In Research Trends in Multidisciplinary Research and Development, Weser Books, pp. 123–142.
25. Krishiviman Drones: Revolutionizing Indian Agriculture. (n.d.). [cited 16 April 24]; Available from: https: //www.krishiviman.com/blog/revolutionizing-indian-agriculture-krishiviman-s-pioneering-drone-technology/.
26. Agras T40 Banana Cultivation. (n.d.). [cited 16 April 2024]; Available from: https: //ag.dji.com/ case-studies/ag-case-study-banana-t40.
27. Di Gennaro, S. F., Vannini, G. L., Berton, A., Dainelli, R., Toscano, P., & Matese, A. (2023). Missing plant detection in vineyards using UAV angled RGB imagery acquired in dormant period. *Drones*, *7*, 349. https: //doi.org/10.3390/drones7060349.
28. In the realm of large-scale commercial sugarcane farming in South Africa, chemical ripening has long been recognized as a crucial factor for successful cultivation. (n.d.). [cited 16 April 2024]; Available from: https: //ag.dji.com/case-studies/ag-case-en-sugarcane.
29. Verma, D., Okhawilai, M., Goh, K. L., & Sharma, M. (2023). Thermomechanical analyses of alkali-treated coconut husk-bagasse fiber-calcium carbonate hybrid composites. *Sustainability*, *15*, 1686. https: //doi.org/10.3390/su15021686.
30. Tamta, S., Vimal, V., Verma, S., Gupta, D., Verma, D., & Nangan, S. (2024). Recent development of nanobiomaterials in sustainable agriculture and agrowaste management. *Biocatalysis and Agricultural Biotechnology*, *56*, 103050. https: //doi.org/https: //doi.org/10.1016/j.bcab.2024.103050.

CHAPTER 4

Sustainable Futures: Smart Agriculture's (Internet of Things) Contribution to Environmental, Social, and Economic Sustainability

PALLAVI JOSHI[1*], PRIYANK VYAS[2], AKANSHA BHANDARI[2], and MUSKAN RAWAT[2]

[1]*Graphic Era Hill University, Dehradun, Uttarakhand, India*

[2]*Graphic Era (Deemed to Be) University, Dehradun, Uttarakhand, India*

*Corresponding author

ABSTRACT

The Internet of Things (IoT) comes as a savior as it can be used to improve agriculture for efficient and effective farming practices and ways. IoT in agriculture is a set of connected devices that collect information and analyze it about aspects of the weather conditions and farming tools that facilitate the work of a farmer. One that is so wide-reaching and relates to the need to tackle issues like climate change, depleting resources, and productivity improvements that must sustain a population of about 10 billion people by 2050. The concept of precision farming is another aspect of smart agriculture model and aims at enhancing efficiency of resource use and minimizing the environmental impacts through the maximization use of data. The application of IoT extends from monitoring crops to monitoring livestock and keeping certain crops as well as improving the supply chain to make sure that the food product can be tracked and reduce loss. Smart agriculture not only increases crop yield and earnings of farmers, but it is also an important tool for ensuring sustainable rural development as maintaining knowledge in the farm and farming community. But at the same time, it is reasonable to expect that the current trends in IoT technology will lead to a significant change in the agriculture industry and the development of a more efficient industry that would be able to solve the problems of food security and environmental problems.

Demystifying Sustainable Farming Using Smart Technologies.
Gunjan Chhabra, Keshav Kaushik, Priya Chugh, Susheela Dahiya, and Aryan Chaudhary (Eds.)

DOI: 10.1201/9781779640932-4

4.1 INTRODUCTION

The infusion of state-of-the-art technology into modern topography is indeed inevitable for the sustainability and efficiency of agriculture practices. One such redefining innovation is the Internet of Things (IoT), which has led to a different paradigm in agriculture. This chapter elaborates on the major smart agriculture systems enabled by the IoT in farming practices, productivity, and resource optimization. The convergence of agriculture and the IoT gives a new meaning to farming by helping connected devices, sensors, and data analytics transform traditional methods [14, 16]. IoT is "a dynamic global network infrastructure with self-configuring capabilities based on standard and interoperable communication protocols" [4]. Understood in the context of agriculture, it means creating a complex system in which various devices can make working decisions, to a large extent, by themselves and communicate with farmers.

This has placed a great deal of pressure on the agricultural sector in terms of climate change, resource scarcity, and the need for increased productivity to feed the multiplying global population [27]. In this sense, the accepted solution is that of so-called IoT-enabled smart agriculture, which provides real-time data on environmental conditions, crop health, and equipment performance [1, 5]. Examples include the use of sensors in fields to obtain real-time data on soil moisture levels, temperature variance, and crop growth patterns. This information gives the farmer the ability to make informed decisions regarding water, fertilizer, and pesticide use, thus optimizing resources to eventually increase crop yield [3, 18].

According to United Nations data for 2019, the global population is projected to rise to 10 billion by 2050, increasing the demand for agricultural products [27]. However, the sector is critically threatened by reduced farmlands, diminishing natural resources, and drastic environmental factors such as global warming, salinization, and flooding. In response, the agricultural sector has been exploring new solutions and technologies, among which the integration of IoT with big data is a rising trend [14, 19].

Several studies and experiments demonstrate how the exploitation of IoT and big data has enhanced efficiency and productivity in agriculture [2, 13]. Cisco forecasts that the number of IoT devices connected to the Internet could reach 500 billion by 2030. IoT, along with big data, offers great potential for realizing smart agriculture, where wireless sensor networks (WSNs) serve as a core framework [3, 4]. WSNs typically include a radio frequency (RF) transceiver, a sensor, a microcontroller, and battery power, enabling environmental monitoring, machine control automation, and traceability [22].

The intensification of agriculture through new technologies has made innovation crucial to improving productivity, which is increasingly achieved via IoT adoption. The main drivers of this shift are breakthroughs in smart farming technologies [10]. Furthermore, IoT is harmoniously embedded in technologies like WSNs, cognitive radio, ad hoc networks, cloud computing, and end-user applications [14]. In smart agriculture, this integration enables automation, decision-support tools, and software systems that empower farmers to increase yields, ensure quality, and maximize profitability.

Often termed "smart agriculture," precision farming reflects the IoT's transformative power. It entails adjusting farming activities based on precise data to enhance productivity and minimize environmental harm [17]. Precision farming aims to increase resource use efficiency, reduce environmental impact, and ensure economic sustainability. The scope of IoT in agriculture extends beyond fields to livestock, crop storage, and supply chains [8, 9]. For instance, livestock fitted with IoT wearables can be monitored for health and behavior, enabling early disease detection and improved farm productivity. Smart storage systems use IoT to maintain optimal conditions for preserving harvested crops. IoT-enabled supply chains enhance traceability, reduce food waste, and ensure fresh produce reaches consumers [12, 13].

The 2019 UN report projected a global population of 10 billion by 2050, heightening the demand for food and putting pressure on production systems [27]. Meanwhile, agricultural lands are shrinking due to resource depletion. Compounding the challenge are erratic environmental threats—such as global warming, salinization, and flooding—that intensify the crisis of global food security.

4.2 TRANSITIONING TRADITIONAL FARMING TO SMART AGRICULTURE: A PROPOSAL FOR SUSTAINABLE RURAL DEVELOPMENT

Traditional farming practices are age-old practices in any community. They exist for generations and serve as the backbone for the community's way of life and economy. These traditional methods of operation, however, are typically excessively labor intensive and extremely inefficient. Therefore, these people have low crop returns and are not profitable. Other than that, another side of this story is the need for the members of a traditional farming community to deal with the problem of the migration of the young generations toward cities in search of better opportunities. This in turn is causing a reduction in the workforce involved in agrarians and loss of expertise and knowledge of farming practices [20]. To curb these challenges and also to spearhead sustainable rural development, we are bound to transition from the traditional agriculture approach to a smart agriculture approach. Smart agriculture, otherwise termed precision or digital farming is farming in which technology and data-driven methods are combined to optimize agricultural practices [14]. With smart agriculture techniques, traditional farmers can achieve higher crop yield and increased profitability without the risks associated with conventional farming methodologies. In other words, farmers can be able to monitor weather conditions, soil moisture, and crop health in real time through technology, for remoteness sensing and the IoT [1, 2]. Thereby, it would ensure more effective decisions and interventions for the targeted quantities of water, nutrients, and pests. In addition, the automated irrigation system would be designed for the optimization of water management with reduced wastage [21]. Smart systems will also assure resource efficiency and sustainability [10, 19]. Such data-driven approaches to modern farming will allow optimization in resource use: water, fertilizers, and pesticides in the right amount, place, and time, reducing waste and making the smallest impact on the environment [15]. Transitioning to smart agriculture can also enable us to face the challenges offered by climate change [27]. Now, with the rising population, frequent changes in climatic conditions, and lots of pressure on resources, the significance of smart technologies for precision farming has increased to facilitate timely, informed decisions [14, 17].

4.3 INTERNET OF THINGS

The introduction of IoT in agriculture has changed the whole approach toward farming practices. The IoT technology, through crop, soil, and livestock health monitoring in real time, has the potential to help farmers make smart decisions for better resource utilization and higher productivity [1, 2]. This makes it farmer-friendly in being able to control irrigation and detect the early onset of diseases in crops, and go on to implement precise strategies in fertilization—all done remotely with the help of sensors, drones, and data analytics [10, 19]. Such precision and efficiency considerably reduce operational costs, which means that sustainability is embedded in agriculture [21]. The worldwide market for IoT in agriculture will be one of the fast-growing markets in the coming years, which will help to explain the new smart technologies in the agricultural sector. As further developed, this potential will change agriculture into a more efficient and sustainable industry [14].

4.4 APPLICATIONS OF IOT IN AGRICULTURE

The applications have been segregated according to their purpose in sections of, monitoring, tracking and traceability, and greenhouse production.

4.4.1 MONITORING

In the agricultural sector, various factors influencing farming and production processes are monitored and collected, including soil moisture, air humidity, temperature, pH level, and more [2, 10]. Different monitoring solutions are applied in different smart agricultural sectors.

4.4.1.1 CROP FARMING

In the context of agriculture, the incorporation of devices like Farm Fox underlines a radical shift toward precision farming, which is a game changer in crop cultivation. Devices like this embed a suite of sensors that continuously monitor key parameters affecting crop growth: air temperature, precipitation, humidity, soil moisture, salinity, solar radiation, and the presence of pests [1, 19]. The internet connectivity of these sensors would help relay data seamlessly to the farmer in order to keep track of and analyze the environmental conditions of the farm. This avalanche of data would enable the farmer to make more intelligent decisions, hence optimizing irrigation, fertilization, and pest management practices [14, 21].

An important instance of the same applied to precision agriculture is the use of IoT devices installed in weather radar, which measure the temperature and humidity of a region, sending an alert to the farmer whenever it crosses a certain predefined limit [2]. This allows the farmer to proactively address environmental issues and sets better crop growth conditions. Web-based Geographic Information System (GIS) IoT systems for monitoring pest activities, issuing early warnings, and producing predictive models of pests and diseases have also been developed. GIS greatly avails chance in the hands of farmers for maintenance and monitoring of pest population activities, applying preventive control actions, and up-grading the strategies of crop protection.

Despite the multiple benefits of precision agriculture through the use of IoT, some hiccups remain, such as the upfront costs of investment in IoT infrastructure in agriculture and the assurance of data privacy, security, and interoperability with other technologies [19]. The potential benefits, on the other hand, are enormous; IoT in agriculture is able to transform the practice of crop and farm management, increasing resource use efficiency and productivity in agriculture while improving sustainability [1]. The future of IoT-enabled precision agriculture harbors continued opportunities for innovation. Major developing trends are seen in progress in sensor technologies, integration of AI, and utilization of blockchain technologies in the use of IoT in agriculture [14, 19]. However, it needs to work on issues such as regulatory frameworks, data standardization, and rural connectivity problems to harness all the benefits that IoT offers to agriculture. In conclusion, the positive transformative effect that IoT-empowered precision agriculture is bringing about underpins an urgent necessity for research, further development, and adoption of IoT technologies to help in the realization of sustainable farming practices [10].

4.4.1.2 AQUAPONICS

Aquaponics farming is a new process that combines aquaculture with hydroponics to capture the found synergy that exists in the fish and plants for the establishment of a self-sustained agricultural ecosystem. In this design, the IoT systems take a central role in monitoring and controlling key parameters necessary for the well-being and productivity of the aquatic organisms and plants. "Aquaponics farming combines aquaculture with hydroponics to establish a self-sustained agricultural ecosystem" [6]. The monitoring solution of IoT allows the aquaponic practitioner to keep track of some important parameters, of which water quality, water level, temperature, salinity, pH, and sunlight are the most vital.

It is in this infusion of IoT technology that these aquaponics farmers can largely automate many of the system management tasks, most notably feeding operations. The IoT-enabled devices automate the feeding process, making it easy to provide the fish with the right amount of nutrients at the right time, therefore reducing waste levels and making the whole aquaponics system more efficient and sustainable. Furthermore, the monitoring of environmental parameters through IoT-based solutions offers insights, leading to data-driven decisions of changes and interventions.

Being an IoT-driven solution, aquaponics holds numerous benefits for farming in terms of productivity, resource efficiencies, and system resilience. The IoT makes sure that the ideal environmental conditions are maintained and that all activities are automated to support the good health and vitality of fish and plants, hence resulting in more yields of fish and good quality of crops. IoT-based monitoring empowers proactive management of aquaponics because problems can be quickly identified and fixed before they escalate, thereby avoiding the various risks and reducing losses. Nevertheless, there are a few challenges despite the benefits conferred by IoT to aquaponics farming, particularly in regard to the integration and administration of such intricate IoT systems. In fact, issues like data security, system reliability, and interoperability with current infrastructure will need to be resolved if IoT-based systems in the aquaponics domain are to be executed properly. In addition, cost efficiency and scalability are two aspects that are crucial for universal adoption and implementation.

In the future, the growth of IoT in aquaponics presents a development and improvement trend. Advanced technologies, such as edge computing, artificial intelligence (AI), and machine learning, which are able to unlock the potential of IoT systems, allow better monitoring and management of aquaponics operations. These, along with data analytics, will further enhance solutions based on IoT sensor technology in optimizing the business in the aquaponics industry for better productivity, sustainability, and profitability. In other words, the deployment of aquaponics farming integrated with IoT systems is one of the steps toward achieving sustainable and efficient food systems.

4.4.1.3 FORESTRY

IoT technology and big data analytics applied to forestry management help in making a big leap in the monitoring and conservation of forest ecosystems. These two methodologies are important in monitoring parameters key to forest health, including soil composition, atmospheric conditions such as air temperature and humidity, and gas concentrations such as oxygen, methane, ammonia, and hydrogen sulfide. It will in capturing and analyzing data available through this, and the forest stewards can get critical insights into the condition of forest ecosystems, which can be used in making the processes of decision-making toward forest governance, environmental monitoring, disaster mitigation, and in-depth assessment of ecosystem vigor.

Integration of IoT devices allows for remote data collection from forest locales in real time and hence overcomes such problems as manual monitoring practices. The real-time data collection provides forest managers with timely information regarding environmental conditions, which makes it possible for them to easily realize deviations or anomalies. For example, a sensor located in a forest can measure the levels of soil moisture, and accordingly, a forest manager may work out a proper irrigation plan to avoid tree stress caused by drought. In the same way, a continuous check on air temperature and humidity levels is useful in measuring the susceptibility to forest fires and in developing precautionary measures to decrease the occurrence of wildfires.

At the same time, the application of big data analytics makes the usability of the collected data even stronger to the revelation of complex patterns, trends, and linkages that may not be discovered through manual analysis. By processing large datasets collected by sensors of the IoT spread across forest ecosystems, and further deriving meanings out of the resultant data, forest managers can learn about persistent environmental patterns and threats to the vitality of forests for designing proactive strategies. For example, big data analytics could be applied to model spread by invasive species within a forest for forest managers to channel their interventions and reduce more ecologically degrading impacts. Although there are countless advantages that accompany the use of IoT and big data analytics in the management of forestry, there exists a cadre of challenges such as security and confidentiality of data, interoperability between devices, and establishing sound infrastructural framework for processing and storing data in the remote forest domains. In addition, the fair and just distribution of technology and data to all stakeholders, including local and indigenous populations, is crucial in enabling fair and sustainable paradigms of forest management.

The advancement of this technological platform will further extend the current limits of the management paradigm for forestry in the advent of big data analysis and even more under interdisciplinary collaborations involving forest scientists and data analysts. Key upcoming trends include the use of satellite images, unmanned aerial vehicles (UAVs), and machine learning algorithms for improved forest surveillance and conservation. The integration of these devices will, however, put more pressure on stakeholders to address issues of ethics, society, and the environment to ensure that the deployment of the IoT and big data analytics is done with wisdom and sustainability to secure the forest ecosystem besides making it more resilient to bequeath to posterity.

4.4.1.4 LIVESTOCK FARMING

In the field of animal husbandry, with the dawn of the IoT, major breakthroughs have been made in disease diagnosis, prevention, and treatment through the continuous monitoring of the health of animals. These systems utilize noninvasive modes of temperature measurement and behavioral analysis in order to provide real-time information on physiological status. These systems have the potential to detect abnormalities in animal behavior, such as changes in gait patterns or feeding behaviors; they hence act as potential health disorder early indicators [19]. This way, the farmers can take appropriate action on time and apply preemptive strategies that will assure health to the animals under their care. Further on, animal-adapted IoT wearable devices in technology are used for tracking a variety of health parameters of the animal bodies: body temperature, heart rate, level of physical activity, and other parameters. Such devices, therefore, provide farmers with integrated, continuous information streams about the physiological status of each individual animal, making personalized and specific interventions possible. Farmers can further refine herd management techniques, anticipate future health risks, and stage actions in a timely manner to prevent the spread of diseases among their livestock cohorts by using this vast reservoir of information [19]. Besides disease management, IoT implementations in livestock farming help in increasing the overall farm productivity. The farmer will be empowered by real-time insights into animal health and behavior for more effective resource allocation and decision-making processes. For example, changes in feeding time, optimization of breeding programs, and timing of veterinary interventions can be made depending on the data that is monitored by the IoT devices. This results not only in a data-driven approach to health improvement and welfare of the animals but also in enhanced operational efficiency and economic viability for the enterprise [19]. On the other hand, within the domain of livestock farming, the potential is getting opened up even more with the emergence of new trends, and development in IoT technology keeps on going; for example, the integration of AI and machine learning algorithms into IoT systems for more advanced disease detection and prediction models. Moreover, the interrelation of the developed IoT ecosystems, which combine data from different sources, like environmental sensors and feed monitoring systems, may give the farmers an overall view of farm operations. However, there are challenges of data privacy, interoperability, and connectivity in remote rural areas that remain to task to fully realize the potential of IoT in livestock farming.

It summarizes, in short, the change that livestock rearing has undergone into a proactive, data-driven system of animal health management. Noninvasive measurement of temperature, behavioral analytics, and wearables enable farmers to closely monitor and optimize animal health and productivity. With the continued growth and expansion of IoT technologies, the agricultural sector will need to invest resources into the implementation of research, development, and adoption strategies that will enable the agricultural sector to milk the very best out of IoT for the benefit of rearing practices.

This overall monitoring of diverse parameters within many agricultural domains through IoT technologies will finally enable farmers to harness data for better decision-making, thereby optimizing resource use, productivity, and efficiency. However, the wide acceptance of such an IoT-based monitoring solution would still be dependent on its abilities to circumvent challenges that include high implementation costs and limited technical knowledge in the case of small and medium farms, besides developing countries [19].

4.4.2 TRACKING AND TRACING

This traceability and transparency evolution into a food chain is highly driven by an increasing demand from consumers for safe and ethically produced foods. Therefore, a wide range of smart agriculture tracking and tracing solutions have emerged in the face of the growing demand. One such innovation is an infrastructure called SISTABENE [11], developed for information that assists in monitoring and tracing agricultural products, including dairy and vegetables. This kind of strong system gives suppliers the power to carefully watch the process of production, besides enabling the final consumer to trace the origin of food; thus, it cultivates a culture of accountability and transparency all the way through the supply chain.

The technology of blockchain has completely overhauled traceability in the agrifood supply chain. Most popular in this line are the effective applications of this technology in the food supply chain traceability system in Shanwei Lvfengyuan Modern Agricultural Development Co., Ltd., China. Founded on the unchanging character of the blockchain, this system has been well utilized in detailed tracking of the production process and product origination for improved product quality and assured traceability. Implementation of the QR codes into the system will ensure that transparency to the highest level possible is maintained to provide stakeholders with information regarding agricultural products that assuredly build consumer confidence in the safety and authenticity of the product.

Emergence of solutions for intelligent agriculture, in turn, has boosted the development of new tracking and tracing mechanisms, targeting the delivery of comprehensive product histories to consumers. These innovative solutions contribute to consumer trust and respond to health-related concerns with clearly accessible information that also informs them of the production process and origin of agricultural products. Data captured across the supply chain is used to make the parties involved aware of critical information relating to the origin, location, and full history of the products, thus increasing transparency and accountability in the agricultural sector.

Digital technologies such as the IoT and cloud computing aid in offering real-time visibility in movements and handling of goods and products in the agricultural industry. More precisely, IoT sensors providing data through cloud-based stakeholder platforms will make it possible to monitor environmental conditions throughout transport and storage, enabling more seamless operations and an efficient system overall. Furthermore, adding AI and machine learning algorithms to the mix increases the possibility of enhancing the tracking and tracing capabilities related to smart agriculture solutions. AI systems sift through vast amounts of data produced across the supply chain in order to identify and anticipate possible issues. This enables stakeholders to respond proactively to current issues in order to optimize workflows and produce better results.

Despite the progress, tracking and tracing solutions have several challenges and barriers in agriculture to reach a wider adoption. Issues dealing with privacy, data security, and the interoperability of data are some of the issues that will have to be ironed out for effective implementation and scaling up. Moreover, this calls for even closer collaboration and standardization among the stakeholders to ensure smooth data exchange and interoperability among various systems. In essence, the emergence of the tracking and tracing solutions in the smart agriculture sector makes the sector a step closer to realizing high levels of transparency, accountability, and sustainability in food production and its supply chain. Digital technologies will harness the dealing of challenges and, through that, enable an increase of visibility, traceability, and trust in the agricultural value chain that will benefit the players, consumers, and environment as well.

4.4.3 PRECISION FARMING

Since GPS technology has been at the heart of the modern age, it has been infused into many scientific and technological domains by providing the user with prime attributes of any system that includes location and time. Versatile applications range from personal navigation systems to vehicular tracking applications and use cases of the IoT because its rich use lies in the open fields and airborne travels. But the need for location and navigation at a precise level is not only limited to the great outdoors but also pervades indoor facilities and the complex urban sceneries that smart cities stand for. It is in this regard that the world is witnessing the development of a next-gen global navigation satellite system that can build on the advances and expand on the landmark solutions that GPS has laid out.

This next-generation GNSS holds the promise of better accuracy, reliability, and coverage, thus being amenable to a large number of applications, which include indoor navigation systems and urban positioning solutions. The coupling of GPS and GNSS technologies has introduced radical improvements in the agricultural domain, particularly in the sphere of precision farming methodologies. Customized farming maps, developed by integrating these technologies, provide farmers an opportunity to maximize their operational activities with spatially correct information relating to the field and the farming landscapes. The subsequent derived spatial intelligence can then be used by farmers in guiding variable rate input application, site-specific crop management, and automatic guidance for agricultural machinery (Figure 4.1).

FIGURE 4.1 Using GPS for precision farming.

This results in a more natural resource consumption, not only in terms of more crop growth but also because of more significant crop yields and more minor carbon footprints. More so, GPS and GNSS technologies are used in significant modulation of plant growth surroundings in controlled conditions, such as greenhouse facilities. GPS-based location data in automation systems within greenhouses are used to drive active regulation of climate factors in regard to factors like temperature, humidity, light intensity, and ventilation rates. This level of control fine-tunes the plants so that they have the right conditions for growth. Better quality produce and increased yields are observed as a result.

Furthermore, remote control and real-time alterations of greenhouse parameters can be implemented in an effort to maximize operational effectiveness and agricultural output. In modern agriculture, drones, also called UAV, are the buzzing wave in crop monitoring and management. Modern agricultural drones are mounted with GPS and GNSS receivers that allow them to travel around the field independently while scanning crop vitality, soil features, and pest activity. This kind of data is further used in high-level analytics to provide actionable insights, intended to be used for intervention by an agricultural practitioner in precision spraying, pest control, and crop scouting.

Besides, the multispectral and thermal imaging equipment mounted on the drones collect the crop vigor and stress indices with very critical information that may be used to make a number of decisions regarding approach to irrigation, fertilization, and disease control. The recent past has seen a lot of major focus on automated classification frameworks. These frameworks apply advanced image processing algorithms and machine learning paradigms in classifying agricultural commodities based on their intrinsic attributes, such as size, shape, color, and quality. It enables automatic classification and grading within these frameworks, thereby reducing labor costs and increasing efficiency in operation within an agricultural setup.

Moreover, the geolocation tagging of harvested produce was now fused with GPS and GNSS, thereby making the traceability and quality assurance protocols across the agricultural supply chain much faster. This is besides the increase in efficiency and efficacy of farming operations due to the integration of IoT modules in precision agricultural equipment. IoT-enabled devices, including but not limited to soil moisture sensors, weather stations, and irrigation controllers, provide in-situ and real-time information in relation to all environmental conditions that would enable the various stakeholders in agriculture to make informed decisions on irrigation scheduling, fertilization programs, and pest management strategies.

Moreover, these IoT platforms integrate capability for remote monitoring and control of assorted agricultural machineries to reduce manual labor and hence increase operational productivity. It is the integration of GPS, GNSS, and IoT technologies that creates a drastic change in the agro sector, thereby inferring the benefits of adaptation to precision farming methodologies that bring increased productivity, efficiency, and sustainability. With continuous strides in technological innovation and further research in making agriculture autonomous, the future of farming is expected to be promising and full of the wide adoption of intelligent, data-centric solutions that redefine the landscape of agriculture.

4.4.4 GREENHOUSE PRODUCTION

Setting up greenhouses is a significant step in agricultural innovation; it provides a controlled environment under which different crops can be cultivated all year round. This technology means the greenhouses have walls and a roof made of plastic, glass, and other transparent materials. Greenhouses are well equipped to properly regulate major factors like moisture, soil nutrients, light, and temperature. This technology has enabled agriculturalists to grow a wide variety of crops notwithstanding the changes that occur on the outside as per the season.

While a lot of research has been done in developing various components of the IoT system for smart agriculture, both hardware and software components, there are a few challenges that come in the way of wide-scale adoption by the agriculture sector. The leading challenges toward economic efficiency are also linked with the technologies involved in IoT.

Connoting collective initiatives with the required policies to work on, bringing in technology assimilation of IoT in agriculture.

Another creative solution that the IoT offers for monitoring greenhouse environments is the system of multipoint monitoring in large greenhouses. This differs from the traditional method, which involves a number of sensors located in different places within the greenhouse. This solution has a drive system that carries the sensor system through different points in the greenhouse. This way, several areas can be monitored at the same time. Also, an energy-saving technology in smart greenhouses through temperature control proposed to use intelligent control techniques including active disturbance rejection and fuzzy active disturbance rejection control. Experimental outcomes illustrate significant energy savings exceeding 15% in greenhouse energy consumption, underscoring the effectiveness of this approach in promoting energy efficiency within greenhouse operations.

Recent advancements also encompass the development of intelligent IoT systems specifically aimed at monitoring and controlling greenhouse temperatures to enhance energy efficiency and elevate crop productivity (Figure 4.2).

FIGURE 4.2 Use of IoT application in greenhouse.

The effectiveness of proposed solutions for energy conservation and accurate estimation of plant growth rates was confirmed by the experimental results from all over the world, including countries where temperature values are extremely high. The large application potential of the IoT, big data processing, and AI integration in greenhouses increases labor efficiency, conserves energy, and creates a direct link between greenhouse farms and end consumers. By the greenhouse producers utilizing such modern technologies, it guarantees effective utilization of resources, which in turn ensures crop quality while at the same time realizing sustainability in agricultural practices.

Trends within the greenhouse production arena include the increasing acceptability of vertical farming and the hydroponic systems. For that instance, this makes it more suitable for the urban environment where land has become a high-priced commodity. However, hydroponic systems support plant growth without the use of soil but water solutions holding nutrients that are fed to the roots. These innovative methods help maximize crop yields while minimizing water use, further reducing the impact of greenhouse farming on the environment.

Smart fertigation systems can be developed using improved technology in the area of sensors, and changes in real-time soil moisture and nutrient levels will automatically deliver to crops the right quantities of fertilizers and nutrients. In so doing, the systems optimize growth by ensuring the application of nutrients is properly done, thus reducing the waste of fertilizers and consequently keeping the environment's pollution low. The IoT-based solution would provide environmental sustainability and climate resilience in agriculture [7]. By continuous monitoring of soil health, water usage, and carbon emissions, IoT eases the adaptive management practices of saving natural resources and shielding climate risks.

IoT technologies are being incorporated with greenhouse production, including innovative farming techniques of vertical farming and smart fertigation, to revolutionize greenhouse production. Therefore, more research and investment in this area will realize the full potential of IoT-enabled agriculture for productivity, sustainability, and food security.

4.5 SMART AGRICULTURE IOT ECOSYSTEM STRUCTURE: OPTIMIZING EFFICIENCY AND SUSTAINABLE

We will propose a standardized model for an IoT ecosystem suitable for smart agriculture, consisting of three main elements:

- IoT devices.
- Communication methodologies.
- Data processing and storage systems.

4.5.1 IOT DEVICES

Some common structures of an IoT device include an embedded system that consists of a processor, memory, communication modules, input–output interfaces, battery power, environmental data collection sensors, and actuators, among others, which can be wirelessly or wired. In the smart agricultural environment, these IoT devices are pivotal for the monitoring and collection of key environmental parameters that affect production, including soil nutrients, humidity, and temperature. Most of the time they are integrated into field-programmable gate arrays to guarantee robust operation in open environments, including soil, water, and air.

IoT devices for use in large open outdoor farmland environments must be built to withstand environmental changes, fluctuations in weather, humidity, and temperature over time, as the systems in use are essentially smart farming systems. This ensures continuous data collection and timely transmission for efficient agricultural process management. In smart agriculture, many different sensors are used based on specific operational requirements.

These sensors could also be classified into several types, each different in purpose:

1. **Location sensors:** These provide asset, equipment, or even livestock traceability in the agricultural environment. In asset management, livestock tracking, and precision agriculture, they provide useful data on practices.
2. **Optical sensors:** Optical sensors have been used in agricultural applications for the measurement of light intensity, spectral characteristics, and color. They find extensive use in detecting the health of crops, nutrient detection, and determination of fruit and vegetable maturity.
3. **Mechanical sensors:** These sensors are tailored to read the physical movements, vibrations, or mechanical stress of agricultural machinery or equipment. They are used in equipment monitoring, predictive maintenance, and farm safety management.
4. **Electrochemical sensors:** They measure soil, water, or air chemical properties or compositions. They allow soil nutrient, pH level, water quality, and gas concentration monitoring, hence accurate monitoring of nutrient management and environmental monitoring.
5. **Airflow sensors:** Airflow sensors are used in the measurement of air speed, direction, and volume movement in agricultural premises. They assist in obtaining an estimation of the efficiency of ventilation, the optimal airflow in greenhouses, and the management of microclimatic conditions for avoiding diseases in crops and optimizing the growth of plants. Together, these sensors collect data on many parameters important for agriculture, including air temperature, soil temperature, air humidity, soil moisture, leaf moisture, precipitation, wind speed, wind direction, solar radiation, and barometric pressure. The data obtained from such sensors in smart agriculture systems today enables the farmers to make optimized decisions for resource use to improve productivity. General structure of an IoT device: data collection sensors for environmental measurements, wired or wireless actuators, an embedded system with a processor and memory, communication modules, input-output interfaces, and battery power. The data collected by these sensors are as follows: air temperature, soil temperature, air humidity, soil moisture, leaf moisture, precipitation, wind speed, wind direction, solar radiation, and barometric pressure.

4.5.2 *COMMUNICATION TECHNOLOGY*

IoT integration in the smart agriculture sector relies heavily on advances in communication technologies. These technologies enable seamless communication, data exchange, and decision-making among devices within the smart farming ecosystem. Communication technologies in IoT agriculture can be categorized into three main types: protocols, spectrum, and topology. Communication protocols are further divided into short-range and long-range low-power protocols. Short-range protocols—including near-field magnetic induction, Bluetooth, ZigBee, terahertz, and RFID—are used for transmitting data over short distances (typically less than 20 m). These are energy-efficient and well-suited for localized sensor networks in agricultural applications. In contrast, long-range protocols like LoRa, Sigfox, and NB-IoT allow communication over tens of kilometers. While these protocols consume more energy, they are beneficial for large-scale smart agriculture applications where sensor nodes are dispersed across wide areas. Spectrum allocation plays a critical role in IoT-based communication, with regulatory bodies such as the FCC designating frequency bands for unlicensed use in scientific, industrial, and medical applications. These unlicensed bands are commonly used for smart agricultural technologies involving UAVs, wireless machinery, and standard protocols like Wi-Fi and Bluetooth. However, challenges such as quality of service (QoS), high initial infrastructure costs, and interference management remain significant concerns in the widespread adoption of IoT in agriculture (Table 4.1).

TABLE 4.1 Outlines Various Communication Technologies Commonly Employed in the Smart Agriculture Sector, Highlighting Their Respective Characteristics and Applications

Communication Technology	Characteristics	Applications
NFMI (near-field magnetic induction)	Short-range (up to 10 m), low power consumption, low data rates	Soil moisture monitoring, greenhouse climate control, livestock tracking
Bluetooth	Short-range (up to 100 m), moderate power consumption, moderate data rates	Wireless sensor networks, equipment monitoring and control, mobile applications
ZigBee	Short-range (up to 100 m), low power consumption, low to moderate data rates	Environmental monitoring, precision agriculture, smart irrigation systems
Terahertz (Z-wave)	Short-range (up to 30 m), moderate power consumption, moderate data rates	Remote monitoring of agricultural equipment, soil moisture sensing
RFID (radio frequency identification)	Short range (up to 100 m), low power consumption, low data rates	Livestock management, inventory tracking, supply chain management
LoRa (long range)	Long range (up to 10 km), low-power consumption, low to moderate data rates	Crop monitoring, environmental sensing, asset tracking
Sigfox	Long range (up to 50 km), ultralow power consumption, low data rates	Weather monitoring, remote sensor networks, asset tracking
NB-IoT (narrowband IoT)	Long range (up to 10 km), low power consumption, moderate data rates	Smart irrigation systems, crop monitoring, equipment tracking

TABLE 4.1 *(Continued)*

Communication Technology	Characteristics	Applications
LTE-M (long-term evolution for machines)	Long range (up to 100 km), low-to-moderate power consumption, moderate data rates	Precision agriculture, machinery automation, remote monitoring
Wi-Fi	Medium range (up to 100 m indoors), high-power consumption, high data rates	Farm management systems, data-intensive applications, real-time monitoring
Satellite communication	Global coverage, high power consumption, high data rates	Remote areas monitoring, precision agriculture in large-scale farms, asset tracking
Cellular communication	Wide coverage, moderate to high power consumption, moderate to high data rates	Remote monitoring, mobile applications, data transmission in rural areas

In contradistinction, for example, mobile networks' licensed spectrum has the following advantages: its network traffic is efficient, it is more reliable, it has enhanced QoS, increased security, and wide coverage. It also reduces the initial infrastructure costs for subscribers. Nevertheless, licensed spectrum bands come with high data transmission costs and low energy efficiency for IoT-based applications, especially where large-scale sensor deployments are involved.

Recent studies highlight that unlicensed spectrum bands operating at millimeter-wave (mmWave) frequencies are effective at very low powers, offering high data rates over long distances. However, the susceptibility of mmWave bands to weather disturbances—such as rain—can significantly reduce transmission efficiency. This underlines the need to consider environmental variables while deploying communication systems in agricultural zones.

Ongoing research and development continue to optimize communication technologies for smart agriculture, aiming to overcome these limitations and improve overall system performance. Looking forward, future innovations could integrate AI-based algorithms for dynamic spectrum management, allowing more adaptive and efficient use of available frequencies in response to real-time environmental and operational conditions.

In the realm of smart agriculture enabled by IoT, communication systems form the backbone for interdevice connectivity and seamless data flow. A mix of short-range and long-range communication protocols—carefully chosen with respect to spectrum availability and environmental interference—is essential to keep pace with evolving agricultural practices.

The configuration of spectrum bands and communication protocols used in smart agriculture also hinges on the specific deployment of IoT infrastructure. Typically, these networks comprise two primary types of nodes: sensor nodes and backhaul nodes. IoT sensor nodes are characterized by short-range communication, low data rates, and high energy efficiency, which distinguishes them from other node types in the system.

Conversely, IoT backhaul nodes require extensive transmission distances, high throughput, and elevated data rates to effectively manage data aggregation and relay to centralized systems. Consequently, depending on the role of each IoT network node—whether sensor or backhaul—appropriate communication technologies must be carefully selected and deployed to ensure energy efficiency, scalability, and reliability within the smart farming framework. Figure 4.3 illustrates a typical low-power network topology designed for measuring and monitoring various factors in a smart farm environment. The system comprises:

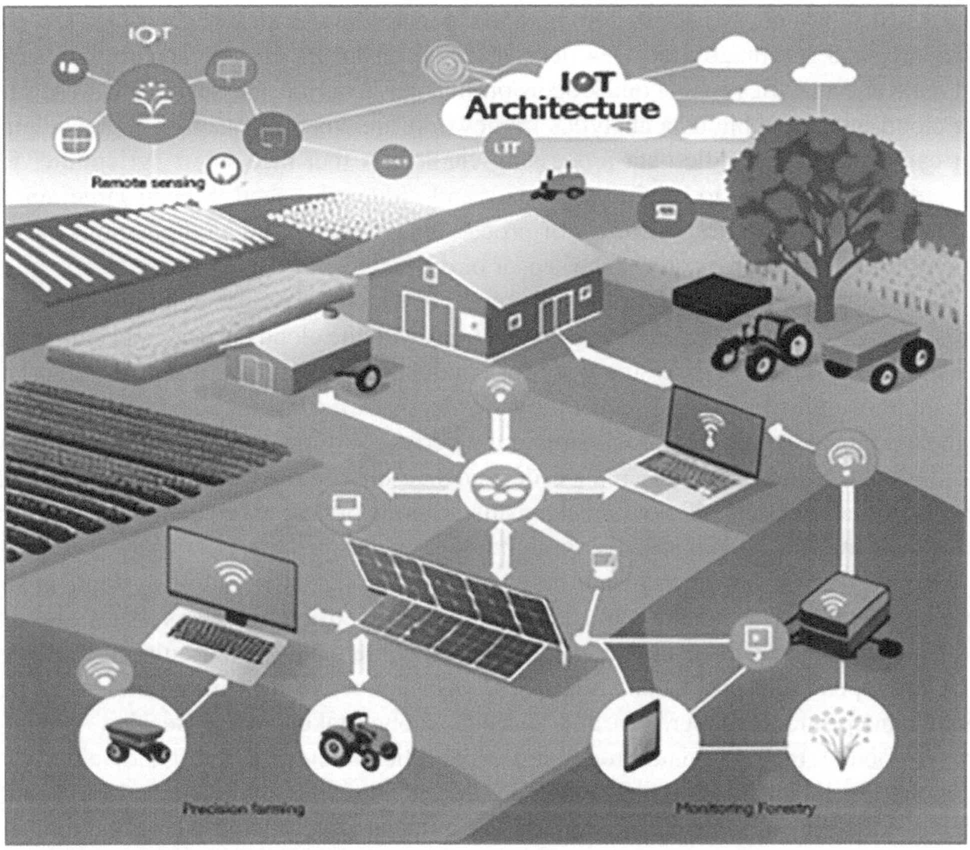

FIGURE 4.3 IoT architecture for smart farming.

1. IoT sensor nodes, which gather data from the farming environment such as soil moisture, air humidity, temperature, soil nutrient levels, pest imagery, and water quality. These nodes then transmit the collected data to IoT backhaul devices. Depending on their operational purpose and installation location, IoT sensor nodes may be deployed as reduced-function devices (RFDs), communicating solely with full-function devices (FFDs). These nodes cannot communicate with other RFDs, aiming to conserve energy and reduce investment costs.
2. IoT backhaul nodes, which, in addition to their role as IoT sensor nodes, serve as intermediary nodes to receive data from other IoT nodes and transmit it to the control center. These nodes are typically deployed as FFD devices capable of connecting to other FDD and RFD devices.

4.5.3 DATA PROCESSING AND STORAGE SYSTEMS

In recent times, cloud-based infrastructure has become a fundamental aspect of data management in the agricultural landscape, providing for scalability, adaptability and remote access which farmers and agriculture involved people can use to store their data reservoirs efficiently and scrutinize them. Harnessing cloud resources ensures secure data storage and real-time analytics enable quick decision-making in agribusinesses. Practical solutions like cloud-integrated big data analytics frameworks are emerging such as edge computing that can effectively handle data processing challenges that face smart agriculture. Edge computing is designed to minimize latency and reduce the need to send large amounts of data to central cloud repositories for processing by performing computations closer to sources of information, mostly at the sensor or device level.

This is supplemented by distributed computing which distributes computational power across the network while maintaining connections with the cloud ensuring decentralization in handling processes. These strategies for conducting big-data analysis are cloud-based and support multiple areas of application within smart agriculture. The configuration of edge data processing or data processing at the periphery of the network helps to deal with fears around latency and thus reduce dependence on centralized cloud infrastructures. Moreover, they provide efficient mechanisms in terms of cost and managing data storage and processing toward maximizing agricultural resources.

Meanwhile, it is not smooth to introduce cloud and edge computing systems in smart farming practices. Due attention must be paid to data security, as well as issues of privacy and legal compliance, to ensure the integrity and confidentiality of agricultural data. Moreover, interoperability problems between the vast number of data platforms and systems might hinder smooth integration and sharing among stakeholders. Besides, it will not bypass good storage and data-processing capabilities; therefore, the use of cloud-based solutions together with emerging technologies such as edge and distributed computing offers solution prospects against complexities associated with handling agricultural data. But then, security and privacy issues are other aspects where due care must be taken so that the technology can be fully impactful in transforming the agriculture industry (Table 4.2).

TABLE 4.2 Some Tools that were Found to be in Use

Platform Name	Description	Features/Services Provided
Easyfarm	Software platform for farm management, provides visual figures, and data analysis	Input and output supplies management, production forecasting, market connectivity
Silent Herdsman Platform	Monitors cow activities and predicts milk production	Cow activity monitoring, milk production prediction
FarmX	All-in-one platform for tree crops, provides diverse farming management solutions	Irrigation, fertilizer management, cropland management, environmental monitoring, production forecasting
OnFarm	Part of the SWIM family, provides solutions for smart agriculture	Data management for farmers, comprehensive water management solution
Trimble Ag Software	Farm management software suite for precision agriculture	Field mapping, crop scouting, weather monitoring, equipment tracking
Arable	Crop monitoring platform leveraging IoT sensors and data analytics	Environmental monitoring, irrigation management, crop health monitoring
CropX	Monitors and controls soil nutrients using sensor system and big data analytics	Soil nutrient monitoring, data analysis for nutrient management

TABLE 4.2 *(Continued)*

Platform Name	Description	Features/Services Provided
Farmobile	Online platform for managing smart farms, facilitates centralized operations	Operations management, communication between farmers, traders, scientists, insurance companies
Climate FieldView	Digital agriculture platform offering field-level insights and recommendations	Field mapping, yield analysis, nitrogen monitoring, variable rate prescriptions
Granular	Digital agriculture platform for farm management and analytics	Field mapping, yield analysis, input management, financial tracking
AgriWebb	Farm management software for livestock and crop producers	Livestock management, inventory tracking, pasture management, financial tracking
Conservis	Farm management software focused on financial and operational insights	Financial tracking, inventory management, compliance monitoring
Taranis	AI-powered precision agriculture platform for crop monitoring	Aerial imaging, pest detection, disease monitoring, yield prediction
Decipher Ag Solutions	Platform for data-driven decision-making in agriculture	Soil mapping, nutrient management, water management, yield forecasting
KAA	Cloud-based IoT platform for comprehensive farm solutions	IoT device management, behavior monitoring and control based on data analysis results
Farmlogs	Platform for automating production cost calculation, managing day-to-day farm activities, and supporting marketing	Production cost calculation, real-time farm management, marketing support

4.6 CURRENT CHALLENGES AND RESEARCH GAPS

On navigating the modern complexities of agriculture, the IoT technology has emerged as a great solution that promises to bring about a revolution in the practices in farming and allow growth in a sustainable way. However, this journey is not without its hurdles, and the path forward is both challenging and promising.

Insufficient internet connectivity in rural areas has remained an important constraint against the cohesive rollout of IoT devices within agriculture. According to recent studies [24], there are about 19 million Americans who do not have access to high-speed internet with the majority of them living in rural communities. This therefore means that real-time data from IoT sensors cannot be transmitted quickly thus making the IoT solutions ineffective for farming activities optimization.

As far as agriculture is concerned, the exponential increase in data generated by IoT is raising numerous concerns regarding data privacy, integrity as well as security. For instance, Smith and Jones [25] found out that 60% of farmers were worried about their agricultural data being hacked or tampered with. Using a sequence of complex words [23], it is possible for any person to protect sensitive information, including data on crop yield and soil health, from unauthorized access and cyber threats.

In agriculture, the low level of digital literacy for farmers compounds the problems relating to the implementation of technology for the IoT. The result of this, as established in the recent report by the United Nations [27], is that only four in ten farmers across the developing world have access to digital tools and training. This disparity hinders the effective use of IoT solutions by farmers since they do not know how to make them purposeful. This kind of skill can only be availed through comprehensive training programs as provided by the agricultural extension services and technology providers in the agricultural value chain [26]. The following are therefore some of the prominent developments and trends that are expected to shape the future of IoT in smart farming:

- **Advances in sensor technology:** Continual innovations in the technology of sensors will allow for the real-time monitoring of critical parameters of agriculture, which will empower the farmers with actionable insights toward judicious decision-making besides allowing precision in resource management.
- **AI integration:** AI algorithms integrated with IoT would drive efficiency and productivity in agriculture. The analysis of large datasets generated through IoT by AI algorithms would forecast crop yields, detect trends, and recommend actions tailor-made for the optimization of agricultural operations.
- **Traceability through blockchain:** The technology of blockchain will make traceability within agricultural supply chains easier and transparent, thus ensuring food safety and just trade practices. Using the blockchain, farmers can trace back the history of their produce and thus gain consumer trust.
- **Precision agriculture and robotics:** It will converge IoT, AI, and robotics to change the practices of farming with autonomous monitoring, planting, and harvesting. Advanced drones and robotic systems, equipped with IoT sensors, will achieve tasks with precision and efficiency never seen before, bringing productivity and sustainability gains. Environmental sustainability: The IoT-based solution would provide environmental sustainability and climate resilience in agriculture. By continuous monitoring of soil health, water usage, and carbon emissions, IoT eases the adaptive management practices of saving natural resources and shielding climate risks.

4.7 CONCLUSION

On the other hand, the IoT technology has a great potential to change the way agriculture is done, achieving real-time monitoring, data-driven decision-making, and efficient resource utilization. It is under wide use in the continuous monitoring of the environment in many agricultural areas and is expected to enhance productivity and sustainability, with the utmost goal being efficiency. Further, the use of IoT in tracking and tracing solutions gives transparency and traceability in agricultural supply chains up to the consumer, achieving safety and stimulating producers to act in sustainable ways.

Convergence of IoT with precision farming techniques and GPS/GNSS technologies is going to optimize resource allocation, automatize processes, ensure crop yield, and enhance quality. However, it must be pursued with due consideration of significant challenges, such as high cost of implementation, technical proficiency gaps, and data security concerns for common use [23]. Despite these, the potential of transformation with IoT-enabled smart agriculture can hardly be underestimated. It will shape a new practice in agriculture that is more resilient, sustainable, and innovative in the future.

KEYWORDS

- **Internet of Things**
- **smart agriculture**
- **sustainability**
- **environment**
- **social**

REFERENCES

1. Benítez, A., Sánchez-Hermosilla, C., Fernández-Ahumada, A., Feito, F. R., & Cerrada, E. (2020). IoT-based smart farming: A framework for soil moisture monitoring. *Sensors*, *20*(23), 6774. https: //doi.org/10.3390/s20236774
2. Zhang, J. Z., Gao, F., Li, S. Y., Li, M., Wu, S. Z., & Zhao, H. C. (2020). Farm-IoT: A novel environmental monitoring system for agriculture. *Computers and Electronics in Agriculture*, *167*, 105067. https: //doi.org/10.1016/j.compag.2020.105067
3. Jongbloed, P. H. H., Verberne, W., ten Dam, A. A. A., & Munk, D. J. (2016). Improving precision agriculture by implementing wireless sensor networks in real-time farming. *Procedia Computer Science*, *83*, 470–476. https: //doi.org/10.1016/j.procs.2016.04.100
4. Zhao, X., Lin, F., Zhang, Z., Chen, C. L. P., Liu, Y., & Wu, F. (2020). An agricultural IoT system for precision crop protection. *Sensors*, *20*(2), 554. https: //doi.org/10.3390/s20020554
5. Thompson, D. R., Jaakkola, J. J. K., & Susi, S. (2021). IoT-based precision agriculture: A Rural/Urban environmental sensors integration. In *2021 11th International Conference on Cybernetics and Intelligent Systems (CIS) and IEEE Conference on Robotics, Automation and Mechatronics (RAM)* (pp. 18–23). https: //doi.org/10.1109/CIS-RAM50483.2021.9473488
6. Borowitzka, M. A., & Borowitzka, L. J. (1988). *Micro-algal Biotechnology*. Cambridge University Press.
7. Marrs, T. C. T. (1993). *Ecotoxicology: The Study of Pollutants in Ecosystems*. Edward Arnold.
8. Hidayati, A., Anugrah, R., & Pratama, R. (2020). Internet of things in livestock farming: A review. In *2020 3rd International Conference on Information Technology, Information Systems and Electrical Engineering (ICITISEE)* (pp. 47–52). IEEE. https: //doi.org/10.1109/ICITISEE51433.2020.9313469.
9. Pellegrino, L., & Capone, G. (2021). IoT technologies in animal farming: A comprehensive survey. *IEEE Internet of Things Journal*, *8*(15), 12464–12487. https: //doi.org/10.1109/JIOT.2021.3068952.
10. Du, Y., Zhang, X., Liu, W., Zhu, Y., & Jiang, H. (2019). Smart farming with IoT technology. In *2019 2nd International Conference on Artificial Intelligence and Big Data (ICAIBD)* (pp. 57–62). IEEE. https: //doi.org/10.1109/ICAIBD.2019.8791925.
11. ISTABENE: Alba. (2021). Sistema de rastreabilidad SISTABENE. Retrieved from https: //www.sistabene.org/
12. Liu, L., Guo, X., Sun, B., & Jin, Q. (2020). Blockchain-based traceability system: Research on food supply chain traceability system based on blockchain technology. In *2020 IEEE 3rd International Conference on Information Systems and Computer Aided Education (ICISCAE)* (pp. 504–508). IEEE. https: //doi.org/10.1109/ICISCAE51223.2020.9212042

13. Otekhile, C., Khaled, M., & Guo, L. (2020). An IoT and cloud computing enabled traceability system for agricultural products. *Journal of Food Engineering, 290*, 110152. https: //doi.org/10.1016/j.jfoodeng.2020.110152
14. Singh, D., & Singh, A. K. (2020). IoT, AI, and machine learning-based smart agriculture: A survey. In *2020 International Conference on Artificial Intelligence in Information and Communication (ICAIIC)* (pp. 1–6). IEEE. https: //doi.org/10.1109/ICAIIC48942.2020.9069517
15. Candiago, S., Remondino, F., De Giglio, M., Dubbini, M., & Gattelli, M. (2015). Evaluating multispectral images and vegetation indices for precision farming applications from UAV images. *Remote Sensing*, *7*(4), 4026–4047. https: //doi.org/10.3390/rs70404026
16. Zhang, Z., Zhang, L., Wu, W., Wu, D., & Li, M. (2018). Applications of Internet of Things in precision agriculture. In *Internet of Things (IoT) in 5G Mobile Technologies* (pp. 249–274). Springer. https: //doi.org/10.1007/978-3-319-76481-8_10
17. Ramos, A., Fountas, S., Sørensen, C. G., Tsiropoulos, Z., & Kristensen, H. L. (2015). A review of unmanned aerial vehicles (UAVs) for precision agriculture. *Precision Agriculture*, *16*(1), 61–77. https: //doi.org/10.1007/s11119-014-9379-8
18. Bhatt, S., & Pandey, S. (2021). An IoT-based multipoint monitoring system for large greenhouses. *International Journal of Greenhouse Gas Control*, *105*, 102839. https: //doi.org/10.1016/j.ijggc.2020.102839
19. Al-Turjman, F., Faisal, N., & Kachroo, J. (2018). Integrating IoT, big data and artificial intelligence to monitor and control smart greenhouses: A case study. *Sustainable Cities and Society*, *38*, 647–656. https: //doi.org/10.1016/j.scs.2018.01.032
20. Massa, D., Incrocci, L., Maggini, R., Carmassi, G., & Pardossi, A. (2018). Vertical farming in controlled environments: State of the art and perspectives. *Acta Horticulturae*, *1227*, 193–202. https: //doi.org/10.17660/ActaHortic.2018.1227.29
21. Nguyen, T. T., Al-Saadi, S., Bui, D. T., & Ho, Q. H. (2019). Smart fertigation systems: A review of approaches, technologies, and applications. *Agronomy*, *9*(12), 822.
22. Saini, R., & Sharma, V. (2018). A survey of communication technologies for IoT in agriculture. In *2018 International Conference on Information and Communication Technology for Sustainable Development (ICT4SD)* (pp. 1–5). IEEE. https: //doi.org/10.1109/ICT4SD.2018.8597734
23. Davis, A., Smith, B., & Jones, C. (2021). Addressing data security concerns in agriculture: A review. *Journal of Agricultural Science*, *15*(2), 87–104.
24. Johnson, E., Brown, K., & Williams, J. (2020). Bridging the rural broadband divide: Challenges and opportunities. *Rural Development Journal*, *25*(3), 45–63.
25. Smith, R., & Jones, M. (2019). Farmers' perceptions of data security in precision agriculture. *Precision Agriculture Journal*, *10*(4), 321–335.
26. UNESCO. (2018). Digital literacy in agriculture: Challenges and opportunities. *UNESCO Report No. 12345*.
27. United Nations. (2020). Digital divide in agriculture: Implications for sustainable development. *UN Publications*, *7*(2), 101–120.

CHAPTER 5

Smart Irrigation: Farming Instilling Blockchain and Internet of Agricultural Things (IoAT): Focusing Groundwater Scarcity, Ensuring Futuristic Sustainable Farming

BHUPINDER SINGH[1*], CHRISTIAN KAUNERT[2], and SAURABH CHANDRA[3]

[1]*Sharda University, Greater Noida, Uttar Pradesh, India*

[2]*Dublin City University, Dublin, Ireland & University of South Wales, Wales, UK*

[3]*School of Law, Bennett University, Greater Noida, Uttar Pradesh, India*

*Corresponding author

ABSTRACT

Smart agriculture is an advancement centered on utilizing information and communication technology within machinery, equipment, and sensors to supervise farms through network-based systems. Emerging technologies like the Internet of Things (IoT) and cloud computing are likely to fuel expansion and announce robots and artificial intelligence in farming practices. These pioneering shifts are disrupting traditional agricultural methods and come with various challenges. The goal of smart farming is to deliver the agricultural industry with the foundation it uses cutting-edge technologies like big data, cloud computing, and the IoT to track, monitor, automate, and analyze activities. Smart farming is distinguished by sensor monitoring and software administration. Its importance is increasing as a result of things like the expanding world population, the need for more climate-resilient agriculture, the adoption and sophistication of information and communication technology, the increasing demand for higher crop yields, and the efficient use of natural resources. This chapter comprehensively explores the applications of blockchain and the Internet of Agricultural Things for smart irrigation and farming, focusing on water scarcity and sustainable development.

5.1 INTRODUCTION

There are many nations that rely heavily on agriculture for their economic growth, which accentuates the need for excellently handling water resources for plants, crops, and

Demystifying Sustainable Farming Using Smart Technologies.
Gunjan Chhabra, Keshav Kaushik, Priya Chugh, Susheela Dahiya, and Aryan Chaudhary (Eds.)

DOI: 10.1201/9781779640932-5

maintaining agricultural land [1]. With remote sensing techniques progressively integrating with Internet of Things (IoT) devices to provide autonomous operations through the communication and data aggregation capabilities of sensors, sensor systems are essential to the implementation of precision agriculture [2]. Real-time sensor situations are utilizing machine learning approaches, expanding the applicability of these techniques beyond agriculture to include transportation, healthcare, military operations, mobile phones, and home appliances [3].

In contemporary agriculture, fields can be monitored by robots and drones, while temperature and moisture sensors can be automated to cater to the specific requirements of crops. There are various systems and devices that contribute to making farms more efficient, environmentally conscious, and economically viable [4]. The adoption of precision agriculture, encompassing these technologies, provides notable benefits. The globe is constantly seeing incredible advancements in technology, and the agricultural sector is no exception. The demand for food and agricultural goods rises in agriculture, becoming a transformative force reshaping traditional agricultural methods and establishing sustainable approaches that are more productive and efficient than traditional cooking methods [5].

5.1.1 BACKGROUND OF STUDY

Farmers benefit from IoT-based solutions that increase productivity and lower yield costs in the modern agriculture. By combining cloud platforms and contemporary wireless communication technologies, smart agriculture is advanced and has the ability to increase production efficiency and product quality. However, a dependable and long-lasting strategy that includes sensing, identification, transmission, monitoring, and feedback capabilities is needed to accomplish agricultural activities successfully [6]. In order to ensure network integrity and carry out genuine tasks in a dispersed fashion, secure technologies are essential. To enable effective and lightweight communication paradigms, agricultural systems must include strong machine learning model functions. Even though agriculture uses over 80% of all the water in the agricultural sector worldwide, traditional irrigation techniques and low water usage efficiency between 35% and 40% from poor management provide significant barriers to sustainable crop production in drylands. This is made worse by the fact that, in comparison to other economic sectors, irrigation in agriculture yields a poorer return per unit of water used [7].

5.1.2 OBJECTIVES OF THIS CHAPTER

The objectives of this chapter are to explore the potential of integrating blockchain and the Internet of Agricultural Things (IoAT) into a smart irrigation system to address groundwater scarcity and promote sustainable farming to:

- overview of the challenges associated with groundwater scarcity in agriculture.
- explain the concept of smart irrigation and its role in addressing the issue.

- explore blockchain and IoAT technologies and their applicability in agriculture.
- discuss the benefits and potential challenges of integrating these technologies.
- insights into the future of sustainable farming practices (Figure 5.1).

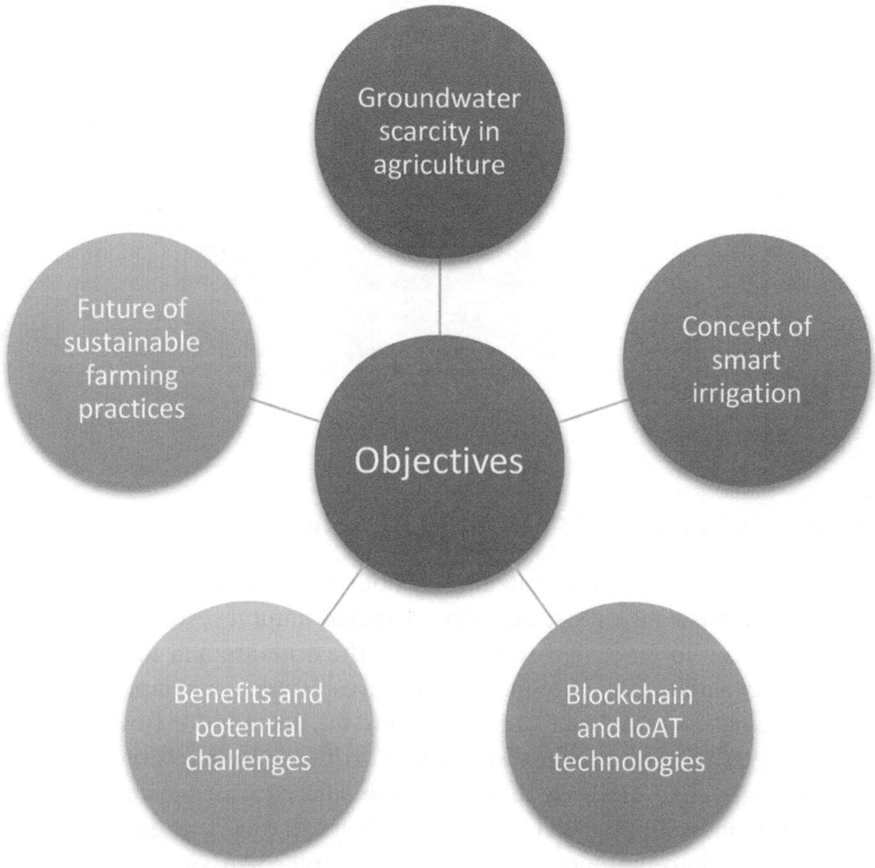

FIGURE 5.1 Objectives of this chapter.

5.1.3 ORGANIZATION OF THE CHAPTER

This chapter comprehensively explores the various dimensions of the smart irrigation for smart farming inculcating blockchain and IoAT concerning Groundwater Scarcity, ensuring futuristic sustainable farming. Section 5.2 elaborates on groundwater scarcity in agriculture. Section 5.3 explains the smart irrigation: solution to groundwater scarcity. Section 5.4 lays down the blockchain and IoAT in agriculture. Section 5.5 specifies the integrating blockchain and IoAT in smart irrigation. Section 5.6 expresses the challenges and opportunities. Section 5.7 discusses sustainable agriculture as global priority. Section 5.8 travels the harnessing IoAT and blockchain for rural farmers. Finally, Section 5.9 concludes the chapter with future scope (Figure 5.2).

FIGURE 5.2 Flow of the chapter.

5.2 GROUNDWATER SCARCITY IN AGRICULTURE

Freshwater is a critical reserve crucial for both ecosystem well-being and human survival, and it stands as the most globally extracted natural resource. The situation is becoming worse due in large part to improper infrastructure and inadequate management of water resources [8]. The uncontrolled freshwater consumption can lead to a reduction in circulation rates, as a situation where the demand for freshwater surpasses its availability [9, 10]. The groundwater stands as the largest reservoir of accessible, fresh water globally, especially in regions where surface water resources are lacking. The demand for domestic, agricultural, and industrial water necessities can only be fulfilled by tapping into the underground water reserves. An analogy provided by the US Geological Survey likens groundwater to money stored in a bank account. Just as drawing money at a faster rate than it is deposited leads to account-supply issues, extracting groundwater at a quicker pace than its replenishment creates analogous problems [11] (Figure 5.3).

5.2.1 CAUSES OF GROUNDWATER SCARCITY

Water scarcity has serious consequences, and the reduced water availability has a significant effect on agriculture, which is the backbone of the country's economy. This leads to lower crop yields and higher food costs. The majority of waterborne illnesses are caused by poor sanitation and hygiene in communities. Also, conflicts over water resources sometimes arise, involving different populations and industries. These difficulties place a great deal of strain on the socioeconomic structures of areas where agriculture is the primary industry [12].

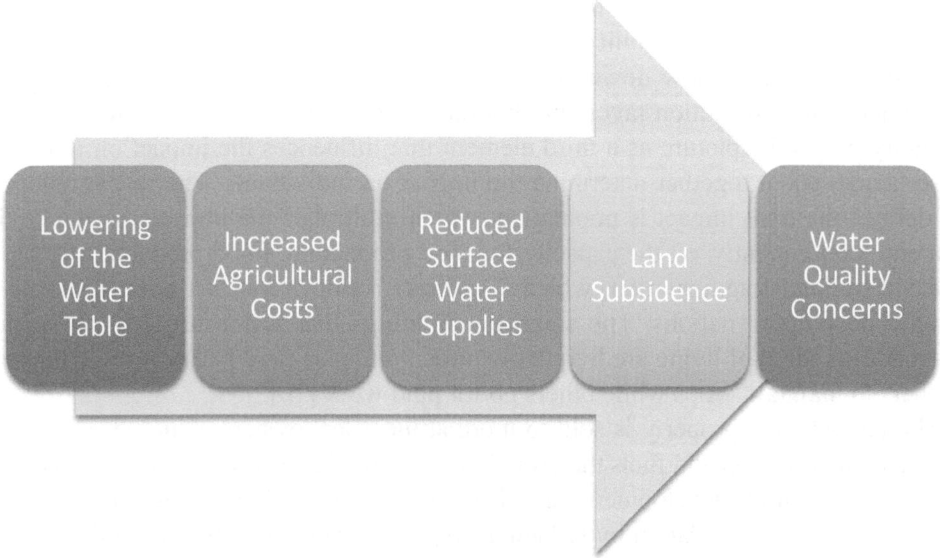

FIGURE 5.3 It highlights the adverse effects—excessive groundwater pumping.

5.2.1.1 OVEREXTRACTION OF GROUNDWATER

The extraction of groundwater entails taking out water that has been stored below the earth's surface. Groundwater is the subterranean water that is essential to human survival, and it is essential for everyday comfort, farming, food production, and survival. Humans rely extensively on groundwater because their rising need for clean water exceeds that of surface water and rainfall. Future food security in drylands may be threatened by the increasing shortage of agricultural water, endangering food production. The lack of water presents a significant obstacle to the achievement of several Sustainable Development Goals (SDGs), including SDG-15 (desertification control), SDG-14 (life below water), SDG-6 (clean water), SDG-7 (clean and affordable energy), and SDG 15.3 (zero hunger). All of these goals depend either directly or indirectly on the availability of water [13].

5.2.1.2 CLIMATE CHANGE AND ERRATIC RAINFALL PATTERNS

The adverse climate change poses a danger to the sustainability of irrigated agriculture and pressurized irrigation systems, and suitable irrigation schedules can lower evaporative or system water loss and increase water productivity, which is defined as the product yield per unit volume of water consumed by the crop. In nations with limited water resources, maintaining irrigation becomes a challenging undertaking [14].

5.2.1.3 POPULATION GROWTH AND INCREASED FOOD DEMAND

There are three primary parts to the equation for the food crisis: First, social structure, wealth, and lifestyle all influence consumption levels. Second, the technologies in use have an effect on the amount of trash produced at a particular consumption level as well as the environmental sustainability of human activities. The adoption of more appropriate technology that might slow down environmental damage may be hampered by poverty [15]. When land distribution favors big holdings over smallholdings or marginal regions, inequality enters the picture as a third element that influences the impact on individuals. These factors taken together determine the impact on individuals. The fourth component that defines the total impact is population, which multiplies the impacts on the environment and, consequently, on the capacity for food production. Food production is strongly impacted by land fragmentation, which is a direct result of the fast population increase in many developing nations. The small landholdings that are insufficient to support a reasonable standard of living are frequently turned into part-time farms, forcing household members to maintain crops while others go for paid work [16].

Alternatively, the property is sold to more affluent landowners, which exacerbates the gap in land distribution and fuels the growth of a sizable labor force lacking access to land. Excessive population development can also lead to unsuitable agricultural methods that cause serious land degradation by reducing vegetation, degrading the soil, misusing agrochemicals, and making it more difficult to manage water resources [17]. The developing nations must include demographic issues, such as growth, distribution, and patterns of movement from rural to urban areas, into their strategic plans in order to meet these difficulties. In addition to sustainable development initiatives addressing soil erosion, deforestation, declining agricultural output, and inadequate water management, community development plans that combine social services and productive resources have to be promoted [18].

The programs for rural agricultural extension that give impoverished farmers guidance, seeds, fertilizer, and financing are essential. So, supporting research on the fusion of established and cutting-edge technologies for food production is also crucial. The future demands must be forecasted, given present population levels and projected changes. The millions of rural poor people can have sustainable livelihoods, stop land degradation, and greatly boost agricultural yields with better resource management. This calls for an equal dedication to the development of human resources, which includes expanding family planning and population control initiatives. Reducing family size and improving community health and well-being should be the goals of national population programs, which should also include accessible family planning services and comprehensive maternal and child health care programs. With working together, it can protect the environment, improve food production, and lessen the hardships faced by the impoverished in rural areas [19] (Figure 5.4).

5.3 SMART IRRIGATION: SOLUTION TO GROUNDWATER SCARCITY

An all-encompassing approach is required to address the shortage of water. The restoration of the water supplies is made possible by using efficient water management techniques, such as watershed management and rainwater collection. The pollution and waste are reduced by improvements in irrigation practices and investments in water treatment systems. So, raising public awareness of water conservation and supporting wise water usage are vital. To further achieve long-term solutions, policies supporting sustainable water allocation and management must be developed [20].

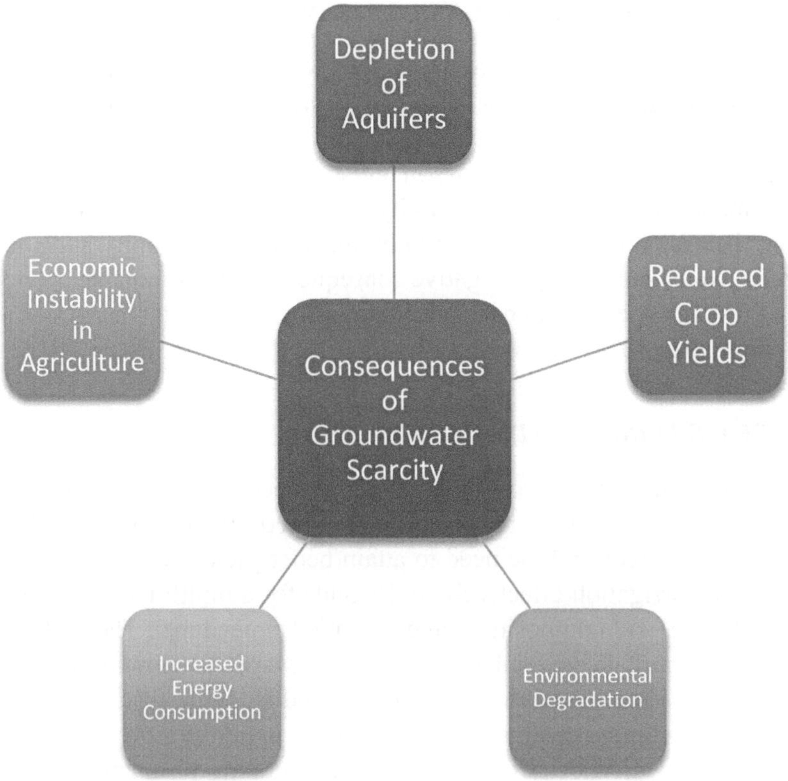

FIGURE 5.4 Consequences of groundwater scarcity.

With accounting for a variety of variables, including differences in soil and climate, soil hydraulic characteristics, plant reactions to water deficiencies, and weather fluctuations, this method improves irrigation decision-making. This method increases yields while also conserving water. The farmers can avoid plant moisture deficits while conserving precious resources by implementing intelligent irrigation systems. It is recommended that smart irrigation be used to manage soil variability and realize financial advantages by addressing each crop's unique irrigation requirements. The smart irrigation system that is properly maintained can provide water and nutrients to plant roots directly [21].

There are not many studies that have been done on the problem of poor water productivity in dryland agriculture and how it may be improved by using intelligent irrigation techniques. Thus, the particular goals are to: (1) find the factors that contribute to dryland agriculture's low water use efficiency; (2) assess traditional irrigation scheduling techniques; and (3) investigate the viability and benefits of smart irrigation systems in improving irrigation management for increased water productivity in water-scarce areas [22, 23].

5.3.1 WATER EFFICIENCY AND CONSERVATION

Water shortage should be taken into account from both an economic and a physical standpoint. The physical water shortage can take two forms: blue water scarcity, which indicates a lack of surface and groundwater availability to fulfill human water requirements, and green water scarcity, which occurs when soil moisture in the root zone is inadequate to meet crop water demands. Economic water scarcity results when water resources are physically available but their exploitation is hampered by institutional capability and socioeconomic circumstances. Water shortage has negative consequences on sustainable economic growth and social cohesion, especially in drylands. It significantly affects the agriculture industry, which uses more than 80% of all fresh water [24].

5.3.2 INCREASED CROP YIELDS

In order to satisfy the future food demands, it is necessary to manage existing water resources effectively through increased water productivity which given the scarcity of accessible water resources and the need to attain better yields with less water usage. It is difficult to manage irrigation effectively in drylands for a multitude of reasons, including crop variety, climate, soil composition, and irrigation procedures. The high temperatures, limited and erratic rainfall, and enhanced potential evapotranspiration are the characteristics of drylands. It is projected that the issues would be made worse by the catastrophic weather events that climate change will bring about. In addition to unfavorable environmental circumstances, irrigated agriculture and the ongoing production of food are seriously threatened by the increasing shortage of water in these areas [25].

5.4 BLOCKCHAIN AND IOAT IN AGRICULTURE

In agriculture, blockchain quickly became an important technology with an extensive range of applications. Blockchain-based IoT systems are being explored in precision agriculture in response to the need for intelligent peer-to-peer (P2P) systems that can validate, secure, monitor, and analyze agricultural data. Blockchain plays a key role in providing a more stable, transparent, decentralized, and unchangeable alternative to conventional techniques for gathering, organizing and exchanging agricultural data [26, 27]. The precision farming's combination of blockchain technology with the IoT is expected to transform smart farms into an internet of smart farms for improving supply-chain network control. It investigates the incorporation of security concepts based on blockchain technology to create a trustworthy system and minimize interference with communication. The simulations and a range of tests have been used to validate the usefulness of the suggested model, which shows that it is a useful tool for improving network parameters [28, 29].

5.4.1 BLOCKCHAIN TECHNOLOGY: POTENTIAL APPLICATIONS IN AGRICULTURE

The disadvantages of conventional agricultural systems include their vulnerability to theft, restricted availability, possibility of loss, and harm to data. The concept of inadequate security, a weak data management system, privacy, and effective transaction record sharing characterize the current system. A cutting-edge, high-performance agricultural knowledge computing system has been created for precision agriculture using computing in order to solve these problems. Smart agriculture uses edge computing, a highly virtualized platform, which places computer units strategically near data sources on the ground to minimize computational burden [30]. Blockchain stands out as a technology with significant potential to revolutionize the agricultural sector [31]. When consumers have access to this level of transparency, they can make informed decisions when purchasing agricultural products. This information is often used to reward farmers and producers adhering to good farming practices [32, 33].

5.4.1.1 *CHALLENGES IN ADOPTING BLOCKCHAIN TECHNOLOGIES*

There are apprehensions about the potential misuse or misapplication of blockchain technology, which could jeopardize food security. Private blockchains, in particular, are considered more vulnerable to hacking and less secure. Since these blockchains operate based on private organizational norms, there is a risk of exploitation by unauthorized individuals. Additionally, small-scale farmers, lacking the necessary size, technological expertise, and scalability, may face challenges in taking advantage of blockchain technology and risk being left behind [34].

5.4.2 INTERNET OF AGRICULTURAL THINGS

Farm management, edge computing, cloud computing, artificial intelligence, and knowledge discovery are all included in the knowledge-based system for smart agriculture. To improve system performance, it proposes a novel edge computing system that moves some of the computing capacity to the edge computing node. Soil moisture levels are monitored by sensors positioned across the field, and this knowledge-based smart agriculture system improves water irrigation. Water irrigation is classified based on recurring analyses powered by system knowledge, and when required, the irrigation water pump is turned on to cut down on water waste [35].

The use of wireless systems based on the IoT has grown significantly in recent years across a variety of sectors. Without needing human interaction, the IoT operates as a network that facilitates communication between physical objects, equipment, sensors, and other entities [36]. Wireless devices are being used by the agriculture industry to integrate current technology and improve farming production and cost management. Smart IoT devices are essential for remotely detecting and tracking crop conditions at various development stages in precision agriculture. Since agriculture is a major economic sector in many countries, it is crucial to manage water resources effectively for plants, crops, and the general maintenance of agricultural land [37].

The spread of wireless technology for the IoAT has significantly advanced smart farming systems. Even though there are many techniques for forecasting crop and soil conditions, developing a high-caliber autonomous system is still a significant research problem [38]. The current methods frequently fail to maintain dependable communication in the face of network congestion. Thus, with an emphasis on dependability and quality, it presents an intelligent optimization model that makes use of machine learning to promote sustainable agriculture. In order to facilitate a consistent process of decision-making for the forwarding scheme, the suggested model first uses intelligent devices to automate data gathering and transmission, and these devices analyze independent performance factors [39].

5.4.2.1 DEVICES AND SENSORS IN SMART AGRICULTURE

IoT-enabled smart sensors can monitor environmental parameters, including humidity, wetness, and temperature with great accuracy. There are certain sensors that can measure the water content and nitrate levels in soil to assess its quality. High-resolution cameras combined with a GPRS system can be used to detect insect pests and plant illnesses. The surveillance using unmanned aerial vehicles helps track crop growth and the topography of farmland. Estimating crop yield is made possible by automated mass flow sensors [40, 41]. With this knowledge, farmers will be able to make well-informed decisions on crop management, which includes irrigation, pest control, and harvest [42]. Wireless sensor technology integration has important ramifications since it enables administrators and farmers to be notified instantly of equipment malfunctions, which starts the troubleshooting process [43]. Data processing can be streamlined, actuation can be improved, and energy consumption may be decreased with automated repair tools. The advent of IoT-enabled unmanned aerial systems is a significant development in agriculture, since it allows farmers to obtain real-time landscape imagery and streamline data processing and storage [44].

5.5 INTEGRATING BLOCKCHAIN AND IOAT IN SMART IRRIGATION

There are critical concerns facing rural agriculture including poor seed quality and insufficient nutrients, as well as irrigation issues made worse by the escalating water crisis. Recent developments suggest that blockchain technology, artificial intelligence, and the IoT will be important developments in the near future [45]. This concept investigates how blockchain technology and IoT may be combined to effectively monitor agricultural crops. Using blockchain technology and the IoT, a system for smart water management and seed quality monitoring has been created to supervise and regulate the use of premium water resources and seeds in local communities [46].

The blockchain network is utilized to protect data and promote community trust. Furthermore, the blockchain network facilitates confidence between systems with limited economic resources that communicate with the blockchain network through a hardware platform [47]. The design of a prototype is also presented, and its performance is assessed through implementation. Blockchain technology eliminates the need for middlemen, such as banks in the cryptocurrency space or middlemen in the agriculture industry, via enabling transparent P2P transactions [48]. By doing away with a central authority, trust is now built through P2P architecture and cryptography instead of a central authority. This change contributes to the restoration of consumer and producer trust and lowers transaction costs in the agrifood sector. Blockchain technology is dependable because it can track transactions between anonymous users and quickly identify fraud and errors [49].

5.5.1 BLOCKCHAIN ENHANCES SMART IRRIGATION

Since the energy sector has historically been reluctant to adopt disruptive technology, a thorough digital transformation strategy incorporating all energy sector players is necessary to successfully transition to a new energy market. Even though blockchain technology is still in its early stages of adoption, pilot projects have demonstrated that it offers enormous promise in changing the future energy market, despite initial resistance, particularly among industry titans [50]. The purpose of this is to provide a way to give small-scale farmers access to alternative irrigation methods. The suggested remedy is for the formation of groups made up of small farmers who would irrigate agricultural land using green energy produced by solar panels [51]. In addition to digital hardware attached to water pumps and solar panels, blockchain technology, which promotes innovation and growth in the energy industry and enables a high degree of automation through smart contracts, may be used to monitor the effectiveness of this system [52]. This utility token accelerates the shift to sustainable energy practices by creating new channels for the trade of energy and water [53].

5.5.1.1 SMART CONTRACTS FOR WATER ALLOCATION

Smart contract integration solves the problem of tracking items throughout the vast supply chain, which is a feature of the intricate agrifood system, by enabling real-time reporting of faults [54]. As a result, blockchain technology addresses the interests of consumers, governments, and other stakeholders while providing answers to issues pertaining to food safety and quality. Blockchain makes sure that everyone is in the know and makes it easier to get reliable data [55]. It has the ability to record each stage of a product's lifecycle, from manufacture to disposal. This trustworthy information on farming operations is crucial for creating data-driven infrastructure and insurance plans, which in turn promotes more intelligent and resilient farming methods [56].

5.5.1.2 DATA INTEGRITY AND TRANSPARENCY

As a decentralized database, blockchain ensures confidentiality and anonymity while recording and sharing data via a distributed computer network [57]. Transactions in this system may only be carried out by the owner who possesses the private key, and other computers on the network operate as validators. Nodes in a public ledger securely record transactions between them without the assistance of a reliable third party [58]. Also, under a centralized cloud method, an item or asset is held and managed by the owner or by a reliable middleman, like a bank or other centralized authority. The server oversees several clients in a centralized computer system; the server has more resources than the client systems. This conventional method restricts processes to one server [59].

In block units, which comprise the hash of the previous block, the hash of the current block, the date, extra information, and the transactions for the blocks, blockchain tracks transactions [60]. Sender nodes broadcast information to other nodes on the network when they start a transaction. Recipient nodes confirm the transaction and offer evidence of their labor [61]. By broadcasting the proof of work to every other node, the node that completes the proof of work successfully links the block to the chain. Any node may validate a transaction since it is signed by the sender and contains the recipient's public key. A linked chain of blocks is created by each block having a hash of the one before it [9].

5.6 CHALLENGES AND OPPORTUNITIES

Until private agricultural data reach legitimate storage and processing equipment, they must be reliably protected from illegal access. IoAT security methods reduce hazards related to sustainable communication while simultaneously offering farmers reliable information on their devices [62]. This is to reduce the additional load on devices and offer a machine learning model for the agricultural network. The proposed architecture includes a security mechanism with several approaches to protect IoT data from dangerous circumstances [63]. The utilization of solar-powered "reverse-osmosis" is being implemented globally to generate irrigation water from saline sources, thereby reducing the reliance on extracting freshwater from underground aquifers [64]. It underscores the significance of research on water scarcity and has the potential to stimulate the development and application of solutions that enhance the efficiency and resilience of agricultural production/consumption [65]. For many in this community, pastoralism and agriculture, which largely depend on natural resources, are the main means of subsistence. This situation mostly affects developing nations, which account for 70% of the world's drylands and are home to people who struggle with issues including marginalization, poverty, food insecurity, and hunger [66].

Water availability and agricultural production are the two main concerns related to drylands and land degradation. Water shortage affects 1–2 billion people globally at the moment, with a large concentration in drylands where water supplies cannot keep up with demand [67]. According to future climate estimates, this issue will only become worse, with an increasing number of people in drylands facing severe water shortages. The decisions on climate change and water management will therefore negatively impact drylands and the people who live there [68]. To satisfy the demands of food security, it is imperative to increase agricultural production in drylands, especially in light of the world's population growth [69]. The adoption of intelligent irrigation techniques appears to be a workable way to maximize the amount of water that is available and raise water production in these areas. Sustainable development in drylands is seriously threatened by the growing worry over water shortages. When the amount of water required equals or exceeds the entire amount of freshwater resources available, water shortage results [70].

5.7 SUSTAINABLE AGRICULTURE AS A GLOBAL PRIORITY

In order to increase the adoption of ecologically friendly methods and decrease the usage of agrochemicals, farmers must find a technical solution to increase crop productivity [71]. This has the major benefit of bringing down the cost of production. Rapid technical advancement and progress have been observed in recent years, surpassing a number of challenges, including skepticism among farmers [72]. Farmers, scientists, and technology manufacturers are working together to discover a methodical way to increase agricultural productivity while cutting expenses [73]. The newest research and development, particularly in the area of innovative sensors and their application, are combined in smart farming [74]. Farmers may choose from a variety of sensors according to their unique agricultural requirements, as sensor networks cover a broad range of farming chores. With the acceptance of new technology, problems with conventional practices like irrigation, greenhouses, organic farming, and organic parks may be resolved [75]. Because of the use of wireless sensor network technologies, the smart farming suggested system is both user-friendly and technically sound [76]. The farmers may rapidly examine environmental conditions using many strategically positioned sensors to measure the real plant conditions using an Android mobile smartphone. Even if the country boasts of its achievements in research and innovation, problems like erratic power supplies or protracted breakdowns nevertheless cause disruptions to everyday life [77].

The term "sustainable agriculture" describes the production of food grains in an eco-friendly way that prioritizes resilience and preservation [78]. In order to ensure economic viability while maintaining soil quality, preventing degradation, conserving water resources, boosting land biodiversity, and nurturing a natural and healthy environment, this approach encourages agricultural techniques that support the well-being of farmers and resources [79]. The maintenance of the natural resources, preventing the loss of biodiversity, and lowering greenhouse gas emissions all depend heavily on sustainable agriculture. Sustainable agriculture focuses on agricultural methods that increase farming productivity while preserving the natural balance and meeting future generation's fundamental requirements [80]. Crop rotation, managing pests and diseases, controlling nutritional deficiencies in crops, recycling, water harvesting, and other smart farming practices are major accomplishments in the field of sustainable agriculture that go toward creating a safer environment generally [80]. The decomposition of dead plants, the use of pesticides and fertilizers, and waste emissions all have a detrimental effect on biodiversity, which is vital to all living things. All living creatures depend on a healthy ecosystem, since the release of greenhouse gases has a negative impact on humans, animals, plants, and the environment [81].

5.8 HARNESSING IOAT AND BLOCKCHAIN FOR RURAL FARMERS

The demand on agriculture to increase food production in a sustainable manner has increased due to the world population's rapid expansion. The distribution and consumption of food are crucial for maintaining limited natural resources and advancing human well-being at the same time [82]. There has been a noticeable desire from politicians, development agencies, civil society groups, and private businesses to learn environment [83].

The agriculture industry is going through a lot of change and facing a lot of social and environmental issues [84]. Due to their lack of direct market access, many farmers continue to rely on conventional farming methods [85, 86]. The challenge of providing end consumers with high-quality food is further compounded by dietary changes, shifting consumer tastes and the effects of climate change [87, 88]. IoT technology is essential to control and diagnostics in smart agriculture [89, 90]. Farmer's productivity and efficiency have grown dramatically as a result of the field's IoT integration, optimizing agriculture. In addition to evaluating and monitoring variables like temperature, soil moisture, and crop diseases, this technology helps determine field variables like plant biomass and soil quality. The tracking of crop growth and yield-influencing variables is made possible by IoT. Based on particular conditions, farmers may choose crops wisely and carry out crop rotation [91].

5.9 CONCLUSION AND FUTURE SCOPE

Through online applications and mobile apps, farmers may engage with the intelligent decision support system to get information about their land from subject matter experts. In order to ensure privacy, security, ease of use, and information availability, the proposed system makes use of blockchain technology. This allows for the efficient use of natural resources and lessens the impact of soil and water pollution caused by the application of pesticides and fertilizers based on data stored in the knowledge base. The Hyperledger platform has been used to develop a knowledge discovery system that effectively stores and shares sensor data while maintaining data confidentiality and privacy. When combined with IoT and edge computing, precision agriculture allows for seamless communication between people and devices in a smart cloud environment, giving farmers access to intelligent services. Smart agriculture was created in response to discussions on agricultural policy related to climate change and sustainable development. Its goal is to offer globally applicable solutions for managing climate change and ensuring agricultural food security. It makes use of cutting-edge technology like big data, cloud computing, and the IoT to monitor field conditions, assess crop development, and provide farmers the information; they need to make wise decisions.

KEYWORDS

- **blockchain**
- **groundwater**
- **Internet of Agricultural Things**
- **sustainable farming**
- **smart irrigation**

REFERENCES

1. Torky, M., & Hassanein, A. E. (2020). Integrating blockchain and the internet of things in precision agriculture: Analysis, opportunities, and challenges. *Computers and Electronics in Agriculture, 178*, 105476.
2. Saba, T., Rehman, A., Haseeb, K., Bahaj, S. A., & Lloret, J. (2023). Trust-based decentralized blockchain system with machine learning using Internet of agriculture things. *Computers and Electrical Engineering, 108*, 108674.
3. Awan, K. A., Ud Din, I., Almogren, A., & Almajed, H. (2020). AgriTrust—a trust management approach for smart agriculture in cloud-based internet of agriculture things. *Sensors, 20*(21), 6174.
4. Mohanty, S. (2021). Internet-of-agro-things (IoAT) makes smart agriculture. *IEEE Consumer Electronics Magazine, 10*(4), 4–5.
5. Bapatla, A. K., Mohanty, S. P., & Kougianos, E. (2021, November). sFarm: A distributed ledger based remote crop monitoring system for smart farming. In *IFIP International Internet of Things Conference* (pp. 13–31). Springer International Publishing: Cham.
6. Noor, N. M., Razali, N. A. M., Sham, S. N. S. A., Ishak, K. K., Wook, M., & Hasbullah, N. A. (2023). Decentralised access control framework using blockchain: Smart farming case. *International Journal of Advanced Computer Science and Applications, 14*(5). http: //dx.doi.org/10.14569/IJACSA.2023.0140560
7. Cheema, S. M., Ali, M., Pires, I. M., Gonçalves, N. J., Naqvi, M. H., & Hassan, M. (2022). IoAT Enabled smart farming: Urdu language-based solution for low-literate farmers. *Agriculture, 12*(8), 1277.
8. Mitra, A., Vangipuram, S. L., Bapatla, A. K., Bathalapalli, V. K., Mohanty, S. P., Kougianos, E., & Ray, C. (2022). Everything you wanted to know about smart agriculture. *arXiv preprint arXiv: 2201.04754.*
9. Leduc, G., Kubler, S., & Georges, J. P. (2021). Innovative blockchain-based farming marketplace and smart contract performance evaluation. *Journal of Cleaner Production, 306*, 127055.
10. Sapra, L., Mathani, A., & Bhatnagar, G. (2022). A thorough analysis of Blockchain's potential for Internet of Things applications in precision agricultural networks. *Mathematical Statistician and Engineering Applications, 71*(4), 4141–4159.
11. Lamtzidis, O., Pettas, D., & Gialelis, J. (2019). A novel combination of distributed ledger technologies on internet of things: Use case on precision agriculture. *Applied System Innovation, 2*(3), 30.
12. Noor, N. M., Razali, N. A. M., Malizan, N. A., Ishak, K. K., Wook, M., & Hasbullah, N. A. (2022). Decentralized access control using blockchain technology for application in smart farming. *International Journal of Advanced Computer Science and Applications, 13*(9).
13. Alruwaill, M., Bapatla, A. K., Mohanty, S. P., & Kougianos, E. (2023, October). FarmIns: Blockchain leveraged secure and reliable crop insurance management system. In *IFIP International Internet of Things Conference* (pp. 381–389). Springer Nature Switzerland: Cham.
14. Elham, M. N., Sabeghi, M. S., Al-Rasbi, F. S., Daud, S. M., Noor, N. Q., Sahak, R., & Mazlan, A. A. (2020, April). A preliminary study on poultry farm environmental monitoring using internet of things and blockchain technology. In *2020 IEEE 10th Symposium on Computer Applications & Industrial Electronics (ISCAIE)* (pp. 273–276). IEEE.
15. Udutalapally, V., Mohanty, S. P., Pallagani, V., & Khandelwal, V. (2020). sCrop: A Internet-of-Agro-Things (IoAT) enabled solar powered smart device for automatic plant disease prediction. *arXiv preprint arXiv: 2005.06342.*
16. Puri, V., Solanki, V. K., & Aponte, G. J. R. (2023, March). Blockchain and federated learning based integrated approach for agricultural Internet of Things. In *The International Conference on Intelligent Systems & Networks* (pp. 240–246). Springer Nature Singapore: Singapore.
17. Mkrttchian, V. (2021). Artificial and natural intelligence techniques as IoP-and IoT-based technologies for sustainable farming and smart agriculture. In: *Artificial Intelligence and IoT-Based Technologies for Sustainable Farming and Smart Agriculture* (pp. 40–53). IGI Global.
18. Pincheira, M., Vecchio, M., Giaffreda, R., & Kanhere, S. S. (2021). Cost-effective IoT devices as trustworthy data sources for a blockchain-based water management system in precision agriculture. *Computers and Electronics in Agriculture, 180*, 105889.

19. Vangipuram, S. L., Mohanty, S. P., Kougianos, E., & Ray, C. (2022). agroString: Visibility and provenance through a private blockchain platform for agricultural dispense towards consumers. *Sensors*, *22*(21), 8227.
20. Ferrag, M. A., Shu, L., Yang, X., Derhab, A., & Maglaras, L. (2020). Security and privacy for green IoT-based agriculture: Review, blockchain solutions, and challenges. *IEEE Access*, *8*, 32031–32053.
21. Waheed, U., Mansoor, Y., & Khan, M. A. (2021). BIoT-based smart agriculture: Food and crops efficiency and improvement in supply chain cycle. In: *Blockchain technology for IoT applications*, pp. 173–189.
22. Arshad, J., Siddique, M. A. B., Zulfiqar, Z., Khokhar, A., Salim, S., Younas, T., Ur Rehman, A., & Asad, A. (2020, February). A novel remote user authentication scheme by using private blockchain-based secure access control for agriculture monitoring. In *2020 International Conference on Engineering and Emerging Technologies (ICEET)* (pp. 1–9). IEEE.
23. Pullo, S., Pareschi, R., Piantadosi, V., Salzano, F., & Carlini, R. (2023). Integrating IOTA's tangle with the internet of things for sustainable agriculture: A proof-of-concept study on rice cultivation. In: *Informatics* (Vol. 11, No. 1, p. 3). MDPI.
24. Latif, S., Idrees, Z., e Huma, Z., & Ahmad, J. (2021). Blockchain technology for the industrial Internet of Things: A comprehensive survey on security challenges, architectures, applications, and future research directions. *Transactions on Emerging Telecommunications Technologies*, *32*(11), e4337.
25. Udutalapally, V., Mohanty, S. P., Pallagani, V., & Khandelwal, V. (2020). sCrop: A novel device for sustainable automatic disease prediction, crop selection, and irrigation in Internet-of-Agro-Things for smart agriculture. *IEEE Sensors Journal*, *21*(16), 17525–17538.
26. Sendros, A., Drosatos, G., Efraimidis, P. S., & Tsirliganis, N. C. (2022). Blockchain applications in agriculture: A scoping review. *Applied Sciences*, *12*(16), 8061.
27. Chen, F., Xiao, Z., Cui, L., Lin, Q., Li, J., & Yu, S. (2020). Blockchain for Internet of Things applications: A review and open issues. *Journal of Network and Computer Applications*, *172*, 102839.
28. Al-Shareeda, M., Ali, M., & Manickam, S. (2023). The blockchain internet of things: Review, opportunities, challenges, and recommendations. *Indonesian Journal of Electrical Engineering and Computer Science*, *31*(3), 1673–1683.
29. Sharma, A., & Singh, B. (2022). Measuring impact of E-commerce on small scale business: A systematic review. *Journal of Corporate Governance and International Business Law*, *5*(1).
30. Kathar, S., Hardel, H., Alam, Z., Phansalkar, S., & Sajjan, R. (2021, December). Auction system for agricultural trade using blockchain technology: A survey with proposed framework. In *2021 IEEE Pune Section International Conference (PuneCon)* (pp. 1–10). IEEE.
31. Upadhyaya, A., Singh, Y., & Anand, P. (2022, May). Blockchain for IoT: A comprehensive review for precision agricultural networks. In *The International Conference on Recent Innovations in Computing* (pp. 787–802). Springer Nature Singapore: Singapore.
32. Bhushan, B., Sahoo, C., Sinha, P., & Khamparia, A. (2021). Unification of Blockchain and Internet of Things (BIoT): Requirements, working model, challenges and future directions. *Wireless Networks*, *27*, 55–90.
33. Fernández-Caramés, T. M., & Fraga-Lamas, P. (2018). A review on the use of blockchain for the Internet of Things. *IEEE Access*, *6*, 32979–33001.
34. Radeva, I., & Popchev, I. (2022). Blockchain-enabled supply-chain in crop production framework. *Cybernetics and Information Technologies*, *22*(1), 151–170.
35. Singh, P., Kumar, A., & Chopra, M. (2023). Real-world applications of BC technology in Internet of Things. In: *Machine Learning Applications: From Computer Vision to Robotics*, pp. 97–122.
36. Shi, X., An, X., Zhao, Q., Liu, H., Xia, L., Sun, X., & Guo, Y. (2019). State-of-the-art internet of things in protected agriculture. *Sensors*, *19*(8), 1833.
37. Sengupta, A., Gill, S. S., Das, A., & De, D. (2021). Mobile edge computing based internet of agricultural things: A systematic review and future directions. In: *Mobile Edge Computing*, pp. 415–441.
38. Prakash, C., Singh, L. P., Gupta, A., & Lohan, S. K. (2023). Advancements in smart farming: A comprehensive review of IoT, wireless communication, sensors, and hardware for agricultural automation. *Sensors and Actuators A: Physical*, *362*, 114605.
39. Akram, S. V., Malik, P. K., Singh, R., Anita, G., & Tanwar, S. (2020). Adoption of blockchain technology in various realms: Opportunities and challenges. *Security and Privacy*, *3*(5), e109.

40. Bathalapalli, V. K., Mohanty, S. P., Kougianos, E., Yanambaka, V. P., Baniya, B. K., & Rout, B. (2021, December). A puf-based approach for sustainable cybersecurity in smart agriculture. In *2021 19th OITS International Conference on Information Technology (OCIT)* (pp. 375–380). IEEE.
41. Vangipuram, S. L., Mohanty, S. P., & Kougianos, E. (2023, October). CroPAiD: Protection of information in agriculture cyber-physical systems using distributed storage and ledger. In *IFIP International Internet of Things Conference* (pp. 375–394). Springer Nature Switzerland: Cham.
42. Haque, M. A., Haque, S., Zeba, S., Kumar, K., Ahmad, S., Rahman, M., Senapathy, M., & Ahmed, L. (2023). Sustainable and efficient E-learning internet of things system through blockchain technology. *E-Learning and Digital Media, 21*(4), 20427530231156711.
43. Tanwar, R., Chhabra, Y., Rattan, P., & Rani, S. (2022, September). Blockchain in IoT networks for precision agriculture. In *International Conference on Innovative Computing and Communications: Proceedings of ICICC 2022, Volume 2* (pp. 137–147). Springer Nature Singapore: Singapore.
44. Amoah, E., & Oh, J. Y. (2020). Blockchain in IoT and project management. *Issues in Information Systems, 21*(3), 268–278.
45. Kose, U., Prasath, V. S., Mondal, M. R. H., Podder, P., & Bharati, S. (Eds.), (2022). *Artificial Intelligence and Smart Agriculture Technology*. CRC Press.
46. Singh, B. (2023). Blockchain technology in renovating healthcare: Legal and future perspectives. In: *Revolutionizing Healthcare Through Artificial Intelligence and Internet of Things Applications* (pp. 177–186). IGI Global.
47. Ordóñez, J., Alexopoulos, A., Koutras, K., Kalogeras, A., Stefanidis, K., & Martos, V. (2023). Blockchain in agriculture: A PESTELS analysis. *IEEE Access, 11*, 73647–73679.
48. Wahid, A., Juliady, I., Zain, S. G., & Parenreng, J. M. (2022). Secure wireless sensor network using cryptography for smart farming systems. *Internet of Things and Artificial Intelligence Journal, 2*(4), 248–262.
49. Hassija, V., Zeadally, S., Jain, I., Tahiliani, A., Chamola, V., & Gupta, S. (2021). Framework for determining the suitability of blockchain: Criteria and issues to consider. *Transactions on Emerging Telecommunications Technologies, 32*(10), e4334.
50. Pikatza, A., Sainz, N., González, I., & Emaldi, M. (2023, July). Development of an open source IoT-Blockchain platform for traceability of fresh products from farm to fork. In: *International Congress on Blockchain and Applications* (pp. 487–497). Springer Nature Switzerland: Cham.
51. Kumari, S., & Muthulakshmi, P. (2023). Artificial intelligence—Blockchain enabled technology for Internet of Things: Research statements, open issues, and possible applications in the near future. In: *Privacy Preservation of Genomic and Medical Data*, pp. 433–480.
52. Panda, S. S., Mohanta, B. K., Dey, M. R., Satapathy, U., & Jena, D. (2020, July). Distributed ledger technology for securing IoT. In *2020 11th International Conference on Computing, Communication and Networking Technologies (ICCCNT)* (pp. 1–6). IEEE.
53. Monteiro, E. S., da Rosa Righi, R., Barbosa, J. L. V., & Alberti, A. M. (2021). APTM: A model for pervasive traceability of agrochemicals. *Applied Sciences, 11*(17), 8149.
54. George, G. M., & Jayashree, L. S. (2022). Fusion of Blockchain-IoT network to improve supply chain traceability using Ethermint smart chain: A review. *KSII Transactions on Internet and Information Systems, 16*, 3694–3722.
55. Karlsson, K., Jiang, W., Wicker, S., Adams, D., Ma, E., van Renesse, R., & Weatherspoon, H. (2018, July). Vegvisir: A partition-tolerant blockchain for the internet-of-things. In *2018 IEEE 38th International Conference on Distributed Computing Systems (ICDCS)* (pp. 1150–1158). IEEE.
56. Alzoubi, Y. I., Al-Ahmad, A., Kahtan, H., & Jaradat, A. (2022). Internet of things and blockchain integration: Security, privacy, technical, and design challenges. *Future Internet, 14*(7), 216.
57. Mitra, A., Singhal, A., Mohanty, S. P., Kougianos, E., & Ray, C. (2022). eCrop: A novel framework for automatic crop damage estimation in smart agriculture. *SN Computer Science, 3*(4), 319.
58. Singh, B., & Kaunert, C. (2024). Integration of cutting-edge technologies such as Internet of Things (IoT) and 5G in health monitoring systems: A comprehensive legal analysis and futuristic outcomes. *GLS Law Journal, 6*(1), 13–20.
59. Farooq, M. S., Javid, R., Riaz, S., & Atal, Z. (2022). Internet of Things (IoT) technologies in greenhouse farming: A systematic literature review. *IEEE Access*, 10, 53374–53397.

60. Voulgaris, S., Fotiou, N., Siris, V. A., Polyzos, G. C., Jaatinen, M., & Oikonomidis, Y. (2019). Blockchain technology for intelligent environments. *Future Internet, 11*(10), 213.
61. Kaur, U., Kaur, M., & Khang, A. (2023). Futuristic technologies in agriculture challenges and future prospects. In: *Handbook of Research on AI-Equipped IoT Applications in High-Tech Agriculture*, pp. 347–364.
62. Gupta, D., Bhatt, P., & Bhatt, S. (2020, December). A game theoretic analysis for cooperative smart farming. In *2020 IEEE International Conference on Big Data (Big Data)* (pp. 2350–2359). IEEE.
63. Siwach, V., Sehrawat, H., & Dahiya, O. (2022, October). Hybrid Blockchain Security Mechanism for protection of data related to Precision Agriculture. In *2022 10th International Conference on Reliability, Infocom Technologies and Optimization (Trends and Future Directions) (ICRITO)* (pp. 1–7). IEEE.
64. Rudrakar, S., & Rughani, P. (2022). IoT based Agriculture (IoTA): Architecture, cyber attack, cyber crime and digital forensics challenges. *Research Square.* doi: 10.21203/rs.3.rs-2042812/v1.
65. Rakhee, M., & Elayidom, S. (2023, January). Blockchain technology as a solution for vulnerabilities in internet of things—A survey. In *2023 Advanced Computing and Communication Technologies for High Performance Applications (ACCTHPA)* (pp. 1–6). IEEE.
66. Kumari, A., & Sharma, I. (2023, October). Augmentation of Internet of Things security and privacy with IOTA tangle network. In *2023 4th IEEE Global Conference for Advancement in Technology (GCAT)* (pp. 1–5). IEEE.
67. Akhtar, M. M., Rizvi, D. R., Ahad, M. A., Kanhere, S. S., Amjad, M., & Coviello, G. (2021). Efficient data communication using distributed ledger technology and iota-enabled internet of things for a future machine-to-machine economy. *Sensors, 21*(13), 4354.
68. Flores, A., Morales, A., Campos, G., & Gelso, J. (2021, December). Energy efficiency using IOTA tangle for greenhouse agriculture. In *Annual International Conference on Information Management and Big Data* (pp. 122–138). Springer International Publishing: Cham.
69. Mistry, I., Tanwar, S., Tyagi, S., & Kumar, N. (2020). Blockchain for 5G-enabled IoT for industrial automation: A systematic review, solutions, and challenges. *Mechanical Systems and Signal Processing, 135*, 106382.
70. Gaba, S., Khan, H., Almalki, K. J., Jabbari, A., Budhiraja, I., Kumar, V., Singh, A., Singh, K. K., Askar, S. S., & Abouhawwash, M. (2023). Holochain: An agent-centric distributed hash table security in smart IoT applications. *IEEE Access, 11*, 81205–81223.
71. Singh, B. (2023). Federated learning for envision future trajectory smart transport system for climate preservation and smart green planet: Insights into global governance and SDG-9 (Industry, Innovation and Infrastructure). *National Journal of Environmental Law, 6*(2), 6–17.
72. Ahamed, N. N., Karthikeyan, P., Velliangiri, S., & Gunjan, V. K. (2023). Blockchain security for the Internet of Things. In: *Modern Approaches in IoT and Machine Learning for Cyber Security: Latest Trends in AI* (pp. 75–92). Springer International Publishing: Cham.
73. Guven, I., & Parlak, M. (2022, November). Blockchain, AI and IoT empowered swarm drones for precision agriculture applications. In *2022 IEEE 1st Global Emerging Technology Blockchain Forum: Blockchain & Beyond (iGETblockchain)* (pp. 1–6). IEEE.
74. Ravindran, M. (2022). Energy efficient clustering for precision agriculture using LEACH protocol. *Neuroquantology, 20*(7), 2702.
75. Strobel, G. (2020). Agriculture in the era of the Internet of Things: An information system architecture for smart agriculture. *CIS 2020 Research-in-Progress Papers*. 84.
76. Vangipuram, S. L., Mohanty, S. P., Kougianos, E., & Ray, C. (2022). G-DaM: A distributed data storage with blockchain framework for management of groundwater quality data. *Sensors, 22*(22), 8725.
77. Chattopadhyay, P., Patel, H. P., & Parmar, V. (2022, August). Internet of Things (IoT) in smart agriculture. In *2022 3rd International Conference on Electronics and Sustainable Communication Systems (ICESC)* (pp. 536–540). IEEE.
78. Lamtzidis, O., & Gialelis, J. (2018, December). An IOTA based distributed sensor node system. In *2018 IEEE Globecom Workshops (GC Wkshps)* (pp. 1–6). IEEE.

79. Bera, S., Dey, T., Mukherjee, A., & Buyya, R. (2023). E-CropReco: A dew-edge-based multi-parametric crop recommendation framework for internet of agricultural things. *The Journal of Supercomputing, 79*, 11965–11999.
80. Dinc, E., Kuscu, M., Bilgin, B. A., & Akan, O. B. (2019). Internet of everything: A unifying framework beyond internet of things. In *Harnessing the Internet of Everything (IoE) for Accelerated Innovation Opportunities* (pp. 1–30). IGI Global.
81. Bodkhe, U., Tanwar, S., Parekh, K., Khanpara, P., Tyagi, S., Kumar, N., & Alazab, M. (2020). Blockchain for industry 4.0: A comprehensive review. *IEEE Access, 8*, 79764–79800.
82. Bayih, A. Z., Morales, J., Assabie, Y., & de By, R. A. (2022). Utilization of internet of things and wireless sensor networks for sustainable smallholder agriculture. *Sensors, 22*(9), 3273.
83. Verma, A., Bhattacharya, P., Madhani, N., Trivedi, C., Bhushan, B., Tanwar, S., Sharma, G., Bokoro, P., & Sharma, R. (2022). Blockchain for industry 5.0: Vision, opportunities, key enablers, and future directions. *IEEE Access, 10*, 69160–69199.
84. Ferrag, M. A., & Shu, L. (2021). The performance evaluation of blockchain-based security and privacy systems for the Internet of Things: A tutorial. *IEEE Internet of Things Journal, 8*(24), 17236–17260.
85. Phang, S. K., Chiang, T. H. A., Happonen, A., & Chang, M. M. L. (2023). From satellite to UAV-based remote sensing: A review on precision agriculture. *IEEE Access, 11*, 127057–127076.
86. Bala, K., & Kaur, P. D. (2022, January). Changing trends of blockchain in IoT: Benefits and challenges. In *2022 12th International Conference on Cloud Computing, Data Science & Engineering (Confluence)* (pp. 324–329). IEEE.
87. Dwivedi, S. K., Roy, P., Karda, C., Agrawal, S., & Amin, R. (2021). Blockchain-based internet of things and industrial IoT: A comprehensive survey. *Security and Communication Networks, 2021*, 1–21.
88. Podder, A. K., Al Bukhari, A., Islam, S., Mia, S., Mohammed, M. A., Kumar, N. M., Cengiz, K., & Abdulkareem, K. H. (2021). IoT based smart agrotech system for verification of Urban farming parameters. *Microprocessors and Microsystems, 82*, 104025.
89. Chauhan, N., & Dwivedi, R. K. (2022, April). Designing a secure smart healthcare system with blockchain: A review. In *2022 1st International Conference on Informatics (ICI)* (pp. 214–219). IEEE.
90. Vilas-Boas, J. L., Rodrigues, J. J., & Alberti, A. M. (2023). Convergence of distributed ledger technologies with digital twins, IoT, and AI for fresh food logistics: Challenges and opportunities. *Journal of Industrial Information Integration, 31*, 100393.
91. Javier Ordonez García, B. J., & Martos Nunez, M. V. (2023). Blockchain in agriculture: A PESTELS analysis. https: //doi.org/10.31220/agriRxiv.2023.0019

CHAPTER 6

Image Processing-Based Applications for E-Pest Control Systems

KALPANA SRIVASTAVA[1], AMBIKA AGGARWAL[1*], ASHUTOSH BHATT[2], and ANCHIT BIJALWAN[3]

[1]*School of Computer Science, UPES, Dehradun, Uttarakhand, India*

[2]*Department of CSE, SST, Swami Rama Himalayan University, Dehradun, Uttarakhand, India*

[3]*School of Computing and Innovative Technologies, British University Vietnam, Hanoi, Vietnam*

*Corresponding author

ABSTRACT

In recent times, image processing technology developments have been at the forefront in opening up new frontiers in diverse fields like agriculture and pest control. These emerging advanced technologies have been adopted due to the increased need for precision and sustainability in pest management, which now provides solutions that were unimaginable before. High-resolution imaging and sophisticated algorithms are used through image processing to improve the efficiency of pest population detection and management. The chapter examines image-based applications of electronic pest control systems by looking at the shortfalls of traditional ways of managing pests while focusing on how improved technology can be used. The chapter gives insight into the basic tenets of image processing and its relationship to agriculture, healthcare, hotel and hospitality, food industry, and research and development with case studies based on real life. Further, applications for the E-pest control technology were also discussed.

6.1 INTRODUCTION

The adage "necessity is the mother of invention" holds particularly true in the ever-evolving battle against pests, where precision and sustainability are key. Imagine a situation whereby pests become a continuous threat to crucial sectors disrupting large and economic losses.

Demystifying Sustainable Farming Using Smart Technologies.
Gunjan Chhabra, Keshav Kaushik, Priya Chugh, Susheela Dahiya, and Aryan Chaudhary (Eds.)

DOI: 10.1201/9781779640932-6

Traditional pest control methods that depend heavily on chemical pesticides usually face challenges of providing the required accuracy while posing significant environmental dangers.

To address these challenges, the combination of image processing technology with electronic pest control systems provides an innovative solution. These systems utilize high-resolution imaging and sophisticated algorithms to accurately detect and identify pests, monitor their populations, and gauge infestation levels. Pattern recognition, feature extraction, and machine learning, among other technologies, have emerged as leading-edge innovations facilitating real-time, efficient pest management.

In 2022, the global pest control services market was valued at USD 22,523.52 million; this figure is expected to expand at a compound annual growth rate of 4.76%, reaching USD 29,767.59 million by 2028. The demand for services in different applications, including insect control, rodent control, termite control, and mosquito control, has been increasing, thus driving their growth in the market. Pest control service providers are increasingly focusing on replacing toxic pesticides with bio-based alternatives to meet stringent government regulations and offer eco-friendly services. Moreover, sectors such as hotels and the mining industry are particularly inclined toward pest control services for effective rat and rodent prevention [1].

Precise and sustainable pest management is necessitated by the growth of this market. Chemical pesticides which are used in traditional ways for control of pests though they have been somewhat effective come with environmental risks, leading to pest resistance. Therefore, the adoption of ecologically benign and biologically based approaches to pest control is on the rise [1].

To extract useful information from images, a sector called image processing has become an important field in pest control. In particular, there are high-resolution imaging techniques as well as advanced algorithms that can identify a specific pest species, track its populations, and monitor infestation levels. These improvements in pattern recognition technology, feature extraction techniques, and machine learning methods offer great tools for real-time pest crop protection.

Through the integration of electronic pest control systems with image processing technologies, more accurate and efficient means of controlling the population size of pests are possible. The system enables detection and identification by combining hardware devices such as cameras and sensors with sophisticated software programs, thus allowing timely interventions. This approach not only enhances the effectiveness of pest control measures but also supports environmental sustainability by reducing reliance on chemical pesticides.

6.2 TRADITIONAL PEST CONTROL METHODS

Figure 6.1 shows different types of traditional pest control methods:

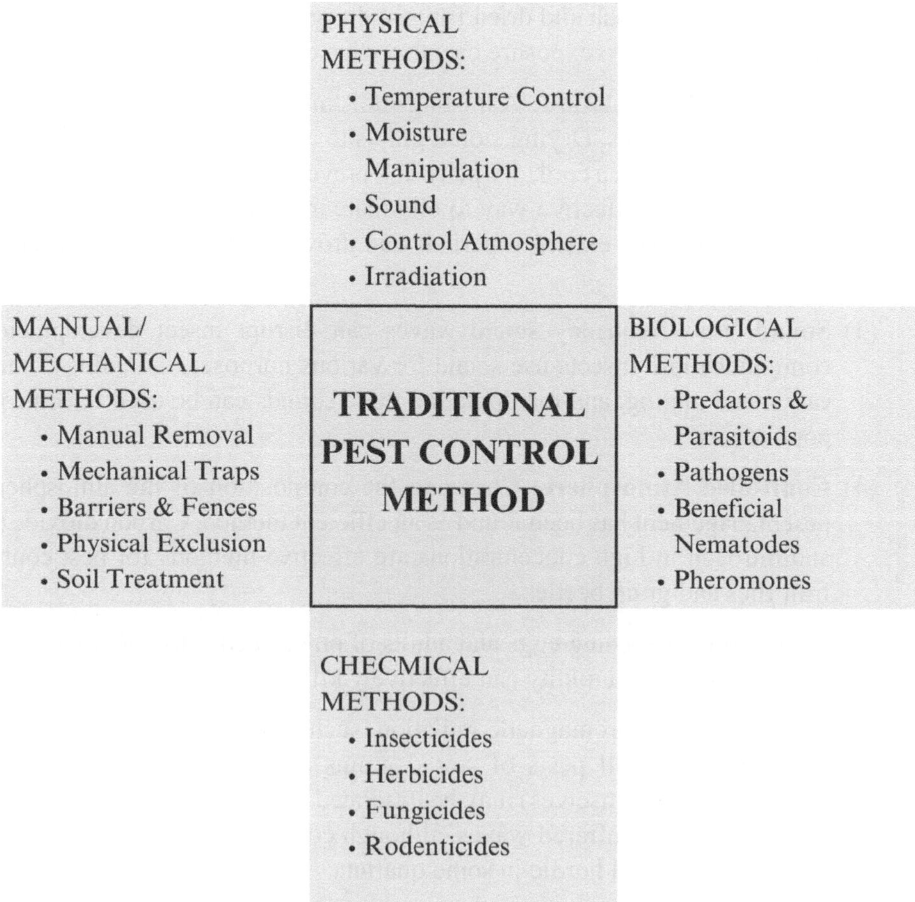

FIGURE 6.1 Traditional pest control methods.

1. **Physical Methods**

Physical methods of pest control employ devices or environmental adjustments to effectively manage and diminish pest populations. These approaches encompass temperature control, moisture manipulation, sound intervention, controlled atmosphere, and irradiation [2, 3].

(1) **Temperature Control:** Raising the temperature can be an effective way to kill pests. During hot months, sun rays can be used to kill pests by exposing seeds and stored commodities, such as the larvae of pink bollworms in cottonseeds. Heating the store to over 50°C for 10–12 hours can kill dormant pests. Steaming: provided by running product through hot water, dry heat, or microwaves to disinfect fruits, vegetables, and woolen clothes. Lowering the temperature works too on the flip side. Cold storage will kill many pests, such as apple maggots: children gather and eat the maggots. This method is used for

a large range of fresh and dried fruits and vegetables. Opening storage in cold climates for subzero exposure can also wipe out grain pests.

(2) **Moisture Manipulation:** Adjusting moisture levels can create unfavorable conditions for pests. Drying stored grains to a moisture content below 8–10% and storing them in a cool, dry place can prevent insect damage. Draining stagnant water is an effective way to eliminate mosquito breeding sites. Soaking logs in water over extended periods can drown boring weevils and larvae of wood wasps.

(3) **Sound:** Low-frequency sound waves can disrupt insect development and communication. Insects use sound for various purposes, including communication and mating, and manipulating these sounds can be used to control pest populations.

(4) **Controlled Atmosphere:** Changing the composition of the atmosphere for pest management has been found as an efficient method. Carbon dioxide (CO_2) and nitrogen in high concentrations are effective methods for pest control of fruit flies and grain beetles.

For example, exposing eggs and adults of pulse beetles to 100% CO_2 at 32°C and 70% relative humidity can effectively kill them.

(5) **Irradiation:** Electromagnetic radiation, such as microwaves or gamma rays, can sterilize or kill pests of stored grains and other products. Insects (or infested grains themselves) may be irradiated with microwaves or nonionizing radiation such as infrared waves, although consumer acceptance of irradiated food remains a real hurdle in some quarters.

2. Manual/Mechanical Methods

Manual and mechanical methods of pest control rely on physical actions or devices to manage and reduce pest populations. These nonchemical techniques are integral to sustainable pest management and include manual removal, mechanical traps, barriers and fences, physical exclusion, and soil treatment [2, 4].

(1) **Manual Removal:** This may be the easiest of all pest control methods, as manual removal simply involves handpicking pests from plants or surfaces. This is a labor-intensive method and suitable for very little infestation in a small garden. Some of these include picking out harmful insects, larvae, and eggs by hand using tongs or gloves and the elimination of infected plant parts to stop pests and diseases from spreading.

(2) **Mechanical Traps:** Designed to capture and kill pests, mechanical traps are widely used in both agricultural and urban settings. Sticky traps coated with a sticky substance capture flying insects like aphids and fruit flies. Snap traps are commonly used for rodent control, employing a spring-loaded mechanism

to capture and kill mice and rats. Live traps, which capture animals alive for relocation, are used for larger pests such as raccoons and squirrels.

(3) **Barriers and Fences:** Physical barriers and fences prevent pests from accessing crops or specific areas. Row covers, made of lightweight fabric, protect crops from insects while allowing light and water to pass through. Netting is used to protect fruit trees and berry bushes from birds and larger insects. Electric fences can keep larger animals like deer and wild boars out of agricultural fields.

(4) **Physical Exclusion:** Blocking entry points to prevent pests from entering buildings, storage areas, or protected environments is another effective method. Techniques include sealing cracks and holes with materials like caulk or foam to keep out pests such as rodents and ants and installing screens on windows, doors, and vents to prevent the entry of flying insects like mosquitoes.

(5) **Soil Treatment:** Soil treatment methods are used to manage pests that live or lay eggs in the soil. Soil solarization, which involves covering the soil with clear plastic to trap solar energy, raises the temperature to levels that kill soil-borne pests and pathogens. Regular tilling disrupts the life cycle of pests by exposing eggs and larvae to predators and environmental conditions they cannot survive.

3. Chemical Methods

Chemical methods of pest control involve using synthetic or natural chemicals to manage and reduce pest populations. These techniques are commonly used as part of integrated pest management (IPM) strategies and include insecticides, herbicides, fungicides, and rodenticides [2, 5].

(1) **Insecticides:** Insecticides are chemicals designed to kill or repel insects. They can be divided into different types based on how they work and how they are applied. Contact insecticides kill insects when they come into direct contact with them, while systemic insecticides are absorbed by plants and kill insects that feed on them. Some examples of insecticides are organophosphates, pyrethroids, and neonicotinoids.

(2) **Herbicides:** Herbicides are chemicals used to control or get rid of unwanted plants, commonly known as weeds. They can be selective, targeting specific types of plants, or nonselective, killing all vegetation. Some commonly used herbicides include glyphosate, atrazine, and 2,4-D.

(3) **Fungicides:** Fungicides are chemicals that prevent or eliminate fungal infections in plants. They can either protect the surface of the plant (contact fungicides) or be absorbed by the plant (systemic fungicides). Examples of fungicides include chlorothalonil, propiconazole, and copper-based fungicides.

iv. **Rodenticides:** Rodenticides are chemicals designed to kill rodents like rats and mice. They can be anticoagulants, which cause internal bleeding, or nonanticoagulants, which kill rodents through other mechanisms. Some examples of rodenticides are warfarin, bromadiolone, and zinc phosphide.

4. **Biological Methods**

When it comes to controlling pests, biological methods offer a natural and eco-friendly approach. These methods involve using living organisms or natural substances to manage and reduce pest populations. They play a vital role in IPM systems and include a range of techniques like predators and parasitoids, pathogens, beneficial nematodes, and even pheromones [2, 6].

(1) **Predators and Parasitoids:** One effective way to control pests is by introducing their natural enemies into the environment. Predators are organisms that chow down on pests, while parasitoids lay their eggs on or in pest hosts, ultimately causing their demise. Examples of these helpful creatures include ladybugs, lacewings, and parasitic wasps, among others.

(2) **Pathogens:** Another biological control method involves using microbial agents like bacteria, fungi, and viruses to infect and kill pests. These pathogens are specific to certain pests and do not harm other organisms. Some examples are Bacillus thuringiensis, Beauveria bassiana, and Nuclear Polyhedrosis Virus.

(3) **Beneficial Nematodes:** These microscopic roundworms are a powerful weapon against soil-dwelling pests. They parasitize and kill a variety of pests such as grubs, termites, and root weevils. Look out for nematodes like Steinernema carpocapsae and Heterorhabditis bacteriophora to get the job done.

(4) **Pheromones:** Insects use chemical signals called pheromones to communicate, and we can manipulate these signals to control pest populations. Pheromones can be used to lure pests into traps or disrupt their mating behavior. Some examples include pheromone traps and mating disruption techniques.

6.3 DISADVANTAGES OF USING TRADITIONAL METHODS

1. **Environmental Impact:** Chemical methods can contaminate soil, water, and air, which ends up harming nontarget organisms and ecosystems. Mechanical and biological methods, while more eco-friendly, can accidentally disrupt local ecosystems and have unintended consequences for nontarget species. Manual methods generally have a lower environmental impact, but large-scale operations can still disturb local habitats [7].

2. **Human Health Risks:** Chemical methods pose significant health risks due to exposure, both acute and chronic. Mechanical methods can result in injuries and exposure to pest-related diseases when handling traps. Biological methods may

cause allergic reactions, and manual methods involve direct contact with pests, increasing the risk of bites, stings, and disease transmission [2].

3. **Labor-Intensive and High Maintenance:** Manual and mechanical methods require a lot of labor and ongoing maintenance, making them impractical for large-scale infestations. Chemical and biological methods also demand significant labor for application and monitoring, adding to the overall workload [2].
4. **Limited Effectiveness and Scalability:** Manual methods usually work best for small-scale infestations, while mechanical methods may not be sufficient for widespread or hidden pest populations. Chemical methods face the challenge of pests developing resistance, and biological methods often depend on specific environmental conditions, which can delay their effectiveness [8].
5. **Cost Considerations:** Setting up and maintaining mechanical methods can be expensive, and the frequent application of chemical methods incurs ongoing costs. Implementing and maintaining biological control agents also involves significant expenses, while the labor costs for manual methods can be high, especially for large-scale operations.
6. **Resistance and Sustainability:** Chemical methods often lead to pests developing resistance, necessitating stronger applications, which can exacerbate environmental and health impacts. Biological methods face the risk of pests adapting, diminishing their long-term effectiveness. Manual and mechanical methods, while not contributing to resistance, face challenges in scalability and overall effectiveness, impacting their sustainability.

In summary, traditional pest control methods have considerable disadvantages, including environmental harm, health risks, labor intensity, limited effectiveness, high costs, and issues with resistance and sustainability. These drawbacks underscore the importance of IPM strategies to achieve more effective, sustainable, and environmentally friendly pest control solutions.

6.4 IMAGE PROCESSING IN PEST-CONTROL SYSTEM

Have you ever considered what would happen if pests in your garden or farm become uncontrollable and traditional methods fail to prevent them from doing so? Suppose that pests are destroying your crops at a fast rate and no matter what you try, you just cannot get rid of them. In this kind of desperate situation, the e-pest control system can save the day by introducing advanced and innovative ways of controlling pest infestation.

Image processing applied in pest control systems employs computer vision techniques to detect, identify, and monitor pests. In addition to enhancing the precision and efficiency of pest management, this technology enables automatic detection while providing real-time data for guiding decisions [9]. Figure 6.2 shows an overview of how image processing could be used in pest control systems; it is technical.

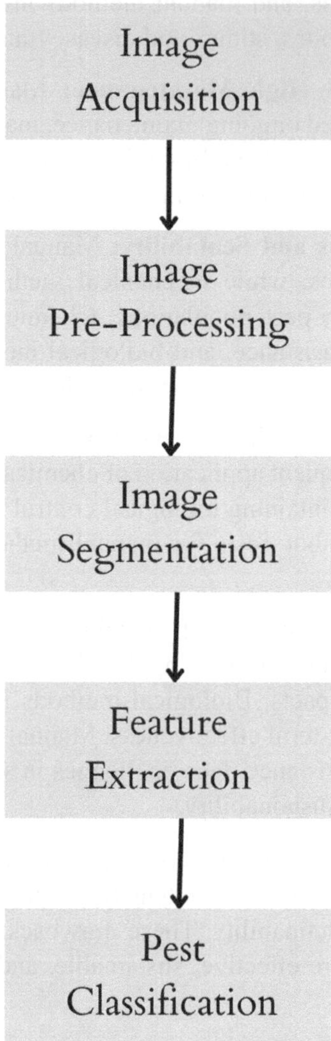

FIGURE 6.2 Flowchart of processing.

The step-by-step visuals are depicted in Figure 6.3.

Image Acquisition: High-resolution cameras or sensors are placed in areas where pests are likely to infest. These cameras capture images or videos at regular intervals, ensuring that no pest activity is missed.

Example: In a vineyard, cameras might be installed to monitor for grapevine moths. These cameras take images every 15 minutes to ensure continuous surveillance.

Image Preprocessing: The captured images often need enhancement to ensure accuracy. Preprocessing improves image quality by adjusting brightness, contrast, and reducing noise.

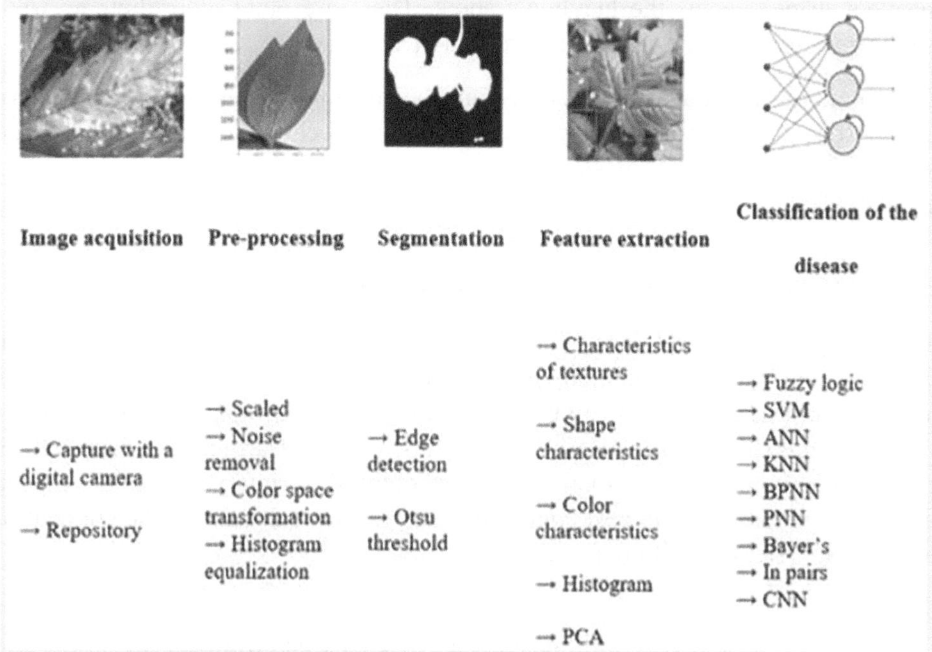

FIGURE 6.3 Steps for classification of pests.

Source: Own elaboration. ANN = artificial neural networks, KNN = k-nearest neighbors, BPNN = back propagation neural network, PNN = probabilistic neural network.

Example: An image taken at dusk might be too dark to see small pests. Adjusting the brightness and contrast makes the pests more visible.

Image Segmentation: Segmentation isolates the pests from the background. This can be done using various techniques like thresholding, edge detection, and region-based segmentation.

Example: A captured image of a leaf infested with aphids can be processed using thresholding to turn the leaf into a plain background while highlighting the aphids as distinct objects.

Feature Extraction: Once pests are isolated, key features are extracted to help identify them. These features include shape, size, color, and texture.

Example: For the Colorado potato beetle, the system extracts features like its elongated shape, distinctive yellow and black stripes, and size.

Pest Classification: The extracted features are fed into machine learning models for classification. Convolutional neural networks (CNNs) are particularly effective for this task because they can learn and recognize complex image patterns.

Example: A trained CNN can differentiate between the Colorado potato beetle and similar-looking insects like the ladybug by analyzing unique features such as the beetle's stripes versus the ladybug's spots. There are other algorithms as well like support vector machine, K- means grouping technique.

This automated approach ensures timely and accurate pest detection, enabling swift and targeted pest control measures. Using image processing technology, farmers and gardeners can effectively manage pest infestations, protecting their crops and reducing reliance on chemical pesticides [10, 11].

6.5 APPLICATIONS OF E-PEST CONTROL SYSTEMS

An e-pest control system refers to using electronic, digital, and automated technologies to manage and control pest populations in crop fields. These systems can monitor, predict, and mitigate pest infestation effectively through various technologies such as sensors, the Internet of Things (IoT), artificial intelligence (AI), machine learning, and data analysis. To treat pests and minimize the use of chemical pesticides, the primary objective is to enhance effectiveness and precision in pest control measures. Figure 6.4 shows the applications of E-pest control systems in different sectors.

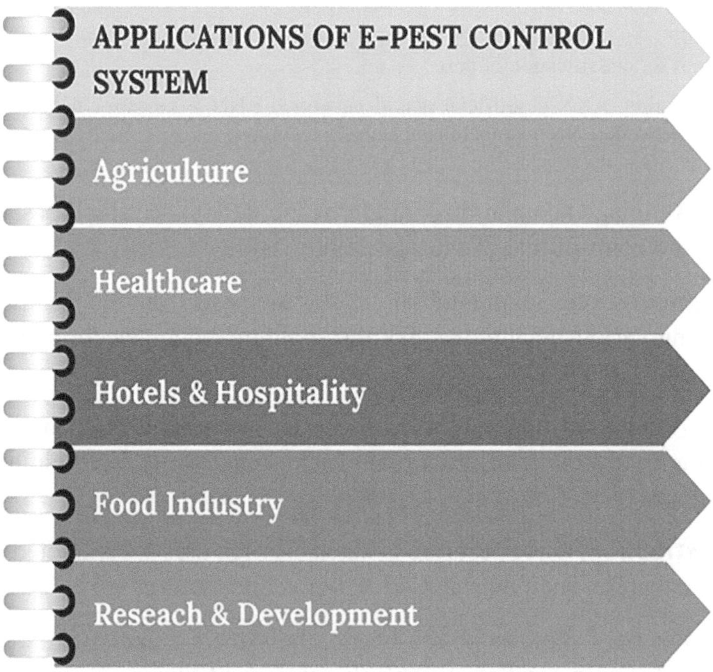

FIGURE 6.4 Applications of e-pest control system.

1. **Agriculture**

   Challenges: Initial costs and the need for technological literacy among farmers are one of the most important challenges in implementing ePath control systems. Moreover, the reliability of the digital infrastructure and the availability of real-time data are crucial to the success of such systems. There are also significant obstacles such as weather conditions, data accuracy, and the adaptability of systems to various crop types and pest species.

Impact: E-Pest control systems' impact on agriculture has been profound in improving crop yield. Farmers can make informed choices quickly, leading to timely intervention to avoid large-scale infestation and crop damage. By guaranteeing stable crop production, this proactive approach is also contributing to food security [12].

Initiative: Initiatives like the eSAP (Electronic Solutions against Agricultural Pests) in India exemplify the efforts to overcome these challenges. eSAP is a tool for farmers to identify pest problems, provide advice on how to solve them, and offer real-time assistance. To provide specific recommendations, it relies on a large number of pest databases, weather forecasts, and agricultural knowledge in the region [13].

Key Insights: The potential for significantly reducing the use of chemical pesticides, thus lowering production costs and minimizing environmental impact, is an essential insight into implementing electronic pest control systems. For example, precision pest control helps to target only the affected areas, reducing the use of pesticides in an indiscriminate manner. This not only protects the beneficial insect population but also prevents soil and water contamination.

Case Study: In the state of Maharashtra, a significant decrease in crop losses due to pests has been achieved by farmers using the eSAP application. By providing farmers with accurate pest identification and control measures, the app helped in timely interventions, resulting in a 20% increase in crop yield and a 30% reduction in pesticide usage within a single growing season. This case highlights the potential of digital solutions to transform traditional agricultural practices, making them more efficient and sustainable [13].

2. **Healthcare**

Challenges: Implementing pest control technology in healthcare settings comes with its fair share of challenges. First off, there is the issue of high initial installation and maintenance costs. It can be quite a financial burden. Another challenge is the need for technical expertise to operate and manage these systems. You got to have the know-how. And let us not forget about integrating the technology with the existing healthcare infrastructure. It is no walk in the park. Oh, and data accuracy and monitoring pests in hard-to-reach areas are also major concerns. Plus, we cannot overlook the importance of maintaining patient privacy and complying with health regulations when it comes to pest control and technology use. It is a lot to consider, that is for sure.

Impact: Now, let us talk about the impact these systems have on healthcare facilities. They work wonders in improving sanitary conditions, reducing the occurrence of HAIs (hospital-acquired infections), and ultimately improving patient outcomes. Not only that, but they also create a more pleasant and hygienic environment for both patients and staff, which adds to the overall quality of care. Plus, by cutting down on the need for chemical pesticides, these systems are doing their part to support eco-friendly pest management practices. Kudos to them! [14]

Initiative: One initiative that deserves a shout-out in this field is the "Hospital Integrated Pest Management System" (HIPMS). Several hospitals in India have adopted this system, and it is making quite a difference. HIPMS uses smart sensors and cameras to detect pest activity, and then it goes a step further by analyzing the data to identify trends and potential problem areas. When it detects something fishy, it sends real-time alerts to the facility management, so they can take swift and targeted pest control measures. Talk about being on top of things! [15]

Key Insights: These systems play a crucial role in maintaining a pest-free environment, which in turn helps prevent the spread of nasty pathogens that pests often carry around. It is all about keeping patients safe and sound. And here is a bonus: the data collected from these systems can be used to improve pest control strategies over time, making them even more efficient and cost-effective. It is a win–win situation!

Case Study: In a large hospital in Mumbai, they implemented HIPMS, and boy, did it make a difference! Staff and patients were no longer bothered by pest sightings and complaints. The real-time monitoring and quick response to pest activity allowed the hospital to maintain a squeaky-clean environment, resulting in a whopping 25% decrease in HAIs related to pests. This case study clearly shows the practical benefits of integrating e-pest control systems into healthcare settings. They truly have the potential to enhance the safety and quality of care provided [16].

3. **Hotels and Hospitality**

   Challenges: The challenges include high costs associated with installing and maintaining the necessary technology, training staff to effectively use and manage these systems, and ensuring seamless integration with existing hotel management infrastructure. Additionally, the reliability and accuracy of the data collected are crucial, as is maintaining guest privacy and meeting regulatory requirements related to pest control and data handling [17].

   Impact: These systems improve the cleanliness and hygiene of hotel properties, contributing to a better overall guest experience. They also help maintain the hotel's reputation by preventing pest-related incidents that can lead to bad reviews and loss of business. Furthermore, the reduction in the use of chemical pesticides supports more sustainable and environmentally friendly pest management practices.

   Initiative: A notable initiative in this area is the "Smart Hotel Pest Management System" (SHPMS), which has been adopted by various luxury hotels in India. SHPMS employs a network of smart traps and sensors that continuously monitor for signs of pest activity. These devices are connected to a central management platform that analyzes the data and provides real-time alerts and recommendations for targeted pest control actions [18].

Key Insights: By ensuring a pest-free environment, hotels can prevent the negative reviews and complaints that often result from pest encounters. Moreover, the data collected through these systems can help hotels identify patterns and areas prone to infestations, enabling more effective long-term pest management strategies.

Case Study: A prominent five-star hotel in Delhi implemented SHPMS and observed a dramatic decrease in pest sightings and guest complaints within a few months. The real-time monitoring and immediate response to pest activity allowed the hotel to maintain high hygiene standards, resulting in a notable increase in positive guest feedback and repeat bookings. This case demonstrates the practical benefits of e-pest control systems in the hospitality industry, highlighting their potential to enhance guest satisfaction and operational efficiency [18].

4. **Food Industry**

Challenges: High initial investment costs and ongoing maintenance expenses can be prohibitive, especially for small to medium-sized enterprises. Additionally, there is a need for skilled personnel to operate and manage these systems effectively. Ensuring the accuracy and reliability of the data collected is vital, as false readings can lead to either complacency or unnecessary interventions. Compliance with food safety regulations and standards, as well as maintaining data privacy, adds further complexity to the implementation of these systems [9].

Impact: These systems enhance the overall hygiene and safety of food products by preventing pest contamination, which is critical for consumer health. They also help maintain the integrity of brands by preventing pest-related incidents that could lead to product recalls and reputational damage. Additionally, the reduction in the use of chemical pesticides promotes sustainable and environmentally friendly practices within the industry.

Initiative: An exemplary initiative in this domain is the "Digital Pest Management System"" (DPMS), which has been adopted by several food processing companies in India. DPMS uses smart traps equipped with sensors that detect and log pest activity. These traps are connected to a central system that analyzes the data and provides actionable insights and alerts to the facility management team. This allows for targeted pest control measures, reducing the reliance on broad-spectrum pesticides [19].

Key Insights: The deployment of e-pest control systems in the food industry significantly reduces the risk of food contamination. Continuous monitoring and real-time alerts enable immediate actions to mitigate pest threats, thus maintaining high standards of food safety and compliance with regulatory requirements.

Moreover, the collected data can help in identifying trends and vulnerabilities, leading to improved long-term pest management strategies.

Case Study: A large food facility in Gujrat implemented DPMS and observed a significant reduction in pest-related incidents. The real-time data and alerts provided by the system enabled the management to take immediate and precise action, preventing potential contamination. This led to improved compliance with food safety standards and a decrease in production downtime due to pest control measures. The facility also reported a 25% reduction in the use of chemical pesticides, underscoring the system's environmental benefits. This case study illustrates the practical advantages of e-pest control systems in the food industry, showcasing their potential to enhance food safety, operational efficiency, and sustainability [20].

5. **Research and Development (R&D)**

Challenges: These include the high cost of deploying advanced technologies, the need for technical expertise to operate and manage the systems, and ensuring the integration of these systems with existing research infrastructure. The accuracy and reliability of data are crucial, as erroneous data can compromise research outcomes. Moreover, maintaining the confidentiality of research data and complying with regulatory standards for pest control and data management are critical concerns [21].

Impact: These systems enhance the precision and reliability of research by maintaining pest-free environments, which is essential for accurate data collection and analysis. They also contribute to the development of more effective and sustainable pest management solutions, benefiting both research and practical applications in agriculture, healthcare, and other sectors. Furthermore, by reducing the need for chemical pesticides, these systems support environmentally friendly research practices.

Initiative: A notable initiative in this field is the "Integrated Pest Management Research System" (IPMRS), which has been adopted by various research institutions in India. IPMRS utilizes a network of smart sensors and traps to continuously monitor pest activity in laboratory and field settings. The data collected are analyzed and used to develop new pest control strategies, improve existing methods, and ensure that experimental conditions remain undisturbed by pest activity.

Key Insights: The detailed information on pest behavior, population dynamics, and environmental interactions gathered through these systems can lead to breakthroughs in pest management techniques. Additionally, real-time monitoring helps maintain the integrity of experimental setups by preventing pest-related disruptions.

Case Study: An institute in Hyderabad implemented IPMRS to monitor and manage pest populations in their experimental fields and greenhouses. The system provided real-time data on pest activity, allowing researchers to take immediate action to protect their experiments. This led to more consistent and reliable research outcomes, as well as new insights into pest behavior and control methods. The institute also reported a significant reduction in the use of chemical pesticides, highlighting the system's contribution to sustainable research practices. This case study demonstrates the practical benefits of e-pest control systems in R&D, showcasing their potential to enhance research quality and environmental sustainability.

6.5 FUTURE WORKS

Smart Traps and IoT devices: Smart traps equipped with image processing capabilities can autonomously identify and count captured pests, sending real-time data to centralized management systems. These IoT devices will streamline pest monitoring and control, making it more efficient and less labor intensive. The data collected will also contribute to broader pest management databases, enhancing overall understanding and strategies [22].

Adaptive Learning Systems: Adaptive learning systems will become more prevalent, where image processing applications continually improve their accuracy and efficiency through machine learning. By learning from new data and past interventions, these systems will adapt to changing pest behaviors and environmental conditions, ensuring sustained effectiveness over time [23].

Precision Agriculture Integration: In precision agriculture, image processing-based e-pest control systems will work in tandem with other technologies such as drones and satellite imagery. Drones equipped with high-resolution cameras can capture images across vast agricultural landscapes, and image processing algorithms can identify pest hotspots. This integration will allow for localized treatments, reducing pesticide use and environmental impact.

User-Friendly Mobile Applications: The development of user-friendly mobile applications incorporating image processing capabilities will empower farmers and pest control professionals to monitor and manage pests using their smartphones. These apps can provide on-the-spot pest identification, suggest control measures, and track pest management activities.

6.6 CONCLUSION

The image-processing base application for e-pest control systems implementation is a giant leap in the era of innovations for pest management. Using advanced algorithms, AI, and real-time data analytics, these technologies provide a reliable, efficient, and sustainable system. It is capable of being integrated with drones, smart traps, and IoT devices, which further escalate monitoring ability and help in taking proactive as well as targeted interventions. They are heralded as a way to reduce pesticide usage and maintain biodiversity—but only if the upscaling can deliver on those promises.

KEYWORDS

- **image processing**
- **pest control system**
- **integrated pest management**
- **biological methods**
- **chemical methods**
- **convolutional neural networks**

REFERENCES

1. *Pest Control Services Market Analysis 2023–2030: Technology Trends and Future Developments*, LinkedIn, 2023.
2. RVJ. (2023). *Pros and Cons of Different Pest Control Manufacturing Techniques*. Deskera.
3. Unit 10 Physical and Mechanical Control Methods. 2021.
4. Traditional Methods of Pest Control. (2021). In eagri.org.
5. PPC-Admin. (2021, December 8). Pest control methods—Chapter 2—Chemical pest control methods. *Panther Pest Control*.
6. Joshi, K., Kumar, A., Anandaram, H., Kumar, A., Aggarwal, A., & Chhabra, G. (2023). Incorporating AI and IoT for smart poultry monitoring and health detection. In *2023 Second International Conference on Augmented Intelligence and Sustainable Systems (ICAISS)* (pp. 76–81). IEEE.
7. Arizona Pest Solutions, 2024.
8. Saxena, R. (2024). The pros and cons of natural pest control methods. *Blogs on Pest Control Treatments, Home Cleaning Tips & Ideas*.
9. Gómez-Camperos, J. A., Jaramillo, H. Y., & Guerrero-Gómez, G. (2022). Digital image processing techniques for detection of pests and diseases in crops: A review. *Ingeniería y Competitividad, 24*(1), e30110973.
10. Chowdhury, M., & Anand, R. (2023). *Automated Pest and Disease Identification in Agriculture using Image Processing*. ResearchGate.
11. Chattu, P., Sathwik, K., Sravani, C., Sai Teja, A., & Samhitha, B. (2024). Using image processing and computational intelligence to detect pest at early stages of crop. In *Journal of Emerging Technologies and Innovative Research (JETIR), Journal of Emerging Technologies and Innovative Research (JETIR)* (Vol. 11, Issue 5).
12. Chodey, M. D. (2020). Pest detection in crop using video and image processing. *International Journal for Research in Applied Science and Engineering Technology, 8*(4), 29–35.
13. University of Agricultural Sciences (UAS), Raichur. (n.d.). e-Pest surveillance in selected cropping ecosystems through eSAP [Pest surveillance and IPM].
14. Dynopest. (2024, April 18). Effective pest control in healthcare facilities: A comprehensive guide. *Dyno-Pest*.
15. Overview and Activities | Directorate of Plant Protection, Quarantine & Storage | GOI.
16. Express News Service. (2019). *Mumbai: Hospital Faces Pest Problem, to Seek Herbal Help. The Indian Express*.
17. Pest Control in the Hospitality Industry According to the Hygiene Doctor.
18. Editor. (2022). *Hotel Property Management Systems: Products and Features*. AltexSoft.
19. Kumar, P., Chaudhary, R., Aggarwal, A., Singh, P., & Tomar, R. (2012). Improving medical image segmentation techniques using multiphase level set approach via bias correction.

20. Parmar, J. (2024). *Restaurants Bugged: Only 10% in Ahmedabad have Pest Control Certificates. The Times of India.*
21. Detection of Pests using Image Processing and Deep Learning. (n.d.). In *IEEE Conference Publication.*
22. Muruganantham, B., Kureshi, F. A., Ishwarya, J., & Murugan, N. (2019). Internet of things (IOT) based electronic pest control using image processing system. *IJARIIE, 5*(2)
23. N, P., & S Murthy, V. (2017). An image processing algorithm for pest detection. In: *Perspectives in Communication, Embedded-Systems and Signal-Processing (PiCES)* (Vol. 1, Issue 3).A

CHAPTER 7

Drone-Based Food Supply Chain: Applications and Challenges

ISHU SHARMA[1*] and JAGDEEP SHARMA[2]

[1]*School of Engineering and Technology, CGC University, Mohali, Punjab, India*

[2]*School of Management and Commerce, RIMT University, Mandi Gobindgarh, Punjab, India & School of Management Studies, CGC University, Mohali, Punjab, India*

*Corresponding author

ABSTRACT

Drones or unmanned aerial vehicle devices integrated with the food supply chain can give enormous benefits in the business for all processes, starting from production to consumption. Food supply chain is the process that combines multiple party's tasks aligned toward a common goal. In the food supply chain, parties like production, postharvest handling, processing and packaging, distribution, retail marketing, and consumption management get involved. The integration of drones for the food supply chain also offers innovative solutions for challenges faced in different areas like monitoring of crops, aerial inspection of fields, etc. This chapter explores the applications and challenges of drone technology in the food supply chain, with an understanding of the technical aspects of drones. The chapter identifies and analyses the challenges associated with the integration of drones in the food supply chain like regulatory hurdles, safety concerns, data management concerns, skills related to operate drones and technical limitations.

7.1 INTRODUCTION

Unmanned aerial vehicles (UAV) or drones are vehicles that can fly and capture images from an aerial view without the need of a human driver. These UAVs can be deployed for multiple purposes, ranging from industry to the surveillance sector [18]. These devices can provide better quality pictures of the targeted area in comparison to satellite images. These devices help in getting real-time data for various areas with requirement of a smaller number of resources. Many industries are investing in building commercial drones for

Demystifying Sustainable Farming Using Smart Technologies.
Gunjan Chhabra, Keshav Kaushik, Priya Chugh, Susheela Dahiya, and Aryan Chaudhary (Eds.)

DOI: 10.1201/9781779640932-7

capturing images or sometimes videos as well. The frequency of capturing images and the area to be captured by drones are based on the application of the deployment of drones. The real-time monitoring features of drones allow administrative managers to take timely action on the situation. UAVs can be used in emergency situations like forest fire, earthquake affected areas, or any other natural disasters to take the necessary action that is best fit for that situation. The areas that are not reachable by humans can also be accessed by these devices like wild life monitoring, sea species investigation, and many other areas. These devices are equipped with sensors, cameras, and global positioning system (GPS) to provide the accurate and timely data of the surroundings [34]. Each drone is equipped with firmware that helps the user to control and interact with the device with a graphical user interface; hence, users can easily use these devices for multiple purposes.

Further, food supply chain is the process that combines multiple party's tasks aligned toward a common goal. In the food supply chain, parties like production, postharvest handling, processing and packaging, distribution, retail marketing, and consumption management get involved. In the production of food, agricultural tasks like crop farming, dairy farming, and drones for monitoring and surveillance can lead to smart resource management, ultimately leading to automation in production. Farmers normally require inputs such as seed, irrigation frequency, irrigation quantity, crop health, pesticides spray. [33]. After the production of crops, products are required to be converted into the form of consumable items like tomato ketchup from tomatoes; this process may include cleaning, packaging, and food preservation. The processed food items are transported to distribution centers from farms or warehouses. The transportation process is managed with efficiency so that processed food can be delivered timely to the destination while maintaining the quality of the food. The end user of the food supply chain is customer, and customers can purchase food items through retail stores like supermarkets, grocery stores, or online platforms. The demand and supply chain of the specific food item can be predicted or computed based on the consumption of the food items. Throughout the entire food supply chain, the need to get real-time data from different units of the food supply chain is critical and can lead to significant achievements in the businesses [7].

With the advent of emerging technologies like artificial intelligence (AI), Internet of Things (IoT), sensor-based data acquisition, blockchain technology, and edge computing, the industries are moving toward Industry 5.0. Drones or UAV devices integrated with the food supply chain can give enormous benefits in the business for all processes, starting from production to consumption [23]. This chapter provides a detailed study on opportunities and challenges for the integration of drones in food supply chain management. Applications in different areas are also elaborated with few case studies. The scope of optimal solutions for tackling the challenges of the integration is also discussed in this chapter.

7.2 TYPES OF DRONES

Drones or UAVs can be of different sizes and different architectures deployed in the battlefield or surveillance domain, and the type of sensors or cameras depends on the application areas. These devices are capable of flying without a flight pilot, and the user can get graphical user interface-based interface to control these devices. The types of different drones are shown in Figure 7.1 and discussed in sections.

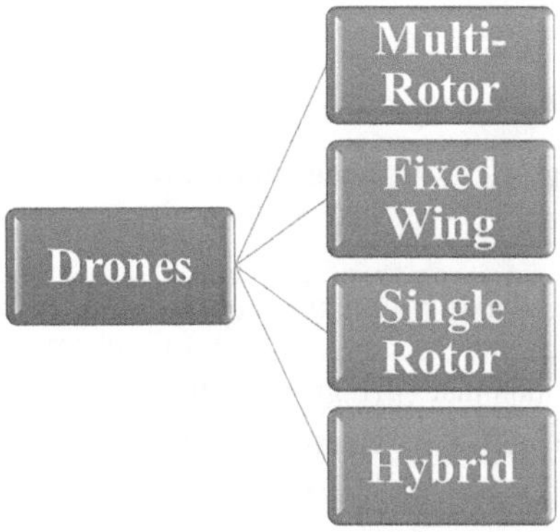

FIGURE 7.1 Generalized classification of drones.

7.2.1 MULTIROTOR DRONE

The multirotor drone does not require a runway to land and is easily operatable by users. This type of drone can also carry lightweight payloads like crop seeds or sprays that make these devices suitable for agricultural and industrial tasks. These drones are manufactured in different types of shapes like "T," "+," and "Y" shape. Table 7.1 contains details about different shapes of multirotor drones.

TABLE 7.1 Different Shapes of Mutirotor Drone

Category	Number of Rotors	Arm Distance (In Degrees)	Shape	Advantages	Limitations
Tri-copter	3	120	Y, T	Easy to use and configure	Limited lifting power
Quad-rotor	4	90	X, H	Enhanced Performance	Expensive in comparison to tri-copter
Hexa-copter	6	60	*	Safe-landing	Motors are working to provide balance in the flight
Octocopter	8	–	**	Can be used for heavy payloads	Size is very big and expensive
Coaxial multirotor drone	Concentrically connected mutirotor	–	**	Small in size; can handle high payload	Expensive

7.2.2 FIXED WING

The design of fixed-wing UAV devices is based on wings that provide lift for the flight. These types of devices are mainly used in research and development tasks for monitoring and experiments [4]. Fixed-wing drones are customized based on applications like military, supply chain, delivery of goods, medicines. Table 7.2 lists the different categories of fixed wing based on the application, and accordingly, the sensor devices or other configurations can be arranged.

TABLE 7.2 Different Types of Fixed-Wing Drone

Application	Mini Fixed Wing	Long-Endurance Fixed-Wing	Vertical Takeoff and Landing (VTOL) Fixed-Wing
Agriculture	RGB Cameras, Sensors—NDVI (Normalized Difference Vegetation Index)	Multispectral cameras, sensors—NDVI, LiDAR (Light Detection and Ranging)	Multispectral cameras, LiDAR, sensors—NDVI
Mapping and surveying	RGB cameras, LiDAR	High-resolution cameras, LiDAR, software	High- resolution cameras, LiDAR, photogrammetry software
Surveillance	Electro-optic infrared cameras, thermal cameras	Electro-optic infrared cameras, surveillance radar	Electro-optic infrared cameras, surveillance radar, automatic identification system receiver
Environmental monitoring	Hyperspectral cameras, gas sensors	Hyperspectral cameras, gas sensors, atmospheric sensors	Hyperspectral cameras, gas sensors, atmospheric sensors
Wildlife conservation	Electro-optic infrared cameras, GPS tracking	Electro-optic infrared cameras, GPS tracking, radio telemetry	Electro-optic infrared cameras, GPS tracking, radio telemetry

7.2.3 SINGLE ROTOR

In single rotor, the horizontal rotor is deployed on the top of the aircraft or helicopter, and antitorque rotor handles the torque generated by aircraft. These types of UAVs or drones are preferred in military applications due to the capability of handling large payload with operations for longer hours. Table 7.3 summarizes different types of single rotors with advantages, disadvantages, application areas, and few notable parameters.

TABLE 7.3 Different Types of Single Rotor

Type of Single Rotor	Advantages	Disadvantages	Application Areas	Notable Parameters
Hiller	Simple design	Limited maneuverability in high winds	Light utility helicopters	Cost of maintenance
No tail rotor	Reduced noise and vibration	Complex system requiring additional maintenance	Civil and military helicopters	Reliability
Fenestron	Reduced noise and improved safety with the device tail	Increased weight and complexity	Civil helicopters for urban areas	Tail protection
Enclosed tail rotor	Improved safety with rotor	Increased weight and complexity	Military helicopters, additional civil applications	Rotor visibility and accessibility in maintenance

7.2.4 *HYBRID ROTOR*

A hybrid rotor is the combination of a rotor aircraft and a fixed-wing UAV with propellers to enhance performance and system reliability [44]. Table 7.4 details the different types of hybrid rotor systems with common applications, pros, and cons of using the hybrid UAV for different applications.

TABLE 7.4 Different Types of Hybrid Rotor

Hybrid Rotor System	Pros	Cons	Common Applications
Increased speed	Ability to achieve higher speeds	Increased complexity and cost	Military inspection and transportation missions
Improved efficiency	Reduced fuel consumption and longer range	Higher maintenance requirements	Civil search and rescue operations
VTOL capability	Ability to operate in confined spaces without runways	Limited payload capacity	Urban air mobility and emergency medical services
Enhanced motility	Improved agility and motility	Higher pilot training requirements	Border patrol and law enforcement assignments

7.3 FOOD SUPPLY CHAIN

The food supply chain is a connection of multiple procedures and with involvement of multiple parties that work together for a common goal to provide food supplies to customers with satisfaction in different terms. Every component of the food supply chain contains an assembly of players and activities [13]. Figure 7.2 represents the components of the food supply chain that are interconnected with each other.

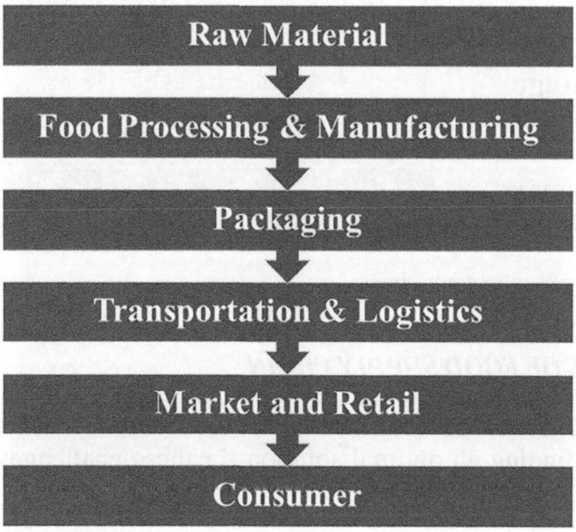

FIGURE 7.2 Components of food supply chain.

7.3.1 FLOW AND UNDERSTANDING OF FOOD SUPPLY CHAIN

The food supply chain consists of several primary components, where agricultural production involves activities such as crop planting, agronomy, and harvesting, where farmers and growers contribute toward producing raw food materials. After having food raw material, food undergoes processing, where it is transformed into various products through cooking, packaging, and other related manufacturing processes [18]. Distribution involves the transportation of food material from processing units to warehouses and other distribution centers. The retail party or component of the food supply chain includes the selling of food products to customers at platforms like supermarkets, grocery stores, restaurants. One of the critical concerns of the food supply chain process is its interdependency, wherein disruptions in one component can affect the complete food supply chain. For example, disruptions in agricultural production due to natural disaster or any other reason can lead to a deficiency of raw materials for food processing, resulting in reduced availability of food products for customers [31].

Figure 7.3 shows the components and parties of the apple supply chain for different stages including farming, processing, distribution, and consumer stage.

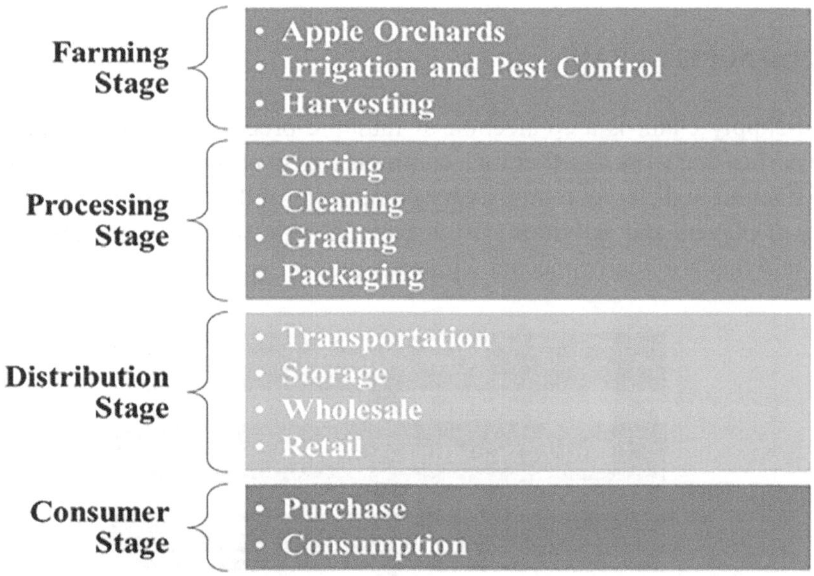

FIGURE 7.3 Example of apple supply chain.

7.3.2 CHALLENGES OF FOOD SUPPLY CHAIN

The food supply chain could face multiple challenges, and different involved parties work on the direction of finding an optimal solution for these challenges. Few challenges are mentioned below:

- Food Safety: For the sake of global health, it is required that food be kept safe in the complete supply chain. Continuous monitoring and regulatory procedures are necessary to address infection threats at every stage from manufacturing to distribution [21].

- Transparency in Supply Chain Operations: Transparency is required in the food supply chain for effective tracing of the food products [21]. There is a need of fulfilling the customer's needs on the parameters of time and quality, otherwise the complete demand and supply chain working in the food supply chain can be affected.
- Climate Change: Crop yields, water availability, and total food production are all impacted by sudden changes in climate due to global warming or natural disasters [24]. Droughts and floods are examples of effects of global warming that may cause hurdles in the supply chain and demand–supply process [30].
- Shortage of Resources: The production of food and other related processes of the food supply chain are vulnerable to resource shortages like water and arable land. Utilizing sustainable farming methods is essential to reducing the loss of natural resources and protecting the environment [1].
- Globalization: Food supply chains have become more linked and complicated with the growth in globalization [9]. The cultural differences, trade exchange rate between foreign countries, and taxation are some concerns that are required to be handled to boost the globalization of food supply chain.
- Food Waste: Food waste is a major concern in the food supply chain at different stages from production to customer that requires timely and right action at every stage. Different people and parties involved in this chain aim to minimize the overall food waste to maximize the profit and throughput [17].
- Infrastructure and Logistics: Inadequate infrastructure may make it more challenging for the delivery of food from the raw material stage to the final stage. Inefficient transportation, poor road connectivity, and a deficiency of cold storage points affect the complete supply chain. On other hand, developed nations can perform better in this sector with advanced and improved resources [25].
- Labor Issues: Migration of labor from one place to another is a constraint in building sustainable manpower. Getting the right labor at right time can be taken as a catalyst in the overall growth and success of the food supply chain [42].
- Technological Disruptions: Technology can increase productivity and reliability of the entire system, and similarly, it can disturb the complete supply chain. For example IoT or drones can provide real-time data or information about the different stages of supply chain with tracking through GPS, cameras or other sensor data, but the same advanced devices can stuck the supply chain due to failures, false information, or security breaches [15].
- Pandemics and Health Crises: Pandemics like COVID-19 can disrupt the complete supply chain because of a hurdle at one single unit, for example, in case of failure of the transportation network, the demand can be increased and the supply cannot be fulfilled that results in a hike of the prices of food items for consumers.

7.3.3 TECHNOLOGICAL INNOVATIONS IN THE FOOD SUPPLY CHAIN

In today's era, the food supply chain is evolving through technological advancements in terms of effectiveness, reliability, sustainability, accuracy, and safety. Blockchain technology, which initially covers the domain of cryptocurrency, has become a trending emerging technology that aims to provide trust and privacy for supply chain management in different domains. Blockchain technology allows multiple parties that are involved in the supply chain to track food items from the production stage to the consumer stage where each party involved can get the real-time information without compromising the data authenticity [12]. The trust and privacy achieved by blockchain can enhance the overall working of the supply chain, and also the tracking of different items can be efficiently improved. The integration of blockchain technology with the IoT allows the gathering of real-time data while maintaining the data privacy. For example, the consumer can check the origin of the food product with assurance that the data are not being manipulated [32]. IoT sensors with radio-frequency identification (RFID) tags are utilized in the supply chain to track different parameters like temperature, pressure, state, location. The use of IoT devices in transportation units can provide the live tracking of different food items with GPS sensors. Similarly for many products, it is necessary to know the temperature and pressure of trucks and other vehicles, and that data can be fetched using multiple sensors. RFID tags, access cards, number locks, facial recognition, fingerprint scanner, etc., can be installed to give secure access to these vehicles that can be further connected to blockchain technology or a centralized database management system [28].

Data analytics and data visualization tools enable different stakeholders of the supply chain to draw meaningful information from extracted data or collected data that help in making future decisions about demand–supply, launch of new product, needs of customers, etc. Consequently, AI algorithms allow predictive analysis for trends in business, sales of the next quarter, where supervised and unsupervised machine learning techniques can be applied to plan the future approaches for the food supply chain. This analysis or projection improves the quality in the food supply chain and the efficiency of different operational tasks. Automation and robots used in production, packaging, or delivery as well have also become powerful technological aspects in the food supply chain, providing an error-free process with less time and disruptions [43]. Robots with a range of sensors can take actions according to the situation or environment where these devices are deployed. Organic farming has boomed in recent years with technical advancement that allows indoor farming in closed surroundings to handle or take care of the needs of special customers. Vertical farming allows reduction in the carbon associated with transportation of food items from one place to another for sustainable growth of food supply chain cycle [45].

On-demand food items can be produced based on dietary requirements and needs with the integration of 3D printing. Drones are also enhancing field-related task in the agriculture sector with approaches like pesticide spraying, soil condition tracking, and aerial observation. Farmers can maximize agricultural outcomes while reducing food waste and other negative effects [29]. Genetic engineering has made it possible to generate crops with improved features to increase the nutritional value of food items that can be customized according to the needs of consumers. Supply chain management software solutions provide each party in supply chain in food management to visualize supply chain processes with efficiency and transparency [37]. Through the use of these emerging technologies, stakeholders can manage all related operations efficiently that can provide optimal resource allocation with giving maximum throughput. This technology stack is also listed in Figure 7.4 for graphical representation.

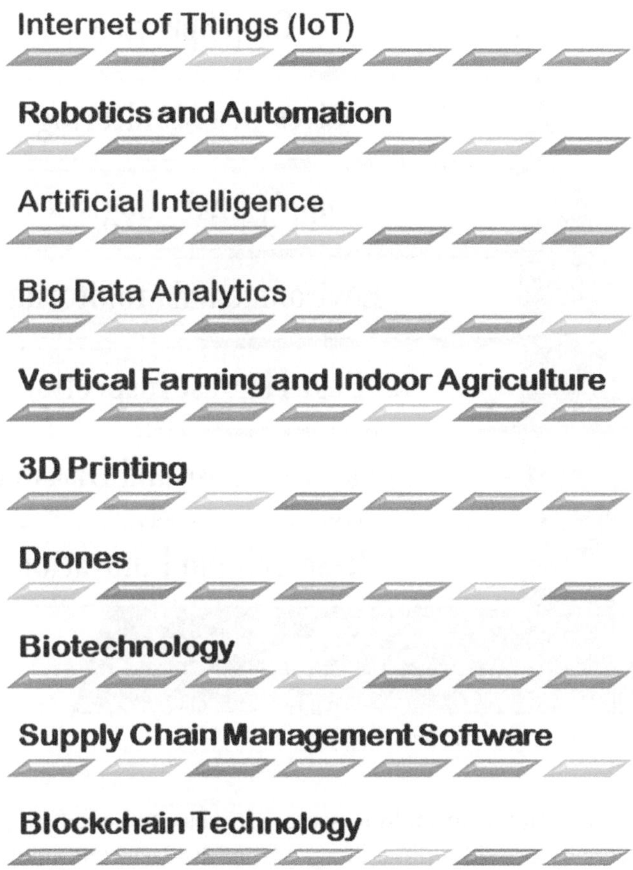

FIGURE 7.4 Technology stack for food supply chain.

7.4 APPLICATIONS OF DRONES IN FOOD SUPPLY CHAIN

With the evolution of emerging technologies and technological advancements, a greater number of stakeholders are moving toward building food supply chain solutions for the food supply chain process. Multiple processes included in the supply chain can be enhanced in terms of efficiency, flexibility, and transparency by including drones for different roles [39]. These applications of drones in the food supply chain process are shown in Figure 7.5.

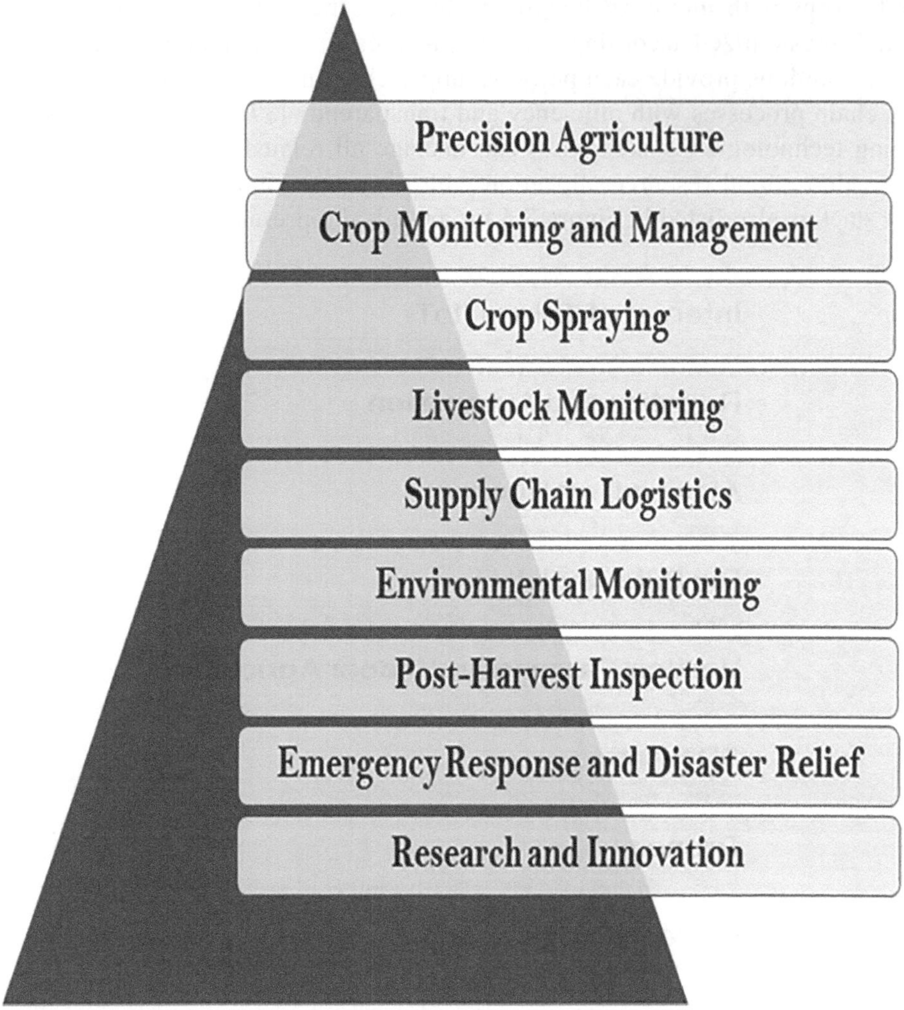

FIGURE 7.5 Applications of drones in food supply chain.

Precision Agriculture: In the agriculture sector, drones can become a good companion of farmers or producers by providing live images from the farm that images can be captured by drones equipped with cameras. Some drones are also capable of capturing videos based on the requirements. Drones with payload capability can be used for pesticide spray or other seed-based requirements [3].

Crop Monitoring and Management: Drones allow farmers to monitor and manage crops from far places rather than going and visiting the actual farm area [5]. The same images captured can be further investigated by AI techniques like crop diseases can be traced at the early stage by advanced image processing methods [35].

Crop Spraying: Advanced spraying systems using drones used for agriculture allow exact deployment of fertilizer, chemicals such as herbicides and pesticides to crops. This type of spraying has multiple benefits over standard spraying methods, like lower chemical use, less environmental pollution, and higher productivity. Drones may also reach places where human presence is not feasible [38].

Livestock Monitoring: Farmers are able to track the health and activity of their animals over a wide area by using drones connected with particular cameras, for instance, those with thermal cameras. UAVs with AI can also be used in such scenarios for recognizing certain patterns with the animals [36]. Through early problem detection and timely action, farmers can boost production, reduce expenditures, and assure the well-being of their animals.

Supply Chain Logistics: Drones with payloads can be deployed for remote delivery tasks without the need of a flight pilot [40]. This is one time investment that can help in minimizing the labor cost in future, and the task can be performed with efficiency avoiding human errors.

Environmental Monitoring: Drones are equipped with different type of sensors to capture the data from environment like temperature, pressure, etc., and these collected data about environment can be utilized for taking numerous decisions like the time of irrigation, crop health, and prediction of crop producing decisions.

Post-Harvest Inspection: Drones can be used for monitoring the crops or food items placed in storage units with aerial images, temperature, and pressure data [14]. These types of sensors or cameras ensure the safety of food items in different types of scenarios.

Emergency Response and Disaster Relief: In some locations, the human interference or reach is not possible, especially in the location affected by natural disasters like flood-affected areas [22]. In such situations, drones can take a flight and give the precise information to the concerned stakeholders about the affected areas.

Research and Innovation: Drones are an important tool for providing timely information for experimental studies for producing hybrid and organic crops. The research and innovation in the field of food supply chain is an ongoing process that can be accelerated by the utilization of drones [2].

7.5 CHALLENGES OF USING DRONES IN THE FOOD SUPPLY CHAIN

Drones provide several advantages for the food supply chain, but there are a few issues with their adoption that need to be resolved as well:

Regulatory Hurdles: Complicated regulatory frameworks make it difficult to integrate drones into the food supply chain. Drone operators often have to manage laws and regulations since drone operating regulations vary depending on the areas. Many firms find it difficult to enter the market due to these laws [16] (Figure 7.6).

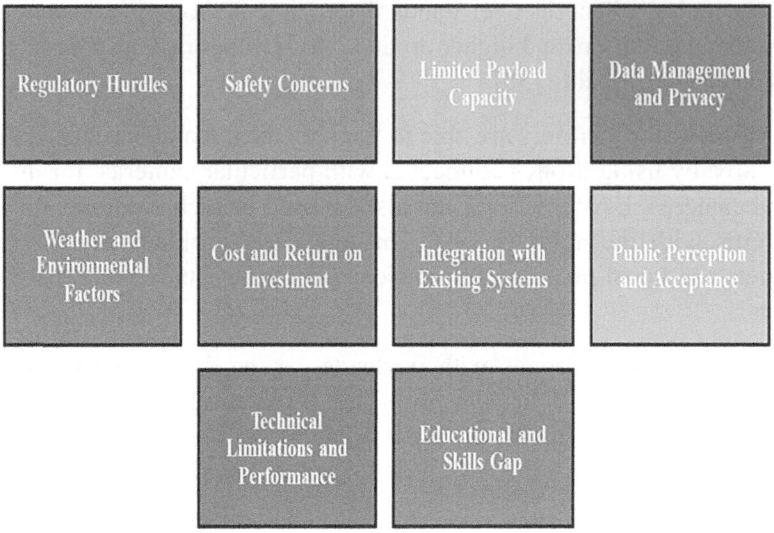

FIGURE 7.6 Challenges of using drones in the food supply chain.

Safety Concerns: Drone use in agriculture requires utmost safety. Drones and human aircraft share airspace, which has the possibility of mid-air accidents [6]. Drone operators must follow safety procedures with continuous training [41].

Limited Payload Capacity: Drones' baggage capabilities are still very small when compared to airplanes with pilots. Drones could find it difficult to transport large tanks of fertilizer over great distances.

Data Management and Privacy: A good amount of data is produced by drones through sensors, and managing this data is very crucial. Further, these data can be misused by unauthorized users in terms of a security breach [19].

Weather and Environmental Factors: Drone operations in agriculture may be greatly impacted by weather and environmental factors. Extreme weather conditions, such as rain, fog, wind, and temperature, may impact drone performance. Drone operations may be hampered by dust, debris, and electromagnetic interference.

Cost and Return on Investment: Drones have a lot of potential applications in agriculture, but some farmers and companies may not be able to afford the initial investment and ongoing operation expenses [10]. Significant initial capital investment is needed to purchase drones, sensors, software, and people training [27].

Integration with Existing Systems: It may be difficult to integrate drones into current agricultural workflows and systems [8]. Careful preparation and communication are needed when coordinating drone operations with other agricultural tasks. When combining drone data with other information systems or farm management software, interoperability problems might occur.

Public Perception and Acceptance: The way the general public views drones in agriculture may have an impact on stakeholders' acceptance. Drone operations may encounter resistance due to privacy concerns, noise pollution, and environmental damage. Overcoming these obstacles requires fostering public knowledge and trust [26].

Technical Limitations and Performance: Drones still have performance and technological limits that affect their efficacy and dependability in agricultural applications. For instance, the operating range and length of drone missions are limited by restrictions in battery life and flying endurance [11]. The quality and dependability of airborne data gathered by drones may also be impacted by issues with sensor resolution, accuracy, and calibration.

Educational and Skills Gap: Drone technology uptake and usage in agriculture require a trained workforce with the skills, knowledge, and experience needed to operate and maintain drones. But there is a big education and skill gap in the agriculture industry right now, and a lot of farmers and agricultural professionals do not have the technical know-how and expertise to use drone technology [20].

7.6 CONCLUSION

With the advent of emerging technologies like AI, IoT, sensor-based data acquisition, blockchain technology, and edge computing, industries are moving toward Industry 5.0. Drones or UAV devices integrated with the food supply chain can give enormous benefits in the business for all processes, starting from production to consumption. AI algorithms allow predictive analysis for trends in business, sales of the next quarter, where supervised, unsupervised machine learning techniques can be applied to plan future approaches for food supply chain. This analysis or projection improves the quality in the food supply chain and the efficiency of different operational tasks. Automation and robots used in production, packaging, or delivery as well have also become powerful technological aspects in the food supply chain, providing an error-free process with less time and disruptions.

KEYWORDS

- **food supply chain**
- **drone technology**
- **unmanned aerial vehicles**
- **single rotors**
- **hybrid rotor systems**

REFERENCES

1. Abatan, A., Obiuto, N. C., Ninduwezuor-Ehiobu, N., Ani, E. C., Olu-lawal, K. A., & Ugwuanyi, E. D. (2024). Integrating advanced technologies for enhanced HSE management in the FMCG sector. *Engineering Science & Technology Journal*, *5*(4), 1270–1280.
2. Ahmad, A., Liew, A. X. W., Venturini, F., Kalogeras, A., Candiani, A., Di Benedetto, G., Ajibola, S., Cartujo, P., Romero, P., Lykoudi, A., & others. (2024). AI can empower agriculture for global food security: Challenges and prospects in developing nations. *Frontiers in Artificial Intelligence*, *7*, 1328530.
3. Ahsen, M. E., & Moshref-Javadi, M. (2024). The role of technologies in supply chain efficiency and resiliency. In: *Impacts of COVID-19 on Supply Chains: Disruptions, Technologies, and Solutions* (pp. 117–144). Springer.
4. Ameer, S. A., Alkhafaji, M. A., Jaffer, Z., & Al-Farouni, M. (2024). Empowering farmers with IoT, UAVs, and deep learning in smart agriculture. *E3S Web of Conferences*, *491*, 4007.
5. Anirudh, K. C., Dixit, M., & Karat, S. (2024). Exploring the applications of wireless sensor networks and the internet of things in agricultural and livestock production and management. In: *Emerging Technologies and the Application of WSN and IoT* (pp. 209–227). CRC Press.
6. Arkeman, Y., Kailaku, S. I., Hidayah, N. J., & Saefurahman, G. (2024). Application of advanced digital technology for green and circular agroindustrial supply chain management. *AIP Conference Proceedings*, *2957*(1), 020002.
7. Chavan, Y., Paul, K., & Kolekar, N. (2024). Food safety and hygiene: Current policies, quality standards, and scope of artificial intelligence. In: *Food Production, Diversity, and Safety Under Climate Change* (pp. 319–331). Springer.
8. Chouhan, S. S., Singh, U. P., & Jain, S. (2024). *Applications of Computer Vision and Drone Technology in Agriculture 4.0.* Springer.
9. Dejanović, M., Popović-Pantić, S., & Kovačević, A. (2023). *The Significance of Digital Transformation in Agriculture for Sustainable Development Conference*, Belgrade, Serbia.
10. Dubey, R., Bryde, D. J., & Foropon, C. (2024). Design and management of humanitarian supply chains for pandemics: Lessons from COVID-19. In: *Annals of Operations Research* (pp. 1–14). Springer.
11. Edirimanne, R., Fernando, W. M., Nielsen, P., Perera, H. N., & Thibbotuwawa, A. (2024). Optimizing unmanned aerial vehicle based food delivery through vehicle routing problem: A comparative analysis of three delivery systems. *Applied Computer Science*, *20*(1), 85–105.
12. El Jaouhari, A., Arif, J., Jawab, F., Samadhiya, A., & Kumar, A. (2024). Unfolding the role of metaverse in agri-food supply chain security: Current scenario and future perspectives. *International Journal of Food Science & Technology*, *59*(5), 3451–3460.
13. Elango, E., Hanees, A., Shanmuganathan, B., & Kareem Basha, M. I. (2024). Precision agriculture: A novel approach on AI-driven farming. In: *Intelligent Robots and Drones for Precision Agriculture* (pp. 119–137). Springer.
14. Gopalan, B., Vijaya, G. S., & Tiwary, A. (2024). Food supply chain management, logistics, and ecosystems in the internet economy. In: *Digital Technologies for a Resource Efficient Economy* (pp. 36–62). IGI Global.
15. Hawashin, D., Nemer, M., Salah, K., Jayaraman, R., Svetinovic, D., & Damiani, E. (2024). Blockchain and NFT-based traceability and certification for UAV parts in manufacturing. *Journal of Industrial Information Integration*, *39*, 100597.
16. Johny, J., & Gurtu, A. (2024). Digital supply chain technologies and sustainability. In: *Environmentally Responsible Supply Chains in an Era of Digital Transformation: Research Developments and Future Prospects* (p. 79).
17. Kantor, C., Eisenback, J. D., & Kantor, M. (2024). Biosecurity risks to human food supply associated with plant-parasitic nematodes. *Frontiers in Plant Science*, *15*, 1404335.
18. Kasyap, C. V. S., & Tiwari, S. (2024a). Crop management using artificial intelligence: A literature survey. *International Research Journal of Modernization in Engineering Technology and Science, 6*(1).
19. Li, W., Li, C., Zhang, D., & Guo, D. (2024). enterprise supply chain optimization algorithm based on big data. *Computer-Aided Design & Applications*, *21*(S21), 116–133

20. Mandal, S., Yadav, A., Panme, F. A., Devi, K. M., & SM, S. K. (2024). Adaption of smart applications in agriculture to enhance production. *Smart Agricultural Technology*, 7, 100431.
21. Mishra, A. C., Das, J., & Awtar, R. (2024). An emerging era of research in agriculture using AI. *Journal of Scientific Research and Technology*, *2*(4), 1–7.
22. Mubarik, M. S., & Khan, S. A. (2024). Digital supply chain and sustainability challenges. In: *The Theory, Methods and Application of Managing Digital Supply Chains* (pp. 133–143). Emerald Publishing Limited.
23. Nath, P. C., Mishra, A. K., Sharma, R., Bhunia, B., Mishra, B., Tiwari, A., Nayak, P. K., Sharma, M., Bhuyan, T., Kaushal, S., & others. (2024). Recent advances in artificial intelligence towards the sustainable future of agri-food industry. *Food Chemistry*, 447, 138945.
24. Odimarha, A. C., Ayodeji, S. A., Abaku, E. A., & others. (2024). The role of technology in supply chain risk management: Innovations and challenges in logistics. *Magna Scientia Advanced Research and Reviews*, *10*(2), 138–145.
25. Oriekhoe, O. I., Ashiwaju, B. I., Ihemereze, K. C., Ikwue, U., & Udeh, C. A. (2024). Review of technological advancements in food supply chain management: A comparative study between the US and Africa. *International Journal of Management & Entrepreneurship Research*, *6*(1), 132–149.
26. Palazzetti, L., & others. (2024). *Optimizing Drone-Based Applications for Delivery and Smart Agriculture*. PhD Thesis.
27. Pungavi, R., & Praveenkumar, C. (2024). Unmanned Aerial Vehicles (UAV) for smart agriculture. In: *Artificial Intelligence and Smart Agriculture: Technology and Applications* (pp. 251–269). Springer.
28. Purtell, C., Hong, S.-J., & Hiatt, B. (2024). Bibliometric analysis on advanced air mobility and drones. *Journal of Air Transport Management*, *116*, 102569.
29. Raj, M., Harshini, N. B., Gupta, S., Atiquzzaman, M., Rawlley, O., & Goel, L. (2024). Leveraging precision agriculture techniques using UAVs and emerging disruptive technologies. *Energy Nexus*, *14*, 100300.
30. Rampalli, N. S., Sri, Y. G., & Bhuvaneshwari, K. S. (2024). Autonomous agriculture and food production: Agritech revolution. In: *The Convergence of Self-Sustaining Systems with AI and IoT* (pp. 40–63). IGI Global.
31. Saidakhmedovich, G. S., Uralovich, M. D., Saidakhmedovich, G. S., & Tishabayevna, R. M. (2024). Application of digital technologies for ensuring agricultural productivity. *British Journal of Global Ecology and Sustainable Development*, *25*, 6–20.
32. Salim, R., & Fajar, A. N. (2024). Object detection of chili using convolutional neural network YOLOV7. *Journal of Theoretical and Applied Information Technology*, *102*(6).
33. Shah, B. R. (n.d.). *Cultivating Connections: Optimizing Efficiency and Sustainability in the Agricultural Supply Chain.*
34. Shah, S. F. A., Mazhar, T., Al Shloul, T., Shahzad, T., Hu, Y.-C., Mallek, F., & Hamam, H. (2024). Applications, challenges, and solutions of unmanned aerial vehicles in smart city using blockchain. *Peer Journal of Computer Science*, *10*, e1776.
35. Shee, H. (2024). Internet of Things: applications and challenges for supply chain management. In: *The Palgrave Handbook of Supply Chain Management*, pp. 1197–1215.
36. Soni, S., Chandra, P., Sharma, P. C., Gangrade, J., & Singh, D. K. (2024). Medical kit delivery using drone: Critical medical infrastructure solution for emergency medical situation. *International Journal of Disaster Risk Reduction*, *108*, 104502.
37. Sorbelli, F. B. (2024). UAV-based delivery systems: A systematic review, current trends, and research challenges. *Journal on Autonomous Transportation Systems*, *1*(3), 1–40.
38. Sulaiman, Y., Khong, B. T. W., Lun, L., Ghani, N. H. A., & Khairi, M. Q. I. M. (n.d.). *A New Adoption Intention Model of Drone Application in Food Delivery Services in Malaysia: A Conceptual Paper*. WSEAS Transactions on Business and Economics. https\\: doi: 10.37394/23207.2024.21.41
39. Sun, X., Kuo, Y.-H., Xue, W., & Li, Y. (2024). Technology-driven logistics and supply chain management for societal impacts. *Transportation Research Part E: Logistics and Transportation Review*, *185*, 103523.
40. Supreetha, S., Sonarthi, H., & Mall, S. (2024). Food processing and management of food supply chain: From farm to fork. In: *Food Process Engineering and Technology: Safety, Packaging, Nanotechnologies and Human Health* (pp. 119–134). Springer.

41. Tadić, S., Krstić, M., Veljović, M., Čokorilo, O., & Milovanović, M. (2024). Risk Analysis of the use of drones in city logistics. *Mathematics*, *12*(8), 1250.
42. Tolooie, A., Sinha, A., & Raftar Nouri, A. K. (2024). Heuristic approach for optimising reliable supply chain network using drones in last-mile delivery under uncertainty. *International Journal of Systems Science: Operations & Logistics*, *11*(1), 2301610.
43. Verma, R. K., & Kishor, K. (2024). Image processing applications in agriculture with the help of AI. In: *Infrastructure Possibilities and Human-Centered Approaches with Industry 5.0* (pp. 162–181). IGI Global.
44. Wahab, S. N., Othman, N., Sahak, M. Y., & Subramaniam, N. (2024). Elevating healthcare resilience: Role of drones in mitigating supply chain challenges in Malaysia. *Environment-Behaviour Proceedings Journal*, *9*(SI19), 55–63.
45. Wang, S., Zheng, C., & Wandelt, S. (2024). Policy challenges for coordinated delivery of trucks and drones. *Journal of the Air Transport Research Society*, 2, 100001.

CHAPTER 8

Development of Prediction Equations for Estimating the Biomass of Standing Trees

DEEPAK KHOLIYA[1*], PRIYA CHUGH[2], AMIT KUMAR MISHRA[3], and RAKESH KUMAR BHADULA[4]

[1]*School of Agriculture, Graphic Era Hill University, Dehradun, Uttarakhand, India*

[2]*School of Agriculture and Technology, Maya Devi University, Dehradun, Uttarakhand, India*

[3]*Department of CS&E, Graphic Era Hill University, Dehradun, Uttarakhand, India*

[4]*Department of Mathematics, Graphic Era Hill University, Dehradun, Uttarakhand, India*

*Corresponding author

ABSTRACT

Assessing the productivity of a forest involves evaluating the biomass of standing trees within the ecosystem. Research on estimating tree biomass in forests is vital for calculating production, understanding soil and plant nutrient dynamics, and assessing the impact of various factors like commercial tree usage, silvicultural practices, and environmental disturbances on forest productivity and stability. These studies provide insights into the performance of tree species in terms of their biological output and guide harvesting practices for mature trees, especially for commercial purposes, that influence total tree biomass. The predictive models developed through these studies are particularly valuable for meeting the needs of forestry and agroforestry industries as tree growers and farmers involved in agroforestry practices. The growth in levels of tree crops as they age has directly impacted the overall biomass of the forest area. In a study involving *Dalbergia sissoo* and *Acacia catechu* trees, which represented all three Eucalyptus hybrid trees, regression equations were developed using models. The above ground biomass (AGB), below ground biomass (BGB), and total tree biomass (AGB + BGB) of all the harvested sample trees were calculated, along with the diameter, height and weight (in kilograms) of each tree component. These plantations vary significantly in terms of age, density, width, height, and total biomass. The fluctuations in biomass are influenced by factors such as tree size, density (spacing), and the quantity of trees present. This approach involved comparing

Demystifying Sustainable Farming Using Smart Technologies.
Gunjan Chhabra, Keshav Kaushik, Priya Chugh, Susheela Dahiya, and Aryan Chaudhary (Eds.)

DOI: 10.1201/9781779640932-8

DBH values with the weights of tree components, from multiple felled trees, to refine the predictive models further.

8.1 GENERAL INTRODUCTION

The stratified mean-tree method is unquestionably the most effective for estimating the biomass of various tree components, but there are several circumstances in which the mean tree for calculating weight may not necessarily be the same as the mean tree for calculating mean diameter. For the prevention of difficulties occurring during cutting down standing green trees in tough locations, where the biomass calculation is more important, their regression equation is necessary. These developed regression equations were created in Punjab Forest of India, by utilizing hybrid plantings (forest and farmers' Agro-forestry plots) that could easily be measured simultaneously.

For agricultural forestry projects in Punjab, *Eucalyptus hybrid* (*E. hybrid*) has been selected as the species that is most suited. In 1966, this species was first seeded. The majority of the timber is used for pulpwood and as fuel. Throughout the 1980s, Punjab saw extensive plantings of eucalyptus trees along canal banks, road edges, and drainage systems. Dogra and Upadhyaya [3] state that the species has been shown to be suitable for planting in semiarid areas.

In the bulk of the Indo-Pakistani subcontinent, *Dalbergia sissoo* (Shisham/Tali) has been cultivated or has naturally grown for a very long time, notably in the Punjabi flood plains and irrigated plantations. New irrigated plantations are being created in Rajasthan, Punjab, and Haryana. Since plantation forestry was established in India, *D. sissoo* has been identified as a promising species to be planted under social forestry, agroforestry, farm forestry, as well as on canal and road borders. This plant has been used by farmers across India's forest divisions, including Punjab.

The past studies also reveal that the amount of biomass produced by agro-forestry species and its other traits have shown similarity, as studied by [5, 7–11]. Specifically for having the least forest cover in Punjab state, this ground-breaking important study involves determining biomass by removing *D. sissoo* trees with commercially demanded *Eucalyptus* and *Poplar* trees, from various places and developing prediction equations for later use to study their growth parameters.

The bulk of Punjab's state woods are confined to the regions close to canals, roads, railways, and sewers because of intensive farming. A single row, many rows, or even 15–30 rows might be used to plant trees. Plantations made of blocks are uncommon. *D. sissoo,* Kikar (*Acacia nilotica*), *E. hybrid*, and Khair (*Acacia catechu*) are the four species that are most common. Khair is cultivated in the sub-montane region of the districts of Hoshiarpur, Ropar, and Gurdaspur [2].

Plantation forests are being cultivated on a massive scale all over the world to meet the increasing need for timber, paper, pulp, and firewood. As more people become aware of the large range of items that may be obtained from forests, the management of such areas is being done with greater rigor. In order to make the most of our limited landmasses and to select appropriate places for commercial forestry, which would enable plantings to be concentrated in the most productive areas, an in-depth study on forest productivity is necessary. For planting reasons, quick-growing species like *D. sissoo* has been introduced, which has helped reduce the supply–demand gap for a number of key forest products.

To reduce the enormous stress upon natural forests inside forest land, "monocultures" of quickly growing, short-rotation forest species were selected for afforestation projects, namely, *Eucalyptus, Poplar, Melia* trees. However, this choice has generated a great deal of debate, particularly in relation to the monoculture eucalyptus farming. By removing allied plants with lesser timber values, other native tree species were also made into monocultures for increased wood production and were branded "kukat" (poor wood), which again caused a great deal of controversy. In India, there have been several studies on *Eucalyptus*, as well as *Poplars*, *D. sissoo*, and Khair. The trees biomass, productivity, and nutrient retention/loss from the sites of all of these commercially demanded species in Punjab have not been studied in the past by any researchers, which makes this study more important for future references.

Keeping the above view in mind, the present study was undertaken in three tree species which are grown as monoculture plantations in Punjab forests, namely, *E. hybrid, D. sissoo,* and *A. catechu* species plantations of different ages at different sites in 10 forest divisions of Punjab, namely Amritsar, Ludhiana, Hoshiarpur, Patiala, Muktsar, Jalandhar, Bhatinda, Ferozepur, Faridkot, Roopnagar/Ropar, representing three agro-climatic zones of Punjab. Map 8.1 [6] makes this point clearly evident.

8.2 STUDY SITES

The marked research trail study sites tree species were studied in all three agroclimatic zones of Punjab, which are depicted in Table 8.1a.

TABLE 8.1a Punjab Topography

Agroclimatic Regions	Climate	Districts Covered	Species Studied
Hot arid (2)	Western plains and Kutch peninsula	Ferozpur, Faridkot, Muktsar, Bhatinda	*E. hybrid* and *D. sissoo*
Hot semiarid (4)	Northern plains and central highlands	Amritsar, Jalandhar, Ludhiana, Patiala	*E. hybrid* and *D. sissoo*
Hot subhumid (9)	Northern plain	Hoshiarpur, Roopnagar (Ropar)	*E. hybrid, D. sissoo, A. catechu,* and *Populus* species

8.3 METHODOLOGY ADOPTED

8.3.1 SPECIES UNDER STUDY

8.3.1.1 EUCALYPTUS HYBRID

More than 3% of Punjab's forest, net transplanted area, covers *Eucalyptus* trees, according to estimates by [2]. Now *E. hybrid* has been transplanted as part of extensive planting and afforestation initiatives in almost every state in India, both to cover bare and deforested areas and to turn sluggish or neglected forest growth into fast-developing monocultures with high yields [4]. Throughout the 1980s, *E. hybrid* was widely planted on woodlands along roadsides, canals, and drains in Punjab barren lands [3] state that this tree species has shown their suitability for being plant in semiarid areas too.

8.3.1.2 D. SISSOO (SHISHAM/TALI)

Throughout the bulk of the Indo-Pakistani subcontinent, it has been grown or has naturally arisen for a very long time, notably in the Punjabi flood plains and irrigated plantations. New irrigated plantations are being created in Rajasthan, Punjab, and Haryana. *D. sissoo* is frequently used for afforestation, with the exception of the nation's coldest, wettest, and driest regions. *D. sissoo* is most likely endemic to the sub-Himalayan and bhabhar tracts. Since plantation forestry was introduced in India, *D. sissoo* has been identified as a promising species to be planted under social forestry, agroforestry, farm forestry, as well as on canal and road borders. Farmers are employed in a number of Indian states.

8.3.1.3 A. CATECHU (KHAIR)

Although khair is grown in many countries, including India, Nepal, Myanmar, Thailand, little is known about the entire life cycle of this species. There has been some study on "Katha" production and other Acacia species, but not much on the biomass, productivity, and nutrient dispersion of this species. Khair is cultivated in the submontane region of the Hoshiarpur, Ropar, and Gurdaspur districts, according to [2].

8.3.2 ESTIMATING TREES' DIAMETER AND HEIGHT CORRELATION AND SELECTION OF SAMPLE TREES FOR HARVESTING

Diameter and height measurements have long been regarded as the most reliable indicators of a tree's growth since there is always a high correlation between the diameter at breast height (DBH), height, and weight of a particular tree to determine its best growth parameters. Thus, it is necessary to measure the DBH and height of some of the sample trees in a research trail plot. A free hand curve or a regression of height and diameter may be used to calculate the remaining trees' heights of the same age groups/classes.

In this study, we collected field data on forest land and farmers agroforestry tree plantations using a tree technique method outlined by [1]. Sample plots of sizes (10 × 10 m, 15 × 15 m, 20 × 20 m, 25 × 25 m, 30 × 30 m, and 50 × 50 m) were established based on the area size in monoculture plantations of *E. hybrid*, *D. sissoo*, and *A. catechu*. During our fieldwork, we measured the DBH of each standing tree in the sample plots. Recorded the heights of trees to cover the full range of diameters in each plot. Subsequently, we analyzed the regression coefficient (R^2) values to understand the relationship between diameter and height.

Following the estimation process, the diameter range, as a whole, was segmented into diameter categories. Within each forestry and agricultural setting with agroforestry plantations, three representative trees from each diameter category close to the DBH of that category were cut down to assess biomass and for other purposes.

8.3.3 CALCULATIONS FOR TREE BIOMASS AND PRODUCTIVITY

The tree's leaves, twigs, branches, bark, and bole were promptly separated when it was chopped down, and their fresh present weights were noted in the field only. From each tree, sample of 100 g of fruit, bark, twigs, leaves, and other representative samples were taken for their laboratory chemical analysis of different macronutrient distribution and for oven dry weight estimation.

The tree trunk part of the sample trees was cut into 2-meter sections to simplify weighing. To determine the volume (including bark) and the fresh and dry weights of the trunk (up to a diameter of 5 cm above bark), a 5-cm-wide disc was removed from the base of each section for measurements. By using the diameter of two discs as a reference, the volume of each segment (with and without bark) was calculated. The volumes of each section were then combined to obtain the volume of the trunk (with and without bark) for each tree sampled.

With the exception of their little rootlets, all of the sample trees' root systems had been completely uncovered. To prevent any further weight loss, the dirt that was sticking to the roots was carefully removed, and new soil was immediately placed. In order to determine its mineral makeup and estimate its dry weight, a representative root sample was also taken for laboratory analysis.

The total biomass of the stand (in tons per hectare) was calculated by multiplying the number of trees in each size category by the weight of the sample trees and summing up the biomass within each size category.

To determine productivity, the overall biomass figures are divided by the number of trees per hectare at research sites. This helps calculate the yield per tree or the productivity per hectare per year for each type of tree, in both forest areas and farmers' agroforestry plantations.

In total, in the areas where the Punjab Forest Department had given permission to do so, we sawed down/felled (23) *E. hybrid* trees, (18) *D. sissoo* trees, and (18) *A. catechu* trees. Even if falling was required or wanted in other locations for better outcomes with different age classes, it was prohibited without authorization under forest laws.

8.3.4 DEVELOPING THE REGRESSION EQUATIONS FOR TREE BIOMASS ESTIMATION

To estimate the amount of wood and growth potential, in forest areas across a region, we have developed a series of equations that link observable characteristics like tree diameter and height with the weight of various parts of the tree (such as leaves, branches, trunk, bark, and roots). These predictive formulas aim to reduce the need for logging or destructive sampling in challenging terrain where such practices may not be practical or feasible, for various reasons.

After studying the data gathered from selected trees that were cut down, several regression equations were developed. These trees, from age groups, showed growth patterns based on their species and locations. The equations, like Log $y = a + \log b\, x$, were used to compare the D^2H values of each tree with the weight of tree parts, across all species that were cut down.

The regression constants are "*a*" and "*b*," where "*y*" is the biomass component given in kilograms and "*x*" stands for the diameter and height values, respectively, of tree. However, D^2H (diameter and height combined) was shown to be the best match for all tree species. The value of "*x*" was also examined for *DBH* and D^2. Regression equations were also created utilizing the 18 trees each for the *D. sissoo* and the *A. catechu* species, as well as the 23 trees that were felled for the *E. hybrid* species. The actual biomass of the sampled trees that were chopped down was compared to the expected biomass values of the trees from each location, each having a different diameter and height.

8.4 OBSERVATIONS

TABLE 8.1b Comparison in Log (D^2H and Total Biomass) for *E. hybrid* Six Trees

$r^2 = 0.9621$									Antilog Predicted Value	Actual value
Tree No.	**Dia**	**Ht.**	**D^2H**	**X log (D2H)**	**Predicted Equation**	***b***	***a***	**Estimated Log Value**	**Biomass (kg/tree)**	**Biomass (kg/trees)**
12	17.6	22.7	7031.552	3.85	$Y = 0.9463\,X - 1.3045$	0.9463	1.3045	2.34	218.78	238.35
13	26.8	29.2	20,972.608	4.32	$Y = 0.9463\,X - 1.3045$	0.9463	1.3045	2.79	616.59	701.35
14	37.7	26.2	37,237.798	4.57	$Y = 0.9463\,X - 1.3045$	0.9463	1.3045	3.02	1047.13	638.32
15	12.7	17.9	2887.091	3.46	$Y = 0.9463\,X - 1.3045$	0.9463	1.3045	1.97	93.32	90.24
16	17.7	22.5	7049.025	3.85	$Y = 0.9463\,X - 1.3045$	0.9463	1.3045	2.34	218.78	213.14
17	31.5	23.9	23,714.775	4.38	$Y = 0.9463\,X - 1.3045$	0.9463	1.3045	2.84	691.83	388.12

Comparison in Log (*D* and Total Biomass) for *E. hybrid* 6 Trees

$r^2 = 0.9329$							Antilog Predicted	Actual Value
Tree No.	**Dia**	**X log (D)**	**Predicted Equation**	***b***	***a***	**Estimated Log Value**	**Value Biomass (kg/tree)**	**Biomass (kg/trees)**
12	17.6	1.25	$Y = 2.6477\,X - 0.9914$	2.6477	0.9914	2.31	204.17	238.35
13	26.8	1.43	$Y = 2.6477\,X - 0.9914$	2.6477	0.9914	2.79	616.59	701.35
14	37.7	1.58	$Y = 2.6477\,X - 0.9914$	2.6477	0.9914	3.18	1513.56	638.32
15	12.7	1.10	$Y = 2.6477\,X - 0.9914$	2.6477	0.9914	1.93	85.11	90.24
16	17.7	1.25	$Y = 2.6477\,X - 0.9914$	2.6477	0.9914	2.31	204.17	213.14
17	31.5	1.50	$Y = 2.6477\,X - 0.9914$	2.6477	0.9914	2.98	954.99	388.12

Comparison in Log (D^2 and Total Biomass) for *E. hybrid* Six Trees

TABLE 8.1b *(Continued)*

$r^2 = 0.9329$								Antilog Predicted	Actual Value
Tree No.	Dia	Dia.²	X log (D^2)	Predicted Equation	b	a	Estimated Log Value	Value Biomass (kg/tree)	Biomass (kg/trees)
12	17.6	309.76	2.49	$Y = 1.3239X - 0.9914$	1.3239	0.9914	2.31	204.17	238.35
13	26.8	718.24	2.86	$Y = 1.3239X - 0.9914$	1.3239	0.9914	2.79	616.59	701.35
14	37.7	1421.29	3.15	$Y = 1.3239X - 0.9914$	1.3239	0.9914	3.18	1513.56	638.32
15	12.7	161.29	2.21	$Y = 1.3239X - 0.9914$	1.3239	0.9914	1.93	85.11	90.24
16	17.7	313.29	2.50	$Y = 1.3239X - 0.9914$	1.3239	0.9914	2.31	204.17	213.14
17	31.5	992.25	3.00	$Y = 1.3239X - 0.9914$	1.3239	0.9914	2.98	954.99	388.12

Comparison in Log (D and Total Biomass) for *D. sissoo* 6 Trees

$r^2 = 0.8941$							Antilog Predicted	Actual Value
Tree No.	Dia	X log(*D*)	Predicted Equation	b	a	Estimated Log Value	Value Biomass (kg/tree)	Biomass (kg/trees)
19	31.8	1.50	$Y = 2.3884X - 0.6985$	2.3884	0.6985	2.89	776.25	876.76
20	36.6	1.56	$Y = 2.3884X - 0.6985$	2.3884	0.6985	3.04	1096.48	963.64
21	40.1	1.60	$Y = 2.3884X - 0.6985$	2.3884	0.6985	3.13	1348.96	1547.35
22	16	1.20	$Y = 2.3884X - 0.6985$	2.3884	0.6985	2.18	151.36	102.23
23	23.1	1.36	$Y = 2.3884X - 0.6985$	2.3884	0.6985	2.56	363.08	464.36
24	42.3	1.63	$Y = 2.3884X - 0.6985$	2.3884	0.6985	3.19	1548.82	1226.94

Comparison in Log (D^2 and Total Biomass) for *D. sissoo* Six Trees

$r^2 = 0.8941$								Antilog Predicted	Actual Value
Tree No.	Dia	D^2	X log (D^2)	Predicted Equation	b	a	Estimated Log Value	Value Biomass (kg/tree)	Biomass (kg/trees)
19	31.8	1011.24	3.00	$Y = 1.1942X - 0.6985$	1.1942	0.6985	2.89	776.25	876.76
20	36.6	1339.56	3.13	$Y = 1.1942X - 0.6985$	1.1942	0.6985	3.04	1096.48	963.64
21	40.1	1608.01	3.21	$Y = 1.1942X - 0.6985$	1.1942	0.6985	3.13	1348.96	1547.35
22	16	256	2.41	$Y = 1.1942X - 0.6985$	1.1942	0.6985	2.18	151.36	102.23
23	23.1	533.61	2.73	$Y = 1.1942X - 0.6985$	1.1942	0.6985	2.56	363.08	464.36
24	42.3	1789.29	3.25	$Y = 1.1942X - 0.6985$	1.1942	0.6985	3.19	1548.82	1226.94

Comparison in Log (D^2 and Total Biomass) for *A. catechu* 6 Trees

$r^2 = 0.9486$								Antilog Predicted Value	Actual Value
Tree No.	Dia	D^2	X log (D^2)	Predicted Equation	b	a	Estimated Log Value	Biomass (kg/tree)	Biomass (kg/trees)
13	12.7	161.29	2.21	$Y = 1.1298X - 0.5476$	1.1298	0.5476	1.95	89.13	81.53
14	17.2	295.84	2.47	$Y = 1.1298X - 0.5476$	1.1298	0.5476	2.24	173.78	175.11
15	20	400	2.60	$Y = 1.1298X - 0.5476$	1.1298	0.5476	2.39	245.47	189.27
16	16.2	262.44	2.42	$Y = 1.1298X - 0.5476$	1.1298	0.5476	2.19	154.88	157.2
17	24.5	600.25	2.78	$Y = 1.1298X - 0.5476$	1.1298	0.5476	2.59	389.05	518.46
18	32.1	1030.41	3.01	$Y = 1.1298X - 0.5476$	1.1298	0.5476	2.86	724.44	732.16

8.5 RESULTS AND DISCUSSIONS

8.5.1 BIOMASS STUDIES

8.5.1.1 E. HYBRID SPECIES

We examined the size of Eucalyptus trees by studying how their diameter and height are related in forest locations such as Nara, Kharkan, Dholbah, Katour (Hoshiarpur Forest Division) Doraha, Ludhiana Reserve Forest (Ludhiana For Div), Kamalpur (Amritsar For Div) Phillaur, Talwandi Bhai (Jalandhar For Div), Mukt Doda Sota Road, Faridkot Nabha Bir Mehas (Muktsar For Div), and Thapar Engineering College (Patiala Forest Div.). The correlations for each site are presented in Table 8.2. The strongest correlation between diameter and height ($r^2 = 0.85$) was found in Dholbah Dehra (Hoshiarpur For Div), indicating a relationship between these two factors. Conversely, Muktsar Doda Sota Road (Muktsar for Div) showed a correlation value of $r^2 = 0.20$, suggesting a link between diameter and height, at that specific location.

TABLE 8.2 Comparison in Log (D^2H and Total Biomass) for *D. sissoo* Six Trees

$r^2 = 0.9118$									Antilog	
Tree No.	Dia	Ht.	D^2H	*X* log (D^2H)	Predicted Equation	*b*	*a*	Estimated Log Value	Predicted Value Biomass (kg/tree)	Actual Value Biomass (kg/trees)
19	31.8	28.2	28,516.97	4.46	$Y = 0.9188X - 1.0399$	0.9188	1.0399	3.05	1122.02	876.76
20	36.6	29	38,847.24	4.59	$Y = 0.9188X - 1.0399$	0.9188	1.0399	3.18	1513.56	963.64
21	40.1	28.9	46,471.49	4.67	$Y = 0.9188X - 1.0399$	0.9188	1.0399	3.25	1778.28	1547.35
22	16	12.8	3276.8	3.52	$Y = 0.9188X - 1.0399$	0.9188	1.0399	2.19	154.88	102.23
23	23.1	18.9	10,085.23	4.00	$Y = 0.9188X - 1.0399$	0.9188	1.0399	2.64	436.52	464.36
24	42.3	22.1	39,543.31	4.60	$Y = 0.9188X - 1.0399$	0.9188	1.0399	3.18	1513.56	1226.94

8.5.1.1.1 Tree Biomass Estimation

Table 8.1 presents a range of trees harvested from forest divisions detailing the diameter, height, and total biomass, per tree in kilograms. The table illustrates variations in age (between 18 and 30 years) density (232–740 trees per hectare), diameter (12.7–49.75 centimeters), height (14.4–41.5 meters), and total biomass (112.98–531.09 t h^{-1}) within the plantations. The differences in biomass are influenced by factors such as tree size, density, the number of transplanted trees per plot, and variations in soil conditions. These observations underscore a connection among all the mentioned parameters.

8.5.1.1.2 Developed Prediction Equations

Prediction equations have been created for the various tree components based on their diameter and height (D^2H), which was demonstrated to be the best option for a dependent variable over D and D^2, based on the sampled trees that were felled down for *E. hybrid.*

The results fo r above ground biomass (AGB), below ground biomass (BGB), and total biomass (AGB + BGB) are particularly substantial; however, all tree component correlation coefficient (r^2) values are most significant. These equations have been demonstrated to be accurate by comparing the expected values of the AGB + BGB of the fallen trees with their actual/calculated biomass as shown in Table 8.1.

Table 8.1 makes it abundantly clear that the biomass of smaller tree diameters differs from that of larger diameters. For sample trees with mean diameters ranging from 13.60 to 35 cm and belonging to different diameter classes, the differences between expected and actual biomass varied from 6.8 to 38.5%. Only five trees showed the smallest percentage variances (6.82–29.5). Only one tree, with a mean diameter of 28.80, had a difference of more than 30%. In general, differences between estimated and actual biomass of 10% to 30% are far acceptable in every aspect.

The D^2H prediction relationship to numerous tree components clearly demonstrates the linear connection between the two parameters, proving their strong linkage. Table 8.1 contains the prediction equations that were thus created and may be used to estimate the biomass of *E. hybrid* plants that are particular to the Punjab. The diameter classes for these studies need to be created with greater attention so that each class may contain more trees. The plot's very few trees that have diameters that are both exceedingly enormous and incredibly small need to be removed.

8.5.1.2 D. SISSOO (SHISHAM)

Various locations such as Nara, Dhirowal, Chaksadhu, Kharkan in the Hoshiarpur Forest Division, Mattewara in the Ludhiana Forest Division, Ajnala and Kamalpur in the Amritsar Forest Division, Bir Dialpura, Bir Talab, Bathinda (MSA) in the Bathinda Forest Division, and Dharamkot (Mixed I), Dharamkot (Mixed II), and Dharamkot (Pure) in the Ferozepur Forest Division were considered for height and diameter correlations. Additionally, sites like Bathinda Muktsar Road, Pathankot Abohar Road, and Barnala Bathinda Road in the Muktsar Forest Division as Bir Mehas and Bir Dosanjh in the Patiala Forest Division were analyzed.

Strong correlations were observed at Dharamkot Mixed I ($r^2 = 0.88$) and Dharamkot Mixed II ($r^2 = 0.82$) sites within the Ferozepur Forest Division. In contrast, the correlation was weakest at Barnala Bathinda Road in Muktsar Forest Division with an r^2 value of 0.05. Two other sites Bir Dialpura (Bathinda Forest Division) and Pathankot. Abohar Road (Muktsar Forest Division) exhibited a lower correlation coefficient of $r^2 = 0.19$, indicating that they are favorable locations for such analysis. Notably, roadside plantings generally displayed correlations where an increase in diameter did not consistently correspond to a rise in height.

8.5.1.2.1 Biomass

In total, 18 sample trees representing the average trees of the various diameter classes (ranging in size from 13.6 to 42.3 cm) were chosen and harvested in accordance with the previously specified technique in order to calculate the actual biomass. Sample trees from various forest divisions have been shown in Table 8.2. You can find the size, height and weight, in kilograms, of every part of the trees, including AGB, BGB, and the AGB + BGB per tree.

Table 8.2 shows that plantation sites exhibit a wide range of variance in terms of their age groups (25–33 years), density (196–453 trees per hectare), mean diameter (13.1–43.1 cm), height (10.95–22.1 m), and biomass (83.56–1226.94 kg $tree^{-1}$). Between 85.42 and 232 t ha^{-1}, there was AGB + BGB. The size of the tree, the density (spacing)/number of trees per plot, and the fluctuation of geographical parameters/soil health/moisture availability are some of the variables that affect biomass.

8.5.1.2.2 Prediction Equations

Prediction equations were created using the actual and estimated biomass of the 12 trees that were felled (Table 8.2). It was discovered that D^2H was a better suitable independent variable and the actual biomass of each component was a more suitable dependent variable. These equations have been confirmed by comparing the predicted biomass of six trees with the actual biomass of each plant part. These six trees represent all the diameter classes of the research plots, with mean diameters ranging from 13.10 to 42.30 cm. As can be shown, the biomass of five trees ranges between 0.54% and 27.82% for both actual and projected values.

The only one tree with a difference in biomass of 48.03% had the lowest diameter of 13.10 cm. The estimated or actual biomass of 18 trees that have been felled is also used to generate prediction equations, producing more significant correlation coefficient values than the equation based on 12 trees. Plotted regression graphs for each of the tree's components show unequivocally that there are linear correlations present in all cases. It is advantageous to use these formulas as a guide when estimating Punjabi biomass in the future.

8.5.1.3 A. CATECHU (KHAIR)

In Punjab, khair plantations are not very common, similar to *D. sissoo* and *Eucalyptus*. The Hoshiarpur and Roopnagar Forest Divisions are where you may find the majority of the plantations. In each of these forest divisions, block plantings with genuine biomass of all tree components have been selected for harvesting. A pair of sample plots measuring 30 × 30 and 20 × 20 m were included at each of the six research sites. Each tree's height and CBH (diameter at breast height) were also measured.

A correlation between height and diameter was discovered everywhere. No significant correlation was found, and their r^2 values ranged from 0.11 to 0.22 at Roopnagar and 0.38 at all locations in Hoshiarpur, showing that the height of these trees does not increase proportionally to their diameter.

8.5.1.3.1 Biomass and Productivity

There have been a total of 18 trees cut down throughout both forest divisions, including 9 from the Hoshiarpur Forest Division and 9 from the Ropar Forest Division. The data on the trees that were removed from these six locations are all listed in Table 8.3, including their age, which varied from 20 to 32 years, the mean diameter of 18 sample trees of each diameter class, which ranged from 10 to 33 cm, and height, which ranged from 7.4 to 17.3 m.

TABLE 8.3 Comparison in Log (D^2H and Total Biomass) for *A. catechu* 6 Trees

$r^2 = 0.9534$										
Tree No.	**Dia**	**Ht.**	**D^2H**	**X log (D^2H)**	**Predicted Equation**	***b***	***a***	**Estimated Log Value**	**Antilog Predicted Value Biomass (kg/tree)**	**Actual Value Biomass (kg/trees)**
13	12.7	9.7	1564.513	3.19	$Y = 0.8261\,X - 0.6923$	0.8261	0.6923	1.95	89.13	81.53
14	17.2	10.7	3165.488	3.50	$Y = 0.8261\,X - 0.6923$	0.8261	0.6923	2.20	158.49	175.11
15	20	10.8	4320	3.64	$Y = 0.8261\,X - 0.6923$	0.8261	0.6923	2.31	204.18	189.27
16	16.2	12.1	3175.524	3.50	$Y = 0.8261\,X - 0.6923$	0.8261	0.6923	2.20	158.49	157.2
17	24.5	14.9	8943.725	3.95	$Y = 0.8261\,X - 0.6923$	0.8261	0.6923	2.57	371.54	518.46
18	32.1	14.8	15,250.07	4.18	$Y = 0.8261\,X - 0.6923$	0.8261	0.6923	2.76	575.44	732.16

The tree biomass varied from 39.4 to 738.98 kg tree^{-1} at each research trail site. The density/number of trees per stand varied from 415 to 640 t ha^{-1}, and the biomass ranged from 77.29 to 223.46 t ha^{-1} (Table 8.3). At Baruhi–Dholbaha, site productivity ranged from 2.57 to 9.31 t ha^{-1} yr^{-1}.

8.5.1.3.2 Developed Prediction Equations

These prediction equations were made using data from 12 trees that were chopped down (Table 8.3). To verify these equations, the estimated biomass of six trees is compared to their actual and calculated biomass (Table 8.3). With discrepancies between actual and projected biomass ranging from 0.81% to 28.45%, there was a strong correlation between D^2H and the overall biomass of Khair trees in Punjab.

The association between D^2H and numerous tree components has been identified, and this connection blatantly illustrates a linear relationship with substantial correlation in each tree component. Prediction equations have also been created based on the 18 trees that were felled (Table 8.3), and they show a strong correlation with D^2H levels. They might be used in future studies of a similar nature. In Punjab, no prediction equations for the Khair tree have ever been developed; therefore, this work is ground-breaking and will help with the biomass evaluation of Khair trees (without falling).

KEYWORDS

- **biomass**
- **diameter at breast height**
- **regression equation**
- **forest productivity**

REFERENCES

1. Art, H. W., & Marks, P. L. (1971). A summary table of biomass and net primary production in forest ecosystems of the world. In: Young, H. E. (Ed.), *Forest Biomass Studies (College of Life Sciences and Agricultural Experiment Station)*. University of Maine: USA.
2. Dogra, A. S., & Sandhu, S. S. (1984). Role of *Eucalyptus* in the economy of Punjab. In: *Eucalyptus in India Past, Present and Future.* Kerala Forest Institute, Kerala.
3. Dogra, A. S., & Upadhyaya, A. (2005). Site index equation for euca*l*yptus hybrid in Punjab. *Indian Forester, 131*(4), 487–498.
4. Qureshi, I. M. (1966) The so-called *eucalyptus* hybrid of Mysore origin. *Indian Forester*, *92*(4), 213–217.
5. Rajvanshi, R., & Gupt, S. R. (1985). Productivity and litterfall in a tropical *D. sissoo Roxb*. *Forest Journal of Tree Science, 4*(2), 73–78.
6. Sehgal, J. L., Mandal, D. K., & Vadivelu, S. , (1990). Agro-ecological regions of India. *Technical Bulletin. NBSS Publ. 24*. (p. 73). National Bureau of Soil Survey and Land Use Planning (ICAR) Nagpur, India.
7. Bisht, S. (2000). Comparative studies on biomass, productivity and mineral cycling in plantation ecosystems of *Dalbergia sissoo* and mixed crops. PhD Thesis (p. 198). FRI Deemed University: Dehradun,.
8. Gupta, S.R., & Rajvanshi, R. (1991). The effect of biotic & abiotic factors on litter decomposition in a tropical D. Sissoo forest ecosystem. In: Verresh, G. K., Rajgopal, D., & Virakta math, C. A. J. (Eds.), *Advances in Management & Conservation of soil fauna. Proceedings of 10th International Soil Zoology Colloinium* (pp. 491–495).
9. Sharma, D. C., Taneja, P. L., & Bisht, A. P. S. (1988). Biomass, productivity and nutrient cycling in a *Dalbergiasissoo* plantation. *Indian Forester, 1114*(5), 261–268.
10. Pal, M. P., & RaturiD P. (1990). Biomass production and organic matter distribution partitioning in *dalbergiasissoo* Roxb. Grown in an energy plantation. *Journal of Tropical Forestry, 6*(11), 120–124.
11. ingh, O., Sharma, D. C., Negi, M. S., & Singh, R. (1994). leaf litter production and decomposition in *D. Sissoo* and *Bombax Ceiba* plantations of Uttar Pradesh. *Indian Forester*, *120*(8), 682–688.

CHAPTER 9

Data-Driven Approaches for Sustainable Farming: Decision Support System

KHUSHBOO JAIN[1], KAMLESH KUMAR RAGHUVANSHI[2], ARUN AGARWAL[2*], and SUBODH KUMAR[2]

[1]*UPES, Dehradun, Uttarakhand, India*

[2]*Ramanujan College, Delhi University, New Delhi, Delhi, India*

*Corresponding author

ABSTRACT

Sustainability within food and fiber farming necessitates a robust integration of knowledge and technology. This chapter underscores the efficacy of leveraging available resources, including data-driven agricultural technology and precision farming methods, over-sole reliance on human observation and experience. By harnessing data from wireless sensor networks and applying artificial intelligence and machine learning techniques, farmers can better comprehend the intricate interplay of natural systems, essential for achieving sustainability and ensuring food security. In the pursuit of sustainable farming practices, this chapter outlines key research domains aimed at addressing decision support systems, critical legal, technical, economic, and social challenges. These include the development of holistic decision-making systems, automated animal intake measurement, low-cost environmental sensors, robot obstacle avoidance, and the integration of remote sensing with crop and pasture models. The integration of emerging technologies is extending the reach of precision farming, presenting opportunities to enhance it further through computer-assisted decision support systems for farm management. Moreover, the research underscores the significance of innovative business models for facilitating data sharing and implementing data regulations that foster trust within the agricultural community. Public funding is identified as imperative for advancing research initiatives, thereby fostering sustainable farming practices and promoting innovation within the agricultural sector.

9.1 INTRODUCTION: BACKGROUND AND DRIVING FORCES

The Food and Agriculture Organization (FAO), dedicated to bolstering global food security, articulates a resolute vision for sustainable food and farming [1]. Central to this vision

Demystifying Sustainable Farming Using Smart Technologies.
Gunjan Chhabra, Keshav Kaushik, Priya Chugh, Susheela Dahiya, and Aryan Chaudhary (Eds.)

DOI: 10.1201/9781779640932-9

is the principle that food should be both nourishing and available to all, while concurrently ensuring that natural resources are accomplished in a manner that preserves environment functions to meet present and future human needs [2]. The FAO's sustainability principles for food and farming encompass a spectrum of objectives, including the enhancement of productivity, employment, and value addition within food systems, the conservation and enrichment of natural resources, and the advancement of livelihoods and inclusive economic growth [3].

However, despite the multifaceted and dynamic nature of sustainability, the terms "sustainable farming" and "regenerative farming" are often interpreted by many as a return to traditional agricultural methods [4]. Conversely, the emergence and proliferation of digital farming technologies represent a burgeoning trend, fueled by the increasing enthusiasm among a new generation of farmers to embrace digital solutions [5].

Today, precision farming is widespread, leveraging computer-assisted decision support systems (DSS) to optimize resource use. However, despite abundant data collected by wireless sensor networks (WSN), analysis often remains basic, missing opportunities offered by advanced data mining techniques [6, 7]. Bridging this gap requires investment in expertise and infrastructure to unlock insights for enhancing agricultural productivity and sustainability.

This posits that the pursuit of sustainability in food and fiber farming necessitates the comprehensive utilization of all available resources, technology, and knowledge, including data-driven sustainable and precision farming methods [8]. Notably, the integration of data gathered by WSN and processed through artificial intelligence (AI) and machine learning (ML) algorithms offers farmers invaluable guidance in the precise and rational application of external inputs, such as water, fertilizer, and pesticides for crops, as well as nutrients and medication for livestock [9, 10]. Moreover, these technologies facilitate the exploration of synergies between natural systems, pivotal for achieving concurrent sustainability and food security. These synergies encompass intricate interactions between plants, the environment, beneficial insects and fungi, grazing animals, nutrient cycling, and soil and crop health [11].

This communication serves to distill the core tenets of sustainable farming, elucidate the transformative potential of data-driven farming in aligning with these principles, delineate prevalent constraints and challenges hindering the adoption of data-driven farm technologies for sustainability, and delineate priority research areas to surmount these challenges [12–14].

Figure 9.1 visually illustrates the evolution and transformation of research focus on data-driven approaches for sustainable farming. Academic literature in this field exhibited consistent yet gradual growth from 2000 to 2015. However, there was a noticeable surge in interest following the impact of DSS on sustainable farming practices, resulting in a gradual increase in the trend from 2018 onwards. Notably, there has been an approximately 300% rise in the number of Scopus-indexed articles and reviews published between 2019 and 2023. This significant growth underscores a marked scholarly inclination toward exploring the applications of data-driven approaches for sustainable farming.

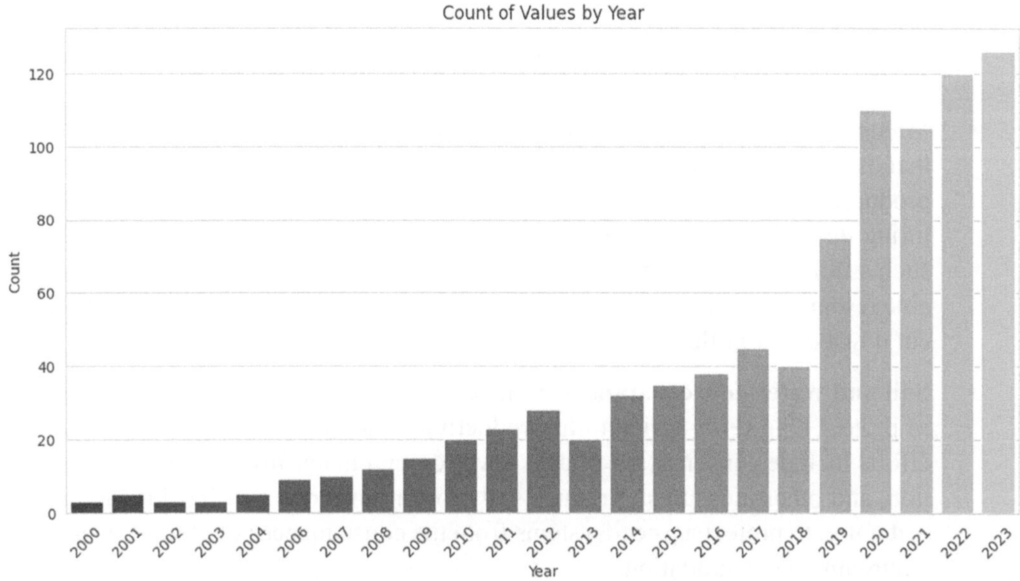

FIGURE 9.1 The evolution and transformation of research on data-driven approaches for sustainable farming.

Moreover, Figure 9.2 illustrates the anticipated impact of public funding directed toward research endeavors addressing these high-yield topics, envisaged to surmount barriers and catalyze the adoption of data-driven approaches for sustainable farming. This will resonate with proponents of sustainable farming across diverse spheres, including agricultural researchers and policymakers, offering insights and perspectives to inform and invigorate the discourse surrounding sustainable farming in the digital era.

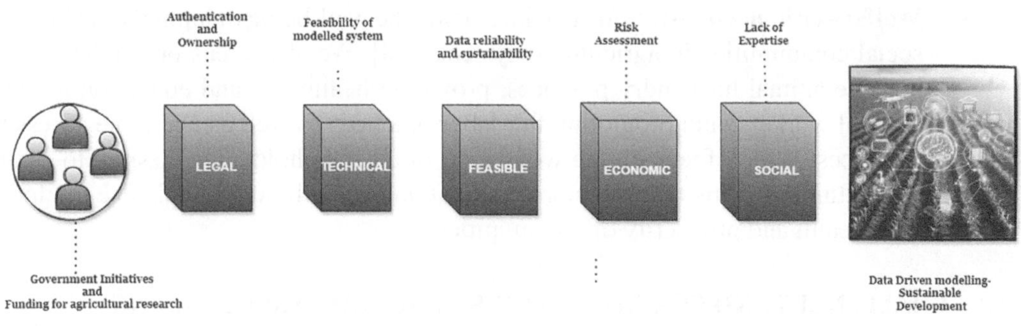

FIGURE 9.2 Unlocking data-driven sustainable agriculture requires substantial public investment in targeted research areas. Funding critical research is expected to yield significant benefits.

9.2 KEY STRATEGIES FOR SUSTAINABLE FARMING

The key strategies for sustainable farming are as follows:

- **Reduced inputs:** Emphasize the reduction of external inputs, such as pesticides, fertilizers, water, and energy to cultivate a more sustainable agricultural system [1, 2, 15]. By employing techniques like integrated pest management, organic farming practices, and precision irrigation, we aim to minimize the ecological footprint of farming while maintaining productivity.

- **Waste recycling:** This might be implemented by enforcing a "zero waste" policy that is targeted at recycling all waste that is organic materials, which can be produced in the agricultural process. Composting programs, bioenergy manufacturing through the use of agricultural residues, and advanced level technologies, such as for instance anaerobic digestion, can really help transform waste into valuable resources. This will likely assist in closing loops along the nutrient rounds and cut down on air pollution.
- **Soil and water conservation:** Prioritize conservation of soil and water resources to protect long-term agricultural productivity and ecosystem health [3, 16]. Our efforts include conservation tillage, cover cropping, agroforestry, precision irrigation, and other measures to reduce soil erosion, improve soil fertility, and reduce water use, to protect agroecosystems from the consequences of climate change and environmental degradation.
- **Biodiversity promotion:** Fostering the conservation and promotion of biodiversity in agricultural landscapes to underpin ecosystem robustness and essential ecological services [1, 7, 17]. This would be done partly by utilizing crop rotation, putting together diversified cropping, enabling actively managed habitats, and establishing links to the preserves framework and species-rich conservation initiatives. Ultimately, they would become a healthy, active, and diverse habitat for the indigenous vegetation and animals and would be as sturdy as possible to pests and diseases.
- **Welfare enhancement:** Prioritize increasing the well-being of animals and human/social communities in agricultural systems [18]. We also focus on implementing humane animal husbandry practices, providing healthcare and education to agricultural workers, and promoting fair labor practices as well as equitable access to resources. With a focus on the well-being of all stakeholders, we aspire to create agricultural systems that are more resilient and socially just to enhance the long-term health and prosperity of communities.

9.3 CHALLENGES AND CONSIDERATIONS OF SUSTAINABLE FARMING

The challenges and considerations of sustainable farming are as follows:

- **Resource trade-offs:** In farming, there is a complicated balance between things like land, workers, and money. Figuring out how to use these resources involves considering things like what people want to buy, new technology, and rules from the government [19]. For example, deciding how to use land might mean choosing between saving nature areas, growing more crops, or building cities. And when it comes to hiring workers and spending money, farmers have to think about things like how skilled the workers are, what is the best value for their money, and if they are doing things in a way that is good for the environment. Finding the right balance helps farms produce a lot while causing as little harm to the environment and society as possible.

- **External influences:** Farms face a lot of challenges from outside factors like changing weather, market prices, and government rules. Bad weather can mess up crops and make it hard to keep food supplies steady. Prices for things farmers need, like seeds and equipment, can go up and down, affecting how much money they make. Also, new rules about things like protecting the environment or keeping food safe can make farming more complicated. To deal with these challenges, farmers have to be flexible, find ways to reduce risks, and work together with others to make sure farms can keep going strong in the long run.
- **Monitoring and evaluation:** It is hard to measure how well farms are doing in terms of being sustainable because some important things change very slowly, like the health of the soil, the variety of plants and animals, and how clean the water is. Also, it is tricky to figure out how different farming methods affect the environment because nature is really complicated, and everything is connected. To understand what is going on and how to make things better, we need to use new technologies like satellites and WSNs and get different experts to work together. By keeping an eye on how farms are doing and trying out new ideas, we can make sure farming is good for the environment and can keep going strong for a long time.

9.4 ADVANTAGES OF UTILIZING DATA-DRIVEN FARMING TO EMBRACE THE TENETS OF SUSTAINABLE FARMING

Agricultural data encompasses information vital to farming practices, farm economics, and environmental impacts. This includes data of varying scales, from small datasets to "Big Data," exceeding traditional processing capacities. Collection methods range from remote sensing imaging to sensor networks and farm documentation. A data-driven approach to sustainable farming leverages available knowledge, technology, and resources to address complex environmental, social, and economic challenges [20]. Data-driven farming facilitates efficient risk management and decision-making throughout the food and fiber production systems (FFPS). Through the adoption of predictive and prescriptive management approaches, it enhances accuracy and has the potential to foster sustainable FFPS.

Data-driven approaches hold significant potential to bolster the sustainability of FFPS across four key areas:

- **Automation of data collection:** This involves developing and deploying practical robotic systems, field and animal sensors, and enhancing aerial communication network for earth observation [21, 22]. Also enables the acquisition of accurate, precise, and high-quality data to inform decision-making.
- **Big data processing and AI and ML integration:** The integration of AI and ML and deep learning techniques into farming focuses on developing learning systems and algorithms to study specific phenomena, revolutionizing scientific

domains [23]. This can be applied in computer vision object recognition, recommendation systems, data mining, informatics, and autonomous control for farming.

- **Development of human–computer interfaces:** This can improve accessibility of insights through text, image, and voice interfaces, aiding farmers in decision support [24]. The challenges in implementing data-driven sustainable farming are complexities in farming data, including volume, variety, velocity, and veracity. Various studies have highlighted challenges in implementing data-driven farming approaches.
- **Role of data curation in farm management:** It serves as the farm's "institutional memory," retaining implicit knowledge from previous decisions [25]. This is essential for retaining generational knowledge as traditional farmers reach retirement, ensuring continuity and sustainability in farming practices.

The widespread integration of data in farming holds the potential to revolutionize farming techniques and lead to a transformative shift in academic research and knowledge dissemination. This way of doing things helps experts from different areas work together to understand complicated connections and changes over space and time. But, it is tough to collect the right information needed for specific questions. We need to focus on getting lots of detailed information, using the same methods, and making it easy for everyone to access.

9.5 METHODS IN DECISION SUPPORT SYSTEMS (DSS)

DSS for sustainable farming uses different ways to help farmers make smart and eco-friendly choices. Some important methods include:

- **Data aggregation:** Data collecting and aggregation from various sources such as weather stations, soil sensors, satellite imagery, and market databases to give farmers all the info they need to make good decisions [26].
- **Modeling and simulation:** Developing simulation programs and mathematical models to forecast crop growth, soil conditions, pest and disease outbreaks, and other variables influencing agricultural sustainability and production [27].
- **Optimization techniques:** Employing optimization algorithms to determine the most efficient allocation of resources such as fertilizers, water, and pest controls to maximize yields while minimizing environmental impacts [28].
- **Risk assessment and management:** Evaluating risks related to factors such as changing weather, market ups and downs, pests, and diseases, then giving advice on how to reduce those risks [29].
- **Scenario analysis:** Looking at different situations and how they might affect, how well a farm does, how much money it makes, and if it is good for the environment. This helps farmers choose the best ways to manage their farms [30].
- **Expert systems:** Integrating expert knowledge and heuristic methods into DSS to provide recommendations based on best practices, local conditions, and farmer preferences [31].

- **Visualization and communication:** Presenting data and information along with recommendations in a user-friendly format, such as interactive dashboards, maps, and charts, to facilitate understanding and communication among stakeholders [32].

By employing these methods, DSS for sustainable farming empowers farmers to adopt practices that optimize productivity, minimize environmental impacts, and enhance long-term sustainability. Table 9.1 presents an overview of various DSS models in data-driven approaches for sustainable farming.

TABLE 9.1 Illustrated the Different DSS Models in Data-Driven Approaches for Sustainable Farming

S. No.	DSS Model	Reference	Benefits	Limitations
1	Probabilistic and optimization models	[33]	• Autonomous software optimizes energy distribution • ML enhances decision-making in energy networks • Real-time ML applications improve energy efficiency • ML aids core energy technologies and distribution utilities • ML optimizes energy efficiency and integrates renewables • ML enables accurate energy forecasting and cybersecurity enhancement	• Dependency on amount and quality of data • Complexity in implementation and maintenance • Resource-intensive training • Interpretability challenges in ML models • Security risks, including adversarial attacks • Regulatory hurdles and initial costs
2	Supervised learning models	[34]	• High model accuracy enhances reliability • Identification of significant environmental contributors aids targeted interventions • Clustering identifies sectors for focused sustainability efforts • Specific sectors like bakery manufacturing offer focal points for improvement	• Potential bias or error in the prediction model • Limited representation of socioeconomic impacts • Interpretation challenges with high variance • Focus on environmental impacts may overlook social or economic factors
3	Time series analysis	[35]	• Addressing global challenges with detailed agricultural monitoring • Yearly national maps facilitate sustainable agriculture planning • Proposed workflow adapts to diverse environmental conditions • Multisensor approach enhances accuracy and spatial consistency	• Potential challenges in data integration and processing • Dependence on satellite coverage and data availability • Interpretation complexity with diverse crop types • Low scalability

TABLE 9.1 *(Continued)*

S. No.	DSS Model	Reference	Benefits	Limitations
4	Genetical programming	[36]	• IoT aids precision agriculture in rural areas • Cheap sensors and drones solve real-time agricultural problems • IoT-based precision agriculture enhances water management and crop yields • Genetic algorithm predicts rainfall, optimizing water supply decisions • Sensor-based system validates rainfall prediction, enabling timely interventions • Enhanced efficiency with integrated atmospheric parameters	• Reliance on sensor accuracy for decision-making • Dependency on consistent data connectivity in rural areas • Implementation complexity and maintenance requirements • Limited scalability and interoperability of the proposed system • Challenges in integrating UAVs for widespread adoption • High cost for small-scale farmers

9.6 OBSTACLES AND LIMITATIONS IN HARNESSING DATA-DRIVEN FARMING FOR SUSTAINABLE PRACTICES

Although the data-driven approach has the potential to do the best in permaculture, it faces both legal and competitive, economic, and social challenges. These problems hinder information sharing and limit the many benefits it can provide. Figure 9.3 illustrates the limitations of using data-driven permaculture practices.

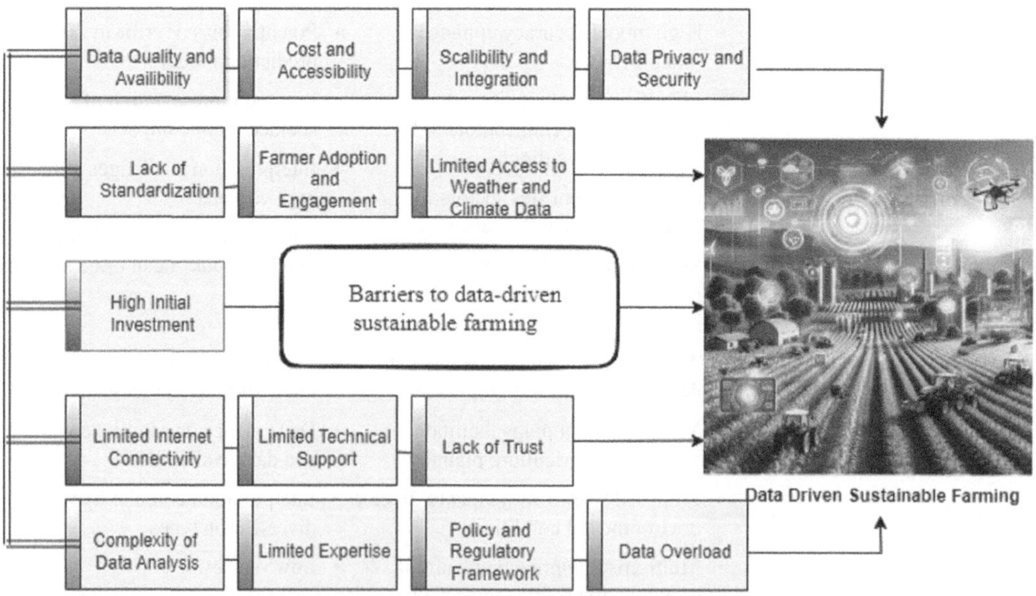

FIGURE 9.3 Barriers in utilizing data-driven sustainable farming practices.

- **Socioeconomic constraints:** Although data ownership has been clarified, privacy concerns addressed, and integration achieved, business constraints still limit the use of agricultural data for further farming. Demonstrating its value can be difficult, especially at multifaceted decision-making levels. Collecting data across farms is important, but faces challenges such as reluctance among farmers due to trust issues and problems in anonymizing data in the field. Trust in decision support is critical to adoption and requires transparency and user involvement in development. Despite technological advances, farmers still prioritize outdoor activities over data interpretation, highlighting the importance of user-friendly solutions for general use [38].
- **Technical obstacles limit widespread agricultural data:** Although the potential of large-scale agricultural data is recognized, challenges persist. Data quality remains a concern, as farmers prioritize tasks over meticulous data collection. Integrating heterogeneous data sources poses another obstacle [39]. Standardizing data formats across various sources is challenging. AI offers solutions but requires robust training datasets, which may be lacking in farming. Additionally, rural internet access limitations hinder the adoption of data-driven technologies in farming.

9.7 CONDUCTING HIGH-IMPACT RESEARCH AIMED AT TACKLING CHALLENGES AND OVERCOMING OBSTACLES

In a short span of time, farming has transitioned from a data-deficient field to one abundant in data. Figure 9.4 illustrates the conceptualizing farming operations and AI and ML-related technologies from various research projects. This wealth of agricultural data presents numerous avenues for high-impact research, including the following areas (listed without prioritization):

- **Soil sensors for cost-effective nutrient information:** On-the-go soil sensing technologies have the potential to significantly reduce the expenses associated with variable rate fertilizer adoption by eliminating the need for manual soil sampling and laboratory testing.
- **Automated animal intake measurement:** Advancing cost-effective and accurate methods for measuring distinct animal intake could revolutionize feed efficiency research in commercial settings, offering opportunities for research in a field traditionally underfunded by feed companies and limited public funding.
- **Robot obstacle avoidance in agricultural settings:** Developing automated systems for livestock and crop robots to navigate and avoid obstacles in the field could substantially reduce the need for human supervision tasks, thus lowering operational costs.
- **Integration of crop models and remote sensing for early pest detection:** Integrating remote sensing data with crop and soil models has the potential to detect plant diseases and pests at an early stage. This approach could reduce the

dependence on scheduled pesticide applications and promote more precise and targeted treatments.

- **Genetics by management by environment (G × M × E) knowledge:** The development of methodologies to utilize on-farm genetic diversity and local knowledge, referred to as Genetics by management by environment (G × M × E), offers a promising avenue for innovative research in agricultural sustainability.
- **Development of extension strategies for data-driven farming:** Additional research is necessary to pinpoint effective extension strategies that encourage data-driven farming. These endeavors are directed toward enhancing food security and mitigating the environmental repercussions of farming methods, underscoring the benefits of aggregating and sharing data.

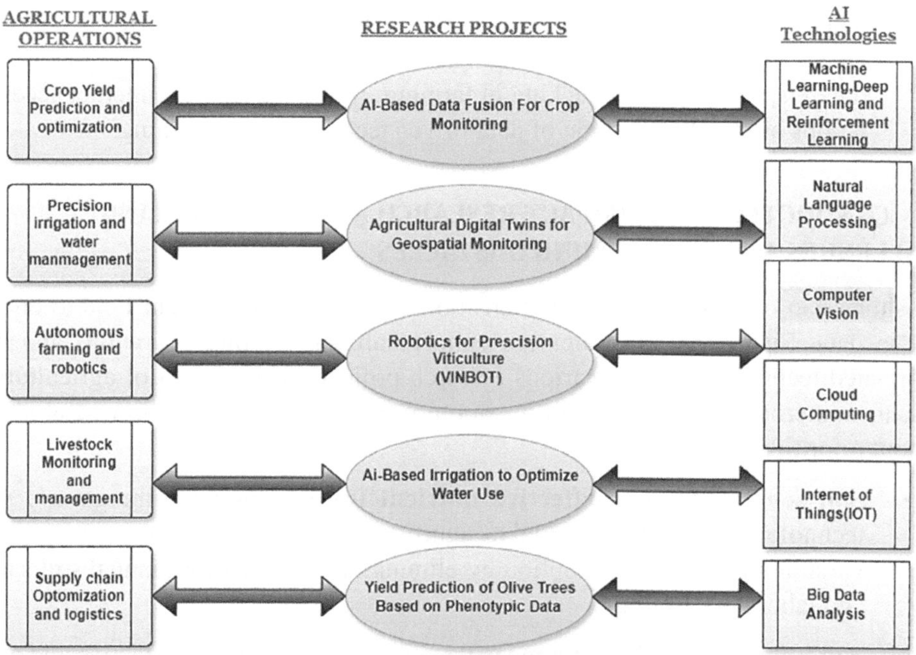

FIGURE 9.4 Conceptualizing farming operations and AI and ML related technologies from various research projects.

- **Business models for farm data gathering and sharing:** Examining feasible business models for gathering and exchanging farm data is crucial for unlocking the complete benefits of data-driven farming. This supports cooperation and data pooling among stakeholders.
- **Regulatory frameworks for farm data management:** Creating regulatory frameworks that enhance trust in gathering, analyzing, and sharing farm data, alongside appropriate business models, is vital for maximizing the advantages of data-driven farming and addressing privacy and security issues.

- **Development of DSS:** It is crucial to develop DSS capable of transforming big data into practical insights, empowering farmers to make well-informed and location-specific decisions for more sustainable and profitable agricultural practices.

9.7 CONCLUSION

The rapid progress in emerging technologies heralds a transformative phase in precision farming, promising significant improvements in farm management through computer-assisted DSS. Achieving sustainability in agriculture necessitates a comprehensive of intricate interactions within biological systems, historically reliant on human cognitive capabilities. However, data-driven technologies now empower farmers, agribusinesses, and researchers to observe, document, and comprehend these interactions to a degree surpassing human cognitive limits. Key areas of impactful research in data-driven agriculture encompass technical domains such as livestock feed intake measurement and soil sensing, alongside innovative methodologies for data acquisition and utilization. Additionally, there is a critical need for inventive management strategies in business models pertaining to data sharing, coupled with the establishment of regulatory frameworks fostering trust. Despite the potential for substantial benefits, inadequate public funding for research in these pivotal domains persists. Addressing this funding shortfall is imperative to unlock the complete potential of data-driven agriculture and realize the pledges of sustainable farming practices.

KEYWORDS

- **sustainable farming**
- **decision support systems**
- **machine learning**
- **food and fiber production systems**

REFERENCES

1. Oweibia, M., Elemuwa, U. G., Akpan, E., Daniel, E. T., Oruikor, G. J., Tarimobowei, E., & Babatunde, A. (2024). Analyzing Nigeria's journey towards sustainable development goals: A *Comprehensive Review from Inception to Present, F1000Reasearch*, 13, 984.
2. Jhariya, M. K., Meena, R. S., & Banerjee, A. (2021). Ecological intensification of natural resources towards sustainable productive system. In: *Ecological Intensification of Natural Resources for Sustainable Agriculture* (pp. 1–28).
3. Bhagat, R., Walia, S. S., Sharma, K., Singh, R., Singh, G., & Hossain, A. (2024). The integrated farming system is an environmentally friendly and cost-effective approach to the sustainability of agri-food systems in the modern era of the changing climate: A comprehensive review. *Food and Energy Security, 13*(1), e534.

4. Le, Q. V., Cowal, S., Jovanovic, G., & Le, D. T. (2021). A study of regenerative farming practices and sustainable coffee of ethnic minorities farmers in the Central Highlands of Vietnam. *Frontiers in Sustainable Food Systems*, *5*, 712733.
5. Júnior, M. R. B., de Almeida Moreira, B. R., dos Santos Carreira, V., de Brito Filho, A. L., Trentin, C., de Souza, F. L. P., Tedesco, D., Setiyono, T., Flores, J. P., Ampatzidis, Y., da Silva, R. P., & Shiratsuchi, L. S. (2024). Precision agriculture in the United States: A comprehensive meta-review inspiring further research, innovation, and adoption. *Computers and Electronics in Agriculture*, *221*, 108993.
6. Avasthi, S., Jain, K., Prakash, A., & Sanwal, T. (2024, February). A Blockchain Architecture for Secure Decentralized System and Computing. In *2024 3rd International conference on Power Electronics and IoT Applications in Renewable Energy and its Control (PARC)* (pp. 415–420). IEEE.
7. Jain, K., Kumar, A., & Singh, A., (2022). Data transmission reduction techniques for improving network lifetime in wireless sensor networks: An up-to-date survey from 2017 to 2022. *Transactions on Emerging Telecommunications Technologies*, *34*(1), e4674.
8. Paul, K., Chatterjee, S. S., Pai, P., Varshney, A., Juikar, S., Prasad, V., & Dasgupta, S. (2022). Viable smart sensors and their application in data driven agriculture. *Computers and Electronics in Agriculture*, *198*, 107096.
9. Raja, A. T., Arunachalam, A. S., & Gobinath, R. Machine learning in smart agriculture. In: *Advanced Technologies for Smart Agriculture* (pp. 129–151). River Publishers.
10. Agarwal, A., Jain, K., & Dev, A. (2021). Modeling and Analysis of Data Prediction Technique Based on Linear Regression Model (DP-LRM) for Cluster-Based Sensor Networks. *International Journal of Ambient Computing and Intelligence (IJACI)*, *12*(4), 98–117.
11. Reicosky, D. (2020). Conservation agriculture systems: Soil health and landscape management. In: *Advances in Conservation Agriculture* (pp. 87–154). Burleigh Dodds Science Publishing.
12. Hoseinzadeh, S., & Garcia, D. A. (2024). Can AI predict the impact of its implementation in greenhouse farming? *Renewable and Sustainable Energy Reviews*, *197*, 114423.
13. Jain, K., & Singh, A. (2022). A two-vector data-prediction model for energy-efficient data-aggregation in wireless sensor network. *Concurrency and Computation: Practice and Experience*, *34*(11), e6898.
14. Agarwal, A., Dev, A., & Jain, K. (2020). Prolonging sensor network lifetime by using energy-efficient cluster-based scheduling. *International Journal of Scientific & Technology Research*, *9*(4), 3410–3415.
15. Pretty, J. (2008). Agricultural sustainability: Concepts, principles and evidence. *Philosophical Transactions of the Royal Society B: Biological Sciences*, *363*(1491), 447–465.
16. Li, Z., Deng, X., Jin, G., Mohmmed, A., & Arowolo, A. O. (2020). Tradeoffs between agricultural production and ecosystem services: A case study in Zhangye, Northwest China. *Science of the Total Environment*, *707*, 136032.
17. Bennett, E. M., Baird, J., Baulch, H., Chaplin-Kramer, R., Fraser, E., Loring, P., & Lapen, D. (2021). Ecosystem services and the resilience of agricultural landscapes. In: *Advances in Ecological Research* (vol. 64, pp. 1–43). Academic Press.
18. Wahyono, N. D., Hasanah, N., Parmawati, R., & Wong, W. K. (2023). Improving economic welfare through capital development: Case study of smallholder dairy farmers in Pujon district. *Sustainability*, *15*(11), 8453.
19. Adolph, B., Allen, M., Beyuo, E., Banuoku, D., Barrett, S., Bourgou, T., & Zongo, A. F. (2021). Supporting smallholders' decision making: Managing trade-offs and synergies for sustainable agricultural intensification. *International Journal of Agricultural Sustainability*, *19*(5–6), 456–473.
20. Rozenstein, O., Cohen, Y., Alchanatis, V., Behrendt, K., Bonfil, D. J., Eshel, G., & Lowenberg-DeBoer, J. (2024). Data-driven agriculture and sustainable farming: Friends or foes? *Precision Agriculture*, *25*(1), 520–531.
21. Jain K., Gupta M., & Abraham A. (2022). Data prediction model in wireless sensor networks: A machine learning approach. In: *Lecture Notes in Networks and Systems* (vol. 419). Springer: Cham.
22. Jain, K., Agarwal, A., & Kumar, A. (2021). A novel data prediction technique based on correlation for data reduction in sensor networks. In *Proceedings of International Conference on Artificial Intelligence and Applications: ICAIA 2020* (pp. 595–606). Springer: Singapore.

23. Javaid, M., Haleem, A., Khan, I. H., & Suman, R. (2023). Understanding the potential applications of artificial intelligence in agriculture sector. *Advanced Agrochem*, *2*(1), 15–30.
24. Tonle, F. B., Niassy, S., Ndadji, M. M., Tchendji, M. T., Nzeukou, A., Mudereri, B. T., .& Tonnang, H. E. (2024). A road map for developing novel decision support system (DSS) for disseminating integrated pest management (IPM) technologies. *Computers and Electronics in Agriculture*, *217*, 108526.
25. Murali, R., Bijoor, A., Thinley, T., Gurmet, K., Chunit, K., Tobge, R., & Mishra, C. (2022). Indigenous governance structures for maintaining an ecosystem service in an agro-pastoral community in the Indian Trans Himalaya. *Ecosystems and People*, *18*(1), 303–314.
26. Salcedo-Sanz, S., Ghamisi, P., Piles, M., Werner, M., Cuadra, L., Moreno-Martínez, A., & Camps-Valls, G. (2020). Machine learning information fusion in Earth observation: A comprehensive review of methods, applications and data sources. *Information Fusion*, *63*, 256–272.
27. Divya, K. L., Mhatre, P. H., Venkatasalam, E. P., & Sudha, R. (2021). Crop simulation models as decision-supporting tools for sustainable potato production: A review. *Potato Research*, *64*(3), 387–419.
28. Li, M., Fu, Q., Singh, V. P., Liu, D., Li, T., & Zhou, Y. (2020). Managing agricultural water and land resources with tradeoff between economic, environmental, and social considerations: A multi-objective non-linear optimization model under uncertainty. *Agricultural systems*, *178*, 102685.
29. Komarek, A. M., De Pinto, A., & Smith, V. H. (2020). A review of types of risks in agriculture: What we know and what we need to know. *Agricultural Systems*, *178*, 102738.
30. Javaid, M., Haleem, A., Khan, I. H., & Suman, R. (2023). Understanding the potential applications of Artificial Intelligence in Agriculture Sector. *Advanced Agrochem*, *2*(1), 15–30.
31. Zhai, Z., Martínez, J. F., Beltran, V., & Martínez, N. L. (2020). Decision support systems for agriculture 4.0: Survey and challenges. *Computers and Electronics in Agriculture*, *170*, 105256.
32. Lea, B. R., & Yu, W. B. (2024). Design an interactive dashboard system to improve university career service: A case study. *International Journal of Business and Systems Research*, *18*(2), 145–169.
33. Ahmad, T., Madonski, R., Zhang, D., Huang, C., & Mujeeb, A. (2022). Data-driven probabilistic machine learning in sustainable smart energy/smart energy systems: Key developments, challenges, and future research opportunities in the context of smart grid paradigm. *Renewable and Sustainable Energy Reviews*, *160*, 112128.
34. Abdella, G. M., Kucukvar, M., Onat, N. C., Al-Yafay, H. M., & Bulak, M. E. (2020). Sustainability assessment and modeling based on supervised machine learning techniques: The case for food consumption. *Journal of Cleaner Production*, *251*, 119661.
35. Blickensdörfer, L., Schwieder, M., Pflugmacher, D., Nendel, C., Erasmi, S., & Hostert, P. (2022). Mapping of crop types and crop sequences with combined time series of Sentinel-1, Sentinel-2 and Landsat 8 data for Germany. *Remote Sensing of Environment*, *269*, 112831.
36. Roy, S. K., & De, D. (2022). Genetic algorithm based internet of precision agricultural things (IoPAT) for agriculture 4.0. *Internet of Things*, *18*, 100201.
37. Wiseman, L., Sanderson, J., Zhang, A., & Jakku, E. (2019). Farmers and their data: An examination of farmers' reluctance to share their data through the lens of the laws impacting smart farming. *NJAS—Wageningen Journal of Life Sciences*, *90*, 100301.
38. Neethirajan, S. (2023). The significance and ethics of digital livestock farming. *AgriEngineering*, *5*(1), 488–505.
39. Šestak, M., & Copot, D. (2023). Towards trusted data sharing and exchange in agro-food supply chains: Design principles for agricultural data spaces. *Sustainability*, *15*(18), 13746.

CHAPTER 10

Smart and Secure Monitoring Devices for E-Agriculture

AMBIKA AGGARWAL[1*], ASHUTOSH BHATT[2], and ANCHIT BIJALWAN[3]

[1]*School of Computer Science, UPES, Dehradun, Uttarakhand, India*

[2]*Department of CSE, SST, Swami Rama Himalayan University, Dehradun, Uttarakhand, India*

[3]*School of Computing and Innovative Technologies, British University Vietnam, Hanoi, Vietnam*

*Corresponding author

ABSTRACT

The integration of smart and secure monitoring devices in e-agriculture is transforming traditional farming practices into data-driven, efficient, and sustainable systems. These devices leverage advanced technologies such as the Internet of Things (IoT), artificial intelligence (AI), and cloud computing to provide real-time monitoring and decision-making support. Key parameters like soil moisture, temperature, humidity, crop health, and pest activity are continuously monitored using sensors and wireless networks. The collected data is analyzed to optimize irrigation, fertilization, and pesticide use, thereby enhancing crop yield and resource management. Security plays a critical role in ensuring the integrity and confidentiality of agricultural data. With increasing connectivity, smart devices are exposed to cyber threats that can compromise operations and data privacy. Therefore, implementing robust cybersecurity measures—such as data encryption, secure communication protocols, and authentication mechanisms—is essential to safeguard these systems. Smart and secure monitoring devices not only improve productivity and reduce environmental impact but also empower farmers with actionable insights. As agriculture faces growing challenges from climate change, resource scarcity, and population growth, such intelligent systems offer a sustainable path forward. This abstract highlights the pivotal role of technology in revolutionizing agriculture through precision monitoring and secure digital integration.

Demystifying Sustainable Farming Using Smart Technologies.
Gunjan Chhabra, Keshav Kaushik, Priya Chugh, Susheela Dahiya, and Aryan Chaudhary (Eds.)

DOI: 10.1201/9781779640932-10

10.1 INTRODUCTION

Overview of e-agriculture: e-Agriculture or digital agriculture is the term used to describe the use of innovative technologies and digital instruments in agricultural activities to improve sustainability, efficiency, and productivity. This section offers a comprehensive understanding of what e-agriculture requires and its significance in shaping the modern agricultural landscape [1].

The term "e-agriculture" refers to a wide range of tools and methods used to enhance several facets of farming and agriculture using digital technology. This includes the use of technologies like Internet of Things (IoT), big data analytics, artificial intelligence (AI), remote sensing, and other innovative technologies for the collection, analysis, and utilization of data for better decision-making and operational efficiency [2].

10.1.1 KEY COMPONENTS OF E-AGRICULTURE

E-agriculture components (Figure 10.1) work together to create a smart, efficient, and sustainable agricultural ecosystem [3].

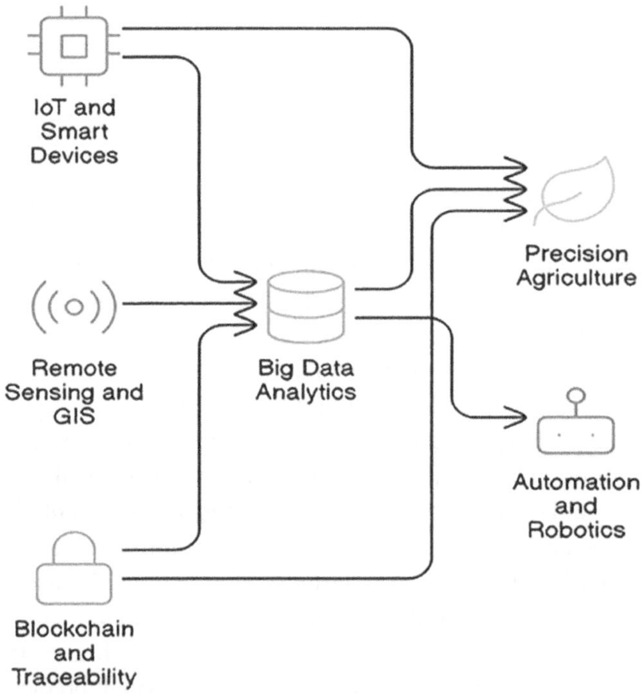

FIGURE 10.1 Key components of e-agriculture.

- **Precision agriculture:** It involves the utilization of data collected from sensors, satellites, and drones to monitor and maintain crop health, soil conditions, and other environmental factors. This allows for precise application of water, fertilizers, and pesticides, reducing waste and optimizing yields.

- **IoT and smart devices:** It involves the deployment of interconnected sensors and devices that collect real-time data on various parameters, including soil moisture, temperature, and humidity. These devices assist in continuous monitoring and enable automated responses to changing environmental conditions.
- **Big data and analytics:** Advanced algorithms and machine learning (ML) approaches are used to examine the massive amounts of data provided by smart devices and other sources. This facilitates trend prediction, problem identification, and well-informed decision-making.
- **Remote sensing and GIS:** Large-scale agricultural lands can be mapped and monitored using spatial data from remote sensing technology and geographic information systems (GIS). This aids in pattern recognition and macrolevel strategic decision-making.
- **Automation and robotics:** It involves the use of robots and automated machinery for performing tasks, such as planting, harvesting, and crop monitoring. This helps in reducing labor costs and increases efficiency.
- **Blockchain and traceability:** It maintains supply chain traceability and transparency by documenting and confirming each stage of the manufacturing and delivery process. This increases consumer trust while improving food safety.

10.1.2 BENEFITS OF E-AGRICULTURE

E-agriculture offers numerous benefits, enhancing various aspects of agricultural practices through technology and digital innovations. Figures 10.2 and 10.3 depict the benefits of e-agriculture.

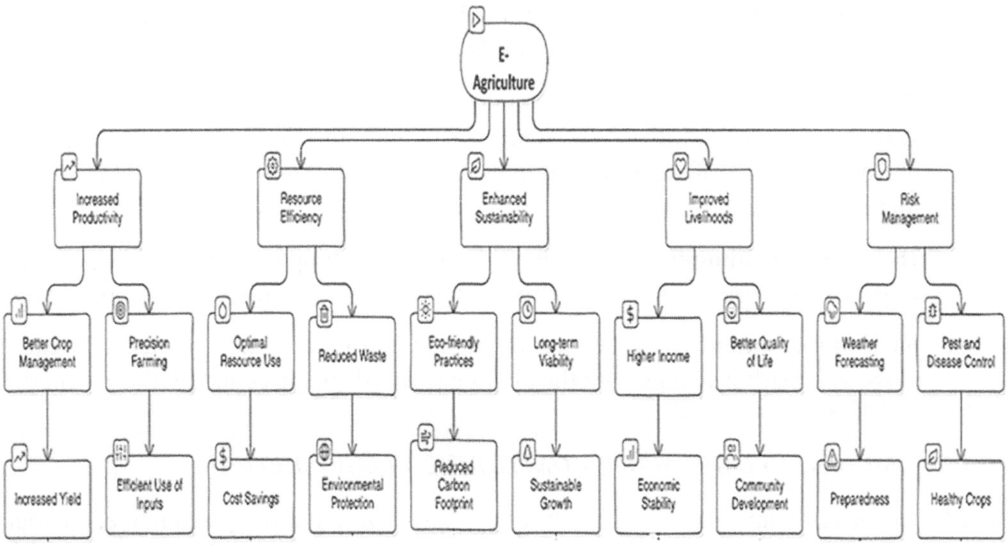

FIGURE 10.2 Benefits of e-agriculture.

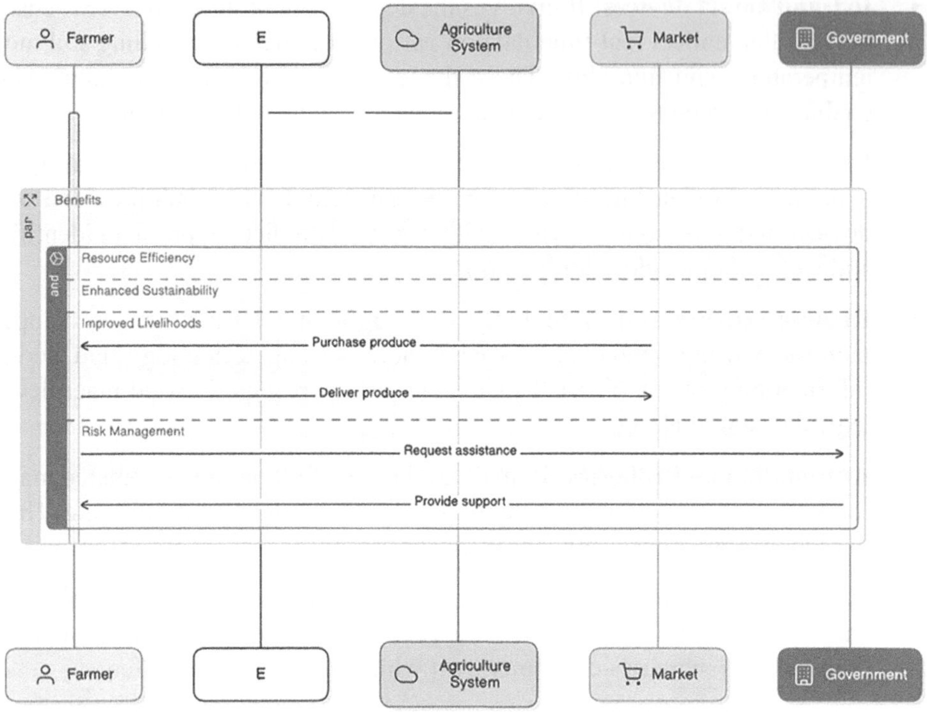

FIGURE 10.3 Benefits of e-agriculture.

- **Increased productivity:** E-agriculture technologies assist farmers in optimizing inputs and procedures, resulting in increased crop yields and improved resource usage, by offering accurate and fast information.
- **Resource efficiency:** More economical use of water, fertilizer, and pesticides is made possible by smart devices and data analytics, which reduces environmental impact and conserves resources.
- **Enhanced sustainability:** Reducing waste and emissions, as well as improving the management of natural resources, are ways to support sustainable farming methods.
- **Improved livelihoods:** E-agriculture can make farming more financially viable by lowering expenses and raising output, especially for smallholders and rural areas.
- **Risk management:** Farmers can anticipate and reduce risks related to weather, pests, and market swings with the aid of advanced analytics and predictive models.

10.1.3 IMPORTANCE OF SMART MONITORING IN AGRICULTURE

In order to convert conventional agricultural methods into effective, sustainable, and data-driven operations, smart monitoring is essential in agriculture [4]. By giving farmers access to real-time data on agricultural conditions through smart monitoring systems, farmers can make more educated decisions about planting, irrigation, fertilizer, and pest control, which leads to healthier crops and higher yields. Weather data and soil moisture sensors work together to optimize watering schedules, guaranteeing crops get the right quantity of water at the right time and reducing water waste. This technique is known as precision irrigation. Monitoring devices that detect soil nutrient levels help farmers apply the right quantity of fertilizer, improving plant health and reducing environmental impact. This improves nutrient management. Smart irrigation systems that accurately assess soil moisture levels and forecast weather patterns improve water conservation. Additionally, by using fewer chemicals and minimizing undesired runoff, the precise application of fertilizers and pesticides is guided by real-time data, which lowers costs. In order to provide prompt action before concerns worsen, smart monitoring technologies identify early indicators of illnesses, pests, and other possible problems. Sensors and imaging technologies that are able to recognize disease symptoms early on allow for timely treatment and stop the progress of the illness. Finally, real-time population monitoring of pests aids in the application of focused pest control strategies and lessens the requirement for extensive pesticide use, both of which are advantageous to pest management.

10.2 THE ROLE OF MONITORING DEVICES IN E-AGRICULTURE

The word "monitoring devices" in e-agriculture refers to a range of electronic tools and sensors that collect and transmit data on agricultural activities [5]. With the use of this technology, farmers and agronomists can keep an eye on crop health, soil quality, weather patterns, and other significant factors that affect agricultural productivity.

10.2.1 TYPES OF MONITORING DEVICES

- **Soil sensors:** Determine the temperature, pH, nutrients, and moisture content of the soil.
- **Weather stations:** Offer current information on temperature, humidity, wind direction, and precipitation.
- **Drones:** With cameras and sensors installed, these aerial vehicles are capable of surveying expansive areas and gathering precise data on crop growth and health.
- **Livestock monitoring devices:** Using wearable sensors, track the location, behavior, and health of animals.
- **Irrigation controllers:** Sensor-based automated devices that maximize water consumption according to soil moisture content and meteorological conditions.
- **Pest and disease detection sensors:** These sensors use a variety of technologies, such as ML techniques, to detect the presence of pests and diseases in crops.

10.2.2 APPLICATIONS OF MONITORING DEVICES IN AGRICULTURE

- **Crop management:** With the use of technology like drones and soil sensors, precision farming maximizes crop yields and minimizes waste by applying precisely the right amounts of pesticides and fertilizer at the appropriate times [6]. Intelligent irrigation systems improve irrigation management by ensuring crops receive enough water, preserving water supplies, and avoiding overirrigation.
- **Soil health monitoring:** Soil sensors are a useful tool for farmers to monitor the amounts of nutrients in their soil. This allows for more accurate fertilizer delivery and supports sustainable soil management. Farmers may optimize irrigation schedules and avoid crop overwatering by using soil moisture tracking, which gives them access to up-to-date information on soil moisture levels.
- **Weather monitoring:** Farmers depend on weather forecasting because it helps them plan planting, harvesting, and other operations more effectively. Weather stations give accurate, localized forecasts. In order to provide more resilient farming practices, monitoring weather patterns also helps with the development of plans for adaptation to extreme weather events and climate change.
- **Livestock management:** Wearable sensors that monitor cattle's vital signs and activities improve health monitoring, lowering mortality rates and facilitating early disease detection. GPS-enabled technology enhances grazing management by regulating and optimizing grazing patterns to guarantee sustainable pasture utilization.
- **Pest and disease control:** Sensors and AI algorithms make it easier to identify disease outbreaks and insect infestations early on, minimizing crop losses and enabling prompt response. Monitoring instruments that offer information on pest populations and environmental variables are beneficial to integrated pest management systems and increase their efficacy.

10.2.3 BENEFITS OF USING SMART MONITORING DEVICES

Automation and real-time data collecting increase agricultural efficiency by decreasing the need for manual labor [7]. Because precision farming techniques use less water, fertilizer, and pesticides, they also save money. Higher crop yields and higher quality food are the outcome of farmers having access to timely and accurate data, which helps them make better decisions. By reducing their negative effects on the environment and optimizing resource consumption, smart monitoring systems help sustainable farming operations. Farmers may anticipate and reduce risks related to weather, pests, diseases, and other factors with the use of ongoing monitoring. By guaranteeing the animals' well-being, health monitoring systems enhance the welfare of cattle by raising output and decreasing losses. Monitoring devices are key to modern agriculture because they offer crucial data and insights that help precision farming, increase productivity, and improve sustainability.

10.3 TECHNOLOGIES USED IN SMART MONITORING DEVICES

- In agriculture, soil moisture sensors are essential, and different kinds have different functions [8]. Resistive sensors track variations in the electrical resistance of the soil as the moisture level varies, whereas capacitive sensors use the dielectric constant of the soil to detect the moisture content. Sensors that use time domain reflectometry measure the length of an electromagnetic pulse to determine the moisture content of soil. When combined, these sensors provide accurate and trustworthy data that is critical for managing irrigation efficiently and achieving ideal crop development.
- Temperature sensors play a vital role in monitoring agricultural environments. Thermistors measure the temperature of soil or air using materials whose resistance changes with temperature. Thermocouples generate a voltage in response to temperature differences between two points. Infrared sensors detect infrared radiation emitted by objects to measure temperature remotely. These sensors provide accurate and essential temperature data for managing crop and soil conditions effectively.
- In order to keep track of the amount of moisture in the air, humidity sensors are crucial. By monitoring changes in capacitance, capacitive humidity sensors calculate relative humidity. Resistive humidity sensors use variations in electrical resistance to measure humidity. Thermal conductivity sensors use the air's thermal conductivity to calculate absolute humidity. In agricultural contexts, these sensors offer vital data for preserving ideal growing conditions.

10.3.1 IOT (INTERNET OF THINGS) INTEGRATION

IoT technology connects various monitoring devices, allowing them to communicate and share data over a network. This integration enables real-time monitoring, data collection, and analysis. Figure 10.4 shows components of IoT integration [9].

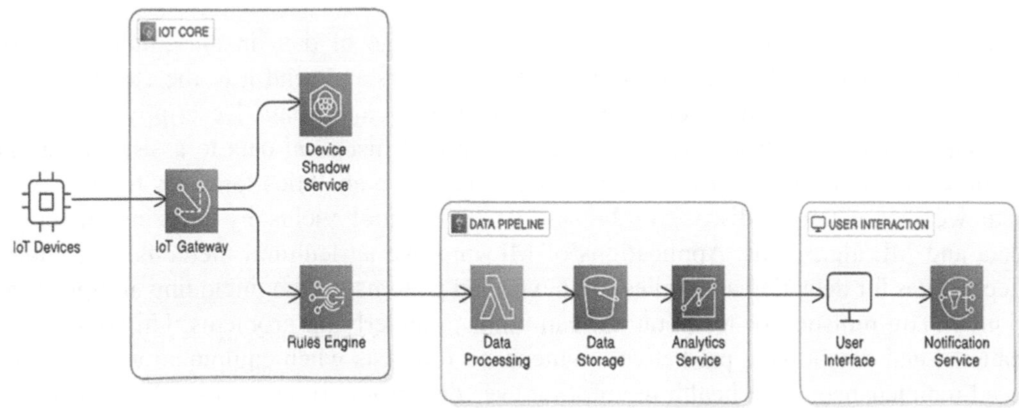

FIGURE 10.4 Components of IoT integration.

10.3.1.1 COMPONENTS OF IOT INTEGRATION

- **Smart sensors:** Embedded with communication capabilities to send data to central systems or cloud platforms.
- **Gateway devices:** Collect data from sensors and transmit it to cloud servers or local data centers.
- **Cloud platforms:** Store and process data, providing analytics and insights accessible via web or mobile applications.
- **User interfaces:** Web-based dashboards and mobile applications that allow farmers to monitor and control their agricultural operations remotely.

10.3.2 WIRELESS COMMUNICATION TECHNOLOGIES

Wi-Fi is an essential agricultural technology that offers a high-speed internet connection and short-range communication within farms or agricultural businesses. It makes it possible for sensors to provide data in real time to central systems, which makes monitoring and controlling agricultural activities more effective [10].

Zigbee is a low-power, low-data-rate wireless communication protocol that is perfect for agricultural short-range networking. It is frequently utilized in mesh networks, which improve connectivity and data collection capabilities by enabling many devices to communicate and expand coverage.

Broad agricultural landscapes are especially well-suited for LoRaWAN (long range wide area network), which allows remote sensors to be connected over several kilometers. It provides an inexpensive solution for IoT applications in agriculture and is made for low-power, long-range communication. This allows for effective data transfer and monitoring over vast areas.

10.3.3 DATA ANALYTICS AND MACHINE LEARNING

IoT devices are essential to the collection and storage of data in agriculture because they continuously collect data from a variety of sensors and send it to the cloud or local servers [11]. Databases are then used to store and analyze this data. Descriptive analytics, a component of data processing, offers a summary of historical data to assist farmers in comprehending historical patterns and trends. Predictive analytics forecasts future events like weather patterns, disease outbreaks, or agricultural yields by utilizing historical data and ML algorithms. Applications of ML improve agricultural methods even more. Techniques for detecting anomalies or unexpected patterns in data, including abrupt variations in soil moisture or temperature, can point to underlying problems. Utilizing usage patterns and sensor data, predictive maintenance forecasts when equipment will break or need maintenance. Crop health monitoring uses data from drones and satellite images to assess crop health, identify illnesses, and suggest treatments. Based on an analysis of crop, weather, and soil data, precision agriculture is a resource-efficient farming technique that offers suggestions for pest control, fertilizer application, and irrigation.

10.3.3.1 BENEFITS OF DATA ANALYTICS AND MACHINE LEARNING

By utilizing cutting-edge technology, smart monitoring devices in e-agriculture facilitate accurate, efficient, and sustainable farming operations. These tools help farmers make well-informed decisions by offering useful insights gleaned from historical and real-time data [12]. Processes for monitoring and processing data are automated to reduce human work and increase efficiency. Furthermore, smart monitoring devices support improved resource management and higher yields through agricultural technique optimization, which raises production. Moreover, risk mitigation can be enhanced by taking proactive measures to prevent or lessen issues as soon as potential threats are identified. All things considered, these advancements guarantee better administration of crops and livestock, assisting farmers in raising productivity, cutting costs, and lessening their adverse effects on the environment.

10.4 SECURITY CHALLENGES IN SMART MONITORING DEVICES

A number of prevalent security risks in e-agriculture can compromise the reliability and efficiency of farming systems [13]. Hacking is the act of breaking into networks or surveillance systems without authorization. It can provide access to vital agricultural systems and result in data tampering, disruptions to operations, or system failures. Weak security protocols, unprotected data storage, and insecure communication techniques are the causes of data breaches, which expose private information, including financial records, farm management data, and crop health data. Ransomware attacks pose a serious threat to monitoring equipment because they encrypt data and demand money for its recovery. Malware infestations can also result in system faults, unauthorized data access, and data corruption. Denial-of-service attacks can impede the collection and transmission of crucial agricultural data by overloading networks and devices with traffic, resulting in system slowdowns or shutdowns. Phishing attacks, a type of social engineering, are designed to trick system administrators or farm operators into giving up important information or login credentials. This could result in unauthorized access and additional security lapses. These dangers demonstrate how crucial strong security measures are to safeguarding e-agriculture data and infrastructure.

10.4.1 VULNERABILITIES SPECIFIC TO AGRICULTURAL MONITORING DEVICES

E-agriculture systems include a number of security flaws that hackers could take advantage of. Devices are vulnerable to unauthorized access due to inadequate authentication methods, such as the usage of default or weak passwords and the absence of multifactor authentication (MFA) [14]. The danger of security breaches is further increased by inadequate remote access authentication methods. Unauthorized access and data interception can result from insecure communication routes between devices and central systems, particularly when data is transferred unencrypted. Vulnerabilities in wireless communication protocols, such as Zigbee and Wi-Fi, may be exploited by attackers. Devices with out-of-date firmware and software may have unpatched security flaws, making them susceptible to known exploits and assaults. This is the result of inadequate firmware and software updates. Surveillance equipment placed in open or remote regions is vulnerable to theft or tampering, which presents a physical security risk. When an attacker has physical access to a device, they can circumvent security measures and seize control. When agricultural monitoring equipment is connected to the same network as other critical systems, inadequate network segmentation might expand the attack surface. Because there is no segmentation, once access is achieved, attackers can move laterally within the network more simply. These flaws demonstrate the necessity of strong security protocols and frequent upgrades to shield e-agriculture systems from possible attacks.

10.4.2 IMPACT OF SECURITY BREACHES ON AGRICULTURAL OPERATIONS

In agricultural operations (Figure 10.5), security breaches can have a big effect on all the stakeholders and interconnected systems [15]. Attacks that damage monitoring equipment can cause operational disruption, which can stop essential operations like irrigation, pest control, and animal monitoring. This can result in agricultural losses, lower output, and monetary losses. Problems with data integrity, such as erroneous information originating from data tampering or corruption, can negatively impact crop health and productivity and may lead to the improper use of pesticides, fertilizers, or water. Financial losses may result from decreased agricultural produce and operational downtime, as well as from the cost of resolving security breaches and data restoration, which may include replacing or repairing compromised devices. Another worry is reputational harm. If a breach reveals private information or interferes with business operations, farms and agribusinesses run the risk of losing their good name, which might have long-term detrimental effects, like losing the trust of partners, stakeholders, and clients.

Cybersecurity incidents involving e-agriculture have the potential to result in severe legal and regulatory ramifications, such as regulatory inquiries and increased monitoring [16]. Penalties and fines may be imposed legally for violating data protection regulations. Compromised safety is a problem as well, since equipment faults resulting from security breaches may put farm personnel in danger, and attacks on livestock monitoring systems might jeopardize the health and welfare of animals. Strong security measures, including encryption, regular upgrades, physical device security, and robust authentication procedures, need to be put in place in order to reduce these security threats. Furthermore, preventing social engineering attempts and ensuring the general security of agricultural monitoring systems can be accomplished by teaching system administrators and farm operators about security best practices.

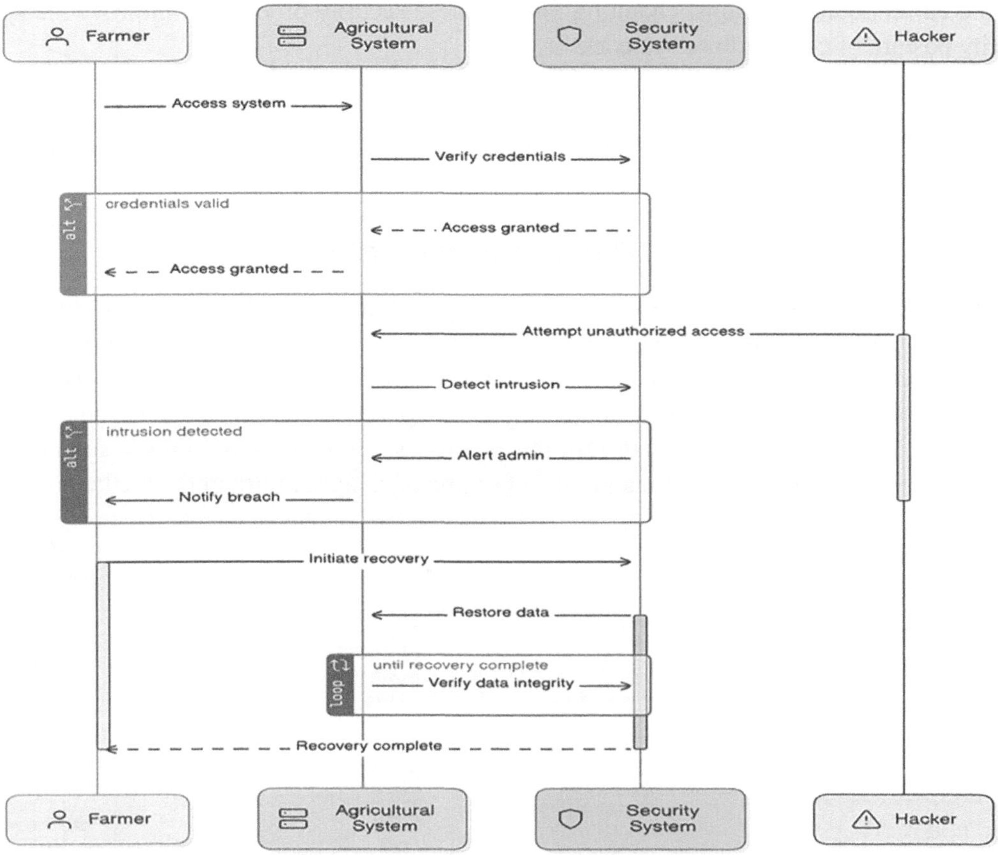

FIGURE 10.5 Impact of security breaches on agriculture operations.

10.5 SECURITY SOLUTIONS FOR MONITORING DEVICES

Protecting agricultural monitoring devices from typical threats and weaknesses requires the implementation of strong security measures [17]. It should be encrypted both while it is in transit and at rest to prevent unauthorized access, even in the event that it is accessed. Data transfer can be protected using secure communication protocols like HTTPS, VPNs, and IoT-specific protocols like MQTT with TLS. Robust authentication protocols like as OAuth and MFA, in conjunction with authorization techniques like RBAC and ABAC, aid in managing access and thwarting unapproved access. Consistent and safe access control across platforms is ensured via secure access management using IAM and API security. Security threats can be found and countered with the aid of anomaly detection systems that use behavioral analysis and ML in conjunction with intrusion detection and prevention systems like NIDS, HIDS, and IPS. The security of agricultural monitoring equipment depends on regular firmware and software upgrades and patches that are applied via OTA and automated updates, as well as efficient patch management and version control. By ensuring data transfer, authorization and authentication, intrusion detection and prevention, and current defense against changing threats, these security procedures improve the security posture of e-agriculture systems.

10.6 CASE STUDIES: SUCCESSFUL IMPLEMENTATIONS OF SMART AND SECURE MONITORING DEVICES IN E-AGRICULTURE

10.6.1 PRECISION AGRICULTURE IN THE NETHERLANDS

The Netherlands is renowned for its advanced agricultural technologies. A case study from Wageningen University & Research demonstrates the successful implementation of smart monitoring devices in precision agriculture [18]. A cloud-assisted IoT-based precision agriculture platform (Figure 10.6) combines cloud computing, IoT technologies, and precision agriculture practices to optimize farming processes. Table 10.1 presents a summary of technology, security procedure, and impact of precision agriculture in the Netherlands.

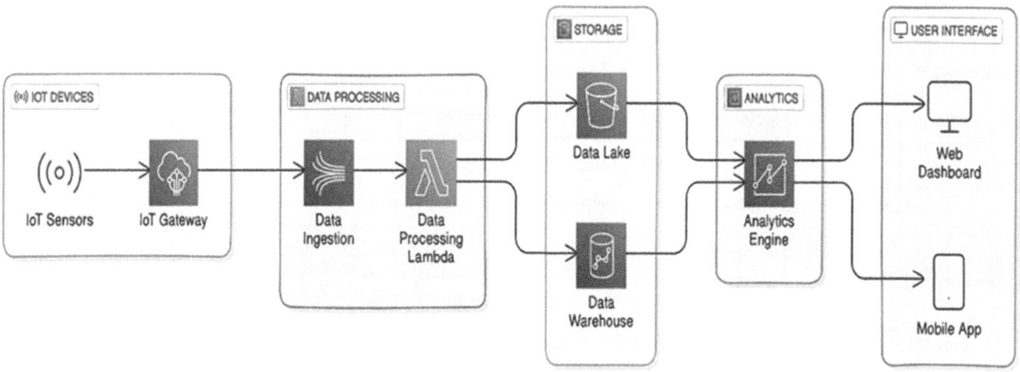

FIGURE 10.6 Cloud-assisted IoT-based precision agriculture platform.

TABLE 10.1 Summary of Technology, Security Procedure, and Impact of Precision Agriculture in the Netherlands

Technologies Used	Security Procedures	Impact
Soil sensors	Data encryption	Higher yields
Drones	Authentication	Cost savings
Internet of Things integration	Frequent updates	Sustainability
Data analytics		

10.6.2 SMART IRRIGATION SYSTEM IN ISRAEL

Netafim, an Israeli company, implemented a smart irrigation system that leverages advanced monitoring devices to optimize water usage in agriculture [19]. Figure 10.7 depicts smart agriculture architecture that integrates various technological components to create an efficient, data-driven agricultural ecosystem. Table 10.2 presents summary of technology, security procedure, and impact of smart irrigation system in Israel.

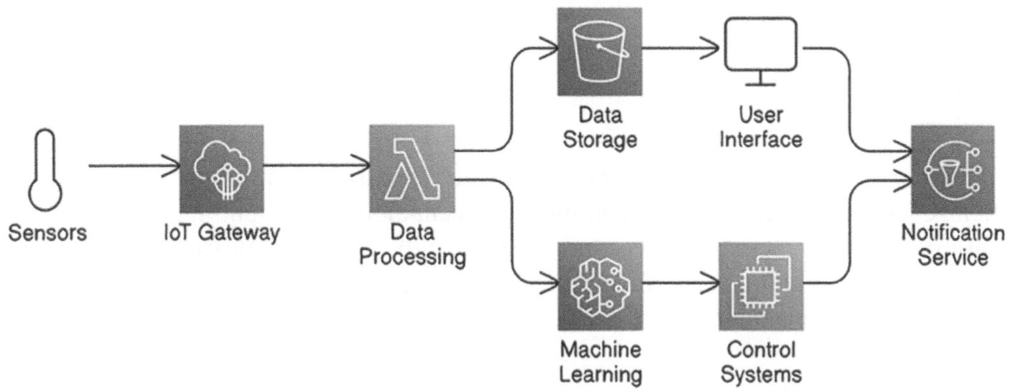

FIGURE 10.7 Smart agriculture architecture.

TABLE 10.2 Summary of Technology, Security Procedure, and Impact of Smart Irrigation System in Israel

Technologies Used	Security Procedures	Impact
Soil moisture sensors	Secure communication	Water conservation
Weather stations	Role-based access control	Enhanced productivity
IoT integration	Intrusion detection	Labor savings
Mobile application		

10.6.3 LIVESTOCK MONITORING IN AUSTRALIA

An Australian livestock farm implemented smart monitoring devices to track the health and behavior of their cattle using IoT and data analytics. Table 10.3 presents a summary of technology, security procedure, and impact of livestock monitoring in Australia [20].

TABLE 10.3 Summary of Technology, Security Procedure, and Impact of Livestock Monitoring in Australia

Technologies Used	Security Procedures	Impact
Wearable sensors	Encryption	Better animal health
GPS tracking	Authorization and authentication	Operational efficiency
Cloud integration	Regular patches	Data-driven decisions
Machine learning		

10.6.4 PEST AND DISEASE MANAGEMENT IN INDIA

In India, a project focused on using smart monitoring devices to manage pests and diseases in cotton crops, leveraging IoT and AI technologies. Table 10.4 presents summary of technology, security procedure, and impact of pest and disease management in India [21, 22].

TABLE 10.4 Summary of Technology, Security Procedure, and Impact of Pest and Disease Management in India

Technologies Used	Security Procedures	Impact
Sensor-equipped pest traps	Secure data storage	Lower crop losses
Weather stations	User authentication	Cost-effectiveness
AI algorithms	Network security	Empowerment of farmers
Mobile application		

These case studies demonstrate how innovative and safe monitoring tools may revolutionize e-agriculture. Utilizing cutting-edge technologies and putting strong security measures in place, these initiatives have significantly increased agricultural operations' production, sustainability, and efficiency. Table 10.5 presents data analysis and outcomes of e-agriculture implementation [23, 24].

TABLE 10.5 Data Analysis and Outcomes of E-Agriculture Implementation

E-Agriculture Implementation	Data Analysis	Outcomes
Precision agriculture in the Netherlands	• Soil sensor data demonstrated that ideal moisture levels were sustained in various soil types, resulting in effective water use	• Targeted interventions based on data-driven insights produced an average 20% improvement in yield
	• Precise fertilizer application, minimal waste, and optimal crop uptake were revealed by nutrient sensors	• Cost reductions of up to 15% were achieved by using resources (water, fertilizers, and pesticides) more efficiently
	• Drones' multispectral imagery gave precise information on crop health and highlighted problem regions	• A smaller chemical runoff and more careful resource management translate into a reduced environmental impact
Smart irrigation system in Israel	• By identifying patterns in soil moisture, ongoing monitoring made it possible to precisely schedule irrigation	• A 30% decrease in water use was achieved, in line with sustainability objectives
	• By correlating localized meteorological data with irrigation requirements, water efficiency was increased	• Crop health was enhanced by consistent moisture levels, which resulted in a 10% increase in yield
	• The data demonstrated a notable decrease in water use while preserving ideal crop health	• By eliminating human labor, irrigation schedule automation resulted in labor cost savings of about 25%
Livestock monitoring in Australia	• GPS data revealed grazing patterns and assisted in optimizing pasture usage	• Early health problem discovery resulted in a 15% decrease in death rates
	• Wearable sensors provide continuous data on cattle health markers, including heart rate and temperature	• Streamlined livestock management with automated monitoring and health alarms, increasing overall effectiveness
	• Early warning indicators of health problems were recognized by machine learning models, enabling prompt interventions	• Improved pasture management brought about by grazing pattern analysis resulted in a 10% increase in feed efficiency

TABLE 10.5 *(Continued)*

E-Agriculture Implementation	Data Analysis	Outcomes
Pest and disease management in India	• Population patterns and hotspots were revealed by real-time data obtained from pest traps • Weather analysis assisted in the prediction of pest outbreaks • Information on the use of pesticides and changes in the number of pests as a result provide insights into the efficacy of various management strategies	• A 25% decrease in crop losses from pests and diseases was achieved by early treatments based on data insights • By using 20% less chemicals, targeted pesticide application decreased expenses and its negative effects on the environment • Farmers' decision-making was enhanced by having access to real-time data and recommendations, which raised their confidence and level of understanding

10.6.5 OVERALL ANALYSIS

In agriculture, precise and timely data collection is essential for making wise decisions. Extensive monitoring of several factors crucial to farming operations is made possible by the integration of diverse sensors and IoT devices [25, 26]. A full picture of agricultural operations is made possible by data integration and analysis, which combine information from multiple sources, including sensors, weather stations, and drones. Proactive management and predictive insights are made possible by advanced data analytics and ML. It is crucial to guarantee the security and integrity of data. To protect sensitive information and uphold trust in the data used for decision-making, data security methods, including encryption, authentication, and frequent upgrades, are required. Additionally, essential to preventing data breaches and illegal access are secure communication techniques. Smart monitoring devices have a big impact on sustainability and productivity. By cutting waste and making the most use of inputs like water, fertilizer, and pesticides, these gadgets contribute to increased sustainability. Additionally, they boost productivity by making precise and effective resource management possible. The long-term advantages of smart monitoring systems—such as reduced expenses, increased yields, and enhanced operational efficiency—outweigh the initial high cost of ownership. Additionally, reducing labor expenses and improving the overall efficacy of farm management are automation and data-driven decision-making. Systems for smart monitoring are flexible and expandable, accommodating different kinds of farms and farming methods. Adoption is facilitated by user-friendly mobile applications and interfaces, which give farmers readily available and practical insights. All things considered, smart monitoring tools revolutionize the agricultural industry, increasing farming's sustainability, efficiency, and data-drivenness. These case studies highlight how innovative and safe monitoring tools may revolutionize e-agriculture. Farmers may make notable gains in productivity, sustainability, and efficiency by utilizing cutting-edge technologies and strong security measures [27, 28]. This will help to transform agriculture into a more resilient and data-driven model in the future.

10.7 FUTURE TRENDS AND INNOVATIONS

Emerging technologies that promise to increase the resilience, sustainability, and efficiency of farming operations are shaping the direction of smart agriculture. Farmers may improve food security, lessen their environmental impact, and streamline their operations by implementing technologies like blockchain, AI, and the IoT [29, 30]. Predictive analytics, fueled by AI and ML models, is one of the major advancements in agriculture yield forecasting. It forecasts weather patterns, pest outbreaks, and agricultural yields by evaluating historical and present data. Based on data analysis, automated decision support systems recommend the best times to plant, when to irrigate, and how much fertilizer to use. With the use of sophisticated image identification techniques, computer vision can diagnose insect infestations, agricultural illnesses, and nutritional deficiencies from photographs taken by drones or satellites. Automation and robotics are important, as self-governing tractors and other equipment carry out tasks like planting, weeding, and harvesting with accuracy and little help from humans. Applications for agricultural robots with cameras and sensors include seed planting, pesticide and fertilizer application, and agricultural health monitoring. In order to obtain precise photographs of crops and learn more about crop health, soil composition, and irrigation needs, remote sensing techniques employ drones and unmanned aerial vehicles. Drones used for precision spraying can minimize waste and adverse environmental effects. By tracking agricultural products from farm to table, blockchain technology improves food safety and quality management by ensuring supply chain transparency. By facilitating safe, automated agreements and transactions between farmers, suppliers, and purchasers, smart contracts lower fraud and boost output. Real-time monitoring is made possible by IoT and edge computing, which continuously track environmental variables, including temperature, humidity, and soil moisture. Edge computing lowers latency and bandwidth consumption by processing data close to the data source, enabling speedier and more efficient analysis. 5G connectivity improves communication by allowing IoT devices in smart agriculture to connect quickly and reliably, especially in remote areas. The precise modification of plant genomes made possible by clustered regularly interspaced short palindromic repeats and gene editing technology increases crop tolerance to pests, illnesses, and environmental stressors.

Hydroponics and vertical gardening allow for precise control over water and nutrient intake while optimizing space efficiency. Biotechnology and bioinformatics methods use helpful microbes and genetic data analysis to increase agricultural yields and resilience. Field simulations and immersive training are provided by augmented reality and virtual reality for farmers to learn new skills.

Reliance on nonrenewable energy sources is decreased by sustainable energy solutions like solar-powered machinery and bioenergy produced from biomass from agricultural waste. Because nanotechnology delivers insecticides and fertilizers directly to plant cells, it reduces environmental effects while increasing output. All things considered, these technologies have the power to completely transform farming practices and provide more effective, efficient, and sustainable agricultural businesses.

10.8 CONCLUSION

The smooth integration and interoperability of diverse technologies, which guarantee that numerous systems and devices operate in harmony with one another, will define the future of smart and secure monitoring in e-agriculture. Data-driven agriculture will be powered by improved data analytics and AI, enabling more accurate and predictive farming methods that boost output and cut waste. By lowering environmental impact and conserving resources, the use of smart monitoring devices will advance sustainability and encourage the use of more sustainable agriculture methods. Global adoption of these technologies will increase as they become more accessible and affordable, which will help both large- and small-scale farmers. Cybersecurity developments will keep agricultural data and systems safe from new attacks. The following actions are necessary for this vision to become a reality: governments and policymakers should support research, funding, and awareness initiatives; educational institutions should integrate smart agriculture into curricula and work with industry; farmers and agribusinesses should invest in smart monitoring technologies and stay informed about advancements; technology developers should innovate and ensure solutions are user-friendly, scalable, and affordable; and consumers and society should advocate for sustainable, technology-driven agricultural practices and support policies promoting these advancement.

KEYWORDS

- **e-agriculture**
- **machine learning**
- **smart monitoring**
- **Internet of Things**
- **soil sensors**

REFERENCES

1. Aulakh, J. S., & Kaur, H. (2020). Smart agriculture monitoring system using internet of things. In *International Conference on Computing, Electronics & Communications Engineering (iCCECE)* (pp. 24–28). Manchester, UK. https: //doi.org/10.1109/iCCECE48183.2020.9162468.
2. Shafiekhani, A., Amro, H. H., Zhang, Y.,& Chen, H C. (2021). IoT-based smart agriculture: A review. *IEEE Internet of Things Journal*, *8*(3), . 1569–1594. https: //doi.org/10.1109/JIOT.2020.3018353.
3. Iftikhar, S., Akbar, M. M., Ahmed, M., & Khan, S. (2020). Smart agriculture: An IoT-based approach to crop monitoring and irrigation management. In *2020 17th International Bhurban Conference on Applied Sciences & Technology (IBCAST),* Islamabad, Pakistan, , (pp. 99–104). https: //doi.org/10.1109/IBCAST48158.2020.9074222.

4. Afzal, A. Ali, M., Ilyas, M., & Malik, K. H. (2020). Smart monitoring and controlling of greenhouse environment using internet of things. In *IEEE 4th International Symposium on Smart Electronic Systems (iSES)* (pp. 126–131). Hyderabad, India. https: //doi.org/10.1109/iSES51748.2020.9264745.
5. Selvaraj, R., Subbiah, P., & Deepa, R. N. (2019). Smart agriculture monitoring and controlling system using IoT. In *2019 3rd International Conference on Electronics, Communication and Aerospace Technology (ICECA)* (pp. 846–850). Coimbatore, India, https: //doi.org/10.1109/ICECA.2019.8822101.
6. Jyothi, N., Sahoo, A. K., Sonawane, A. R., & Balamuralidhar, R. (2020). IoT based smart agriculture monitoring system, 2020 *International Conference on Communication and Signal Processing (ICCSP)* (pp. 0137–0141). Chennai, India https: //doi.org/10.1109/ICCSP48891.2020.9111510.
7. Arasu, A., Kumar, K. B., Raj, M. S., & Rajkumar, P. P. (2020). A survey on IoT based smart agriculture monitoring system. In *2020 International Conference on Recent Trends in Electrical, Electronics & Computing Technologies (ICRTEECT)* (pp. 1–6). Kanchipuram, India, https: //doi.org/10.1109/ICRTEECT48253.2020.9154183.
8. Alkawsi, A. A. M. S., Al-Nuaimy, W. Q. I., Al-Jumaily, K., & Abid, M. T. (2020). A smart IoT-based system for monitoring and controlling agricultural environmental parameters. *IEEE Access, 8*, 58010–58021. https: //doi.org/10.1109/ACCESS.2020.2983550.
9. Mishra, A. Singh, A., & Singh, R. (2020). Design and implementation of IoT based smart agriculture monitoring and controlling system. In *2020 International Conference on Electrical, Computer and Communication Technologies (ICECCT)* (pp. 1–5). Coimbatore, India. https: //doi.org/10.1109/ICECCT48541.2020.9190732.
10. Reddy, M. R. K., Chiluka, N. R., & Guntupalli, A. K. (2021). Smart agriculture monitoring and controlling system using IoT and raspberry pi, In *2021 International Conference on Smart Electronics and Communication (ICOSEC)* (pp. 1–6). Trichy, India. https: //doi.org/10.1109/ICOSEC51453.2021.9423518.
11. Li, J., Hu, X., Wang, W., Li, C., & Li, X. (2020). Smart agriculture monitoring system based on IoT, In *2020 12th International Conference on Intelligent Human-Machine Systems and Cybernetics (IHMSC).* (pp. 235–238). Nanjing: China. https: //doi.org/10.1109/IHMSC50020.2020.9130284.
12. Gupta, M. K. Sharma, R. G., Katiyar, S. K., & Verma, A. (2020). Smart agriculture monitoring system using internet of things. In *2020 4th International Conference on Intelligent Computing and Control Systems (ICICCS)* (pp. 1061–1064). Madurai, India. https: //doi.org/10.1109/ICICCS48265.2020.9120793.
13. Deshmukh, S. V., Borole, D. A., & Mhetre, N. A. (2021). IoT based smart agriculture monitoring and controlling system. In *2021 International Conference for Emerging Technology (INCET)* (pp. 1–4). Pune, India. https: //doi.org/10.1109/INCET53433.2021.9633982.
14. Alahakoon, N., Fernando, H. W. C., Ekanayake, R. & Lakmali, G. D. (2020). A smart agriculture monitoring system based on internet of things, In *2020 5th International Conference on Electrical Engineering and Information & Communication Technology (ICEEICT)* (pp. 1–6). Dhaka, Bangladesh. https: //doi.org/10.1109/ICEEICT49484.2020.9198827.
15. Salam, S. K. (2020). Smart agriculture monitoring and controlling system using IoT. In 2020 *International Conference on Inventive Computation Technologies (ICICT)* (pp. 1669–1673). Coimbatore, India. https: //doi.org/10.1109/ICICT49183.2020.9110437.
16. Kalubarme, S. V., & Wankhade, P. P. (2021). Smart agriculture monitoring and controlling system using IoT. In *2021 International Conference on Inventive Research in Computing Applications (ICIRCA)* (pp. 1369–1373). Coimbatore, India. https: //doi.org/10.1109/ICIRCA53282.2021.9655866.
17. Asrani, K. S., Adhau, S. P., & Sontakke, P. J. (2021). Smart agriculture monitoring and controlling system using IoT. In *2021 International Conference on Advanced Computing and Intelligent Technologies (ICACIT)*, Mysore, India.
18. Kumar, S. Verma, R. & Jha, M. K. (2021). IoT-based smart agriculture: a review. In *2021 4th International Conference on Computing, Communication and Security (ICCCS)* (pp. 1–5). *Patna, India*. https: //doi.org/10.1109/ICCCS52272.2021.9731236.
19. Kuniyal, P., Singh, A., & Bhatt, A. (2023). Detection techniques used for foodborne pathogens and its significant financial and human behavioral effects. *Journal for ReAttach Therapy and Developmental Diversities, 6*(2s), 181–184.

20. Mohammed, S. S., Abdelgawad, A. M. E., Elhoseny, M., & Abd Elaziz, G. M. Smart and secure monitoring devices for e-agriculture: A survey. *IEEE Access, 9*, 134636–134654. https: //doi.org/10.1109/ACCESS.2021.3118003.
21. Akter, H., Hasan, M. J., & Mamun, M. A. A. (2021). A review on smart agriculture monitoring system using IoT and machine learning techniques. In *2021 2nd International Conference on Smart Systems and Data Science (ICSSDS)* (pp. 1–6). Dhaka, Bangladesh. https: //doi.org/10.1109/ICSSDS51289.2021.9635805.
22. Alvi, A. A., Al-Mahmood, M. O. A., & Rahman, S. M. A. (2021). A smart IoT-based system for monitoring and controlling agricultural parameters. In 2021 *IEEE International Conference on Consumer Electronics—Taiwan (ICCE-Taiwan)* (pp. 1–4). Taipei, Taiwan. https: //doi.org/10.1109/ICCE-Taiwan53267.2021.9639480.
23. Saini, D. K. J. B., Ahammad, S. H., Das, P., Bhatt, A., Malusare, P., & Pande, S. D. (2023, September). Grading of fruit ripeness using arduino in IoT. In *2023 International Conference on Sustainable Emerging Innovations in Engineering and Technology (ICSEIET)* (pp. 184–188). IEEE.
24. Khan, A. F. R., Islam, N., & Islam, M. M. (2021). Smart agriculture monitoring system using IoT: a review. In *2021 International Conference on Intelligent Systems Engineering (ICISE)* (pp. 1–6). Dhaka, Bangladesh, https: //doi.org/10.1109/ICISE53272.2021.9641891.
25. Hasan, M. K., Hasan, M. J., & Mamun, M. A. A. (2021). Smart agriculture monitoring system using IoT and machine learning: a review. In *2021 5th International Conference on Computing, Communication and Security (ICCCS)* (pp. 1–5). Patna, India. https: //doi.org/10.1109/ICCCS52272.2021.9731148.
26. Chaudhary, R., Singh, P., & Agarwal, A. (2012). A security solution for the transmission of confidential data and efficient file authentication based on DES, AES, DSS and RSA. *International Journal of Innovative Technology and Exploring Engineering, 1*(3), 5–11.
27. Mamun, M. A. A., Hossain, M. M., Hasan, M. K. & Hasan, M. J. Smart agriculture monitoring system using IoT: A review. In *2021 IEEE International Conference on Industrial Engineering and Engineering Management (IEEM)* (pp. 1–6). Singapore. , https: //doi.org/10.1109/IEEM45035.2021.9732566.
28. Joshi, K., Kumar, A., Anandaram, H., Kumar, A., Aggarwal, A., & Chhabra, G. (2023, August). incorporating AI and IoT for smart poultry monitoring and health detection. In 2023 Second *International Conference on Augmented Intelligence and Sustainable Systems (ICAISS)* pp. 76–81. IEEE.
29. Hasan, M. J., Hasan, M. K. Mamun, M. A. A., & Hossain, M. M. (2021). Smart agriculture monitoring system using IoT and machine learning: A review. In *2021 2nd International Conference on Electrical, Communication and Computer Engineering (ICECCE)* (pp. 1–5). Dhaka, Bangladesh. https: //doi.org/10.1109/ICECCE52269.2021.9462070.
30. Hossain, M. M., Mamun, M. A. A., Hasan, M. K. & Hasan, M. J. (2021). A review on smart agriculture monitoring system using IoT: Challenges and future directions. In *2021 International Conference on Robotics, Electrical and Signal Processing Techniques (ICREST)* (pp. 1–6). Dhaka, Bangladesh, https: //doi.org/10.1109/ICREST53259.2021.9636570.

CHAPTER 11

Mathematical Algorithms in the Convergence of AI and IoT in Agriculture

RASHMI SINGH[1], NEHA BHARDWAJ[2*], and ARYAN CHAUDHARY[3]

[1]*Amity Institute of Applied Sciences, Amity University, Noida, Uttar Pradesh, India*

[2]*Department of Mathematics, Sharda School of Basic Sciences & Research, Sharda University, Greater Noida, Uttar Pradesh, India*

[3]*BioTech Sphere Research, Kolkata, West Bengal, India*

*Corresponding author

ABSTRACT

The merging of AI-controlled IoT devices into the execution of the agricultural industry and food production is revolutionizing the industry by providing innovations in the application of precision farming, resource management, and improved food production yields. This change pertains primarily to computational methods for decision-making or patterns, for data processing, maximization, or prediction. The main algorithm that forms the base of integration of mathematical algorithms with AI and IoT in agriculture is also discussed in this chapter from the aspect of optimization technique, learning approach, and predictive analysis. These innovations are able to help the farmers make the right decision with regards to irrigation, feeding, and pest control in real time thus promoting efficiency and sustainability at the farm level. Further, the chapter explores the implementation issues of using these algorithms in the agriculture industry and gives directions to the future studies. Through the use of AI and IoT in agriculture, there is a possibility of feeding the increasing population, with less and less negative environmental impacts.

11.1 INTRODUCTION

The agriculture sector is feeling the pressure as the global population continues to grow and accelerate, creating immense amounts of pressure on sustainable food production. The world's population is expected to reach 9.7 billion by 2050—and the United Nations says the only way to meet the urgent challenge of food security is to innovate. In response to this problem becomes a new breakthrough of integral application of artificial intelligence

Demystifying Sustainable Farming Using Smart Technologies.
Gunjan Chhabra, Keshav Kaushik, Priya Chugh, Susheela Dahiya, and Aryan Chaudhary (Eds.)

DOI: 10.1201/9781779640932-11

(AI) and Internet of Things (IoT) into modern agriculture. Most notably, AI technology including machine learning, data analytics, and computer vision makes possible the processing of big data generated from IoT devices, like sensors and drones [1, 2]. Such devices capture real-time data on soil health, crop conditions, and environmental variables, and these are critical insights for farmers to make data-driven decisions to optimize their farming, mathematically algorithms underpinning this integration make it an efficient data analysis, efficient farming practices, and predictive modeling. This chapter examines the critical part played by mathematical algorithms to drive the convergence between AI and IoT in agriculture by using them to advance productivity and sustainability in the practices of farming.

According to Figure 11.1, these are the key steps that go on to depict the need for the convergence of agriculture with AI and the IoT.

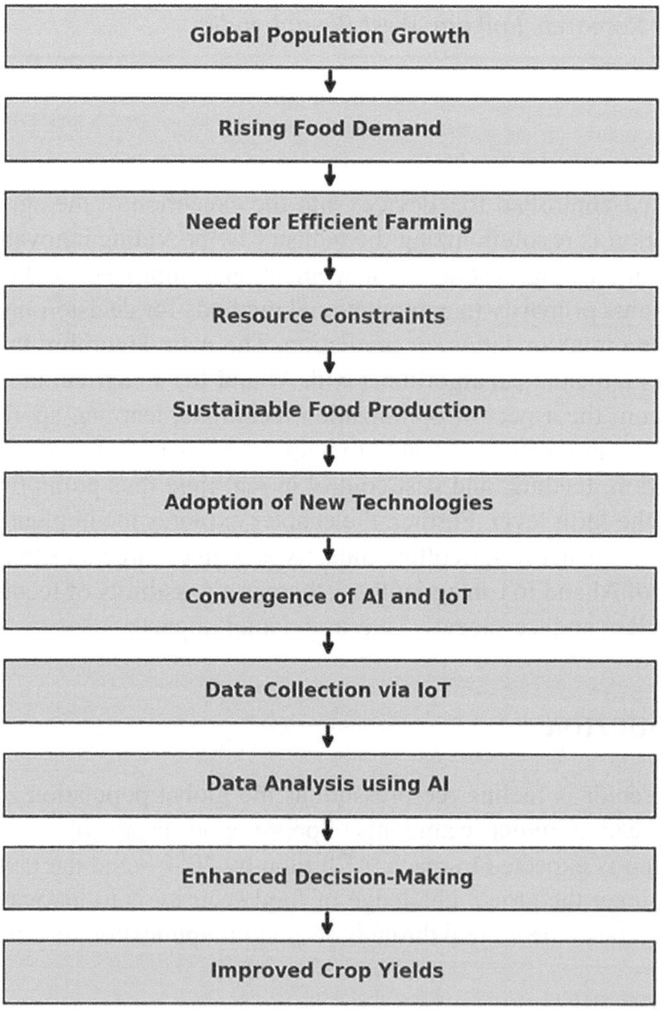

FIGURE 11.1 Interconnectedness of various factors influencing the agricultural landscape.

11.2 AI AND ALGORITHMS IN AGRICULTURAL APPLICATIONS

The impact of AI is changing agriculture with innovative solutions leading to boosting productivity and strengthening efficiency as well as sustainability [3]. Advanced algorithms doing analysis of huge amounts of data are already being used in several AI applications in farming to automate processes and to help farmers make informed decisions. In this section, we explore a few important AI-driven applications in the agricultural industry and what the algorithms powering these advancements. The complete picture of these AI applications in agriculture is given in Table 11.1 that provides a succinct overview of the contributions of each.

TABLE 11.1 Summary of AI Applications in Agriculture

Application	Description	Associated Algorithms
Weather forecasting	Predicts weather patterns to assist in better agricultural planning	Machine learning models, data integration
Soil and crop health monitoring	Evaluates soil quality and monitors crop health for optimal growth	Remote sensing, image recognition
Precision agriculture	Enhances the efficient use of resources like water and fertilizers [4, 5]	Optimization algorithms, VRT
Pest and disease management	Detects and controls pests and diseases to protect crops	Predictive analytics, computer vision
Yield prediction	Forecasts crop yields based on environmental and growth data	Regression analysis, neural networks
Resource management	Maximizes the efficient use of water, fertilizers, and energy	Smart irrigation systems, fertilizer optimization [6]
Supply chain optimization	Improves the efficiency of agricultural logistics and supply chains	Demand forecasting, logistics management
Agri-robotics	Automates tasks like harvesting and weeding to increase productivity	Robotics algorithms, AI-based decision-making

Note: VRT: variable rate technology, AI: artificial intelligence.

11.2.1 WEATHER FORECASTING

By providing accurate weather forecasting, farmers can better understand their planting, irrigation, and harvesting decisions and can improve crop yields and resource management. Using advanced algorithms, AI systems can take a massive amount of weather data and produce highly precise forecasts for just agricultural uses. Figure 11.2 shows how different data sources fit together to improve the overall accuracy of the weather forecasts for the agricultural sector.

11.2.1.1 ROLE OF MACHINE LEARNING MODELS

AI-powered weather forecasting is based on machine learning models that use historical weather data to find the patterns and trends into how weather will look in the future. Key features of these models include the following:

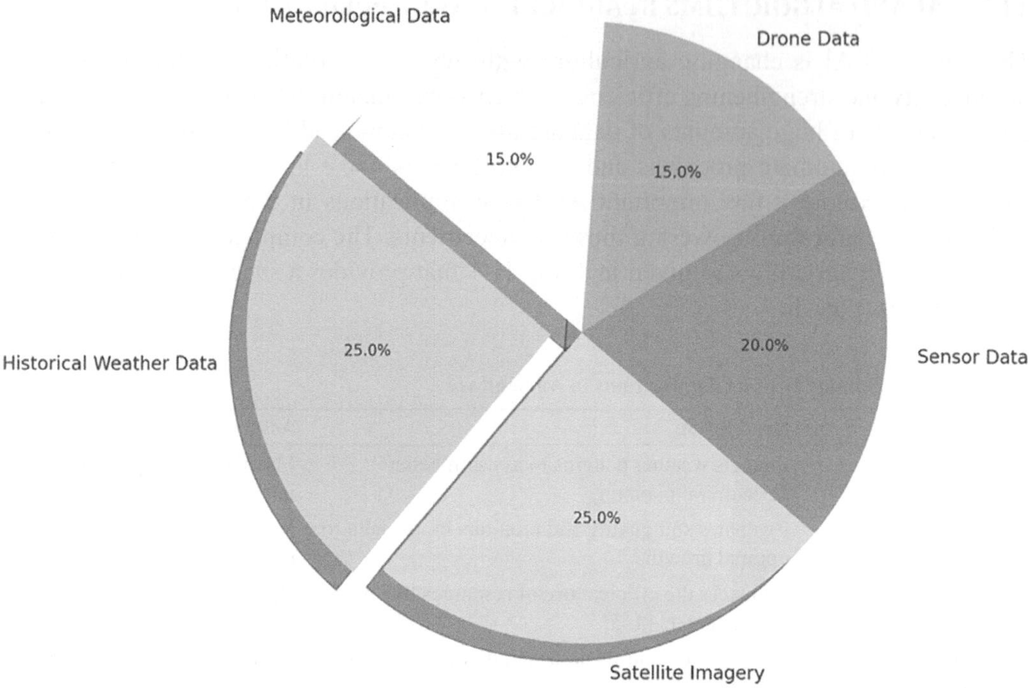

FIGURE 11.2 Contribution of data sources to AI-driven weather forecasting.

- **Pattern recognition:** Machine learning algorithms are better at teasing out the intricate Japanese in which historical data tells us things like temperature variations, precipitation levels, and wind trends. The models also can make more accurate forecasts of what upcoming weather events will look like.
- **Predictive analytics:** But these models give actionable insights to help farmers plan critical things like planting and harvesting. For example, if a dry period is forecasted, farmers may change their planting schedule or adjust their irrigation plans in order to achieve desired outcomes.
- **Adaptive learning:** With more data, there are continually more machine learning models refining their accuracy. These models get better and better with each new input, improving their predictions until they are ever more reliable.

11.2.1.2 DATA INTEGRATION

Adding weather data from different sources make the AI systems work better to create accurate weather forecasts. This data integration includes the following:

- **Satellite imagery:** Up-to-date information from satellites about cloud cover, temperatures, and precipitation patterns is available. This real-time data lets AI models make more accurate, more reliable weather predictions.

- **Sensor data:** Local conditions, such as humidity, temperature, and soil moisture, are recorded from ground-based sensors. Real-time sensor data is analyzed with historical records by these AI algorithms to produce highly localized weather forecasts [7].
- **Drone data:** Weather sensors that are attached to drones take specific area data on microclimates and soil health. Then these hyperlocal data are integrated into AI models to provide even more precise forecasts for specific locations.
- **Meteorological data:** But traditional weather data from meteorological stations is still essential. These weather predictions are further refined by AI systems that use this data with satellite and sensor inputs.

11.2.2 SOIL AND CROP HEALTH MONITORING

Modern agriculture recognizes soil and crop health monitoring as a critical tool in growing conditions optimization and productivity maximization. Traditional methods, like manual soil sampling and visual inspections, cannot keep up with the new data-driven agriculture practice here, which demands accurate actions. In this regard, the need for AI integration into soil and crop monitoring has consequently risen. There is analysis of large data sets from different sources and valuable insights to make the decisions. Different AI applications for monitoring the soil and crop health and their impact on the agricultural products are given in Table 11.2 along with the sources of information.

TABLE 11.2 AI Applications in Soil and Crop Health Monitoring

AI Application	Data Sources	Impact
Soil moisture monitoring	Sensors, satellites	Optimizes irrigation practices, reduces water wastage
Nutrient level monitoring	Remote sensors	Improves soil fertility, ensures balanced nutrient distribution
Crop health monitoring	Drones, Satellites	Detects early signs of stress, nutrient deficiency, disease, or pest damage
Disease detection	Image recognition	Identifies and helps manage diseases early, reducing crop loss
Pest infestation detection	Image analysis	Enables precise pest control, reducing reliance on pesticides

11.2.2.1 ROLE OF AI IN SOIL AND CROP HEALTH MONITORING

Monitoring soil and crop health is about as important—if not more important—as monitoring any other living organism; such monitoring, however, becomes more critical if, perhaps, being not only more efficient and precise but also more healthful for its subject. With the use of advanced algorithms and machine learning models to process highly complex data collected from sensors, drones, and satellites, AI can offer to farmers and agronomists key information about the health of their fields. By capturing these insights, real-time interventions can occur to maximize growing conditions, reduce the waste on resources, and prevent crop loss. What is notable is that AI can detect early signs of stress in crop, nutrient deficiency, or pest infestation so it can be adjusted before damage to yield.

Remote sensing technologies, image recognition algorithms, and predictive models form key parts of AI in this area, which contribute to soil and crop health monitoring by monitoring different aspects of the farm ecosystem from a comprehensive view [8].

11.2.2.2 REMOTE SENSING TECHNOLOGIES

One of the most commonly used technologies in AI-driven monitoring of soil and crop health is remote sensing. With drones, satellites, and ground-based sensors you can use them to gather data pertaining to different parameters affecting the soil and crops. All these data are then processed by AI algorithms to give you actionable insights.

11.2.2.3 SOIL MOISTURE LEVELS

It is known how much we depend on soil moisture for effective irrigation management. Real-time moisture that runs throughout large areas of farmland is monitored by AI systems using remote sensors. It means that we are able to practice more specific irrigation and conserve water at the same time.

11.2.2.4 NUTRIENT LEVELS

Crop health and productivity depend on the content of soil nutrients. Remote sensing of nutrients with AI-enabled technologies can help farmers identify areas of nutrient deficiency. It lets us use the fact better and more effectively and takes targeted fertilizer practices.

11.2.2.5 CROP HEALTH INDICATORS

Images of crops are captured by drones and satellites which are analyzed by AI algorithms against plant health indicators like chlorophyll content and overall biomass. These indicators give you clues as to crop stress, its growth patterns, and whether leaf lesions are caused by disease or nutritional deficiencies. Early detection of these issues with the help of AI analysis can rectify them quickly.

11.2.2.6 IMAGE RECOGNITION ALGORITHMS

AI-based image recognition algorithms are an important part of detecting particular issues related to the health of crops. Using AI, images from the field can be analyzed and will be able to identify some of the subtle visual cues that may be an indication of an underlying problem, nutrient deficiencies, plant diseases, and pest infestations ([("nutrient deficiencies," ("plant diseases," ("pest infestations")))]).

11.2.2.7 NUTRIENT DEFICIENCIES

Image dataset-trained AI algorithms can identify early signs of nutrient deficiencies, like yellowing leaves or stunted growth. Identifying these symptoms early on means specialist and farmers can adjust their fertilization practices quickly, preventing unnecessarily costly crop loss due to nutrient imbalance.

11.2.2.8 DISEASE DETECTION

It is known that AI models can detect the visual signs of various plant diseases before the naked eye can see the disease. These models are trained to recognize the patterns and symptoms of certain diseases so farmers can apply targeted treatment. It reduces the spread of diseases and minimizes crop damage.

11.2.2.9 PEST INFESTATIONS

Pests and their damage to crops are also identified by image recognition algorithms. Passive monitoring of images of affected plants can be useful toward the detection of pests early enough for timely and specific pest management techniques. Specifically, this approach cuts down on the necessity for broad-spectrum pesticides, which makes pest control more sustainable [9, 10].

11.2.2.10 PREDICTIVE MODELS

Apart from real-time monitoring, AI's capabilities to predict future trends and potential risks using predictive models. The combination of historical data with current real-time input is used to predict and forecast crop performance and to identify anomalies that can cause problems soon [11].

11.2.2.11 FORECASTING CROP PERFORMANCE

Predictive models are being combined with controlled field experiments, precision agricultural data, and weather data to identify what is working and what is not and forecast crop yield based on previous growing seasons, using changing weather patterns or soil health trends as factors. It can help farmers anticipate challenges and therefore take preventive measures to optimize crop performance [12].

11.2.2.12 ANOMALY DETECTION

By looking at soil and crop health data for anomalies it can see if a problem might occur soon, it can be a nutrient shortage, water stress, or a pest invasion. By through detecting these anomalies early, AI can let farmers treat problems before they crescendo, thereby allowing better management of the resources and higher crop yields [9, 13].

11.2.3 PRECISION AGRICULTURE

Through the power of AI and optimization algorithms, precision farming provides farmers with data-driven insights to use their resources better—water, fertilizers, and pesticides—efficiently. For sustainable agriculture this approach elevates crop yield without waste [14].

11.2.3.1 OPTIMIZATION ALGORITHMS

Mathematical tools known as optimization algorithms allow a farmer to spend water, fertilizer, and labor just as optimally as possible. Some key algorithms used in precision agriculture are as follows:

1. **Linear programming (LP):** By taking into account such constraints as land area, budget, and resource available, this method is used to increase the efficiency of resource allocation. LP assists in developing more efficient use of limited resources among various areas of a farm [15].
2. **Genetic algorithms (GA):** They come out of natural selection and are used to solve problems where a lot of variables work with each other, such as what is the best crop mix or irrigation strategy. While genetic algorithms are a form of evolutionary algorithm, they evolve solutions based on testing different inputs for farming to find the most suitable farming strategy [16].

11.2.3.2 VARIABLE RATE TECHNOLOGY

Variable rate technology (VRT) is an AI-driven tool that defines different rate for the input given (water, fertilizers, and pesticides) at different places of a field. The amount of input is just right, and the quality of it is good, a "just rightness" that leads to reduced waste and higher yields. This method adjusts inputs based on:

- Soil quality differences
- Moisture levels
- Crop growth stages

These variable applications are informed by the data collected and processed by the AI system from sensors, drones, and satellites.

Figure 11.3 shows a pie chart showcasing the distribution of different AI technologies used in precision agriculture: 30% optimization algorithms, 40% VRT, 20% remote sensing, and 10% predictive analytics. Based on the chart, we can see what each AI technology contributes to precision agriculture, and how much they are used: VRT is the most widely used to optimize input usage across fields. The major benefits of key components of precision agriculture are represented in Figure 11.4. It illustrates the approximate contribution of each component in yield improvement, resource efficiency, and cost reduction in precision agriculture [8].

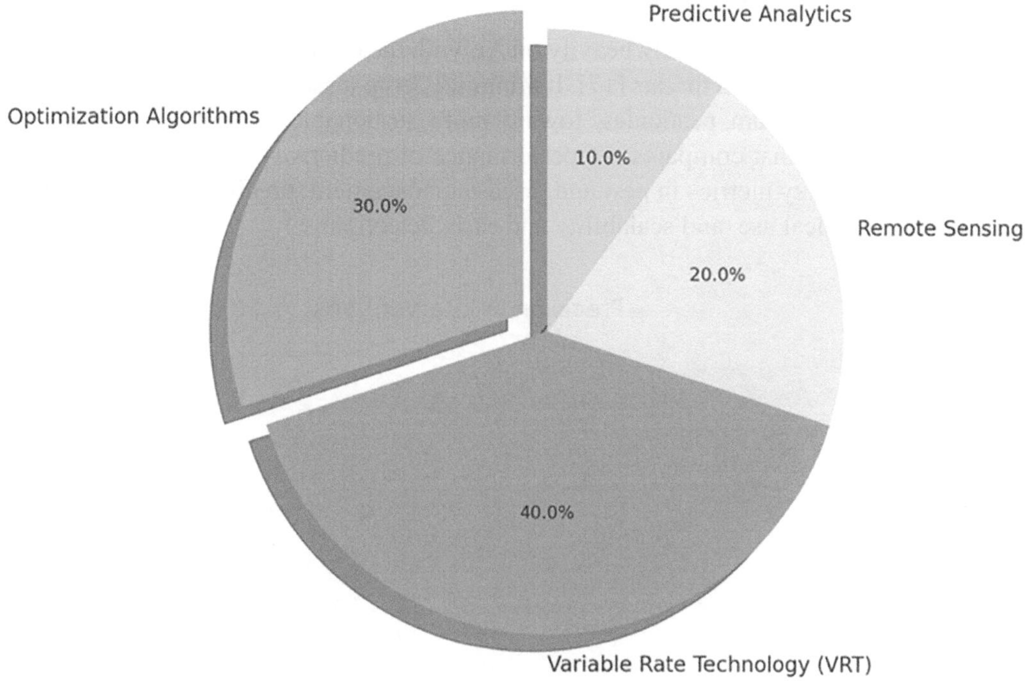

FIGURE 11.3 Distribution of AI tool in precision agriculture.

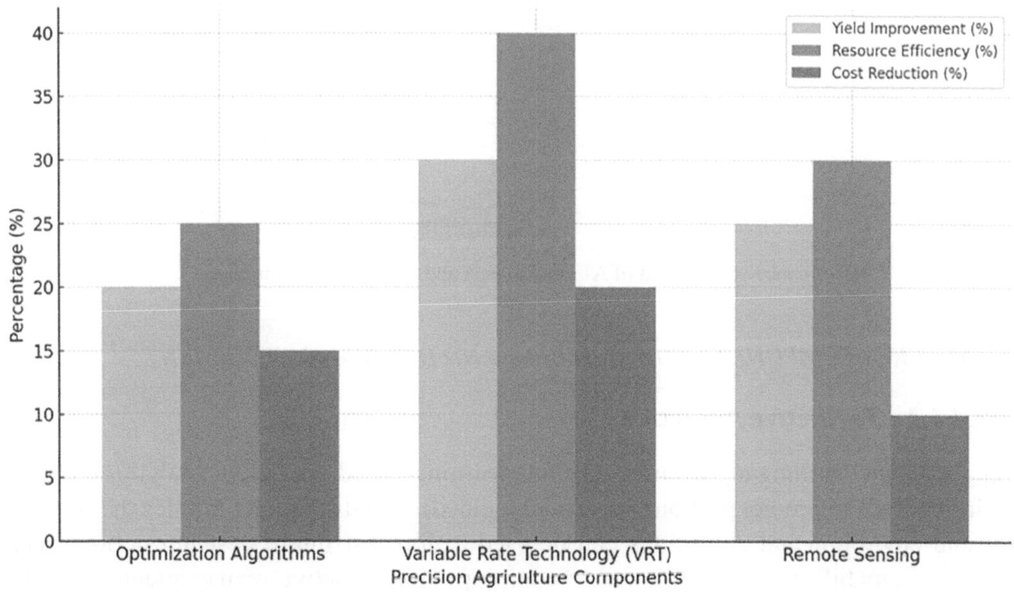

FIGURE 11.4 Benefits of precision agriculture components.

11.2.4 PEST AND DISEASE MANAGEMENT

Pest and disease management rely heavily on AI, with the earliest signaling to early detection and control of potential threats [17]. It minimizes crop losses while minimizing dependency on broad-spectrum pesticides, toward more sustainable agriculture. Figure 11.5 shows a radar chart that compares the performance of predictive analytics and computer vision across four key metrics in pest and disease management: precision in interventions, reduction in chemical use, and scalability and early detection.

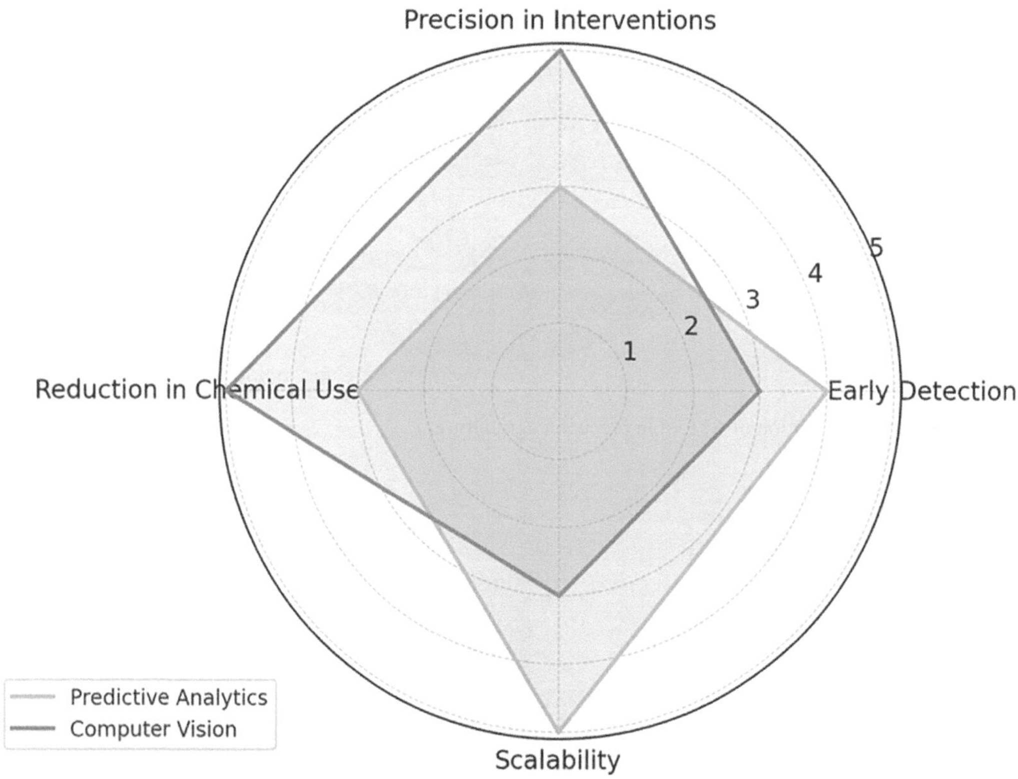

FIGURE 11.5 Performance comparison of AI tools in pest and disease management.

11.2.4.1 KEY COMPONENTS OF AI IN PEST AND DISEASE MANAGEMENT

11.2.4.1.1 Predictive Analytics

Using machine learning algorithms, machine learning-based predictive analytics tools use AI-driven tools to predict pest outbreaks and incidences of disease. AI mimics the way that past and environmental conditions can be used to forecast future pest infestations. That predictive capability enables farmers to come up with proactive management strategies, like early pesticide application, and earlier bio-control measures that can limit the damage significantly.

11.2.4.1.2 Computer Vision

Thus, the image analysis of the computer vision technology is applied to the AI system, which detects pests and diseases in the crop. AI models can process high-resolution images from drones or stationary cameras and identify visual signs of pest infestations or crop diseases also including spots on leaves, discoloration, or wilting. It is about early identification that allows for targeted intervention. The system instead treats only where the problem is, cutting chemical use and environmental impact.

11.2.5 YIELD PREDICTION

An accurate yield prediction is important to modern agriculture for planning and determining market supply. This allows farmers, agronomists and policy makers to make informed decisions on crop production, resource allocation, and financial planning. Yield prediction is helpful in supply chain management in controlling food shortages and maximizing profits. Accurate yield prediction is greatly aided by AI that can analyze a number of inputs. Machine learning algorithms of AI use historical data, particularly environmental data like soil quality and weather conditions to predict future crop yields with super high precision.

11.2.5.1 REGRESSION ANALYSIS

- For yield prediction, one of the fundamental algorithms is a multiple linear regression (MLR). To predict the future yield as the dependent variable, it analyzes historical data along with a number of independent variables, which include soil quality, weather conditions, and so on, and past yield. This is a relatively simple and commonly used method in an agricultural setting for its interpretability and modeling of linear relationships between multiple factors.
- Example: Data relating to the last 10 years, such as rainfall, temperature, or soil quality, might be used by MLR to predict the yield for a given season on the basis of these variables. It will work fine for simple relationships but may have a hard time dealing with complex, nonlinear ones.

11.2.5.2 NEURAL NETWORKS

- One of the fundamental algorithms used for yield prediction is MLR. In this, historical data as well as a series of independent variables, for example, the quality of the soil, the weather conditions, and past yields are analyzed to predict future yield as a dependent variable. However, this method is relatively simple and is used fairly widely in agricultural settings because it is easy to interpret and can model linear relationships of several factors.

- Example: If we had data from the past 10 years, containing rainfall, temperature, and soil quality, MLR could predict the yield for that season. It can be quite effective to make basic predictions out of it but may not be so good when dealing with complex but nonlinear relations in data.

11.2.6 YIELD PREDICTION

Accurate yield prediction is as important as it is essential for efficient planning and market supply management [18]. Helping farmers, agronomists, and policymakers make informed decisions on crop production, resource allocation, and financial planning. Yield prediction helps not only in managing the supply chain, preventing food shortages, but also helps optimize profits. By analyzing multiple data inputs, AI is used to improve the accuracy of yield prediction. Machine learning algorithms incorporated within AI systems can use historical data, including soil quality and weather conditions, to predict possible crop yield, with a level of precision well beyond what humans can achieve.

11.2.6.1 KEY AI TECHNIQUES IN YIELD PREDICTION

11.2.6.1.1 Regression Analysis [19]

- MLR is one of the fundamental algorithms used for yield prediction. It involves analyzing historical data and various independent variables, such as soil quality, weather conditions, and past yields, to predict the future yield as a dependent variable. This method is relatively simple and widely used in agricultural settings for its interpretability and ability to model linear relationships between multiple factors.
- Example: MLR might use data from the last 10 years, including rainfall, temperature, and soil quality, to predict the yield for a particular season based on these variables. While effective for basic predictions, it may struggle with complex, nonlinear relationships in the data.

11.2.6.1.2 Neural Networks

- Neural networks, especially deep learning models, have revolutionized yield prediction by handling more complex and high-dimensional datasets. These models consist of layers of interconnected nodes (neurons) that mimic the human brain's structure, allowing them to learn intricate patterns in the data [20].
- Deep learning algorithms can analyze nonlinear relationships in large datasets, including satellite images, soil nutrient levels, moisture content, and climate forecasts, to provide highly accurate yield predictions. Neural networks excel at

handling a large number of variables and can significantly improve prediction accuracy over time as they are exposed to more data.

- Example: A neural network model could take satellite imagery data combined with on-the-ground weather sensor data to predict the health of crops and forecast their yield. This process involves recognizing patterns such as plant growth stages, moisture stress, and pest infestations, which traditional linear models might miss.

As depicted in Table 11.3, the following observation may be drawn:

TABLE 11.3 Comparison of Regression Analysis and Neural Networks in Yield Prediction

AI Technique	Data Complexity Handling	Accuracy	Scalability	Interpretability
MLR	Low	Moderate	High	High
Neural networks (deep learning)	High	High	Moderate	Low

Note: MLR: multiple linear regression.

- Data complexity handling: Neural networks excel in handling complex, high-dimensional data, while MLR is better suited for simpler, linear relationships.
- Accuracy: Neural networks generally offer higher accuracy due to their ability to capture nonlinear patterns in the data.
- Scalability: MLR is highly scalable and easier to implement across large datasets, while neural networks require more computational resources.
- Interpretability: MLR models are easy to interpret and explain, whereas neural networks function as a "black box," making them harder to explain.

11.2.7 RESOURCE MANAGEMENT

In agriculture, the use of water, fertilizers, and energy is crucial for resource management to improve the use thereof and hence optimize crop productivity and environmental sustainability. As climate change complicates the processes around farming and increases global demand for food, AI is being used more than ever to aid in the sustainability of farming. Agricultural substrates make more precise, data-driven decisions, using AI to maximize yield and minimize resource waste [21].

11.2.7.1 AI-DRIVEN RESOURCE MANAGEMENT TECHNIQUES

11.2.7.1.1 Smart Irrigation Systems

Smart irrigation systems based on AI with integrated soil moisture sensors, weather forecasts, and crop water needs are programmed so as to decide the best time for irrigating [22]. These systems, which analyze real-time data to prevent under- or over-irrigation, also reduce water waste to ensure crops get the exact amount of water they need. The systems, however, are particularly useful in areas where water supply problems are a serious issue.

Example: So a smart irrigation system may use data from sensors in the soil to sense how wet the soil is and, along with weather forecasts, work out how much water to apply and when. This approach not only works to be as efficient with water as possible while also using it only when it is needed but also for crop health.

11.2.7.1.2 Fertilizer Optimization

- Fertilizer application is another area where AI is making significant strides. AI algorithms analyze soil nutrient content, crop growth stages, and environmental conditions to determine the most efficient fertilizer application strategy. This minimizes nutrient waste, optimizes fertilizer usage, and reduces the negative environmental impacts such as soil degradation and runoff into water bodies [23].
- Example: An AI system can process soil data and recommend specific types and amounts of fertilizers to apply at different times during the crop growth cycle. The system continuously adapts its recommendations based on ongoing data collection, ensuring that crops receive the right nutrients without excess.

11.2.8 SUPPLY CHAIN OPTIMIZATION

From on-farm practices onward, AI revolutionizes on-farm practices themselves, and also optimize the agricultural supply chain by delivering goods from farm to consumer efficiently. Harvesting, transportation, storage, and market distribution are the steps of the agricultural supply chain. Since AI-driven solutions provide valuable insights and recommendations, to aid in making the best decisions, reducing waste and timely delivery of agricultural products. It improves supply chain efficiency by optimizing both demand forecasting and logistics management.

11.2.8.1 KEY AI APPLICATIONS IN SUPPLY CHAIN OPTIMIZATION

11.2.8.1.1 Demand Forecasting

Machine learning algorithms run on huge data sets, such as market trends, historic sales data, weather patterns, and customer preferences to forecast demand for agricultural products. It assists in such planning of the production and distribution for the farmers and distributors [24].

Example: With AI-based demand forecasting, farmers find it easier to change crop production levels in line with predicted demand so that they do not overproduce or underproduce. It allows for no excess crop wastage, and because it brings supply closer to consumer demand, this ultimately adds to profitability [25].

Impact: Better inventory management, reduced wastage, and better production planning are all the result of improved demand forecasting.

11.2.8.1.2 Logistics Management

AI-based logistics systems enable optimized transportation routes, vehicle schedules, and inventory tracking so that agricultural products are delivered with efficiency and in time. The identification of the most efficient routes and corresponding schedules, as well as real-time schedule adjustments, reduces delays and spoilage, and lowers transportation costs for these systems [26].

Example: Real-time traffic conditions and weather forecasts, along with other inputs like seasonal availability of products may be monitored by AI-based route optimization software which dynamically updates the transportation plans such that the products can reach the markets faster, thereby minimizing spoilage, especially for perishable goods like fruit and vegetable.

Impact: The use of AI for logistics management results in little transportation delays, higher freshness of agricultural products, and lower transportation and storage costs.

11.2.9 AGRI-ROBOTICS

With AI, power comes to agri-robotics that are able to automate a multitude of laborious tasks including, planting, harvesting, and weed control [27]. These robots give the farmer the ability to undertake agricultural tasks with far greater precision, efficiency, and less dependence on human labor. These robots are able to perform complex tasks like detecting ripe crops, pulling weeds, or working on workflows within the field thanks to the use of AI algorithms.

11.2.9.1 KEY AI APPLICATIONS IN AGRI-ROBOTICS

11.2.9.1.1 Automated Harvesting

Information about crop harvesting with AI-driven robots both spotting and harvesting ripe crops using advanced image recognition and machine learning algorithms. It is equipped with routers and robots with sensors and cameras to sense color, size, and ripeness of fruits and vegetables. It reduces the labor demands with a reduction in human labor especially during labor demands and a reduction in postharvest losses as a result of the speed of harvesting operations [28, 29].

Example: The strawberry-picking robot runs AI so that its tiny arms scan the strawberry field for ripe berries, then pick the berries for a long distance without hurting the plant. It makes the crop more efficient since the only berries that are harvested are fully ripened, improving crop quality.

Impact: The major advantage of automated harvesting is that it drastically reduces labor cost making harvesting more efficient and faster.

11.2.9.1.2 Weed Control Robots

The robot aims to weed control and relies on computer vision and machine learning algorithms to distinguish crops from weeds. Without artificially disturbing crops, the robots can then precisely weed out without the use of chemical herbicides. The targeted approach of this helps maintain soil health, and in a more sustainable farming practice by reducing overuse of herbicides [30].

Example: The sensors and AI algorithms take over from a weed-control robot that moves through a field. After weeds are identified, the robot uses mechanical means (e.g., weeding tools) or a supplied directed herbicide spray to kill the weeds, reducing the use of chemicals.

Impact: These robots diminish dependence on herbicides, cut costs, and create more precise weed removal, all without triggering herbicide resistance in weeds.

11.3 CHALLENGES AND LIMITATIONS

1. Data collection and integration: Gathering accurate, high-quality data is one of the biggest challenges in integrating AI and IoT in agriculture. In order for AI algorithms to be able to process the collected data from various IoT devices, these data should be reliable, timely, and relevant. Decisions on farms can be hampered by inaccurate predictions and recommendations from inconsistent or incomplete data.
2. Scalability and infrastructure: For making IoT devices and AI solutions work with agriculture, you often need a good infrastructure: internet access, power supplies, and so on. However, in many rural areas, these resources are limited which makes it hard to scale AI and IoT solutions properly. The cost of deploying advanced technologies can also be a barrier to small and medium-sized farm(s).
3. Data privacy and security: Supporters of AI and IoT technologies regarding farms have risen as more farms adopt AI and IoT technologies. For example, farmers tend to be reluctant to give the data to large corporations because they fear misuse or loss of control over their data. It is imperative to ensure that data are protected and there is transparency about how data is used in order to gain trust in AI and IoT solutions.
4. Adaptability to diverse agricultural conditions: What has always made this inherently diverse: different crops, climates, and soil conditions all require different approaches. These variations mean that the AI algorithms have to be adaptive enough as well. Unfortunately, the current AI systems have difficulty adapting to the highly variable environment, limiting such systems' usefulness in a variety of different regions.

11.4 CONCLUSION

Combining AI and the IoT has the power to transform agriculture. Mathematical algorithms, with the help of AI-driven technology, can help farmers use real-time data from the IoT gadget to make sensible choices in gaining knowledge of crop health, irrigation, pests administration, and yield predictions. While the potential for AI and IoT in agriculture is great, so are its challenges. Barriers for widespread implementation exist in issues such as data collection and integration and in problems of scalability, data privacy, adaptability to different conditions, and others. Understanding the challenges posed by these technologies and keeping up ongoing research and development, infrastructure improvements, and the application of tailored AI solutions in various agricultural settings is essential to enable these technologies to help secure sustainable and efficient farming the world over. As the world population grows, the successful application of AI and IoT in agriculture will be a fundamental component to feeding the growing population in a sustainable way.

KEYWORDS

- **artificial intelligence**
- **Internet of Things**
- **mathematical algorithms**
- **precision agriculture**
- **predictive analytics**
- **crop yield prediction**
- **resource optimization**
- **sustainable farming**
- **smart agriculture**
- **data-driven decision-making**

REFERENCES

1. Jha, K., Doshi, A., Patel, P., & Shah, M. (2019). A comprehensive review on automation in agriculture using artificial intelligence. *Artificial Intelligence in Agriculture*, *2*, 1–12. https: //doi.org/10.1016/j.aiia.2019.05.004
2. Singh, R. (2024). Application of AI and IoT in agriculture. *Reshaping Intelligent Business and Industry: Convergence of AI and IoT at the Cutting Edge,* Wiley. https: //doi.org/10.1002/9781119905202.ch28
3. Singh, R., Bhardwaj, N., & Islam, S. M. N. (2023). The role of mathematics in data science: Methods, algorithms, and computer programs. In: *Advanced Mathematical Applications in Data Science* (pp. 1–23). Bentham Science Publishers. https: //doi.org/10.2174/9789815124842123010004
4. Neupane, J., & Guo, W. (2019). Agronomic basis and strategies for precision water management: A review. *Agronomy, 9*, 87. https: //doi.org/10.3390/agronomy9020087
5. Jiang, W., Quan, L., Wei, G., Chang, C., & Geng, T. (2023). A conceptual evaluation of a weed control method with post-damage application of herbicides: A composite intelligent intra-row weeding robot. *Soil and Tillage Research*, *234*, 105837. https: //doi.org/10.1016/j.still.2023.105837
6. Torres-Sanchez, R., Navarro-Hellin, H., Guillamon-Frutos, A., San-Segundo, R., & Domingo-Miguel, R. (2020). A decision support system for irrigation management: Analysis and implementation of different learning techniques. *Water, 12*(2), 548. https: //doi.org/10.3390/w12020548

7. Goldstein, A., Fink, L., Meitin, A., Bohadana, S., & Ravid, G. (2018). Applying machine learning on sensor data for irrigation recommendations: Revealing the agronomist's tacit knowledge. *Precision Agriculture, 19*, 421–444. https: //doi.org/10.1007/s11119-018-9564-8
8. Sagan, V., Peterson, K. T., Maimaitijiang, M., Sidike, P., Sloan, J., Greeling, B. A., Maalouf, S., & Adams, C. (2020). Monitoring inland water quality using remote sensing: Potential and limitations of spectral indices, bio-optical simulations, machine learning, and cloud computing. *Earth Science Reviews, 205*, 103187. https: //doi.org/10.1016/j.earscirev.2020.103187
9. Bioresources and Bioprocessing. (2023). *AI and Machine Learning for Soil Analysis: An Assessment of Sustainable Agricultural Practices. Springer.*
10. Talaviya, T., Shah, D., Patel, N., Yagnik, H., & Shah, M. (2020). Implementation of artificial intelligence in agriculture for optimization of irrigation and application of pesticides. *Artificial Intelligence in Agriculture, 4*, 58–73. https: //doi.org/10.1016/j.aiia.2020.04.002
11. MDPI Sensors. (2024). *Recent Methods for Evaluating Crop Water Stress Using AI Techniques: A Review. MDPI.*
12. Frontiers in Agronomy. (2023). *Modeling and Artificial Intelligence (AI) for Climate-Smart Agronomy. Frontiers.*
13. Bai, B., Wang, J., Tian, S., & Peng, X. (2021). Analysis on the competitiveness of global agricultural remote sensing research based on AI technology. *PLoS One, 16*(1), e0230522. https: //doi.org/10.1371/journal.pone.0230522
14. Singh, R., Bhardwaj, N., & Islam, S. M. N. (2024). Applications of mathematical techniques to artificial intelligence: Mathematical methods, algorithms, computer programming, and applications. In: *Advances on Mathematical Modeling and Optimization with Its Applications* (pp. 152–169). CRC Press. https: //doi.org/10.1201/9781003387459-11
15. Alotaibi, A., & Nadeem, F. (2021). A review of applications of linear programming to optimize agricultural solutions. *International Journal of Information Engineering and Electronic Business, 15*(2), 11. http: //dx.doi.org/10.5815/ijieeb.2021.02.02
16. Roy, S. K., & De, D. (2022). Genetic algorithm based internet of precision agricultural things (IopaT) for agriculture 4.0. *Internet of Things, 18*, 100201. https: //doi.org/10.1016/j.iot.2020.100201
17. Donatelli, M., Magarey, R. D., Bregaglio, S., Willocquet, L., Whish, J. P., & Savary, S. (2017). Modelling the impacts of pests and diseases on agricultural systems. *Agricultural Systems*, 155, 213–224. https: //doi.org/10.1016/j.agsy.2017.01.019
18. Gharde, Y., Singh, P. K., Dubey, R. P., & Gupta, P. K. (2018). Assessment of yield and economic losses in agriculture due to weeds in India. *Crop Protection, 107*, 12–18. https: //doi.org/10.1016/j.cropro.2018.01.007
19. Ansarifar, J., Wang, L., & Archontoulis, S. V. (2021). An interaction regression model for crop yield prediction. *Scientific Reports, 11*(1), 1–14.
20. Attri, I., Awasthi, L. K., Sharma, T. P., & Rathee, P. (2023). A review of deep learning techniques used in agriculture. *Ecological Informatics, 77*, 102217. https: //doi.org/10.1016/j.ecoinf.2023.102217
21. Pant, L. P., & Hambly-Odame, H. (2009). Innovations systems in renewable natural resource management and sustainable agriculture: A literature review. *African Journal of Science, Technology, Innovation and Development, 1*(1), 103–135.
22. Darshna, S., Sangavi, T., Mohan, S., Soundharya, A., & Desikan, S., 2015. Smart irrigation system. *IOSR Journal of Electronics and Communication Engineering (IOSR-JECE), 10*(3), 32–36. https: //doi.org/10.1016/j.ifacol.2022.10.125
23. Ojeniyi, K., Ngonidzashe, C., Devkota, K., & Madukwe, D. (2024). Optimizing Split-Fertilizer Applications for *Enhanced Maize Yield and Nutrient Use Efficiency in Nigeria's Middle-Belt. Heliyon. 10*, https: //doi.org/10.1016/j.heliyon.2024.e37747
24. Chelliah, B. J., Latchoumi, T. P., & Senthilselvi, A. (2024). Analysis of demand forecasting of agriculture using machine learning algorithm. *Environment, Development and Sustainability, 26*(1), 1731–1747. http: //dx.doi.org/10.1007/s10668-022-02783-9

25. El Bilali, A., Taleb, A., & Brouziyne, Y. (2021). Groundwater quality forecasting using machine learning algorithms for irrigation purposes. *Agricultural Water Management, 245*, 106625. https: //doi.org/10.1016/j.agwat.2021.106625
26. Talaviya, T., Shah, D., Patel, N., Yagnik, H., & Shah, M. (2020). Implementation of artificial intelligence in agriculture for optimisation of irrigation and application of pesticides and herbicides. *Artificial Intelligence in Agriculture*, *4*, 58–73.
27. Tianjing, Y., & Mhamed, M. (2024). Developments in automated harvesting equipment for the apple in the orchard. *Smart Agricultural Technology*, 100491. https: //doi.org/10.1016/j.atech.2024.100491
28. Zeeshan, S., Aized, T., & Riaz, F. (2024). In-depth evaluation of automated fruit harvesting in unstructured environment for improved robot design. *Machines*, *12*(3), 151.
29. Jiang, W., Quan, L., Wei, G., Chang, C., & Geng, T. (2023). A conceptual evaluation of a weed control method with post-damage application of herbicides: A composite intelligent intra-row weeding robot. *Soil and Tillage Research*, 234, 105837.
30. Genze, N., Bharti, R., Grieb, M., Schultheiss, S. J., & Grimm, D. G. (2020). Accurate machine learning-based germination detection, prediction, and quality assessment of three grain crops. *Plant Methods, 16*, 157. https: //doi.org/10.1186/s13007-020-00686-y

CHAPTER 12

Precision Farming Technologies: Transforming Agriculture for a Sustainable Future

RATNESH KUMARI THAKUR[1*], PRIYA CHUGH[2], and RICHA JASWAL[3]

[1]*College of Basic Sciences and Humanities, PAU, Ludhiana, Punjab, India*

[2]*School of Agriculture and Technology, Maya Devi University, Dehradun, Uttarakhand, India*

[3]*Department of Soil Science, CSKHPKV, Palampur, Himachal Pradesh, India*

*Corresponding author

ABSTRACT

Precision farming is one of the promising methods to reduce the waste of resources and helps in improving farming practices worldwide. The scenario of agriculture has been transformed due to interventions of modern technologies. The use of sensor technology, satellite imaging, global positioning system (GPS), remote sensing, drones, and so on enables the farmer to get real-time information about the field. The infusion of these technologies with traditional agriculture practices makes easy management of field operations. The available sensor helps the farmers to accurately measure soil traits, nutrient levels, and moisture content. It allows target-specific fertilizers and irrigation. The use of drones and unmanned aerial vehicles helps the farmers to get updated on pest attacks and disease infestation and the health of crops. Moreover, these practices utilize geographic information system and GPS technologies to map and monitor field conditions. These technologies also optimize seed planting, spray distributions, and irrigation under field conditions. Overall precision farming reduces cost and increases profitability by reducing the wasteful use of resources. In conclusion, traditional farming has been transformed with the help of modern technological practices, focused resource allocation, and improved crop productivity, enabled by advanced technologies.

12.1 INTRODUCTION

The present estimate of 7.6 billion people on the earth will give way to 11.2 billion by the year 2100. To maintain dynamic ecosystem services and protect the planet's wildlife, we

Demystifying Sustainable Farming Using Smart Technologies.
Gunjan Chhabra, Keshav Kaushik, Priya Chugh, Susheela Dahiya, and Aryan Chaudhary (Eds.)

DOI: 10.1201/9781779640932-12

need a system that can sustain an extra 3.6 billion people while, ideally, consuming as little land and leaving as small an environmental footprint as possible. Modern technology is therefore expected to be introduced and adopted in the agriculture industry. A key tactic in the global movement toward environmentally friendly farming methods and sustainable farming practices is precision agriculture (PA). Precision farming raises average yields by carefully calculating the amount of inputs utilized, as opposed to traditional gardening approaches. A comprehensive system like this one was created to maximize production by applying essential information, technology, and management components in order to boost output, improve product quality, optimize crop chemical use, preserve energy, and protect the environment [30]. Because of this, precision farming is a compelling idea, and its principles inevitably give rise to optimism that farming inputs may be used more wisely, boosting profitability and causing less environmental harm. The technological advancements in precision farming today can supply the means for tomorrow's environmentally sustainable agriculture. Precision farming offers the potential to significantly increase yields while utilizing less external inputs, particularly for small farmers in developing nations [11]. The use of modern technologies such as drones and AI-driven analysis helps in mitigating the effect of climate change, water scarcity, fertilizers, and conserving soil health [23, 33].

12.2 TECHNOLOGIES INTERVENTION IN PA

The PA is the key to addressing various world challenges like population growth, climate change, scarcity of resources, and environmental conservation. Implementation of state-of-the-art technology has become critical, allowing us to produce more food than we ever have and with considerably less negative influence on the planet. The agriculture industry can achieve more with these technologies and meet the increasing demand for food while also reducing its negative environmental impact; as a result, the future can appear more sustainable and fruitful. To address the gaps and promote the implementation and integration of technology in PA, stakeholders, specifically researchers, industries, and policymakers, should collaborate to share ideas and do research and development.

12.2.1 IMPROVE PRODUCTION LEVELS AND FARMING EFFICIENCY

One of the key drivers of propulsion precise technological deployment in agriculture is output and operational efficiency. Due to cutting-edge disciplines such as remote sensing (RS), satellites, drones, and GIS, farmers can acquire data best-describing crop growth and health, soil moisture, and climatic differentiation in real time. The data, in turn, allow them to make more informed business decisions, such as optimal use of synthesized pesticides, organic farming, and irrigation, which results in high resource utilization and profit [22] have published proofs showing an increase in crop production and a decrease in productivity input as a result of deploying presiding farming technology. The study, which was exemplified by experiments comparing the effect of digital soil mapping and variable-rate nitrogen applied in corn fields, achieved up to 8% yield increase compared to the traditional corn farming method.

12.2.2 *ENVIRONMENTAL CONSERVATION AND RESOURCE SUSTAINABILITY*

Precise agriculture technologies promote environmental sustainability and resource conservation. They ensure that products are used with minimal wastage and ensure that resources are not negatively influenced by their application. Using technology such as precision application of chemical fertilizers, pesticides, herbicides, and water based on site-specific data plays a key role in managing waste products and conserving the quality of water and soil. Due to precise sensor technology, Internet of Things (IoT), and data analytics, farmers can identify particular fields requiring intervention and apply precision and target management methods. In a current study on actual data, conducted by [9], scholars evaluated how precision irrigation has impacted the environment. The results show that inappropriately applying irrigation increases farm productivity while reducing water consumption by at least 30%.

12.2.3 *REDUCED PRODUCTION COST AND ECONOMIC SUSTAINABILITY*

PA technology helps farmers to minimize outputs and maximize income, which promotes economic sustainability. With the aid of leftover inputs such as seeds, fertilizer, and pesticides, farmers may reduce their variable costs. In addition, the technology is reducing labor costs, making use of autonomous tools and autonomous robot systems, improving operational efficiency, and eliminating human errors. A recent investigation by [44] assessed the financial impact of precision agricultural techniques on total production in the grains sector. Results suggest a marked increment in the net export per hectare is possible by employing precision farming technologies versus typical farming; this is realized through improved work performance and management, the least cost of input versus high yield.

12.3 PA: DEFINITION AND COMPONENTS

PA is a relatively young field arising at the intersection of agriculture and technology; it is currently revolutionizing traditional farming methods. Developed using newer developments like drones, bezel sensors, AI-driven analytics, and robotics, the field has honed in on close examination of resourcefulness supervision and sturdiness as methods to enhance ingrain yields. For example, by leveraging real-time data collection and AI-inspired bodies of info to optimize irrigation, fertilizer, and pesticide utilization in specific spear of areas and multiple fields, drones have made crop production more economical. According to the latest technology text, PA is a fresh and unique technology with a bright prospect that has the potential to increase resources effectively and inspire the highest sustainable grilling.

12.3.1 IOTS AND SENSORS

IoT and sensors are staple of the ever-evolving precision farming sector that revolutionizes agribusiness. According to Delgado et al., "Advancements in IoT-enabled Precision Agriculture: Enhancing Crop Monitoring and Resource Management," IoT sensors are central to this study in that the real-time data-collection tools offer unique revelations into soil moisture and crop health and much more. When incorporated into a farm's rich tapestry of farming, IoT sensors permit the farmers to distinguish mediocre treatments and make effective data-based decisions, such as adequately covering fertilizer, irrigating, scheduling specialist tending attention, and swiftly recognizing areas of concern practice. The Delgado et al. study offers an in-depth look at how IoT devices have become vital components for farmers to promote output, limit utilization, and back sustainable agrosystems.

Drones and unmanned aerial vehicles (UAVs) represent revolutionary technologies that lead modern agriculture. Velusamy et al.'s research of 2022 named "Unmanned Aerial Vehicles (UAV) in Precision Agriculture: Applications and Challenges" implies the critical role of precision farming drones equipped with a high-resolution camera and multispectral sensors [42]. As described by Smith et al., these flying devices collect data on crop health, pest infestations, and soil quality highly accurately and quickly. UAVs provided partly with artificial intelligence (AI) help farmers get the full picture and respond timely by applying pesticides to the affected areas, changing irrigation patterns, and isolating the first signs of a disease, thus helping Smith et al. better prove that the introduction of drones significantly increased yields.

12.3.2 AI AND MACHINE LEARNING

AI and machine learning are two rapidly evolving technologies that have reshaped agriculture, making a significant impact on precision farming. It is all due to the advent of sophisticated technologies such as AI algorithms, machine learning models, to name a few. Recent AI-driven trends are related to optimized resource allocation, disease identification, and prediction of yield. For instance, computer vision and machine learning in analyzing drone-based images helped detect a plant disease at the early stage and around the vicinity of the disease, which makes a reduction in damage to the total yield possible. Additionally, with AI-powered predictive analytics, farmers can now make crucial irrigation and fertigation and optimal harvesting decisions based on multiple models that take into account many factors in woodland, like weather forecast data, soil conditions, past farm performance, and other sources. It fosters optimal yield, profitability, and resource utilization.

12.3.2.1 MONITORING AGRICULTURAL LANDS

12.3.2.1.1 Satellite Photography and Global Positioning System Technology

Satellite photography and global positioning system (GPS) technology have revolutionized PA by providing farmers with critical data and tools to improve crop management.

12.3.2.1.2 Imageries from Space

Satellite imagery refers to high-resolution images appropriate for agricultural land captured by satellites rotating the earth. Information such as crop health, soil conditions, and moisture content is garnered from the images. The satellite can identify crop health, pests affecting the crops, soil variability, and even predict crop yields through its different spectral bands and sensors.

12.3.2.1.3 Global Positioning System

A satellite-based GPS system can pinpoint an exact location trace in the dust of the planet. Its application in agriculture is that in conjunction with satellite images, it is utilized for PA. GPS receivers in tractors, drones, and other farming equipment play a role in accurately navigating the farm to apply resources such as fertilizer, herbicides, and water. Thus, only the needed resources get appropriately used in targeted locations.

12.4 ADVANTAGES OF PA

1. Enhanced resource management: Farm-generated maps produced using satellite imagery and GPS allow farmers to allocate resources more efficiently, reducing waste, and minimizing the overall environmental impact.
2. Early warning and prioritization: The technologies can recognize crop health indicators, such as levels of nutrition and hydration, which signal concerns like nutrient deficits, pest outbreaks, or diseases.
3. Forecasting and planning: Analyzing historical information and the present environment allows farmers to create educated planting, watering, and harvesting timelines and procedures according to the optimized environment.

12.4.1 RS AND GISS

RS and geographic information systems (GISs) are two technologies taken together. These help us to explore geographical space and objects such as the Earth's surface. According to Refs. [2, 4, 6, 16], these technologies have a wide applicability. RS and GIS are used in agriculture, environmental research, urban planning, and natural resource management. Here, we shall introduce the basic concepts of RS and GIS and point out some of these technologies' applications and advantages. RS is a process of getting information about a thing or an event without putting the sensors onto the thing or going to the event. To achieve this, information is collected by remote sensors, which can be satellites, airplanes, drones, and ground-based sensors. The remote sensors take pictures or get digital signals from a target area, and then the information is processed to derive new types of information from it.

Thanks to RS, data may be consistently collected at fixed intervals over extensive areas. Due to their comprehensive and precise observation of the Earth's surface, remote sensing technologies enable scientists, researchers, and decision-makers to monitor and assess changes in land use, vegetation cover, natural resources, and environmental conditions. It is now an essential instrument in climate change research, resource management, disaster surveillance, and PA.

12.4.2 GEOGRAPHIC INFORMATION SYSTEMS

GIS combines spatial data—information linked to specific geographic locations—with characteristics, or nonspatial data to develop, manage, analyze, and visualize geographically aligned information. GIS enables users to address complex problems related to spatial interactions and patterns by organizing, storing, managing, and analyzing spatial datasets within an integrated framework of hardware, software, data, and expert knowledge. GIS incorporates diverse data sources—including GPS, satellite imagery, aerial photography, and survey data—and employs spatial analysis tools to generate maps, perform spatial modeling, extract actionable insights, and support informed decision-making. Applications of GIS span a wide range of fields, including disaster response, natural resource management, environmental monitoring, urban planning, and transportation management.

RS and GIS: These are essential to modern-day PA as they aid the farmer in managing resources and maximizing crop yields and decisions as well. Some important applications include the following:

1. Crop monitoring and management: The satellite or drone-obtained spectral data are used to check the health of crops. In the case of attack, sickness, or nutrient shortages, certain arguments can be made.
2. Yield prediction: Using modeling and GIS, a predictive map for crop yield is created after data analysis. Assistance in market forecasts for monitoring and harvest planning and labors.
3. Soil mapping and analysis: GIS and RS technology simplify the process of soil analysis by detecting the kind, moisture contents, and nutrient contents of the soil. These data assist in developing variable rate irrigation and fertilizer prescriptions.
4. Precision irrigation: GIS helps to develop irrigation schedules based on topography, soil composition, and crop water needs. RS data aid in optimizing water use through real-time plant health and soil moisture monitoring.
5. Pests and disease management: Before being visible to the naked human eye, areas in a farm with irregularities that were caused by pests or diseases could be identified using RS. GIS can be used to map and control the spread of these problems.
6. Variable rate application: GIS technologies allow the variable rate application (VRA) of inputs, such as fertilizers, herbicides, and pesticides. VRA is made possible by farmers who use these technologies to save resources and reduce costs and environmental impact by mapping field variability.

The following can be considered as benefits of RS and GIS:

1. Spatial decision-making: Support spatial data and information from RS and GIS are data-driven decision procedures. Spatial relations and interactions may be seen and understood to help stakeholders improve choices regarding resource allocation, land development, and other spatial uses with increased spatial relations and interactions.
2. Efficiency and cost-effectiveness: They help gather vast amounts of spatial data over huge areas rapidly and efficiently. When these technologies are combined with contemporary analysis approaches, the cost and time of data collection are reduced, making them cheap alternatives for usage across various scopes.
3. Better resource management: RS and GIS technologies help to manage natural resources more effectively water, soil, and forest resources that play an important role in the agricultural sector. Concerned agencies can develop sustainable ways founded on frequent observation and evaluation, minimizing wastage and enabling. Utilizing RS technologies will help stakeholders ensure optimal usage of limited resources.
4. Improved visualization and communication: A strong platform for conveying and visualizing spatial data is offered by GIS. This makes it simpler to communicate results and insights to stakeholders, decision-makers, and the general public by facilitating the understanding and interpretation of complex information.

12.4.3 HYDROPONICS AND VERTICAL FARMING

PA benefits greatly from the creative solutions provided by hydroponics and vertical farming, which maximize crop development in regulated conditions. Water, nutrition, light, and temperature are just a few of the factors they carefully control and monitor using technology [20, 21, 32, 34, 41]. The goals of PA include higher yields, lower resource use, and increased efficiency. Contributing to this objective are soilless farming techniques like hydroponics and vertical farming, which stacks layers of crops in a regulated environment.

1. Resource efficiency: Water, fertilizers, and space can all be used precisely with hydroponics and vertical growing. They are especially important in places where there is a shortage of water or arable land since they reduce waste and maximize resource allocation.
2. Controlled environments: These techniques allow environmental parameters to be adjusted. Sensors, IoTs, and automation technologies monitor and adjust variables like humidity, pH, and light intensity to provide crops the optimal growing circumstances.
3. Year-round production: These techniques make it easier to cultivate year-round and lessen reliance on seasonal fluctuations by managing environmental conditions. Supply networks can be stabilized and food security enhanced by this steady output.

4. Maximum yield in minimal space: Soilless hydroponic systems and stacked vertical farming structures make the most utilization of the given area. This efficiency is especially helpful in urban settings where land is limited.

12.4.4 BLOCKCHAIN TECHNOLOGY

As of late, blockchain is the only novel technology that can be applied to PA to mitigate major issues, particularly when combined with IoT technology [27, 38]. The application of blockchain technology in the precision agricultural and food supply chain sectors is expected to rise from $41.9 million in 2018 to $1.4 billion by 2028, according to a new market intelligence analysis by BIS Research [1]. PA can benefit greatly from a number of new uses for blockchain technology. For instance, take into account data assurance and security, supply chain tracking and monitoring [3, 18, 28], and financial management [7]. The government's determination to rearrange its laws and rules to incorporate blockchain technology into its economic strategies is partially motivated by the increasing adoption of public blockchain technology in food markets.

Following are some areas where blockchain helps in achieving precision farming.

1. The blockchain ensures that in precision farming data collected from various sources is tamperproof and transparent. It helps in accurate decision-making.
2. It helps in keeping track of products from farm to end customer.
3. It ensured the exchange of information between farmers, researchers, and agronomists for conductive decision-making. It maintained data privacy and ownership by enabling secure data exchanges across a wide range of agriculture.
4. Smart contracts for agreements: In PA, related contracts and transactions can be done through programming contracts on a blockchain. For instance, smart contracts can interconnect contracts between a farmer and equipment provider on usage or maintenance, or it can interconnect payment upon meeting the agreed standards such as volume and quality.
5. Verification of practices and inputs: Blockchain can assure and authenticate the use of certain inputs, such as residential areas, fertilizers, and water intake. Such transparency may help validate the use of ethical farming and compliance with specific principles on sustainability.

In decision support systems (DSS), and PA, precision farmers utilize techno-instruments to make data-driven informed decisions to maximize their investment in crop production, resource management, and over-farm productivity. This does a healthy synthesis: First, data are integrated, where DSS shall combine multiple sources of data to facilitate precision farming. These include weather stations, soil sensors, satellite photographs, and farmer's previous records in one platform. Second, data integration is analyzed and modeled, where several analytics and modeling techniques are used to analyze integrated data. This integrated data has information about the predicted yield, disease patterns and recommendations, soil study for soil fertility analysis, and predictive modeling for weather patterns. Third, DSS offers a decision support tool, whereby depending on the analysis carried out on the data, DSS analysis is presented in a board, and a user-friendly interface helps the farmer grasp passive and active options from the data. This system outlays recommendations and alternatives depending on the data analysis carried out. Fourth, these systems give customized advice, which is data analyzed such that it is expressly tailored for the farmer considering crop type, soil content, weather prediction, and previous data. Such advice may be on the timing of applying the insecticides, the exact watering schedules, or even the timing to seed. Fifth, these systems can also afford real-time monitoring and control, where a few DSS can monitor and control in real time, and farmers can watch their farm's performance where they can control it. For example, one can watch the performance of the spraying machine or control the watering systems. Lastly, it offers integration with other precision technologies, where DSS works with integrated other pseudo-technologies in precision farming such as aerial surveillance, variable rates technology (VRT) for variable fertilization or seeding, and G prairie seed for location precision canning—guided equipment for real-time monitoring, for of water canning, fertilizers, and pesticides.

12.4.4.1 VARIABLE RATES TECHNOLOGY

VRT is a fundamental component of PF since it enables the application of input elements like seed, water, fertilizer, and pesticides to various parts of a field, which require unique amounts of elements based on their needs. Farmers can do better due to the introduction of variable rate technology, a significant achievement of PA. VRT makes it possible to apply input to different areas of the field or block and quantify input values based on spatially continuous data. According to Ferguson et al. [10], instead of the traditional uniform application approach, VRT technology allows farmers to treat one field area as distinct from another for treatment purposes by configuring the spreading rates to a reopening place on a per hectare basis. VRT operates as follows:

1. Data collection and analysis: Soil maps, yield maps, satellite images, and sensors are sources of data used by VRT. The obtained data are then analyzed for spatial differences in elements such as soil fertility, moisture content, organic matter content, and historical performance yield.
2. Mapping and prescription generation: Farmers or agricultural consultants create maps based on the analyzed data. The created map indicates the various areas within the field, which details the precise input needs for each area. For example, an area with a high soil fertility level can do with less fertilizer, and those with low soil fertility levels will require more.
3. Precision application equipment: VRT equipment, such as variable rate spreaders or sprayers, utilizes technology that changes the input application rate as the equipment moves across a field in real time. GPS-guided equipment ensures exact application to prescription map specifications.
4. Optimized input use: Due to precise application and allocation of inputs, resource use is optimized. In addition, it increases overall sustainability, reduces input costs, and mitigates the environmental footprint while increasing crop yield.

5. Adaptability and record-keeping: With VRT systems, farmers can monitor the effectiveness of various inputs across the field over time. Systems can adapt to different field conditions and keep data on source rates applied.

12.5 APPLICATION OF TECHNOLOGIES FOR PA

In the mid-to-late-1980s, research on PA began in the United States, Australia, Canada, and Western Europe. Only a small percentage of farmers have embraced any kind of PA technology, despite substantial research efforts. The primary method of implementing PA has been to use readily available field equipment and add GPS and controllers to allow for spatially variable applications.

12.5.1 MONITORING OF CROP GROWTH STAGES

Crop growth can be tracked with PA. During the mature period, agricultural issues including nutrition and water stress, illness, bug, and weed infestation can all be identified with RS and GIS. Various sensors can be used to collect data, which can then be correlated with GPS to generate field management strategies for water, fertilizer, and crop production.

12.5.2 CONTROL OF WEED, INSECTS, AND DISEASES

The mapping of hotspots for weeds, pests, and diseases is comparable to postdisaster management, which is permitted only in cases where the crop is rendered virtually unusable. Mapping disease incidences is made possible by the system that is created when RS and GIS are integrated. The experts can determine the underlying causes of crop illnesses by mapping. By mapping the locations of disease occurrences, agro-ecological zones can be spatially integrated, aiding in the prediction of additional hotspots for associated infestation in the future. One of the main benefits of the GIS and RS is that because of their analytical powers, farmers may simulate and get the early warning system in addition to mapping.

12.5.3 AGROCHEMICAL APPLICATION OF VRT

When it comes to maximizing the use of fertilizers, precision farming is essential when it comes to the three main nutrients: N, P, and K. These nutrients are evenly sprayed to fields at specific periods of the year in conventional farming. This causes some areas to receive excessive nutrient treatment while others receive insufficient application. Excessive use of fertilizer leads to the leakage of nitrogen and phosphorus into surface and ground waterways as well as other unwanted areas, directly increasing environmental expenses. Nutrients can be applied more precisely when using PA, and the application can be optimized using a spatial and temporal component. For VRT applications, a number of firms are currently manufacturing herbicide applicators, sprayer systems, controllers, and air spreaders.

12.5.4 ACCURATE HARVESTING

By switching around where their animals spend the night, farmers may better control the manure that is applied to planted areas. They can carefully apply manure to the "bad spots" that require the greatest nutrients and organic matter application by using this precision manuring technique. Small farmers frequently use the deliberate application of manure, compost, and other nutrients to low-yielding crops [40]. The technique is especially beneficial to small farmers, who do not have enough property to overlook areas with diminishing soil richness [36].

12.5.5 ACCURATE WATERING IN PRESSURIZED SYSTEMS

Recently, innovations that use GPS-based controllers to control the movement of irrigation devices are being made available for commercial usage in sprinkler irrigation. Sensor technologies and wireless communication are being developed in addition to movement control to monitor soil and ambient conditions as well as the irrigation machines' operating parameters (flow rate and pressure) in order to achieve high water application efficiency and crop consumption [25]. These technologies have a lot of promise, but more work needs to be done before it's possible to buy them commercially.

12.5.6 ACCURATE CONSERVATION AGRICULTURE

By customizing techniques to local conditions, precision conservation agriculture helps farmers implement conservation agriculture successfully [17]. Three basic principles characterize conservation agriculture:

1. Reducing the amount of soil disturbance.
2. Applying associations and/or crop rotations.
3. Maintaining crop residue layers on the soil [12].

Among the PCA land preparation techniques are hand-dug planting holes, meticulous lime application in each plant's root zone, and precise plant placement.

12.5.7 CROP SUITABILITY AND CLIMATE COMPATIBILITY

Including reflectance from the visible, near-infrared, and mid-infrared regions of the electromagnetic spectrum, hyperspectral and multispectral images are useful for tasks including stress mapping, irrigation management, and pesticide and fertilizer application. Physical characteristics including crop cover, crop health, and soil moisture can all be used to interpret results [31]. Hyperspectral and multispectral RS data have also been used to examine the nutrient content of many crops, including rice [35], wheat, and sorghum [43].

12.5.8 MONITORING OF YIELD

Yield monitors are made up of multiple parts. They are made up of various sensors and parts, such as a data storage device, a task computer located in the combined cab that controls how these components integrate and work together, and a user interface with a keyboard and monitor. The sensors measure the separator speed, ground speed, and grain flow volume. Grain yield is continuously measured in combination by calculating the force of the grain flow through the grain elevator's sensible plate. The latest development in mass flow sensors measures the amount of energy that is reflected back after hitting the seed stream going down the chutes and emitting microwave beams. All of the yield monitors' yield maps and location-based yield statistics are created using GPS receivers.

12.5.9 SYSTEMS FOR ROBOTIC HARVESTING

Japanese researchers designed an autonomous vehicle and a sophisticated, automated follow-up vehicle system for various farm tasks. Commercial harvesting robots had already been produced for cherry tomatoes, tomatoes, cucumbers, grapes, watermelons, and strawberries. Early in the twenty-first century, more robotic harvesting systems are expected to go on sale. In the horticultural industry, robotic harvesting has been successfully tested. Companies like Agrobot and Harvest Croo (FAO) have used robotics in strawberry harvesting.

12.5.10 METICULOUS MANAGEMENT OF LIVESTOCK

Radiofrequency identification tags are already in use in the dairy industry to identify calves in robotic milking equipment and computer-controlled self-feeders [14]. Calves can have their milk supplement customized, their body temperature and weight measured, and reports can be generated by automatic milk feeders. Dairy operators can work less and milk cows on a schedule thanks to milking robots. These robots can also be designed to perform online analyses of the content of milk, including cell counts, lactose, protein, and fat—a crucial determinant of sanitary condition [24]. Individual animal feeding is made possible by knowledge of the quantity and quality of milk [24]. When used in conjunction with additional sensors, Global Navigation Satellite System receivers enable behavior and welfare tracking of individual animals in outside environments.

12.6 THE ADVANTAGES OF PRECISION FARMING

12.6.1 BENEFITS ASSOCIATED WITH CROPS

Increased crop yields are the outcome of precision agricultural techniques, which improve water and nutrient use efficiencies [8]. The appropriate amount of fertilizer is administered at the appropriate time and location for the stage of crop growth [13]. Soil structure is improved by PA activities like conservation agriculture. Reducing tillage increases soil biodiversity, adds organic matter, and permits irrigation and rainwater to permeate the soil [37]. Timely and efficient weed removal eliminates competition for nutrients and moisture in the soil, freeing up resources for the crop to thrive. Because PA makes optimal use of input resources like labor and fertilizers, farm output rose dramatically [19].

12.6.1.1 ECONOMIC BENEFITS

According to Jensen et al. [15], PA can be the cornerstone of productive agroecosystem production, resulting in large savings on agricultural inputs. According to Rider et al. [29], PA can save up to 60% on agrochemicals like pesticides and nearly 30% on fertilizers. Using PA techniques also lessens the chance of applying fertilizer and agrochemicals incorrectly or excessively. Certain studies indicate that PA has good potential for site-specific nitrogen management, particularly in small-scale farming systems [5]. Thus, accurate application of fertilizers, herbicides, and insecticides minimizes input costs for smallholder farmers by reducing losses [39]. Because precision agricultural techniques increase the output from labor invested in farming operations, they raise labor productivity in the commercial farming industry. Ultimately, the higher production per unit input results in higher agricultural income.

12.6.1.2 ADVANTAGES FOR THE ENVIRONMENT

In commercial farming, where crop production requires substantial input consumption (fertilizer, herbicides, and insecticides), precision is particularly crucial. A significant worldwide concern is the contamination of surface and ground-water resources by agrochemicals [26] state that PA practices are essential to reducing the inappropriate use of environmentally harmful substances. By increasing output in a unit area, PA can reduce the amount of new land available for agricultural production [8]. On and near farms that use PA methods, soil erosion, and degradation are also reduced. Thus, precision farming is essential for protecting the environment.

12.6.2 OPPORTUNITIES AND CHALLENGES OF TECHNOLOGY IN PA

12.6.2.1 OPPORTUNITIES

Enhanced efficiency: Thanks to technologies including IoT sensors, drones, and GPS-guided machinery, farmers may monitor crops, make the greatest use of resources, and even automate farming tasks. This results in cost-cutting and increased productivity.

- Data-driven decision-making: Farmers who have access to up-to-date information on crop health, soil health, and weather are better equipped to make well-informed judgments. Crop management is aided by AI and predictive analytics, which improve quality and output.

- Resource optimization: PA reduces resource waste by accurately applying inputs like fertilizer, pesticides, and water. This reduces the impact on the environment while also cutting expenditures on resource-efficient.
- Sustainability improvement: The adoption of technology allows for the adoption of sustainable agricultural practices. Precision farming makes it easier to monitor soil, use beds of chemicals, and protect vital resources.
- Quality assurance and new markets: Technology-enabled traceability and certification guarantee product security. People are becoming increasingly curious about where their food originated and producing methods. High-quality new market opportunities for ethically sourced products abound.

12.6.2.2 CHALLENGES

- High initial investment: When implementing PA technologies, equipment, sensors, and even software are frequently acquired with a lot of money upfront. Farmers may find this to be a barrier in a small scale.
- Data management and integration: Managing vast amounts of data from numerous sources can be tough. Data security and interoperability across platforms must be ensured.
- Access to education and technology: Farmers do not have comparable access to technology for a variety of reasons like geography, infrastructure, and technical expertise. The digital divide needs to be closed to make it inclusive.
- Infrastructure and technical assistance: Reliable broadband internet access and ongoing technical support are essential to ensure the effective implementation of these technologies in remote areas. Infrastructure should be developed for wide adoption.
- Privacy and data ownership: Since the farmers give data away to the technology companies, there is a problem of privacy and data ownership. It will require regulations for the two issues.
- Skill gap and adoption hurdles: In order to put sophisticated technologies in place, the farmers will need training and to have new skills. There might be a challenge of adoption because of the lack of information and something new to some.

12.7 CONCLUSION

In conclusion, PF technologies have made a difference and transformed agriculture by providing innovative solutions to modern-day agriculture challenges and creating the path to a more sustainable future. These allow the farmers to use resources such as pesticides, fertilizers, and water roll in the farming process as possible. The use of data-driven insights and real-time monitoring enables farmers to make better decision-making and effective use of resources. In summary, precision farming technologies provide a way forward to a more sustainable and resilient future for agriculture. By using data, technology, and teamwork, we can feed a growing population while ensuring agriculture is socially, economically, and environmentally sustainable. To achieve the maximum, we will need to have a farmer, policy, research, and technology developer to hire parties to join hands and take sustainability to the next level.

KEYWORDS

- **precision farming**
- **global positioning systems**
- **unmanned aerial vehicles**
- **geographic information systems**

REFERENCES

1. BIS Research. (2018). *Global Blockchain in Agriculture & Food Market—Analysis and Forecast 2018–2028.*
2. Blaschke, T. (2010). Object-based image analysis for remote sensing. *ISPRS Journal of Photogrammetry and Remote Sensing, 65*(1), 2–16.
3. Bordel, B., Lebigot, P., Alcarria, R., & Robles, T. (2018). Digital food product traceability: Using blockchain in the international commerce. In: Antipova, T., & Rocha, A., (Eds.), *Digital Science. DSIC18. Advances in Intelligent Systems and Computing* (vol 850). Springer: Cham. https: //doi.org/10.1007/978-3-030-02351-5_27
4. Campbell, J. B., & Wynne, R. H. (2011). *Introduction to Remote Sensing*. Guilford Press.
5. Cao, Q., Cui, Z., Chen, X., Khosla, R., Dao, T. H., Miao, Y., et al. (2012). Quantifying spatial variability of indigenous nitrogen supply for precision nitrogen management in small scale farming. *Precision Agriculture, 13*(1), 45–61.
6. Chang, K. T. (2018). *Introduction to Geographic Information Systems*. McGraw-Hill Education: New York.
7. Chinaka, M. (2016). *Blockchain Technology—Applications in Improving Financial Inclusion in Developing Economies: Case Study for Small Scale Agriculture in Africa* (p. 43). PhD thesis, Massachusetts Institute of Technology (MIT): Cambridge, MA, USA.
8. Dobermann, A., & Nelson, R. (2013). *Opportunities and Solutions for Sustainable Food Production.* Sustainable Development Solutions Network: United Nations.
9. Elhakeem, A., Ramadan, M., Toma, Y., & Youssef, M. A. (2021). A novel IoT/cloud-based datalogger for precision irrigation systems. *IEEE Internet of Things Journal, 8*(1), 751–760.
10. Ferguson, R., Dobermann, A., & Schepers, J. (2007). *Precision Agriculture: Site-Specific Nitrogen Management for Irrigated Corn* (pp. 1–7). University of Nebraska Lincoln Extension. Bulletin.
11. Fountas, S., Ess, D., Sorensen, C. G., Hawkins, S., Pedersen, H. H., Blackmore, S., & Deboer, L. J. (2004). Farmer experience with precision agriculture in Denmark and US Eastern corn belt. *Precision Agriculture, 6*, 121–141.
12. Giller, K. E., Corbeels, M., Nyamangara, J., Triomphe, B., & Affholder, F. (2011). A research agenda to explore the role of conservation agriculture in African smallholder farming systems. *Field Crops Research, 124*(3), 468–472.

13. Ibrahim, A., Pasternak, D., & Fatonji, D. (2015). Impact of depth of placement of mineral fertilizer micro-dosing on growth, yield and partial nutrient balance in pearl millet cropping system in the Sahel. *Journal of Agricultural Science*, *153*, 1412–1421.
14. Ipema, C., Lokhorst, P. W. G., & Koerkamp, G. (Eds.), (2009). In: *Precision Livestock Farming* (pp. 135–142), Wageningen Academic Publishers: Wageningen, Netherlands.
15. Jensen, H. G., Jacobsen, L. B., Pedersen, S. M., & Tavella, E. (2009). Socioeconomic impact of widespread adoption of precision farming and controlled traffic systems in Denmark. *Precision Agriculture*, *13*(6), 661–677.
16. Jensen, J. R. (2007). *Remote Sensing of the Environment: An Earth Resource Perspective*. Pearson Education India.
17. Jerich, M. (2011). Potential of precision conservation agriculture as a means of increasing productivity and incomes for smallholder farmers. *Journal of Soil and Water Conservation*, *66*, 171–174.
18. Lin, J., Shen, Z., Zhang, A., & Chai, Y. (2018). Blockchain and IoT based food traceability system. http://dx.doi.org/10.1145/3265689.3265692
19. Knight, B., & Malcolm, B. (2009). A whole-farm investment analysis of some precision agriculture technologies. *Australian Farm Business Management Journal*, *6*(1), 41.
20. Kumari, R., Bedi, S., Dhatt, A. S., & Sethi, V. P. (2021). Effect of proportion of NO_3^- and NH_4^+ in nutrient solution and growing seasons on yield and quality in coriander. *Journal of Agricultural Science and Technology*, *23*(4), 891–902.
21. Kumari, R., Bedi, S., & Dhatt, A. S. (2020). Effect of nitrogen forms on physiological and biochemical constituents of lettuce grown in soilless conditions during different seasons. *Indian Journal of Horticulture*, *77*(4), 663–670.
22. Li, Z., Meng, Y., Rui, Y., Xie, H., Xie, S., & Wang, D. (2021). Effects of digital soil mapping and variable-rate nitrogen application on yield and nitrogen use of maize. *Agronomy*, *11*(1), 176.
23. Liu, Z., Qiu, C., & Zhu, L. (2022b). Dynamic role of green energy efficiency and climate technologies in realizing environmental sustainability: Fresh insights from China. *Frontiers in Environmental Science*, *10*, 908470. https: //doi.org/10.3389/fenvs.2022.908470.
24. Maltz, E., Antler, A., Halachmi, I., & Schmilovtich, Z. (2009). In: Lokhorst, C., & Groot Koerkamp, P. W. G., (Eds.), *Precision Livestock Farming* (pp. 17–23), Wageningen Academic Publishers: Wageningen, Netherlands.
25. O'Shaughnessy, S., & Evett, S. (2010). Canopy temperature based system effectively schedules and controls center pivot irrigation of cotton. *Agricultural Water Management*, *97*, 1310–1316.
26. Oliver, M. A., Bishop, T. F. A., & Marchant, B. P. (2013). Precision agriculture for sustainability and environmental protection. In: *Earth Scan Food and Agriculture*. Routledge: Abingdon.
27. Prashar, D., Jha, N., Jha, S., Lee, Y., & Joshi, G. P. (2020). Blockchain-based traceability and visibility for agricultural products: A decentralized way of ensuring food safety in India. *Sustainability*, *12*(8), 3497. https: //doi.org/10.3390/su12083497
28. Casado-Vara, R., et al. (2018). Howblockchain improves the supply chain: Case study alimentary supply chain. *Procedia Computer Science*, *134*, 393–398.
29. Rider, T. W., Vogel, J. W., Dille, J. A., Dhuyvetter, K. C., & Kastens, T. L. (2006). An economic evaluation of site-specific herbicide application. *Precision Agriculture*, *7*(6), 379–392.
30. Shibusawa, S. (2002). Precision farming approaches to small farm agriculture. *Agro-Chemicals Report*, *2*(4), 13–20.
31. Singh, D., Sao, R., & Singh, K. P. (2007). A remote sensing assessment of pest infestation on sorghum. *Advances Space Research*, *39*, 155–163
32. Siregar, R. R. A., Seminar, K. B., Wahjuni, S., & Santosa, E. (2022). Vertical farming perspectives in support of precision agriculture using artificial intelligence: A review. *Computers*, *11*(9), 135.
33. Smith, P., Martino, D., Cai, Z., Gwary, D., Janzen, H., Kumar, P., McCarl, B., Ogle, S., O'Mara, F., Rice, C., Scholes, B., Sirotenko, O., Howden, M., McAllister, T., Pan, G., Romanenkov, V., Schneider, U., Towprayoon, S., Wattenbach, M., & Smith, J. (2008). Greenhouse gas mitigation in agriculture. *Philosophical Transactions of the Royal Society B: Biological Sciences*, 363 (1492), 789–813. https: //doi.org/10.1098/rstb.2007.2184.

34. Soojin, O., & Chungui, L. (2023). Vertical farming—Smart urban agriculture for enhancing resilience and sustainability in food security. *The Journal of Horticultural Science and Biotechnology*, *98*(2), 133–140. https: //doi.org/10.1080/14620316.2022.2141666.
35. Stroppiana, D., Boschetti, M., Brivio, P. A., & Bocchi, S. (2008). Plant nitrogen concentration in paddy rice from field canopy hyperspectral radiometry. *Field Crop Research*, *111*, 119–129.
36. Taddesse, G., Peden, D., Abiye, A., & Wagnew, A. (2003). Effect of manure on grazing lands in Ethiopia, East African highlands mountain. *Research and Development*, *23*, 156–160.
37. Thierfelder, C., Rusinamhodzi, L., Ngwira, A. R., Mupangwa, W., Nyagumbo, I., Kassie, G. T., et al. (2014). Conservation agriculture in Southern Africa: Advances in knowledge. *Renewable Agriculture and Food Systems*, *30*, 328–348.
38. Tripoli, M., & Schmidhuber, J. (2018). *Emerging Opportunities for the Application of Blockchain in the Agri-food Industry* (pp. 1–29). FAO: Geneva, Switzerland.
39. Twomlow, S., Hove, L., Mupangwa, W., Masikati, P., & Mashingaidze, N. (2008). Precision conservation agriculture for vulnerable farmers in low potential zones. In *Proceedings of the Workshop on Increasing the Productivity and Sustainability of Rain-fed Cropping Systems of Poor, Smallholder Farmers* (pp. 22–25). Tamale, Ghana.
40. Vanlauwe, B., Bationo, A., Chianu, J., Giller, K. E., Merckx, R., Mokwunye, U., et al. (2010). Integrated soil fertility management: Operational definition and consequences for implementation and dissemination. *Outlook on Agriculture*, *39*, 17–24.
41. Velazquez-Gonzalez, R. S., Garcia-Garcia, A. L., Ventura-Zapata, E., Barceinas-Sanchez, J. D. O., & Sosa-Savedra, J. C. (2022). A Review on hydroponics and the technologies associated for medium- and small-scale operations *Agriculture, 12*(5), 646.
42. Velusamy, P., Rajendran, S., Mahendran, R. K., Naseer, S., Shafiq, M., & Choi, J.-G. (2022). Unmanned Aerial Vehicles (UAV) in precision agriculture: Applications and challenges. *Energies*, *15*(1), 217. https: //doi.org/10.3390/en15010217.
43. Zhao, D., Reddy, K. R., Kakani, V. G., & Reddy, V. R. (2005). Nitrogen deficiency effects on plant growth, leaf photosynthesis and hyperspectral reflectance properties of sorghum. *European Journal of Agronomy*, *22*, 391–403.
44. Zhou, L., Lin, S., Feng, X., Hu, C., Han, H., & Cao, W. (2020). Financial evaluation of precision agriculture: Benefits and uncertainties in grain production. *Precision Agriculture*, *21*(3), 460–477.

CHAPTER 13

Internet of Climate Change Things (IOCCT) for Sustainable Agricultural Production

M. S. SADIQ[1*], I. P. SINGH[2], M. M. AHMAD[3], and I. K. NAZIFI[4]

[1]*Department of Agricultural Economics and Extension, FUD, Dutse, Nigeria*

[2]*Department of Agricultural Economics, SKRAU, Bikaner, Rajasthan, India*

[3]*Department of Agricultural Economics and Extension, BUK, Kano, Nigeria*

[4]*Information Technology Unit, Federal Inland Revenue Service, Nigeria*

*Corresponding author

ABSTRACT

Due to their effects on the physical and biological components of the environment, the problems of environmental pollution and climate change have gained international attention. Precision agriculture (PA) is a solution that can be used to address the problem of low agricultural yields and losses caused by recent unanticipated and severe weather occurrences. The development of sensors for frost prevention, remote crop monitoring, fire hazard prevention, precise nutrient control in soilless greenhouse cultivation, solar energy autonomy, and intelligent feeding, shading, and lighting control to increase yields and lower operating costs are all results of technological advancements over time. PA reduces environmental pollution and labor expenses while delivering higher yields at cheaper input prices during a period of rising food demand. The use of the most advanced computer and electronic technologies is anticipated to increase significantly in modern food production and PA.

13.1 INTRODUCTION

Because of ongoing climate change, the entire planet is presently in danger [43]. Global warming is the term used to describe the observed rise in global average temperature over the last 100 years and the resulting changes. The released data have persuaded many scientists that this change is manmade and the result of high levels of global greenhouse gas (GHG) emissions [10]. Thermal radiation is absorbed and emitted by gases like ozone, carbon dioxide (CO_2), methane, nitrous oxide (N_2O), and water vapor. The amount of solar energy that is preserved changes proportionally as a result of these changes in the relative

Demystifying Sustainable Farming Using Smart Technologies.
Gunjan Chhabra, Keshav Kaushik, Priya Chugh, Susheela Dahiya, and Aryan Chaudhary (Eds.)

DOI: 10.1201/9781779640932-13

amounts of the GHGs [13, 50]. The steady increase in the global mean temperature is currently considered as a sign of global warming [22, 46]. This concept is intended to take into account the possibility of some isolated exceptions to this rise. The requirement for average temperature arises from the possibility that one area may experience cooling while the entire world may see an increase in temperature [1]. The GHGs' ability to trap more heat in the atmosphere is a major worry because it impacts both the climate and localized weather patterns. The number of unfavorable weather occurrences, such as storms, heat waves, cold snaps, droughts, and fires, increases as a result [5]. With global warming of 1.5°C and additional increase at 2°C, climate-related threats to human health, livelihoods, food security, water supply, human safety, and economic growth are anticipated to rise [11]. Furthermore, it is predicted that by the end of the century, 1.5°C will be less risky for global aggregated economic growth than 2°C. The most significant contribution to global warming comes from CO_2 [14, 37]. The situation is much worse than previously thought, with three-quarters of the gas likely to remain for a time in the area of up to 1000 years and the remaining portion lasting for an indeterminate period of time [7]. It was stated that the current effects of humanity on the environment can undoubtedly result in a long-term issue [35].

Globally, countries have maintained fairly strict goals for lowering their GHG emissions [6]. Significant energy consumption reductions in the city are necessary to achieve these targets. Over half (55%) of the world's population lives in urban areas, a number that is expected to rise to 68% by the middle of this century [16]. The largest local economies, the highest levels of energy use, and gas emissions all belong to urban regions. In order to lower their levels of gas discharge, metropolitan areas must cut back on consumption and use renewable energy sources whenever possible. To combat GHG emissions, smart cities frequently use digital sensors to detect and send data about the current amounts of GHGs in the city [16]. Thus, the effectiveness of such a system depends on the network that is used to compile and evaluate the data gathered as an existing network. Given that designing and implementing a novel system would be more expensive; the mobile telecommunications networks provide an easy way to fulfill this aim. Smart cities will undoubtedly play a crucial role in achieving these lofty goals, as is widely acknowledged [3].

A universal foundation for information lovers, the Internet of Things (IoT) enables advanced services by combining (physical and virtual) things based on evolving interoperable Information Communication Technologies (ICT) [2]. In 1999, Kevin Ashton coined the phrase "IoT" [24]. Almost all agricultural operations, including animal monitoring, conservation monitoring, and plant and soil monitoring, can be used to growing IoT trends. As a result, by cutting back on expenses and time spent on agriculture, it will help to improve both product and process efficiency. The IoT integrates satellite technology and uses a variety of sensors to collect data [30]. Farmers and producers can make better judgments by using forecasting services provided by IoT services after the acquired data has been processed.

Environmental issues like climate change are receiving a lot of attention right now, and environmental monitoring, modeling, and management help us understand how the environment works naturally. Changes in land and water management, which are the primary forces behind change, are anticipated to be directly impacted by climatic variability and change in the agricultural sector (probably the main channels of change). It is anticipated that there would be changes in the frequency and severity of floods, storms, and droughts. Although the effects of climate change will differ greatly depending on the locality, they can have a major impact on agricultural production. Because of lower agricultural yields, susceptible areas can experience a loss of agricultural output. Input shortages and drought have caused difficulty for Indian farmers since 2015 [34]. Climate change has contributed to the remarkable increase in natural disaster incidents over the years, which has cost us not just significant losses but also damage to our assets and infrastructure. Natural calamities cannot be prevented, but contemporary technology makes lifesaving more efficient. During a natural disaster, communication networks can be very helpful in preserving lives in the affected areas [8]. Training and implementing the use of information and communication technologies to increase agricultural output may be one of the greatest strategies [34].

13.1.1 AGRICULTURAL PRODUCTION AND GLOBAL CLIMATE CHANGE

It is possible to observe the change in climate that is either directly or indirectly linked to human activities that alter the composition of the atmosphere in contrast to natural climatic fluctuations across comparable time periods [11]. Crop cultivation has a very high sensitivity to both short-term and long-term climatic change. Agricultural productivity is directly impacted by temperature and precipitation. In particular, the main crops of Sri Lankan agriculture, such as rice, are affected by the rise in atmospheric CO_2 concentrations, the rise in temperature and the development of precipitation, and the evapotranspiration regimes [34]. For instance, an increase in temperature shortens the duration of most cereal crops by accelerating their phenological development. The development of agricultural yields and blooming could both be adversely affected by short-term high temperature variability, hence the effects of short-term climate variation and severe weather events should be taken into consideration with caution. Reports on whether the negative effects of rising temperatures could be offset by the impact of higher CO_2 emissions on agricultural harvest are contradictory.

13.2 INTERNET OF THINGS

One of the main causes of man-made climate change has been the Industrial Revolution and the technology that supported it [47]. However, it appears that a new industrial revolution and a new set of technologies may help humanity avoid the worst consequences of man-made global warming. The IoT and smart connectivity will play a crucial role in enabling the aforementioned solutions, allowing for the deployment of green microgrids in the event that nonrenewable utility networks fail or are unavailable [31]. On the one hand, green technology and renewable energy sources will have the most positive impact. A time when the future of our environment depends on it, the adoption of the IoT and smart gadgets will also help energy networks and customers become more energy efficient generally.

A variety of businesses stands to benefit from the IoT increased efficiency, but if the technology is handled improperly, negative consequences could result. The two words for the IoT are Internet and Things. Several IoT devices with unique identities that have the ability to remotely sense, act upon, and monitor a certain sort of data are referred to as "things" [30].

The IoT can be characterized as an all-encompassing framework for information-loving individuals that enable innovative services by fusing (physical and virtual) things based on existing and emerging interoperable information transmission technologies [24]. With the aid of device identification, data gathering, processing, and communication abilities, IoT clearly ensures the use of specific devices to deliver services to all sorts of applications and fulfillment of security and privacy standards (Figure 13.1). The IoT can be seen as a concept with both technological and social significance in a larger sense. The IoT represents a faster-developing stage of computing communications. IoT's technology revolution will have an effect on how people work, think, and live [32]. The IoT, big data, and nearly rapid expansion of cloud computing technologies have improved flood monitoring. The responsibility for real-time data collection falls to industrial IoT. Real-time data analysis is done effectively via big data analysis. After big data and high-performance computing (HPC) converged, fog computing reduced latency, and cloud computing provided the required IT infrastructure [18]. IoT integrates services, technology, and people by connecting digital devices to work together to detect the environment, produce results, and convey results to people or other digital devices. Every sector of the economy, including manufacturing, urban planning, finance, education, healthcare, and emergency services, has been impacted by the growth of IoT. Countries all around the world are frantically looking for a solution to the ongoing and escalating threat of climate change as they teeter on the edge of disaster. However, with innovations like the industrial IoT (IIoT) already cutting carbon emissions and improving energy efficiency across industries, the overall situation may start to appear a little more positive.

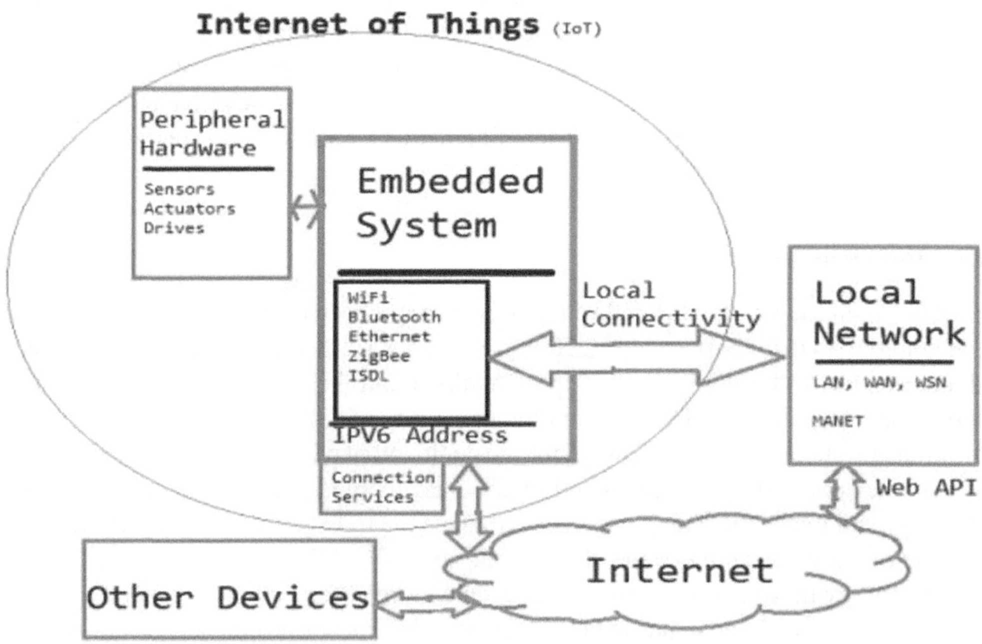

FIGURE 13.1 (a) Internet of Things (IoT).

Source: Adapted from Refs. [34] and [38].

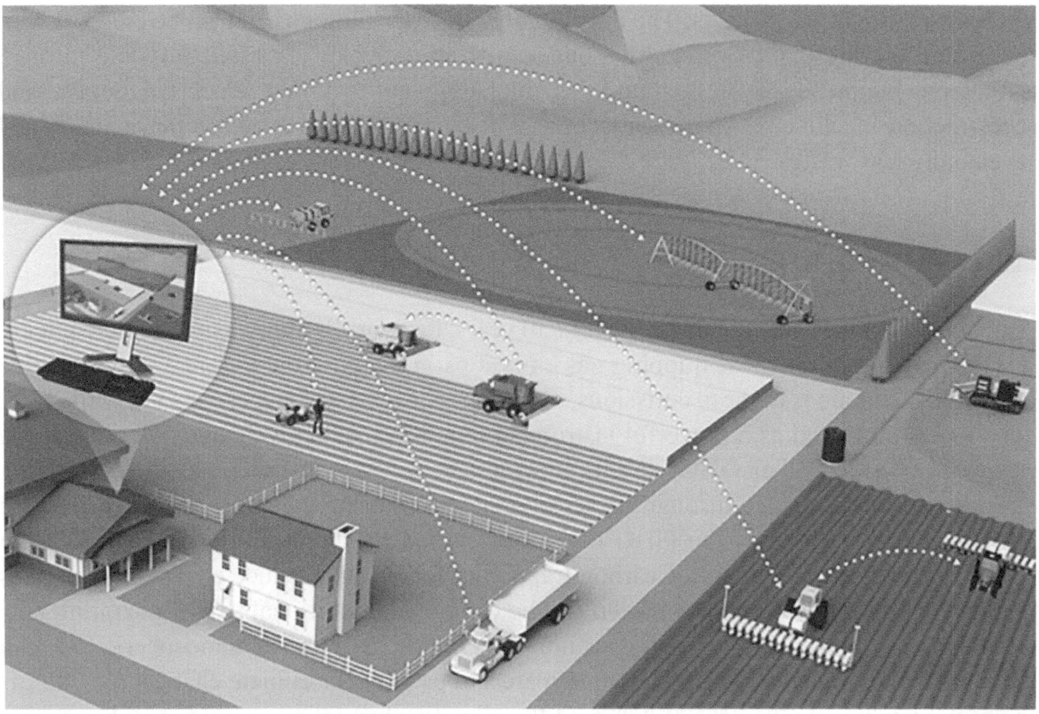

FIGURE 13.1 (b) Pictorial chain of IoT.

13.2.1 WHERE IS IOT TECHNOLOGY ADDRESSING ENVIRONMENTAL PROBLEMS?

To use the least amount of water, fertilizer, and pesticides possible, farmers adopt precision agriculture (PA) [29]. The technology keeps an eye on the moisture, temperature, and soil minerals. By doing so, the usage of resources and land is reduced while yields are improved and increased. The management of farm resources can be enhanced by combining algorithms, or "grow recipes," with IoT sensors in the soil and surrounding environment [30]. The negative effects of GHGs produced by livestock could be mitigated with the aid of industrial IoT. "Livestock health monitoring, such as keeping track of an animal's temperature and nutrition in order to identify and treat sick animals, has the potential to lower methane emissions from ruminants, which helps to lower GHG emissions." Sensors are employed in the manufacturing sector to automatically arrange products and increase production. The energy use of manufacturing equipment is also being tracked by industrial IoT, allowing operators to spot wasteful machinery.

"Logistics firms may shorten delivery routes through intelligent route planning, tracking can cut delivery times through new insights into the supply chain, products can be located, fewer products are lost, which ultimately has a beneficial impact on the climate by lowering total direct and indirect energy consumption." "More accurate monitoring of external elements can reduce errors, resulting in less waste and more effective use of materials." Smart street lighting is being employed in the energy and utilities sector to save consumption, while leakage sensors are being installed in water and gas pipes to find and fix losses brought on by leaks. The development of IoT solutions can also be credited with extending the lifespan of products. Longer utilization cycles are made possible through predictive maintenance of commodities, including automobiles, electrical goods, and construction machinery. As a result, there will be fewer failures and replacements, and eventually less waste.

13.3 CLIMATE IOT

Because of human activity during the past 50 years, the world's climate is changing quickly [1]. Depending on the heat-trapping gas emissions in the environment and how sensitive the Earth's climate is to these emissions, this change is expected to persist for a while [14]. The IoT is promoted as a powerful instrument to address climate change. By sensing the levels of CO_2 and other GHGs in our atmosphere, it offers insights into the underlying causes of climate change through its sensing and monitoring capabilities [48]. Real-time monitoring is possible of the GHG emissions caused by the burning of fossil fuels. In order to enhance the amount of carbon trapped in forests and reduce emissions, the rates and processes of carbon sequestration can be examined. Additionally, by integrating new atmospheric "things" and technologies into the climate IoT (CIoT), atmospheric CO_2 can be permanently reduced. In order to anticipate and prepare for climate change, the CIoT is helpful. Insights into the precise nature of the climatic changes are provided by its sensing and communication technologies in combination with prediction systems and models [12].

The climate decision-making tools that are IoT enabled can forecast how the climate will change and how the ecosystem will probably react to the shift and other influencing factors. Empirical studies on the effects of increasing GHGs have been made possible by IoT technology [15]. Additionally, it facilitates simulations of ecological responses under various climatic situations (both current and future). The procedures for managing the environment and the atmosphere are adjusted in light of this new information, and new management strategies might be created as a result. The CIoT has the ability to fulfill short- and long-term goals and application needs by employing modern scientific and technology breakthroughs. Better knowledge and insights into the world's ecosystems are provided by this design, which also encourages wise decision-making. Our capacity to evaluate water supplies, forecast climate, and anticipate weather patterns may all be improved by a better understanding of the Earth system at all scales, from the global to the local. The effect these variables have on our community defines the requirement for the creation of applications useful to society (Figure 13.2).

13.3.1 CHANGING CIOT FOR SUSTAINABLE ENVIRONMENTAL PRACTICES

The land and ocean bio-geochemical processes, ecosystems, air quality, hydrosphere, lithosphere, biosphere, and atmospheric chemistry make up the global environment [42]. The essential contextual elements for the IoT's capabilities in environmental sustainability and climate change are outlined in the following climate change elements (Figure 13.3).

FIGURE 13.2 Precision farming in combating climate change.

Source: https: //agfundernews.com/why-precision-agriculture-is-essential-in-combating-climate-change

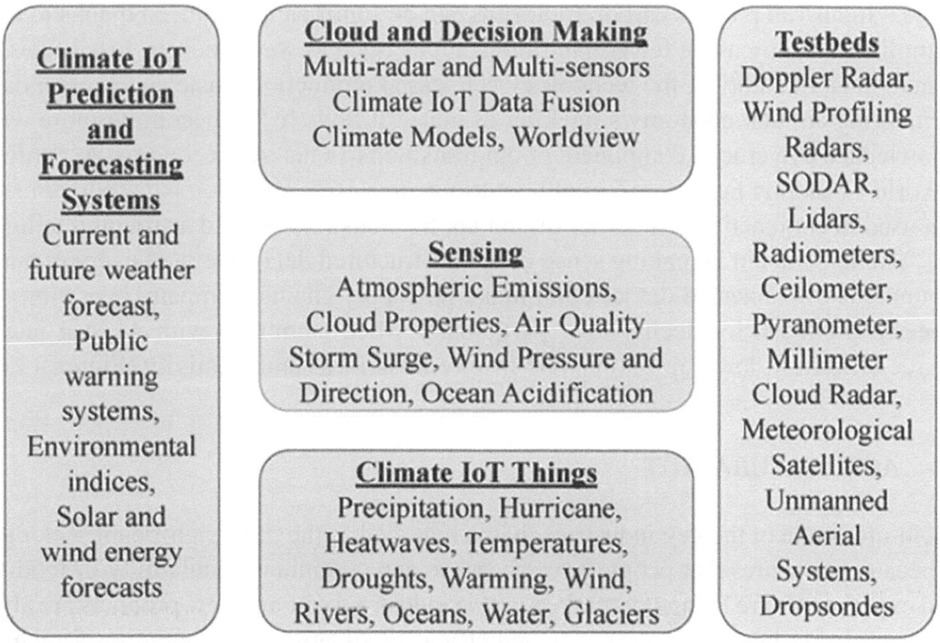

FIGURE 13.3 The architecture of environmental sustainability and climate change IoT.

Source: Reprinted with permission from Ref. [42]. Copyright © 2020, Springer Nature Switzerland AG

- Hurricanes, heat waves, and severe precipitation.
- Storm surge, in-land flooding, rising sea levels, and flooding around the coast.
- The modification of marine habitats and ocean acidification.
- The growth in population and changing land-use patterns have exacerbated the decline in water supply and increased water competition.
- Rising levels of CO_2.
- Climate change brought on by wildfires, droughts, and rising temperatures.
- Increased water demand for energy and water.
- Changes in stream flow timing brought on by snowmelt.
- Thawing permafrost and shrinking glaciers.
- Depleting sources of fresh water and water scarcity.

13.3.2 IOT'S ROLE IN PREVENTING CLIMATE CHANGE

Through sensors that can monitor a wide range of elements, including everything from air and water quality to monitoring pollution levels near industries, rivers, and towns, industrial IoT measures the influence of industrial processes and human activities. By detecting factors like flood and river levels, wind speed, land erosion, the activity of bees and beehives, and tracking animals or vegetation in damaged areas, the device can also detect the more indirect effects of climate change. "The IoT is our planet's digital skin." "We become aware of existing problems and may follow them through time when we apply specific actions to battle these problems by assessing the real status of the world through sensors." Industrial process carbon footprints can be significantly reduced thanks to IoT. It accomplishes this by using fewer natural resources, such as water, power, fossil fuels, and raw materials. In addition, the technology can lessen production waste and is essential for the emerging circular economy's tracking of material flow. IoT is becoming more widely acknowledged as a crucial component of our transition to a net-zero economy. According to the World Economic Forum, IoT might contribute to a 15% reduction in carbon emissions when used in conjunction with other digital applications like 5G and artificial intelligence (AI). "The IoT can aid in making sense of the unstructured data generated every minute by the hundreds of connected devices that make up supply chains, company operations, and connected products. IoT technology, particularly when combined with AI, can increase resource efficiency, lower pollution, and inspire fresh ideas and creativity (Figure 13.4).

13.4 AGRICULTURAL IOT

Agriculture is one of the key industries that is regarded as the most significant sector globally because it ensures that people have access to and continuous availability of food. IoT-based experiments are being undertaken in this industry daily, and new products are always being unveiled to improve the efficacy and efficiency of agricultural activities for a higher output (Figure 13.5). Indian farmers, for instance, are today facing severe challenges and disadvantages in terms of farm size, business, technology, government policy, weather,

FIGURE 13.4 IoT and augmented reality (5G and AI).

Source: Adapted from siilc. https: //www.siilc.edu.in/the-rise-of-smart-farming-how-iot-and-ai-are-revolutionizing-agriculture/

and so on. There is no denying that ICT-based solutions have solved some issues, but they are insufficient for effective and secure production. IoT, sometimes known as "ubiquitous computing," is a modern evolution of ICT [25]. The expansion of IoT into a variety of sectors, including business, private residences, and urban areas, has created enormous potential to make everything smart and sensible. Today's adoption of IoT technologies by the agricultural industry has given rise to the "Agricultural IoT" [40].

Due to the inadequate integration of data-driven decision support systems (DSSs), traditional agriculture is characterized by superfluous human contact, which leads to higher labor costs & susceptibility to extreme weather occurrences. Through automated farm operations, mechanization to address crop monitoring concerns, and energy- and water-saving techniques based on AI and machine learning, IoT offers a reprieve [34]. In view of the negative impacts of global warming and climate change on food production and food safety worldwide, the significance of technology in farming in the future cannot be discounted. In the US and other countries, unusual climate occurrences like excessive precipitation and temperature have been linked to a sharp decline in agricultural output. Given that the higher expenses would be passed on to consumers, the influence of climate-related variables on agriculture would not be limited to farmers. The advanced economies in the western hemisphere were not the only ones dealing with this issue. Similar difficulties were identified in Greece, where the area of arable land had shrunk, according to [19]. Severe food shortages and malnutrition have been caused by the disturbance of weather patterns brought on by climate change in combination with sociodemographic factors. The variety of problems faced by industrialized and developing countries emphasizes the need for creative and technology solutions to lessen the disruptions to agriculture brought on by climate change.

By 2050, the prices of agricultural production are expected to rise by 1–29%, according to the Intergovernmental Panel on Climate Change [30]. The optimization of greenhouse settings using IoT could somewhat counterbalance the difficulties caused by climate change in agriculture; as a result, cutting-edge technologies would be essential to global food supply chains [19]. The four crucial areas of the application of IoT systems in smart agriculture and greenhouses-maintaining an optimum microclimate for ideal plant growth, improved irrigation and fertilization methods, infection control, and increased security-would be the focal point of any prospective IoT-based solutions for current farming difficulties [39]. Infrared cameras, unmanned aerial vehicles/systems for remote monitoring, optical monitors, infrared, and thermal sensors placed strategically across large farms to mitigate crop losses due to invasion by wildlife (such as birds, goats, and buffaloes) are all ways to increase security in agricultural production [30]. Cost factors also influence the deployment of intelligent security solutions because robotics, big data, and remote monitoring sensors are more cost-effective and productive than hiring human labor. With the help of technology, it was possible to save costs and boost productivity, as shown by the commercially accessible sensors. Minimizing manual interventions, increasing yields, and maximizing the use of resources and agrochemicals are important future objectives. The optimization of certain parameters and the integration of precise sensors for tracking water and moisture content as well as plant physiology are necessary for the realization of the potential benefits connected with sensors.

IoT technologies provide considerable promise for PA, according to preliminary data. Farmers were informed, for instance, when it was best to apply fungicides by weather data-driven DSSs [4]. Fungicides that were applied in a timely manner served to reduce the danger of late blight, which resulted in direct cost savings of roughly $500 per acre [4]. The use of electrical capacitance sensors for soil–water balance and soil–water content in conjunction with DSSs driven by weather data was shown to have similar advantages. Wheat farmers saved 25% on irrigation costs because to the design. The use of optical sensors to measure plant chlorophyll concentration during the application of fertilizers has also shown the technology to be beneficial (a predictor of plant nitrogen levels). More nitrogen-rich fertilizer was applied to the plants that had a more severe nitrogen shortage, which led to better fertilizer use and higher yields [4]. Greenhouses in Saudi Arabia produced 252,824 tons of fruits and vegetables in 2016, supporting the empirical finding that greenhouse-based agricultural production increases yields by 10–12% on average [30].

The performance of sensors in desert locations remained a worry because current IoT systems for PA were less suited to semiarid regions [54]. The drawback is that PA and technology adoption in underdeveloped countries are not well documented. The consistently low yields, which contribute to the gaps in food security between industrialized and emerging countries, can be explained by the poor uptake of new technology in developing countries. Despite weather/climate-related technology limitations, the impressive success seen in Greece, Saudi Arabia, and other nations is evidence of the need for smart and intelligent agriculture to increase crop yields in industrialized and rising countries with wide temperature swings [41]. The potential increases in yields strengthened the argument for tailored solutions and interventions. The normal yield increases vary from 10% to 12%, although better performance can be obtained by optimizing plant growth factors, making sensors more reliable technologically, and reducing costs [55]. For instance, farms with dispersed greenhouses may be forced by cost considerations to select a dispersed rather than a concentrated deployment of sensors [44]. According to studies by [54] on the relationship between PA and cost, the worries expressed over the expense of IoT in agriculture were valid. The majority of the time, IoT in agriculture was costly and out of the price range of smallholders, who make up the foundation of the global agricultural system. The lack of set parameters regarding energy and water allocations and energy use, as well as the current uncertainty regarding the selling prices of agricultural produce, are factors that may prevent smallholder farmers from investing in IoT technologies [51]. Contrary to major commercial producers that can easily obtain IoT systems, smallholder farmers have little incentive to invest in new technologies like IoT because of the narrow production margins. Cost was a significant barrier to the general adoption of IoT infrastructure, as shown by the comparative analysis of the two [51]. Madushanki et al. [28] supported the difficulties mentioned by Villa-Henriksen et al. [51]. Given the potential of IoT infrastructure to catalyze smart farming and urban greening, it was hypothesized that the advantages exceeded the hazards in the latter situation.

FIGURE 13.5 Agriculture IoT.

13.5 PRECISION AGRICULTURE

As an integrated crop management system, PA, according to [34], integrates information technology with the agricultural sector and aims to supply inputs and their types in accordance with the actual needs of farming in small farms that are a part of larger farms. PA is additionally viewed as an information technology-based agricultural management system that can be used to recognize, assess, and manage changes on a farm to maintain profitability, sustainability, and ideal conservation. This technology considerably benefits the environment and aids in cost reduction. By decreasing the excess and underuse of inputs like fertilizers and pesticides, this approach can increase the manufacturer's profitability and lower the risk of agricultural chemicals contaminating ground or surface water (Figure 13.6).

FIGURE 13.6 (a) Precision farming.

FIGURE 13.6 (b) Irrigation in precision farming.

13.5.1 EMERGING SMART TECHNOLOGIES AND PA: A CHAIN REACTION

After computers and the internet, technology has made an enormous contribution to modern civilizations, which serves as the foundation for the IoT. The integration of intelligent machinery, actuators, sensors, unmanned aerial systems, radio frequency identification (RFID) devices, big data analytics, AI, and satellites could be the key to the IoT's enormous contribution to agricultural and commercial greenhouses [26]. Its widespread use in numerous agricultural and nonagricultural applications, such as intelligent farming and frost prevention in greenhouses, intelligent control of greenhouse structures, fire hazard prevention, the transition to agriculture 4.0, precise nutrient control in soilless greenhouse cultivation, smart cities, emission monitoring, distributed/decentralized energy storage, solar-powered sensors, smart feeding, shading, and lighting, has been made possible by this [12, 30]. The creation of very effective communication protocols like MQTT (Message Queuing Telemetry Transport) Protocol, which has gradually phased out Hypertext Transfer Protocol, has contributed to the broad adoption of IoT in smart greenhouses and PA [30]. Because MQTT can operate with less bandwidth, it has less overhead than other protocols. Despite the extensive usage of IoT systems in smart greenhouses, there is still a lack of knowledge on how the technology can best regulate greenhouse environments, particularly in tropical areas with extreme temperature swings.

The future of PA is expected to be substantially impacted by a number of developing technologies. Cloud computing, edge computing, fog computing, embedded software, embedded systems, cyber-physical systems, wireless sensor networks (WSN), big data gateway, machine-to-machine, human-to-machine, LoRa Protocol, ZigBee/Z-Wave RFID, gateway general packet radio service (GPRS), application programming interface, and advanced encryption standard are the technologies that would have the most significant impact [30]. Figure 13.7 depicts how diverse ICTs, cloud computing, WSN, geo-location, satellites, and computer-to-human interfaces interact. The long-term cost-benefits in greenhouses are briefly explored, as well as the prospective significance of these technologies in water and energy saving.

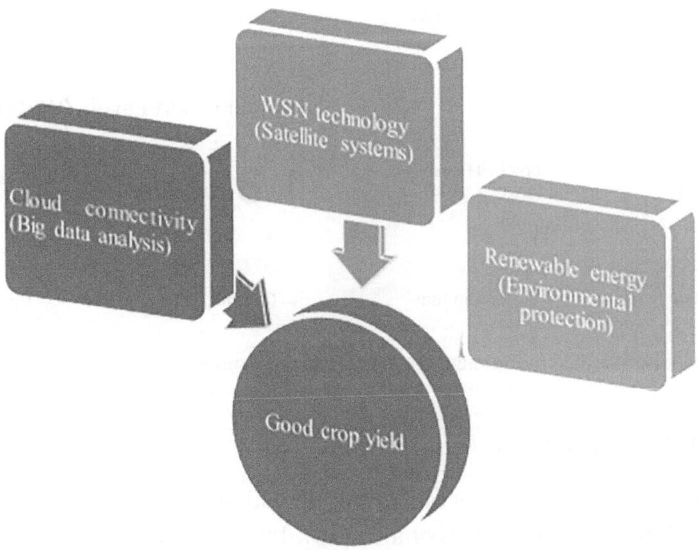

FIGURE 13.7 Cloud computing.

PA and smart greenhouses have both demonstrated the value of cloud computing [12] (Figure 13.8). It has resulted in significant cost savings and increases in yields to save sensor data in the cloud and integrate it with smart technologies for remote monitoring of plant water levels, nutrient content, soil pH, humidity, and temperature. There are numerous strategies to increase productivity. To develop supply and demand trends across various product markets, for instance, the farmers can obtain historical predictive analytics data from institutions. In addition, the predictive data gives farmers up-to-date weather information, reducing the negative consequences of climate change and global warming. The prolonged use of cloud computing in places with poor network coverage and slow internet speeds, however, has serious downsides.

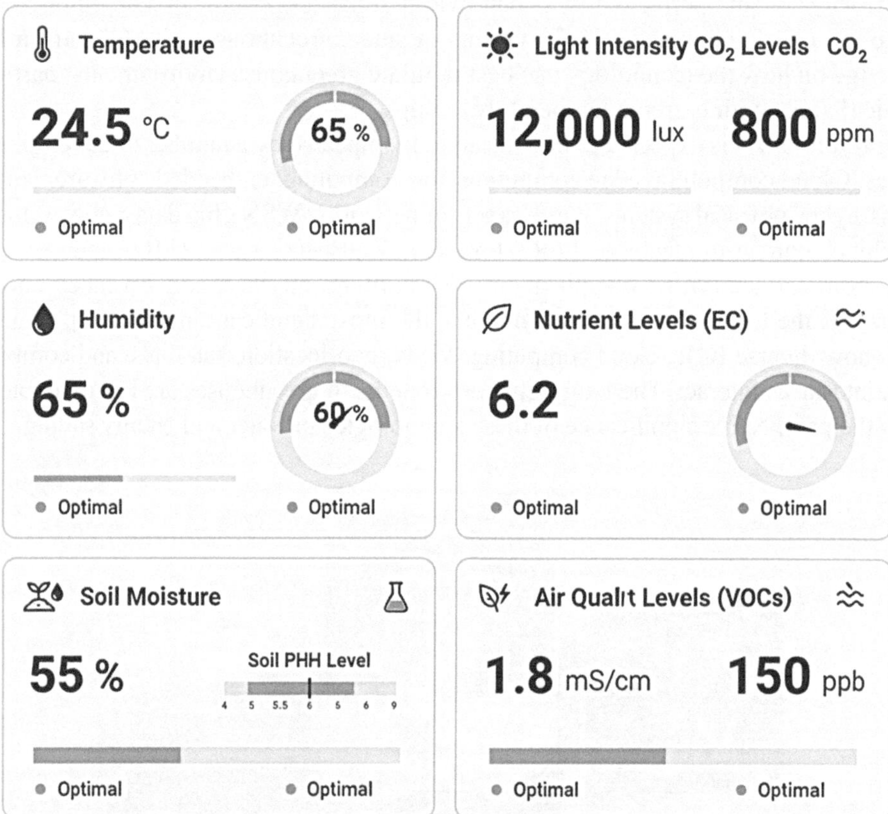

FIGURE 13.8 Monitoring the environmental condition of a greenhouse.

13.5.1.1 BLOCKCHAIN AND AGROVOLTAICS

Another new IoT-related technology breakthrough with potential for applications in sustainable energy-food production is agrovoltaics [30]. Integrated electrical and thermal energy generation systems and foldable PV modules and solar tiles for light transmittance and extended service life are two recent advancements. The emergence of agrovoltaic systems is a significant negative. The sole operational agrovoltaic system existed in Belgium as of 2020 [53]. Given the difficulties in adopting new technologies, agrovoltaics' infancy is a constraint. The most recent discoveries gleaned from this review are useful in commercial agriculture. At the All-Russian Research Institute of Electrification of Agriculture, additional advancements have been made in the commercialization of PV modules and solar tiles [36]. It is projected that as agriculture switches to renewable energy, demand for agrovoltaics would increase rapidly outside of Russia and Belgium. In addition to agrovoltaics, the incorporation of blockchain into IoT-based farming systems is a developing field that merits more R&D focus. Blockchain could decentralize solar energy production and consumption while enhancing security in agro-systems.

13.5.2 OPTIMIZATION OF IOT SYSTEMS FOR INTELLIGENT FARMING AND SMART GREENHOUSES

In order to optimize IoT systems for greenhouses, experts generally agree that careful sensor selection, data collecting, optimization, identification of the appropriate settings, and rule-based management are necessary (Figure 13.9) [21, 49]. There is a need for sophisticated data collecting methods employing Kalman filter prediction and related approaches that can anticipate future environmental conditions using historical data since the conditions must be optimized for various horticulture crops. In fact, controlling every parameter in smart greenhouses—including temperature, moisture, pH, pesticides, humidity, UV radiation, rain, CO_2, and pressure-can be difficult, especially when there is a lack of sufficient historical data [30].

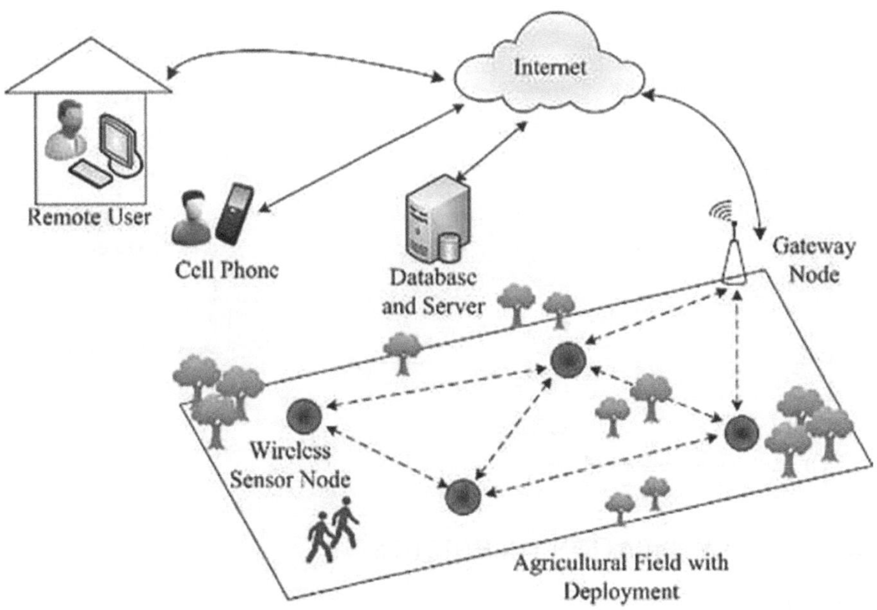

FIGURE 13.9 IoT-based system for smart greenhouses.

13.5.3 SMART GREENHOUSES WITH IOT-BASED SENSORS

There is now a large range of IoT-based sensors for smart greenhouses, including plant growth sensors, temperature and humidity sensors, insect detection sensors, soil temperature, pH, and moisture sensors, as well as solar radiation, air pressure, wind speed, and CO_2

(and other gas) sensors [25]. These sensors rely on Bragg, piezoelectric, electrochemical, electromagnetic, and fiber-optic technologies for an accurate assessment of the relevant parameters. Different light wave lengths, photocurrent, fluorescence intensity, the fluorescent signal produced by plant chlorophyll, optical density, and the electrochemical signal produced by an enzyme-catalyzed redox reaction (SHA principle) are among the metrics of importance [52]. Electromagnetic sensors have been created to analyze the chlorophyll content and nitrogen levels in plants; this method depends on light reflection and pulsating laser diodes and was made possible by advancements in research and design [39]. Additionally, the method has been shown to be effective in assessing plant physiology in real-time, including a plant's vegetative index, dietary needs, electrical conductivity, and magnetic susceptibility and conductivity (quad-phase) (Figure 13.10).

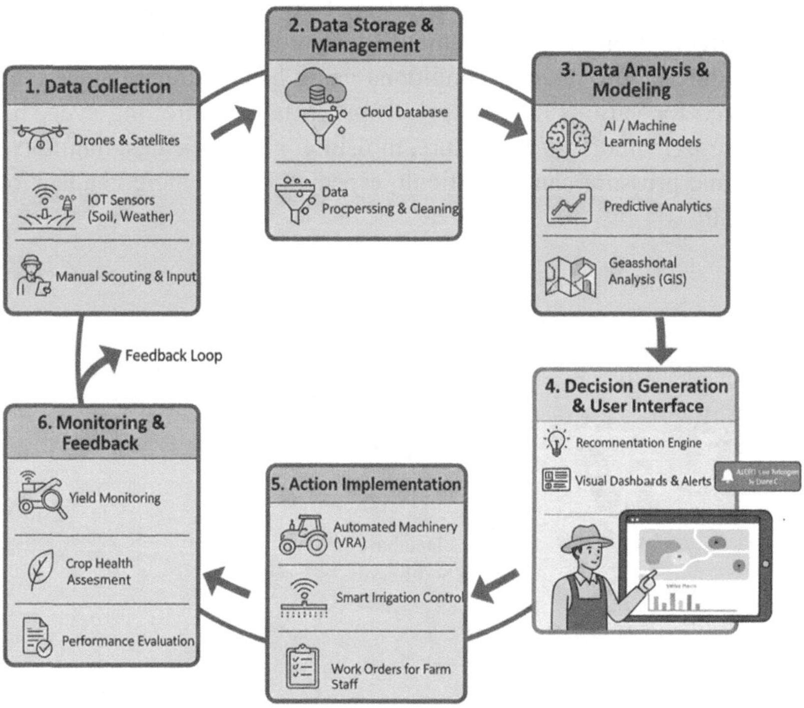

FIGURE 13.10 IoT-based sensor network.

13.6 IOT-BASED MITIGATION TECHNIQUES

13.6.1 FLOOD

Flooding reduces agricultural productivity, which leads to misery in rural regions. Floods, which are typically the results of major weather phenomena such prolonged rainfall, snowmelt, and so on, could be influenced by geography and human activity. Farm land may be significantly affected when flooding occurs at a key period of the crop cycle. At the moment, there is no reliable generic system for estimating how severe the floods' socio-economic effects will be. For scientists and researchers, effectively modeling, managing, and monitoring floods is crucial. Some academics have devised various strategies for flood management systems or water management systems.

In order to demonstrate the potential and effectiveness of industrial IoT [8], as reported by [34] provided a Big Data-HPC-convergent industrial IoT service for flood management. This service included a thorough explanation of the architecture, algorithms, experimental methods, results, and studies. The flood management service that is being offered can help with the convergence of IoT, big data, and HPC and give more accurate flood analysis and prediction, for instance by leveraging the business intelligence methodology and contemporary prediction and data visualization tools. In order to prevent unwanted access, it is crucial to keep all results and data secure. However, as more data is collected, managing security and management issues becomes more challenging. When there are natural disasters that cause damage, the capability of data recovery and the continuation of services without interruption become essential. Because ubiquitous access, storage, and analysis of such a vast set of sensory data is a complicated and resource-intensive procedure, hydro-meteorological methodologies for sensory data analysis were at danger. There are certain commonalities across watersheds, or regions like ponds, rivers, and lakes where water collects from various sources like precipitation or drainage. Since watersheds with comparable characteristics would react with the same flood generating responses, a universal system for global flood monitoring can be created. This is because most research and analysis can be reused.

Additionally, a computerized water management system was created to regulate the field's moisture level. The soil moisture sensor-connected motor will turn on to water the field when the moisture content falls below the acceptable level. A second motor linked to the soil moisture sensor will transfer any extra moisture from the field to a well or water bed in the event of severe floods. This water management system is automated using an Arduino microcontroller [34]. A more effective method to save many lives is real-time flood detection and rescue services. IoT-based rescue services make it possible for victims and survivors of natural disasters to communicate with one another and with their loved ones in order to provide information about protection and their current situation. The IIoT services related to warning about and predicting the consequences of natural disasters like floods can use this method. However, if the location and spread of floods are to be predicted, a real-time flood forecasting application must be sensitive to delay. Due to the push and pull of data between the cloud and the devices, latency may occur with cloud-based IoT. Most of the study has focused on spatial closeness, such as in the Geographic Information System, which only allows for real-time flood detection. Digital elevation model and Synthetic Aperture and Radar imaging models are important, but not as important as an early warning system.

13.6.2 DROUGHT

A frequent and ongoing effect of climate change is drought. A drought in agriculture happens when the amount of soil moisture that vegetation can access has dropped below the necessary level, which has an impact on crop yields and consequently, agricultural output. In accordance with [34], agricultural drought is one of the most frequent disasters, has a substantial impact on the population, and causes large yield losses in both rain-fed and irrigated farming. As a result, the crop production need for expanding population is called into question. Agricultural drought causes economic losses, environmental deterioration, and pollution, according to [29]. The harvest benefits from more rain that falls consistently throughout the growing season. Droughts could be brought on by low precipitation (below-average precipitation), which affects agricultural productivity and reduces water availability, which affects irrigation and the water supply for farm animals. Continuously bad droughts make it difficult for there to be abundant precipitation and restrict the dispersion of wells during the growth season. So it represents the creation of a thorough drought monitoring system for drought avoidance and IoT aids farmers in overcoming scarcity and droughts. With the IoT, plans and practices can be more accurate, and groundwater drought management and watershed sustainability can be better understood. Groundwater protection, utilization, and efficiency can all benefit from IoT.

In a platform based on IoT technologies, hybrid programming, and parallel computing, [27] as related by [34] created a synthetic system that allows monitoring and forecasting of drought and prediction of irrigation volume. Most of China's northern regions have adopted this technology, which has the ability to provide an integrated service for tracking and forecasting droughts as well as irrigation volume. In order to properly manage and irrigate the farmlands and conserve the available water, an Automatic Smart Irrigation DSS was developed. Estimating weekly irrigation requires knowledge of soil characteristics, current weather, and forecasted weather. To implement this new system, various sorts of sensors are used. A soil sensor is utilized to monitor various crops and environmental factors, and a Global System for Mobile Communications/GPRS modem is used to gather data at various locations. The inputs for the system include environmental factors like precipitation, humidity, needed water depth, and so on.

13.6.3 DISEASE IN PLANTS

Agriculture experiences severe damage and financial loss as a result of plant diseases and nutritional deficiencies. Temperature, light, and water can affect the development of plant diseases and crop growth, and they can also affect the kind and health of the host plant. The development and output of plants are significantly impacted by high CO_2 concentrations and temperatures in the atmosphere, changing rainfall patterns, and the frequency of extreme weather events. Because of this, the likelihood of developing a disease varies. Crop physiology and resistance will also vary as a result of climate change, as will pathogen development stages and growth rate. Agriculture productivity may be considerably impacted by plant disease management techniques. One management activity to which Agricultural IoT may offer support to raise crop productivity is integrated pest management. The farmers could precisely administer chemicals to the identified problem regions with the help of IoT equipment like sensors, robots, and drones. The entire growing region uses cutting-edge methods, such as field sensors, unmanned aerial vehicles (UAVs), or remote sensing satellites, to take raw photos and process them to identify pests and diseases. An IoT-based agricultural greenhouse environmental monitoring system that allows for real-time remote monitoring of the greenhouse's environmental conditions was proposed by Ref. [23] as cited by Ref. [34]. The WSN can also be used to track and manage variables that have an impact on plant performance and growth. This technique can be used to determine the best time to harvest, plant diseases, control equipment, and other factors. It proved successful in controlling the potato late blight disease with the use of DSS. By determining the precise quantity of fungicide needed, DSS increases effectiveness, lowers costs, and has a smaller negative impact on the environment. In order to forecast late blight, this prediction uses a weather model. But with the rise of the IoT, it is now possible to collect reliable weather data on a broad scale using cheap, low-power sensor nodes on agricultural land. With the help of a newly emerging service like Cloud-IoT, remote sensors can be connected to and utilized. The gathered information can also be sent over the internet to the Cloud IoT Framework. A mechanistic model of the disease and a DSS for late blight are used to anticipate the disease and provide real-time (during the season) support for blight control.

A smartphone was used to manage the irrigation system, and as reported by Refs. [17, 20, 34] focused on irrigation systems that employ WSN to collect environmental data. As a long-term, sustainable solution for farm automation for plant data or for environmental measurements, Ref. [17] as cited in Ref. [34] developed a portable measurement technique. A soil moisture sensor, an air humidity sensor, and air temperature sensors are all part of the system. In order to improve water utilization and obtain baseline data for the study of soil water infiltration fluctuations and intelligent precision irrigation [9], as cited in Ref. [34] presented a multilayer measurement system to assess soil temperature and water content of cultivation area with WSN. By examining the crop's health and applying smart pesticide sprays in accordance with monitoring, Ref. [45] as reported in Ref. [34] devised a method to assist smart irrigation and smart pest management. A drone was utilized to determine the affected area when the system was administered to a grape vineyard. Data on the relative humidity, temperature, and ultra-violet radiation were taken every 15 min. The system's components included remote sensing technologies, IoT, cloud servers, intelligent systems, and agricultural experts (Figure 13.11).

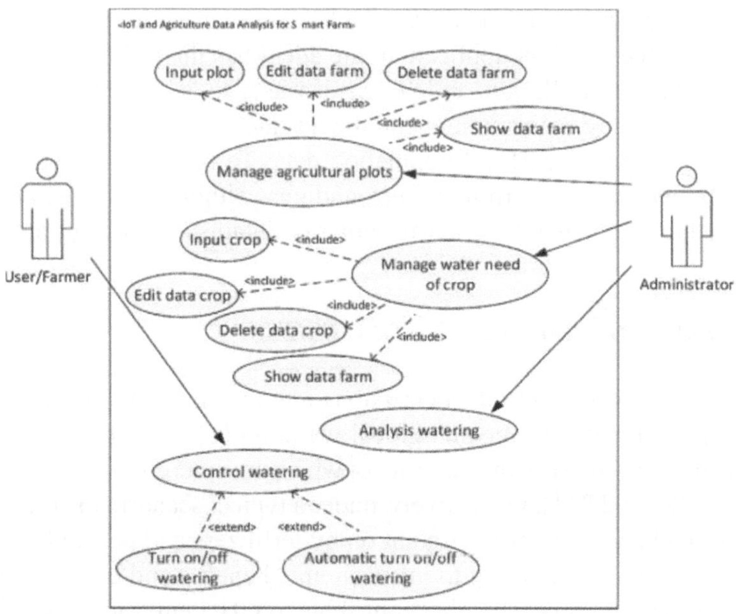

FIGURE 13.11 A web-based and mobile application for plant disease control.

Source: Adapted from Refs. [33] and [34].

13.6.4 IOT-BASED AGRICULTURAL POLLUTION MITIGATION

The biosphere and the atmosphere around it alter owing to various contaminants, which is known as environmental pollution (chemicals and energy). Climate is primarily impacted by pollution. Monitoring the pollution is necessary to guarantee the preservation of the environment. The level of contamination ought to be within acceptable bounds. In recent years, urban air quality has become a major global concern. In order to ensure the viability of the cities, the urban air quality index must be continuously checked. With the use of numerous sensors, an IoT system was created to measure the environmental factors that contribute to pollution, such as temperature, air pressure, UV radiation, air quality, nitrogen dioxide, and carbon monoxide (CO) levels in the atmosphere, noise level, and smoke, among others. The UVI-01 sensor, for instance, was used to generate an output based on the quantity of UV rays observed, while the two in one Temperature and PH sensor was in charge of monitoring water quality. To the cloud, the collected data will be sent. A report will be created after the collected data has been examined. Actions to control pollution will be taken in accordance with the report's findings [34]. With the use of IoT for smart cities, another air quality monitoring system was created. It is possible to detect atmospheric parameters like temperature, smoke, humidity, CO level, liquid petroleum gas level, and particulate matter levels like PM2.5 and PM10 that pose health risks. A global android application will provide access to the real-time information on air quality.

Global worry exists over rising environmental contamination. The increase in ocean acidity brought on by this pollution harms marine life. By studying environmental parameters, a project was devised to reduce pollution and enhance agricultural. At numerous locations throughout the field, the air, water, humidity, and oxygen levels have all undergone routine testing. Drones were used to take precise digital pictures, which were then uploaded to a cloud server and made available to the agricultural expert at any time. Utilizing a smartphone or tablet application, the adept has the right to access cloud-based photographs. To track air pollution, an IoT sensor was also exhibited. CleanSpace was the sensor's name. Users may now comprehend the air quality in real time thanks to machine learning, which provides hyperlocal population data. To recharge the battery, the sensor captures radio frequency energy from its surroundings. Although it currently only monitors CO, it is claimed to be a general marker for other pollutants (including transport).

13.6.5 GHG EMISSION REDUCTION STRATEGIES

Agriculture is a major source of GHG emissions into the atmosphere, which significantly accelerated current climate change and global warming [5]. By 2050, GHG emissions in Asia, Africa, and Latin America, the majority of which are emerging countries, are predicted to rise by 37%, 32%, and 21%, respectively, under a typical scenario [34]. The reduction of GHG emissions and contamination brought on by fertilizers and pesticides should also be taken into consideration, in addition to the economic benefits and environmental benefits. The principal agricultural processes that contribute to GHG emissions, which lead to N_2O emissions, are soil nitrification and denitrification. Methane and N_2O emissions originate from the decomposition of manure. According to recent studies, increased GHG emissions are a significant contributor to climate change, which includes rising temperatures, reduced and erratic precipitation in certain regions, greater flooding and hurricane spread in others, and more frequent tornadoes [1].

With the goal of coordinating agricultural processes and maximizing productivity, precision farming aids in minimizing GHG emissions. To provide farmers a more thorough understanding of the current conditions in the producing area and/or set up automated devices to optimize energy consumption, water consumption, and the application of pesticides and fertilizers, quick, reliable, and widespread measures are required. At the most fundamental level, after gathering data from numerous heterogeneous systems, well-evaluated scientific knowledge can be arranged in the form of intelligent algorithms to enable a better comprehension of current processes, justify the existing conditions, and determine projections based on heterogeneous data, warn about dangers to variety at an early stage, and strengthen automated control signals in relation to crop reactions. An analysis of the CO_2 emissions from automobiles, forest fires, and industries uses IoT. The CO_2 levels were measured using an MG811 sensor, and a Raspberry Pi served as the control module. The algorithm detects the location with the most pollution as well as the amount of CO_2 emissions. The device might be used to detect wildfires before they spread. This can help save a lot of lives. If a significant amount of CO_2 is emitted in a specific location, a notification will be delivered to a mobile phone through Simple Notification Service [34].

13.7 CONCLUSION

One of the most exciting and forward-thinking concepts for achieving environmental objectives is the IoT. IoT technologies now exist that make it feasible to examine the ecological state of various regions of our world at this time. For sustainable health, IoT must be integrated with tools for sensing, communicating, monitoring, and decision assistance. Accordingly, DSSs have been created during the past few decades to give farmers the professional knowledge they need for managing their agricultural operations.

KEYWORDS

- **climate change**
- **Internet of Things**
- **agriculture**
- **smart technologies**
- **greenhouses**
- **mitigation**

REFERENCES

1. Abbas, S. (2022). Climate change and major crop production: Evidence from Pakistan. *Environmental Science and Pollution Research*, *29*(4), 5406–5414.
2. Adamides, G. (2020). A review of climate-smart agriculture applications in Cyprus. *Atmosphere*, *11*(9), 898.
3. Adamides, G., Kalatzis, N., Stylianou, A., Marianos, N., Chatzipapadopoulos, F., Giannakopoulou, M., Papadavid, G., Vassiliou, V., & Neocleous, D. (2020). Smart farming techniques for climate change adaptation in Cyprus. *Atmosphere*, *11*(6), 557.
4. Antony, A. P., Leith, K., Jolley, C., Lu, J., & Sweeney, D. J. (2020). A review of practice and implementation of the internet of things (IoT) for smallholder agriculture. *Sustainability*, *12*, 3750
5. Arunrat, N., Sereenonchai, S., Chaowiwat, W., & Wang, C. (2022). Climate change impact on major crop yield and water footprint under CMIP6 climate projections in repeated drought and flood areas in Thailand. *Science of the Total Environment*, *807*, 150741.
6. Bhardwaj, M., Kumar, P., Kumar, S., Dagar, V., & Kumar, A. (2022). A district-level analysis for measuring the effects of climate change on production of agricultural crops, i.e., wheat and paddy: Evidence from India. *Environmental Science and Pollution Research*, *29*(21), 31861–31885.
7. Chandio, A. A., Shah, M. I., Sethi, N., & Mushtaq, Z. (2022). Assessing the effect of climate change and financial development on agricultural production in ASEAN-4: The role of renewable energy, institutional quality, and human capital as moderators. *Environmental Science and Pollution Research*, *29*(9), 13211–13225.
8. Chang, V. (2014). The business intelligence as a service in the cloud. *Future Generation Computer Systems*, *37*, 512–534.
9. Chen, K. T., Zhang, H. H., Wu, T. T., Hu, J., Zhai, C. Y., & Wang, D. (2014). Design of monitoring system for multilayer soil temperature and moisture based on WSN. In *2014 International Conference on Wireless Communication and Sensor Network* (pp. 425–430). IEEE.
10. Čirjak, D., Miklečić, I., Lemić, D., Kos, T., & Pajač Živković, I. (2022). Automatic pest monitoring systems in apple production under changing climatic conditions. *Horticulturae*, *8*(6), 520.
11. Dawadi, B., Shrestha, A., Acharya, R. H., Dhital, Y. P., & Devkota, R. (2022). Impact of climate change on agricultural production: A case of Rasuwa District, Nepal. *Regional Sustainability*, *3*(2), 122–132.
12. Gikunda, R., Jepkurui, M., Kiptoo, S., & Baker, M. (2022). Quality of climate-smart agricultural advice offered by private and public sectors extensionists in Mbeere North Sub-County, Kenya. *Advancements in Agricultural Development*, *3*(1), 32–42.
13. Hamidov, A., Khamidov, M., & Ishchanov, J. (2020). Impact of climate change on groundwater management in the northwestern part of Uzbekistan. *Agronomy*, *10*(8), 1173.
14. Hasegawa, T., Wakatsuki, H., Ju, H., Vyas, S., Nelson, G. C., Farrell, A., Deryng, D., Meza, F., & Makowski, D. (2022). A global dataset for the projected impacts of climate change on four major crops. *Scientific Data*, *9*(1), 58.
15. Huang, K., Shu, L., Li, K., Yang, F., Han, G., Wang, X., & Pearson, S. (2020). Photovoltaic agricultural internet of things towards realizing the next generation of smart farming. *IEEE Access*, *8*, 76300–76312.
16. Huseien, G. F., & Shah, K. W. (2021). Potential applications of 5G network technology for climate change control: A scoping review of Singapore. *Sustainability*, *13*(17), 9720.
17. Kaewmard, N., & Saiyod, S. (2014, October). Sensor data collection and irrigation control on vegetable crop using smart phone and wireless sensor networks for smart farm. In *IEEE Conference on Wireless Sensors (ICWiSE)* (pp. 106–112). IEEE.
18. Kalyani, Y., & Collier, R. (2021). A systematic survey on the role of cloud, fog, and edge computing combination in smart agriculture. *Sensors*, *21*(17), 5922.
19. Kavga, A., Thomopoulos, V., Barouchas, P., Stefanakis, N., & Liopa-Tsakalidi, A. (2021). Research on innovative training on smart greenhouse technologies for economic and environmental sustainability. *Sustainability*, *13*(19), 10536.
20. Khriji, S., El Houssaini, D., Jmal, M. W., Viehweger, C., Abid, M., & Kanoun, O. (2014). Precision irrigation based on wireless sensor network. *IET Science, Measurement & Technology*, *8*(3), 98–106.

21. Khudoyberdiev, A., Ullah, I., & Kim, D. (2021). Optimization-assisted water supplement mechanism with energy efficiency in IoT based greenhouse. *Journal of Intelligent & Fuzzy Systems*, *40*(5), 10163–10182.
22. Kumar, P., Sahu, N. C., Kumar, S., & Ansari, M. A. (2021). Impact of climate change on cereal production: Evidence from lower-middle-income countries. *Environmental Science and Pollution Research*, *28*(37), 51597–51611.
23. Li, S. L., Han, Y., Li, G., Zhang, M., Zhang, L., & Ma, Q. (2012). Design and implementation of agricultural greenhouse environmental monitoring system based on Internet of Things. In: *Applied Mechanics and Materials* (Vol. 121, pp. 2624–2629). Trans Tech Publications Ltd.
24. López-Morales, J. A., Martínez, J. A., Caro, M., Erena, M., & Skarmeta, A. F. (2021). Climate-aware and IoT-enabled selection of the most suitable stone fruit tree variety. *Sensors*, *21*(11), 3867.
25. Loukatos, D., & Arvanitis, K. G. (2021). Multi-modal sensor nodes in experimental scalable agricultural iot application scenarios. In: *IoT-Based Intelligent Modelling for Environmental and Ecological Engineering* (pp. 101–128). Springer: Cham.
26. Lova Raju, K., & Vijayaraghavan, V. (2020). IoT technologies in agricultural environment: A survey. *Wireless Personal Communications*, *113*(4), 2415–2446.
27. Luan, Q., Fang, X., Ye, C., & Liu, Y. (2015). An integrated service system for agricultural drought monitoring and forecasting and irrigation amount forecasting. In *23rd International Conference on Geoinformatics* (pp. 1–7). IEEE.
28. Madushanki, A. R., Halgamuge, M. N., Wirasagoda, W. S., & Ali, S. (2019). Adoption of the Internet of Things (IoT) in agriculture and smart farming towards urban greening: A review. *International Journal of Advanced Computer Science and Applications*, *10*(4), 11–28.
29. Maheswari, R., Azath, H., Sharmila, P., & Gnanamalar, S. S. R. (2019). Smart village: Solar based smart agriculture with IoT enabled for climatic change and fertilization of soil. In *2019 IEEE 5th International Conference on Mechatronics System and Robots (ICMSR)* (pp. 102–105). IEEE.
30. Maraveas, C., & Bartzanas, T. (2021). Application of internet of things (IoT) for optimized greenhouse environments. *AgriEngineering*, *3*(4), 954–970.
31. Martinho, V. J. P. D., & Guiné, R. D. P. F. (2021). Integrated-smart agriculture: Contexts and assumptions for a broader concept. *Agronomy*, *11*(8), 1568.
32. Morkunas, M., & Balezentis, T. (2021). Is agricultural revitalization possible through the climate-smart agriculture: A systematic review and citation-based analysis. *Management of Environmental Quality: An International Journal, 33*(2), 257–280.
33. Muangprathub, J., Boonnam, N., Kajornkasirat, S., Lekbangpong, N., Wanichsombat, A., & Nillaor, P. (2019). IoT and agriculture data analysis for smart farm. *Computers and Electronics in Egriculture*, *156*, 467–474.
34. Narmilan, A., & Niroash, G. (2020). Reduction techniques for consequences of climate change by internet of things (IoT) with an emphasis on the agricultural production: A review. *International Journal of Science, Technology, Engineering and Management—A VTU Publication*, *2*(3), 6–13.
35. Ozdemir, D. (2022). The impact of climate change on agricultural productivity in Asian countries: A heterogeneous panel data approach. *Environmental Science and Pollution Research*, *29*(6), 8205–8217.
36. Panchenko, V., Izmailov, A., Kharchenko, V., & Lobachevskiy, Y. (2021). Photovoltaic solar modules of different types and designs for energy supply. In: *Research Anthology on Clean Energy Management and Solutions* (pp. 731–752). IGI Global.
37. Pickson, R. B., He, G., & Boateng, E. (2022). Impacts of climate change on rice production: Evidence from 30 Chinese provinces. *Environment, Development and Sustainability*, *24*(3), 3907–3925.
38. Priyadharsnee, K., & Rathi, S. (2017). An IoT based smart irrigation system. *International Journal of Sciences & Engineering Research*, *8*(5), 44–51.
39. Ratnaparkhi, S., Khan, S., Arya, C., Khapre, S., Singh, P., Diwakar, M., & Shankar, A. (2020). Smart agriculture sensors in IoT: A review. *Materials Today: Proceedings*.
40. Rayhana, R., Xiao, G., & Liu, Z. (2020). Internet of things empowered smart greenhouse farming. *IEEE Journal of Radio Frequency Identification*, *4*(3), 195–211.
41. Sagheer, A., Mohammed, M., Riad, K., & Alhajhoj, M. A.(2020). Cloud-based IoT platform for precision control of soilless greenhouse cultivation. *Sensors*, *21*, 223.

42. Salam, A. (2020). Internet of things for environmental sustainability and climate change. In: *Internet of Things for Sustainable Community Development* (pp. 33–69). Springer: Cham.
43. Shahbaz, P., & Boz, I. (2022). Linking climate change adaptation practices with farm technical efficiency and fertilizer use: A study of wheat–maize mix cropping zone of Punjab province, Pakistan. *Environmental Science and Pollution Research, 29*(12), 16925–16938.
44. Sharma, A., Singh, P. K., & Kumar, Y. (2020). An integrated fire detection system using IoT and image processing technique for smart cities. *Sustainable Cities and Society, 61*, 102332.
45. Srilakshmi, A., Rakkini, J., Sekar, K. R., & Manikandan, R. (2018). A comparative study on internet of things (IoT) and its applications in smart agriculture. *Pharmacognosy Journal, 10*(2), 260–264.
46. Subahi, A. F., & Bouazza, K. E. (2020). An intelligent IoT-based system design for controlling and monitoring greenhouse temperature. *IEEE Access, 8*, 125488–125500.
47. Symeonaki, E., Arvanitis, K. G., Loukatos, D., & Piromalis, D. (2021). Enabling IoT wireless technologies in sustainable livestock farming toward agriculture 4.0. In: *IoT-based Intelligent Modelling for Environmental and Ecological Engineering* (pp. 213–232). Springer: Cham.
48. Symeonaki, E., Arvanitis, K., & Piromalis, D. (2020). A context-aware middleware cloud approach for integrating precision farming facilities into the IoT toward agriculture 4.0. *Applied Sciences, 10*(3), 813.
49. Ullah, I., Fayaz, M., Aman, M., & Kim, D. (2022). An optimization scheme for IoT based smart greenhouse climate control with efficient energy consumption. *Computing, 104*(2), 433–457.
50. Van den Berg, J., Greyvenstein, B., & du Plessis, H. (2022). Insect resistance management facing African smallholder farmers under climate change. *Current Opinion in Insect Science, 50*, 100894.
51. Villa-Henriksen, A., Edwards, G. T., Pesonen, L. A., Green, O., & Sørensen, C. A. G. (2020). Internet of Things in arable farming: Implementation, applications, challenges and potential. *Biosystems Engineering, 191*, 60–84.
52. Wang, K., Khoo, K. S., Leong, H. Y., Nagarajan, D., Chew, K. W., Ting, H. Y., Selvarajoo, A., Chang, J.-S., & Show, P. L. (2021). How does the Internet of Things (IoT) help in microalgae biorefinery. *Biotechnology Advances, 54*, 107819.
53. Willockx, B., Herteleer, B., & Cappelle, J. (2020). Combining photovoltaic modules and food crops: First agrovoltaic prototype in Belgium. *Renewable Energy & Power Quality Journal (RE&PQJ), 18.*
54. Zamora-Izquierdo, M. A., Martı, J. A., & Skarmeta, A. F. (2018). Intelligent systems for environmental applications smart farming IoT platform based on edge and cloud computing. *Biosystem Engineering, 177*, 4–17.
55. Zhang, Y., Geng, P., Sivaparthipan, C. B., & Muthu, B. A. (2021). Big data and artificial intelligence based early risk warning system of fire hazard for smart cities. *Sustainable Energy Technologies and Assessments, 45*, 100986.

CHAPTER 14

Technologies for Precision Farming

SAVITA TAMTA

Department of Vegetable Science, College of Horticulture, VCSG Uttarakhand University of Horticulture and Forestry, Bharsar, Pauri Garhwal, Uttarakhand, India

ABSTRACT

Precision farming, alternatively referred to as precision agriculture, can be described as a new groundbreaking paradigm in modernizing agricultural practices using advanced technologies, improving efficiency, and reducing negative environmental impacts. This chapter explains different aspects of such a method, focusing on the various technologies that facilitate its functioning. More specifically, the Global Positioning System, remote sensing, variable rate technology, automated machinery, data analytics, the Internet of Things, and nanotechnology are thoroughly examined, and their specific applications and relevance for realizing precision farming are explained. It is highlighted in their roles in improving crop management, maintaining soil health, and monitoring livestock, as well as their general significance in making farm work more efficient. Consequently, for the reasons outlined above, precision farming is a method through various technologies; farmers can make more informed choices and act more attentively, adapt their facilitation for specific field conditions, automate some processes, and eventually promote more sustainable agriculture. This abstract summarizes the various directions of precision farming and the complex technology stack implemented in modern agriculture.

14.1 INTRODUCTION

Modern agriculture faces severe challenges due to rapid population growth, climate change, and natural resources continue to shrink. According to [21], global food demand has risen significantly and is expected to increase by 59–98% by 2050. Furthermore, conventional farming methods cannot be developed further to help food producers avoid damaging the environment or achieving economic sustainability [5, 14, 65]. At the same time, there is evidence that inefficiencies within the food themes are associated with the food insecurity [6, 66, 83]. To manage the constraints on agricultural inputs, more economical technology is needed for farmers and growers.

Demystifying Sustainable Farming Using Smart Technologies.
Gunjan Chhabra, Keshav Kaushik, Priya Chugh, Susheela Dahiya, and Aryan Chaudhary (Eds.)

DOI: 10.1201/9781779640932-14

Precision farming has emerged as a new vision, providing an opportunity to completely reshape farming by integrating new cutting-edge technological tools and techniques [39]. It helps to maximize the crop yield and minimize the environmental impact, as the usage of pesticides and fertilizers is optimized [63]. The crops are controlled by monitoring almost all indicators of their state of health, the amount of water and mineral or organic fertilizers, and the presence of infections. It controls the state of health, amount of water and fertilizers, possible infections, and many other indicators of crop planting [92]. The presence of all resources alone is not the same as the result of access to resources due to scarcity, inadequate technology, lack of field agrarian knowledge and awareness, and the acquisition of obsolete techniques [11].

Precision agriculture leverages cutting-edge data technologies and expertise to optimize crop yields in a geographically-targeted manner. Some economically viable strategies for boosting harvests involve implementing wireless sensor networks to monitor local conditions [37]. While the field continues advancing, few farmers have incorporated precision agricultural practices successfully due to specific terrain challenges and upfront expenses. However, those who adopt precision agriculture can recoup costs through increased outputs over time. Identifying, locating, and tracking plants, pests, and weeds as well as calibrating fertilization, pesticide, and fungicide application rates, mapping plantings, and monitoring crop health and development are all aspects of precision agriculture's integrated approach. Meanwhile, researchers continue exploring novel techniques like precision livestock farming and variable-rate irrigation to enhance productivity and sustainability [16].

Modern agriculture faces severe challenges because of the rapid population growth, climate change, and the fact that the world´s resources become less and less available. [21] argue that the global food demand has increased to the greater extent and will climb even more—by 59–98% in 2050. It is impossible to imagine that the conventional ways of the farming can be further developed so that they above all prevent the food producers from harming the environment or being economically sustainable. However, there is sufficient evidence to suggest that inefficiencies in the food themes themselves exacerbate the food insecurity problem. A more economically farmers and growers technology should be introduced to manage the constraints on agricultural inputs. The new vision is precision farming, which allows us to radically change farming by integrating new cutting-edge technological tools and techniques. As a result, crops produced by farmers with this new technology fully address the demands for crop yield and simultaneously minimize the environmental impact since the use of pesticides and fertilizers is optimized. The crops are monitored on almost all indicators of their state of health and, most importantly, on the amount of water and mineral or organic fertilizers and for the presence of infections. The presence of all resources alone is not the same as the result of access to resources due to scarcity, inadequate technology, lack of field agrarian knowledge and awareness, and acquisition of obsolete techniques.

14.2 PROBLEMS IN AGRICULTURE

Agriculture is one of the oldest human activities. Still, it faces several problems that have existed since its inception, plus others that have arisen or strengthened in connection with such factors as climate change, globalization, and scientific and technological progress. These are the following problems of agriculture:

1. **Climate change**: Altering weather conditions, increased extreme events, and a temperature rise are the most severe challenges to agricultural productivity. Unstable precipitation, prolonged droughts, and rising floods disrupt crop cycles and reduce yields while compromising food safety and security.
2. **Soil depletion**: Erosion, a lack of necessary nutrients, salinization, and desertification threaten to reduce agriculture's capacity. Overusing a limited number of plants and excessively high tillage methods leads to soil degradation, lowering fertility and reducing the ability to grow food over time.
3. **Water scarcity**: Agriculture is responsible for a large proportion of water use globally, and water shortage is fast becoming a reality through overextraction, contamination, and competition between various sectors. Water management practices like drip irrigation and rainwater collection are integral to sustainable agriculture.
4. **Loss of biodiversity**: Under intensification, cultivation practices often focus on certain high-yielding species, which reduces biodiversity.
5. **Pesticide overuse and resistance**: Due to the lack of genetic variety, crops are more vulnerable to pests, diseases, and the effects of climate change. Therefore, this practice negatively impacts food security and the ability of ecosystems to recover after shocks. For example, excessive pesticide and herbicide use kills benign life forms and pollutes the environment while promoting the development of the pesticide's resistant species. So, instead, people should use integrated pest management and promote agroecology.
6. **Food waste**: Although enough food is produced, a large quantity is lost or wasted at various postfarm stages. Reducing food waste entails enhanced distribution and marketing and fostering better consumer behavior. It also involves improvements in infrastructure, storage facilities, and transportation networks, among other postfarm practices.
7. **Market volatility and price fluctuations**: Farmers face financial uncertainty from fluctuating prices, insecure input markets, and trade policies. Farmers experience food price instability that affects their livelihoods and discourages them from investing in the sector. The instability of food prices may result in more poverty since most smallholder farmers depend on it for their income.
8. **Resources and technology**: Smallholders, especially in developing countries, lack important resources such as land, credit, inputs, and technical knowledge of modern agriculture. These imbalances must play a prominent role in boosting productivity and raising rural people's income.
9. **Labor shortages and rural depopulation**: There are many cases of labor shortages in agriculture, localized in various regions that are explained by rural-to-urban

migration, aging farming populations, and other socioeconomic factors. Mechanization, automation, and investment in rural infrastructure and services are needed to make farming more attractive for younger generations.

10. **Policy and governance issues**: In certain jurisdictions, inadequate policies, bad institutions, and corruption or lack of political will can impede agricultural development and worsen existing problems. Governance, policy coherence, and stakeholder engagement are prerequisites for developing an environment that enables sustainable farming.

These issues are complex and interconnected and require a systems perspective that will address them holistically and address the ecological, social, and economic dimensions of agriculture while creating opportunities for innovation, collaboration, and resilience-building on local, national, and global scales.

14.3 NEED OF AGRICULTURE

First, the importance of agriculture is the elimination of hunger and overcoming malnutrition. At the same time, this area is often pursued in rural areas. This means that it has food access, provides livelihoods, employs thousands of people, brings in money and, as a result, develops people's power. Furthermore, the raw materials price is used in many industries, including textiles, biofuels, and pharmaceuticals, while critical resources for production include sugar cane, cotton, and timber. Livestock products, including wool, leather, and other manufacturing processes, are also available. Agroecological approaches and sustainable practices can help sustain ecosystems and preserve biodiversity, reducing climate risks, preserving and maintaining carbon ecosystems, and storing carbon in the soil.

Agriculture is the foundation of the rural sector. It is important for the perspective of both economic and social because of the increased access to social services, infrastructure, and the quality of life in rural places. It also has an impact on human health and nutrition through the provision of a few different kinds of food items, which are often more beneficial when combined than when eating alone. Moreover, agriculture advances the provisioning of crucial ecosystem processes such as pollination, pest management, water regulation, and nutrient cycling. Further, it facilitates global trade and exchange through the exportation of agricultural commodities that make economic development a reality. Lastly, agriculture research and development that investigates new versions of crops, farming methods, and practically sustainable agriculture has to advance productivity, resilience, and competitiveness.

14.4 TECHNOLOGIES FOR PRECISION FARMING AND THEIR APPLICATION

Precision farming, also known as precision agriculture, is a farming management concept that uses technology to ensure optimal efficiency and profitability while minimizing waste and environmental impact (Figure 14.1).

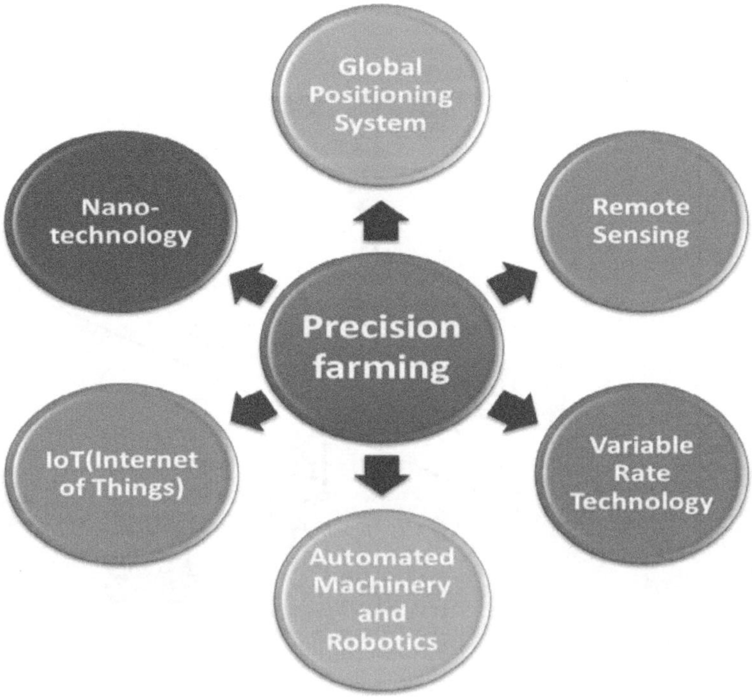

FIGURE 14.1 Technologies in precision farming.

Several technologies have emerged to enable precision farming, revolutionizing traditional agricultural practices. Here are some key technologies and their applications.

14.4.1 GLOBAL POSITIONING SYSTEM

The first global navigation system was called global positioning system (GPS). GPS technology enables farmers to map their field boundaries precisely, monitor crop health, navigate farming tools more accurately, and develop their careers. Afterward, as technology, freelance planting, spraying, and harvesting can be used, which ensures the independence of all operations and no overlap in the use of resources. GPS is more effective than aerial triangulation, especially in large areas [33]. Furthermore, broadcast equipment creates detailed maps of various parts of the field, so much better decisions can be made (Figure 14.2). Deep learning technologies such as mask R-CNN could be beneficial in automatic weed removal from the field [18].

- **Accurate field mapping**: GPS allows farmers to accurately map their fields and identify spatial variability in soil properties, topography, and crop health. With a 95% likelihood, the determined position will be within 10–15 m of the actual location [26]. Farmers can also achieve site-specific, or field-scale, management. They can make detailed field maps and apply management practices that fit specific parts of the field.

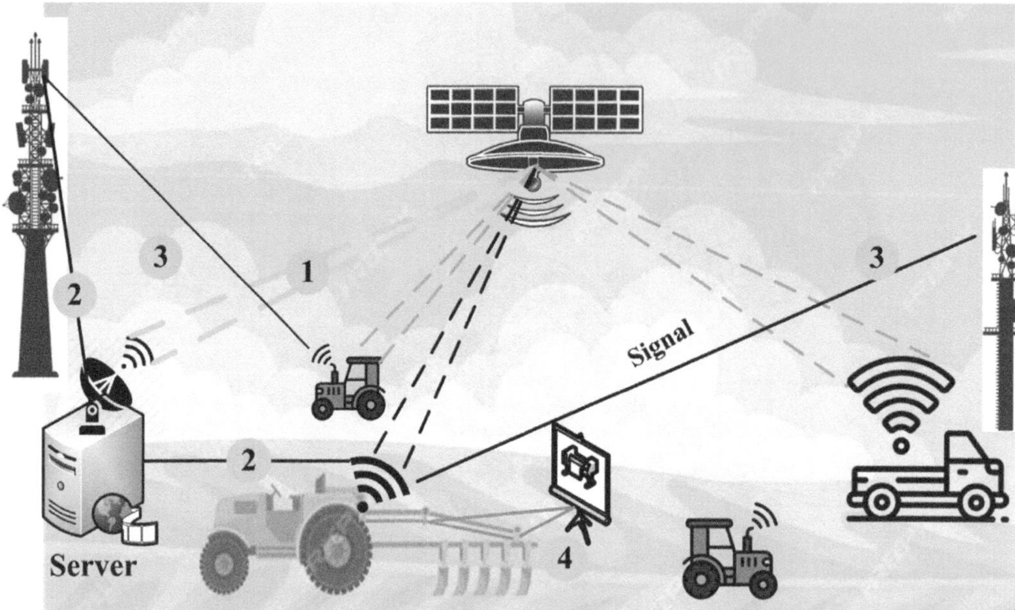

FIGURE 14.2 Global positioning system.

- **Site-specific management**: Farmers can use GPS to develop site-specific management practices, which is also called variable rate application (VRA). Basically, farmers develop a prescription map with field data, like soil sampling and remote sensing, to apply their input exactly where it is needed [68]. They could vary their inputs, including fertilizers, pesticides, and irrigation water, according to individual pieces of ground. This technique ensures that a farmer only spends inputs where they will have the greatest impact instead of wasting—reducing the environmental effect.
- **Precision planting**: Accurate positioning and guidance for seed placement have been made possible by GPS technology. According to [89], GPS-guided planting equipment assists farmers in planting seeds with uniform spacings and precise seeding depths for maximum seed germination and seedling growth. The system determines the best possible placement opportunities for innumerable seeds.
- **Auto-steering and guidance systems**: Moeller et al. [54] point out that the GPS signals farm machinery with GPS-based auto-steering and guidance systems. It allows tractors and sprayers to be managed precisely without the need for human interaction. In terrain that was ragged and potato fields that were uneven, GPS-guided implement steering contraptions (like the precision machine from Massey Ferguson called Auto-Guide, John Deere's AutoTrac, Case IH's Advanced Farming package, Mahindra Novo Guidance, Swaraj Tractor's Auto Guidance software, the AGS from YTO Group, and the Lovol 504 Guidance tool) could function and give exact direction to farming implements (as reported in Ref. [62]). These devices lessen where work overlapped and get rid of gaps by taking advantage of GPS

signals to maintain pathways that were straight and continuous all through agricultural activities happening in the land. Systems for auto-steering boost productivity, decrease fatigue for the person operating equipment, and guarantee fertilizer, herbicides, and pesticides are applied accurately and consistently.

- **Yield monitoring and mapping**: GPS-enabled yield monitoring systems enable farmers to measure accurately and map crop yields across their fields [3]. Yield data can also be analyzed along with other field information such as soil maps and management practices to ascertain various factors contributing to yield variance and allow farmers to decide how to address them based on data to boost productivity.
- **Remote sensing integration**: GPS technology can be used in combination with remote sensing technologies, like drones and satellite imagery. This will allow for real-time data on crop health, pest infestation, and field conditions. When merged with GPS data, remote sensing can inform farmers where in their fields there is cause for concern and what areas to take proactive action in, before yields are affected [12].
- **Data management and decision support**: Precision agriculture data management and decision support are achieved by GPS technology. Based on the collected data about variability in fields, input use, and crop performance, farmers subsequently decide a series of decisions related to their farming practice. Using such data helps the farmer minimize the variability of his work, improve his efficiency, and thereby increase his total profit [44].

GPS technology is used in precision agriculture for field mapping, site-specific management, precision planting, auto-steering, and guidance systems, yield monitoring, integration of remote sensing with GPS, DM, and decision support. The information and data from this technology will enable the farmer to gain more yields using less input and not deplete the environment and consequently be able to develop sustainable, effective, and productive agriculture.

14.4.2 REMOTE SENSING

Remote sensing is of utmost importance in the field of precision agriculture since it is used to assess the status of crops, soils, and pertinent environmental conditions, especially in the extensive agro-ecological landscapes [80]. The uses of remote sensing are the following:

Crop monitoring and management: Remote sensing technology utilizes satellite imagery and aircraft drones to obtain more precise data on crops' condition and progress [41]. Multisensor or hyperspectral imagery assessment allows producers to detect variation in the vigor of canopies and issues with crop influence by soil factors, particularly nutrient scarcity and water shortage, diseases, and pest attacks and promptly rectify policy to address the conditions. The efficient treatment of critical inputs, such as fertilizers, pesticides, and irrigation water, can help to more appropriately match inputs to certain field demands with respect to each area's optimal resources for the cultivation of crops [27]. While drones are more frequently used tower frame lightweight UAS for IWM, they are also well-suited to mapping weeds in spatially variable contexts such as those found on min-field edges in small-holder farms. As a remote sensing platform, drones have also been used for data synthesis from field-deployed nodes to boost crop and its management [25].

Early detection and monitoring of pest and diseases: Besides, remote sensing allows for the early detection of stress and pest infestations on crops, enabling farmers to take action before substantial yield losses occur. Farmers may notice small changes in plant health indicators such as chlorophyll content or leaf temperature that indicate seniority by regularly monitoring crops from above. The cause of plant stress can be identified and managed through time intervention to reduce damage and prevent the infestation and disease of pests from spreading to surrounding areas. Another classification is that sensing systems are also based on their characteristics and maturity in remote sensing of plant diseases and pests. They may include visible and near-infrared spectral sensors; fluorescence and thermal sensors; and synthetic aperture radar and light detection and ranging systems [93]; spectroradiometer is the sensor used in early detection and monitoring in the remote sensing sensor for the internal aspects and hidden signs of *Bactrocera zonata* in fruits and vegetables [32].

Yield prediction and mapping: The collected remote sensing data is used in making maps that show the return of the crop. Yield maps are important tools for making management practice decisions that contribute to sustainable resource application [38]. This can be made possible by sending signals to the farmers from satellite or drone images on the vegetation indices for both biomass accumulation and canopy coverage, which will help in estimating the final crop yield in his or her field per square meter. Such information would enable a farmer to be conversant with the more and less productive areas and be better positioned to apply ideas relevant to making wise management decisions where profit is to be maximized. Besides the biodiversity aspect, it may involve the use of remote sensing data obtained from different satellites, with the help of machine learning regression models such as the Random Forest Regressor, XGBRegressor, and Gradient Boosting Regressor, together with other models that exist to predict the yield of almonds, amongst others [28].

Soil mapping and management: Remote sensing can be used to harness information for examples of soil properties as well as their variability patterns across agricultural landscapes. These would help determine the status of the moisture content, texture, and organic matter percentage of the soil from the satellite or aerial images. The software compiles that into soil maps, and it can facilitate decision-making in soil preparation progress, drainage, and nutrient management practices that are appropriate for different soil types and under varying conditions [80].

Environmental monitoring and sustainability: This technology allows monitoring of land use changes, availability of water, and dynamics of vegetation, among other environmental factors, through remote sensing [49]. Eventually, it will enable farmers to track their land cover and land use change and facilitate decision making regarding sustainable land use practices. Consequently, the farmers are able to track changes brought about by agricultural activity within the natural resources and biodiversity of a given piece of land. Remote sensing also provides information on areas that are suitable for conservation, more sensitive to erosion, and the effectiveness of conservation efforts.

Data integration and decision support: Lastly, the integration of remote sensing data can assimilate other data repositories like GPS data, weather data, soil data, and so on, that will finally be able to help precision agriculture in terms of decision support [74]. Several layers of data are integrated, giving an overall picture of the agronomic situation to the farmer who will then make more knowledgeable decisions to enhance their management practices and efficiency with better yields.

As a result, remote sensing is one of the very essential technologies in precision agriculture because it helps to capture variations and changes in crop and soil, as well as identify the environmental impacts and potential in the creative support of productivity throughout the various areas. Because of this remote sensing technology, farm managers can enhance input use, lessen their environmental footprint, and ramp up the productivity of modern agriculture.

14.4.3 VARIABLE RATE TECHNOLOGY

Variable rate technology is one of the key technologies of precision agriculture, which allows farmers to apply fertilizer, pesticides, and irrigation water to their fields in a more optimal way. Here is an overview of Variable Rate Technology and its significance in precision agriculture:

Customized input application: The system that enables farmers to change the application rate of inputs, such as seeds, fertilizers, or pesticides depending on the field conditions. Instead of applying the same rate throughout the field, variable rate technology provides the ability to change the rate of inputs depending on spatial variability within a field [31, 52, 73, 91]. In addition, variable rate technology (VRT) is used to develop prescription maps based on soil data, yield maps, remote sensing imagery, and other relevant information on a specific field to ensure optimal crop production and minimum input costs taking into consideration environmental impacts [61].

Optimized resource use: VRT allows farmers to apply inputs only in the areas that require them enabling better uses of resources such as fertilizers, pesticides, and water. For instance, in vegetable farming VRT allows for more efficient application of fertilizers without applying more where the nutrient is abundant or less where the nutrients are deficient; pesticides can also be reduced or diverted to a more critical area as opposed to the uniform application which many leads to wastage and environmental pollutions [15]. VRT can be used to prepare the field where potatoes grow, by first mapping the level of soil variation, then determining essential fertilizing rates per square meter for the known area, and then feeding the data to a computer that will regulate the application of fertilizers based on the optimal required level per area. The process will eliminate extinction in other areas and hence no more wastage [15].

Crop improvement: VRT helps optimize crop performance and yield potential by applying nutrients, water, and other inputs as per on the amount required, at the appropriate time they needed, and location where they are needed without much spillage on the other areas. Farmers, by utilizing their VRT system, apply inputs in areas of the field with different nutrient requirements in response to soil fertility, soil pH levels, and moisture availability, promote optimal crop growth and production. VRT systems resulted in more efficient use of inputs and better soil fertility, therefore healthier crop growth [35].

Reduced environmental impact: VRT is a mode of reducing environmental impact of agriculture where VRT mitigates excess application of inputs hence minimizing spillage of the said inputs on the farms which in most cases runoff and are spilled into the water sources [30]. The VRT technology helps farmers to apply suitable rates to the different areas of the field, in areas with high nutrients apply lower nutrients that suits the crop requirement hence reducing the pollution, topsoil loss, and groundwater contamination and when it was utilized by [47] it enabled minimized N being used but kept the productivity and reduced environmental pollution.

Cost savings: Despite the need for an upfront investment in equipment and technology, VRT can bring substantial long-term cost savings by minimizing input costs and enhancing crop yields [76]. As inputs are applied with greater efficiency, farmers can save on input costs while securing yield or even increasing them. Because of this, the overall profitability of farm operations is increased.

Integration with precision agriculture technologies: VRT and other agricultural machinery technologies have gained popularity among farmers worldwide due to the numerous benefits they have to offer. First, the technology is frequently combined with GPS, geographic information system, and remote sensing, where seamless data collection, analysis, and utilization are feasible. This way, farmers can make better informed decisions in potato management and preparation to boost the productivity of the crop [54]. Furthermore, by integrating VRT with real-time monitoring and data analysis, the management of farm activities can adapt to the prevailing field conditions as they change [84]. In the future, given the high pressure on agriculture for production knowledge intensification in the limits of available resources and to preserve minimum harm to the environmental ambition, VRT on the results mentioned above and many others will secure an unfading place in agriculture.

14.4.4 AUTOMATED MACHINERY AND ROBOTICS

Precision agriculture is also increasingly affected by automated machinery and robotics, which provide farmers with tools to further increase the efficiency of their operations, save on labor, and optimize production. Here is an overview of the role of automated machinery and robotics in precision agriculture.

Autonomous tractors and equipment: Through the use of autonomous tractors that have advanced sensors and GPS, a farm can use them to perform various field activities with limited human input [75]. For example, autonomous tractors can navigate the fields, plough the fields, plant the seeds, apply fertilizers and pesticides, mow the grass among other duties, and follow preprogrammed paths or through real-time monitoring. The use of guidance systems is critical for enabling a significant level of automation in many types of agricultural machinery. The guidance systems are used to acquire data on the actual position of the machines, the speed of the machine, the direction of movement, and the current steering wheel angle (Figure 14.3). Consequently, through the guidance systems' influence, automated machinery and robotics can perform its tasks with minimal human intervention [75, 77].

FIGURE 14.3 Electric field robot for seeding and weeding was used in the experiment with additional sensors.

Planting and seeding precision: Another key application technology is planting and seeding, where robotics ensures accurate and proficient planting or deployment of the seeds at the required depths and gaps [56]. When it comes to planting, automated systems can establish the most suitable speed and configuration depending on the soil conditions, the landscape, or demanding plants for high levels of germination and production. Robotics help achieve plans uniformity and reduce established crop levels.

Weed and pest control: Robotics technology has been used more and more to exhibit automated weeding and pest control in the field of precision agriculture [64]. The robotic system uses cameras, sensors, and artificial intelligence algorithms to detect and eradicate weed or selectively apply chemical agents to pests. Automating these tasks reduces the use of herbicides and pesticides, avoids crop damage and promotes crop health. Although UAV systems are primarily used for weeding, sowing, and chemical spraying, these applications are needs less due to constraints in control, respectively, consumption cost, energy consumption, and autonomy [48, 72].

Harvesting and crop management: Robotics in agriculture, particularly for harvesting, has seen significant advancements. Robots are now capable of performing delicate and complex tasks that were traditionally done by human labor in Table 14.1 [72]. Automated machinery and robotics could transform existing labor-intensive procedures in precision farming harvesting and crop management [29]. Autonomous harvesters with sensors and computer vision systems can recognize ready crops and harvest fruits or vegetables selectively, sorting them according to size, color, and quality [87]. Automated manual working systems for observation crops and after-planting and emerging crops are also being developed to enhance output and quality while using fewer workers.

TABLE 14.1 Robotics in Harvesting and Post-Harvest Handling

S. N.	AI Technology	Functions	References
1	Robotic arms and manipulation systems	Used for sowing of seeds, fertilizer application, irrigation, weed control, and harvesting of crops	[60, 70]
2	Machine learning	Used in post-harvest handling operations and cost-effective grading system of tomato	[36]
3	Robotics	Ensure consistent cutting height and high-quality cuts, which is crucial for crops like sugarcane and lettuce	[13]
4	Robotics	Used for harvesting operations such as fruit detection and picking efficiency and rate useful for delicate fruits like strawberries and tomatoes	[53]
5	Robotics	Suitability and development of harvesting tools	[55]
6	Harvesting robot	Used to identify and detect mature fruit or fruit bunches	[50]
7	Robotics	Used for harvesting in low light conditions	[34]
8	Robotics	Used for harvesting of cotton, tomato, cherry tomato, and kiwi fruit	[22, 24, 42, 81, 86]
9	Robotics	Used for development of gripper to minimize damage to produce	[94]
10	Robotics	Smooth and safe harvesting support in the field	[90]
11	Robotics	Used for harvesting of fruits	[71]

Data collection and analysis: Automated machinery and robotics ensure that farmers have access to agricultural and environmental data. Precision agriculture uses robotic platforms that have sensors, cameras, and other on-board devices to collect data such as soil properties, crop health, and environmental conditions involved. Such automated systems enable farmers to adopt informed and accurate decision-making processes dependent on monitoring critical components of aggregate field performance and large data volume collection systems. In conclusion, automated machinery and robotics have significantly contributed to the future of sustainable and profitable field activities. Farmers will have access to more advanced automated tools that enable them to utilize and optimize resource use to enhance production.

14.4.5 INTERNET OF THINGS

The Internet of Things (IoT) has also become a technology in precision agriculture. Consequently, IoT has redefined how farmers monitor, manage, and optimize farming operations. The IoT is defined as a network of interconnected gadgets and sensors that collect, archive, transport, and analyze data in real-time. Under this definition, IoT affects precision agriculture through the following ways:

Sensor networks: Sensor networks have been put into place in agriculture fields to measure environmental parameters such as soil moisture, temperature, humidity, rainfall, and air quality [9]. These sensors measure data on a continuous basis and transmit the same to one centralized platform using wireless communication and information that is used for real-time interpretation (Figure 14.4).

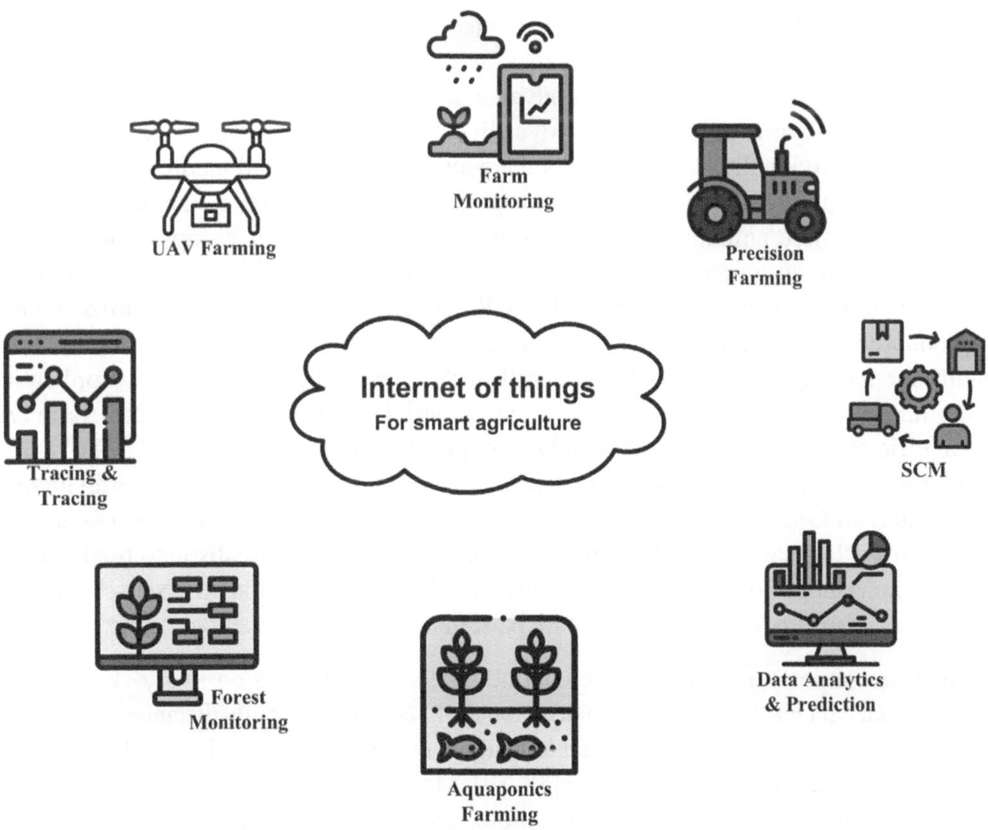

FIGURE 14.4 Use of internet of things (IoT) in agriculture.

Precision irrigation: IoT-based irrigation systems are equipped with soil moisture sensors and weather forecasts, which help in determining the efficient water usage for irrigation scheduling. By monitoring the moisture level of the soil in the real-time and changing the irrigation schedule, farmers can supply crops from the desired quantity of water at the desired time. As a result, water is remarkably wasted, and crops are given more yields. The agricultural field irrigation schedule is predicted with the WSN acting as an observant environmental condition, such as pressure, humidity, temperature, soil moisture, soil salinity, and soil conductivity [4].

Smart crop monitoring: To track crop development and individual health in real-time, the IoT places sensors and cameras throughout fields monitoring growth [10]. Detecting stress, nutrient levels, and harmful conditions in fruits allows ranchers to make timely adjustments before losses occur. Quality, health issues in fruits, detecting weeds, and optimizing water usage are now achievable thanks to a plethora of sensor-gathered multimedia data from installations across orchards and vineyards [17, 69, 78]. Sentinel sensors vigilantly scan each plant, cross-referencing readings with a plant's unique fingerprint enabling a discerning eye on the entire operation.

Livestock monitoring and precision livestock farming: Many of the technologies that precision farming relies in livestock systems are paramount in livestock monitoring and precision livestock farming. Sensors, RFID tags, and monitoring systems are some of the primary tools, and they include sensors that measure animals' health, and behavior, among other characteristics. One of the main advantages of this type of technology in farming is that it gives the farmer precise control over livestock, and it becomes possible to monitor each animal individually. This enhances the accuracy of feeds and medication, as well as allows better optimization and breeding programs. In addition to that, the more controlled environment where animals are mostly kept alone instead of in groups, leads to fewer outbreaks of animal-human diseases. Above all, this is meant to increase productivity and efficiency, more than most developments do. Precision farming seeks to maximize the level of productivity and minimize the level of input cost while at the same time minimizing environmental degradation. Tool use is still in its infancy and will find new applications in the future. New technology will have the potential to completely transform farms and create efficient, resilient, and sustainable food production systems. Currently, the two key technological trends are the rise of big data, advanced analytics capabilities, aerial imagery, feeding and milking robots, and intelligent sensors [95]. Heat and nutrient sensors and GPS trackers use RFID tags. Although several years ago, it seemed unrealistic to track the vital signs of thousands of dairy cows. It is quite possible now due to biometric sensors and GPS technology. Growers re-observe the same patterns and make quicker decisions.

Real-time animal trackers provide continuous location updates on livestock along with useful spatial data interpretations from decision tools. Metrics like daily distances traveled and terrain usage reveal patterns with valuable insights into animal well-being, such as detecting immobility or uneven herd distributions [58, 59]. Motion sensors likewise analyze behavioral activity, shedding light on factors, for example, grazing duration which can offer a view of forage availability and potential supplementary feeding needs [8]. When the circumstances are properly understood, a well-calibrated rain gauge may provide invaluable information regarding both the quantity and quality of available fodder, painting a holistic picture of the environmental conditions. Through the employment of wearable collars that leverage cues to deter boundary breaches of predefined virtual fences, virtual fencing is becoming a viable option for effectively managing livestock dispersion [7, 57]. This approach works particularly well in situations where traditional methods such as erecting physical barriers, herding animals manually, or strategically positioning water sources are not feasible due to financial or practical limitations. However, we must proceed with care and caution, closely observing their responses to ensure their well-being remains the top priority. Though promising in its aims, we still have much to learn to optimize this technology for the benefit of all [85]. While numerous sensor tools exist, validation efforts still require significant initiatives even as use remains limited due to infrastructure needs, moderate to high skill requirements for interpretation, financial constraints especially for small herds, scarce availability, and other causes primarily in underdeveloped and developing regions [40].

Equipment monitoring and maintenance: IoT sensors installed on farm machinery and equipment can monitor performance metrics such as fuel consumption, engine temperature, and operating hours. By analyzing this data in real-time, farmers can identify potential issues, schedule preventive maintenance, and optimize machine usage, decreasing downtime and boosting operational productivity. IoT helps farmers become more efficient in various ways, for instance, tracking and implementing fleets electronically. IoT technology will permit farmers to remotely monitor their gear, from tractor lineups to grain conveyor lines. Tractors nowadays are fitted with sensors that constantly monitor the machine's performance. These internet-linked detectors assess whether the heavy-duty vehicle is functioning at its top potential. The observant technology will promptly alert the farmer about any needed servicing to restore peak effectiveness. Addressing issues immediately can lengthen the tractor's usable lifetime by averting unanticipated breakdowns that reduce productivity in the fields. Furthermore, cutting-edge grain transportation lines have their own intellectual detectors. These perceptive sensors continuously measure and report metrics such as belt deterioration. When abrasion becomes too excessive, a notification will be dispatched to operating staff advising that extra lubrication be applied to the conveyor to ease the friction and prevent damage or malfunctions from occurring [43].

Supply chain management: IoT technologies can greatly improve traceability and transparency in agricultural supply chains by closely observing the movement and environment of farm products from planting to plates. Radio frequency identification tags, GPS monitors, and temperature detectors can track where agricultural goods are located, how they are stored, and what routes they travel, helping to ensure quality and safety benchmarks are consistently met throughout the supply chain. Wireless sensor networks and radio-frequency identification are often viewed as two primary building blocks for sensing and connecting technologies central to the IoT, allowing vast amounts of real-time data to be collected and analyzed for more effective oversight and operations management [23, 51]. During drier periods, an array of optical instruments are applied to monitor the condition of crops like papayas, catching any issues at the earliest stages [82].

Data integration and analytics: IoT generates vast amounts of data that can be integrated with other sources of information, such as weather data, soil maps, and market trends, to provide actionable insights and decision support for farmers. Advanced analytics techniques such as machine learning and predictive modeling can analyze this IoT data to identify complex, hidden patterns, trends, and anomalies, enabling farmers to optimize management practices and remarkably improve decision-making [46]. By leveraging big data approaches and sophisticated algorithms, analytics of integrated IoT and other agricultural data sources afford discerning previously obscure connections and multifaceted inter-dependencies that facilitate more informed choices.

14.4.6 NANOTECHNOLOGY

Nanotechnology proposes promising prospects for cultivating crops while restricting resource intake and environmental impact. By incorporating nanomaterials and nanoscale instruments, accuracy agriculture can revolutionize various facets of crop administration, from proficient nutrient transport to real-time tracking of plant health [2, 79].

At the nanoscale, materials have many unmatched benefits related to nutrient administration. Nanoparticles and nanostructured fertilizers increase the effectiveness of absorption and reduce nutrient leaching, thereby decreasing environmental contamination. For example, aspects such as controlled release derived from nanoscale encapsulation of fertilizers ensure that there is a timely supply of nutrients to the crop plants [19]. Second, nanoparticle formulations could be designed to supplement a lack of specific nutrients in the daily diet, which would improve both plant development and productivity.

Precision delivery systems: Nanotechnology ensures many twists and turns in the targeted distribution of agrochemicals, meeting the overall benefits needed for crops with a minimal environmental load [67]. With this development are nanoscale sensors that continuously analyze the plant's health, together with host microstructures that carry nutrients to their target sites, into which roots can take them up [88]. Today, this precision is unrivaled in the elimination of unwanted pests and weeds without killing other living beings. In a different context, nanoscale carriers ensure that the active ingredients are not destroyed, thus prolonging their action and reducing intervals between applications. Such enhanced delivery enhances the protection in field applications while decreasing the environmental doubts related to conventional chemical use. Taken together, nanoscience-enabled applications support high-yield harvests by providing efficient, effective solutions bearing respect for ecological health.

Sensor technologies for real-time monitoring: Another significant operation in agriculture is the provision of real-time information to farmers such that valuable knowledge can be transferred for improved decision-making relative to crop conditions. The nanoscale sensors often turn out to be a favorite option for on-site monitoring applications because of the characteristics of high sensitivity, specificity, and scalability. These advanced sensors incorporate unique capabilities to monitor numerous parameters relevant to both healthy plant growth and soil quality, such as moisture levels, nutrient contents, pH balances, and signs of illness. In addition, coupling such nanosensors with wireless networks allows for remote monitoring from any place so that real-time responses to issues emerging may be put in place and resources economically allocated in cases where challenges arise. Continual data gathered through nanoscale monitoring devices about soil conditions, plant wellness, and surrounding environmental factors allows maximum yields with water conservation and lessening fertilizers while reducing the environmental effect [20, 45].

Nanobiotechnology in disease management: Nanobiotechnology offers new, promising ways in the creative handling of the crop in terms of disease. For example, targeted delivery of antimicrobial treatments in a controlled manner by the use of nanocarriers helps in effective removal of the pathogen menace without hurting beneficial microbes and the environment. The role of nanomaterials as nucleic acid carriers also facilitates proper transportation of RNA interference molecules that are effective in the silencing of pathogen virulence or stimulation of host resistance in plants. Targeted interventions of such a nature offer sustainable solutions in the fight against plant illnesses and decrease the reliance on chemical treatments. These experiments on Ag-NP treatments applied as a plant resistance inducer in squash plants infected with CMV show promise for developing this new alternative technique in the management of plant viral diseases without pesticide application [1].

Although nanotechnology offers a perspective development in precision agriculture, several issues continue to persist and have to be resolved under the multistakeholder context. The concern of regulations, environmental impact, and economic feasibility of cooperative evaluations by researchers, lawmakers, and stakeholders will lead to useful and safe applications of practical nanomaterials. Future effort may involve maximal synthesis methods to maximize compatibility with existing methods, or investigate new avenues altogether, such as enquiring about fully new possibilities, like the automation of everyday work through nanorobots to enhance productivity. Adaptive models with social and scientific oversight hope to fulfill the potential of nanotechnology in terms of both ecological and humanitarian priorities.

14.5 CONCLUSION

Precision farming uses new technologies and data analysis to give farming a new facade. It helps farmers, therefore, use their resources most efficiently by zeroing in on particular sections of each field with the help of GPS, remote sensing, variable-rate technology, self-driving equipment, analytical software, the IoT, precision livestock monitoring, and finally, nanotechnology. This will save resources, increase yield, and support environmental sustainability to be improved security for generations to come. The ultimate goal is to hand down prosperous land and abundant harvests to our children.

However, there are important hurdles to its application. Some of the obstacles are high upfront costs, challenges of managing large datasets, and lack of enabling regulatory environments; this will have to be dealt with collectively by industry stakeholders to increase access to more innovative solutions, reduce complexity in handling the data, and create well-thought-out laws and policies. The future looks bright for precision agriculture.

These include constant new innovations in sensor tech, AI, and robotics, but most importantly blockchain: they will disrupt farming, food production, and environmental outcomes in a dramatic way in the coming years. Many big barriers have impeded the way, and important solutions are badly needed for the general adoption of precision agriculture. All these integrate a huge initial investment, the ability to maintain huge databases, and missing policy support. To overcome major barriers, industry stakeholder groups have to cinch practical regulation, harmonizing of data procedures, and shutting down the inaccessibility of leading-edge technologies. Still, precision agriculture holds enormous promise in the near future. Continuous development within areas such as blockchain, artificial intelligence, robotics, sensing technology, and others places crop production, food manufacturing, and environmental outcomes in good stead for many years yet to come. In other words, the climate-smart tools provide options to reduce vulnerabilities of unpredictable weather changes through a combination of adaptation and mitigation strategies. The farmer of the future under precision agriculture will be able to produce more with less, making a prosperous life for people and the planet possible if the challenges are effectively tackled.

KEYWORDS

- **global positioning system**
- **geographic information system**
- **variable rate technology**
- **remote sensing**
- **Internet of Things**
- **nanotechnology**

REFERENCES

1. Abdelkhalek, A., El-Gendi, H., Alotibi, F. O., Al-Askar, A. A., Elbeaino, T., Behiry, S. I., Abd-Elsalam, K. A., & Moawad, H. (2022). Ocimum basilicum-mediated synthesis of silver nanoparticles induces innate immune responses against cucumber mosaic virus in squash. *Plants*, *11*(20), 2707. https: //doi.org/10.3390/plants11202707.
2. Abd-Elsalam, K. A. (2024). Agricultural nanotechnology. *Plants*, *13*(4), 489.
3. Ahmad, U., & Sharma, L. (2023). A review of best management practices for potato crop using precision agricultural technologies. *Smart Agricultural Technology*, *4*, 100220.
4. Akhter, R., & Sofi, S. A. (2022). Precision agriculture using IoT data analytics and machine learning. *Journal of King Saud University—Computer and Information Sciences*, *34*(8), 5602–5618.
5. Akter, S., Voumik, L. C., Rahman, M. H., Raihan, A., & Zimon, G. (2023). GDP, health expenditure, industrialization, education and environmental sustainability impact on child mortality: Evidence from G-7 countries. *Sustainable Environment, 9*(1), 2269746.
6. Ali, A. Z., Rahman, M. S., & Raihan, A. (2022). Soil carbon sequestration in agroforestry systems as a mitigation strategy of climate change: A case study from Dinajpur, Bangladesh. *Advances in Environmental and Engineering Research, 3*(4), 1–15.

7. Anderson, D. M., Estell, R. E., Holechek, J. L., Ivey, S., & Smith, G. B. (2014). Virtual herding for flexible livestock management—a review. *The Rangeland Journal*, *36*(3), 205–221. https: //doi.org/10.1071/RJ13092.
8. Asher, A., & Brosh, A. (2022). Decision Support System (DSS) for managing a beef herd and its grazing Habitat's sustainability: Biological/Agricultural basis of the technology and its validation. *Agronomy*, *12*(2), 288. https: //doi.org/10.3390/agronomy12020288.
9. Atalla, S., Tarapiah, S., Gawanmeh, A., Daradkeh, M., Mukhtar, H., Himeur, Y., Mansoor, W., Hashim, K. F. B., & Daadoo, M. (2023). Iot-enabled precision agriculture: Developing an ecosystem for optimized crop management. *Information*, *14*(4), 205.
10. Balaji, G. N., Nandhini, V., Mithra, S., Priya, N., & Naveena, R. (2018). IoT based smart crop monitoring in farm land. *Imperial Journal of Interdisciplinary Research (IJIR)*, *4*(1), 88–92.
11. Bhargava, K., Kashyap, A., & Gonsalves, T. A. (2014, February). Wireless sensor network based advisory system for Apple Scab prevention. In *2014 Twentieth National Conference on Communications (NCC)* (pp. 1–6). IEEE.
12. Bilotta, G., Genovese, E., Citroni, R., Cotroneo, F., Meduri, G. M., & Barrile, V. (2023). Integration of an innovative atmospheric forecasting simulator and remote sensing data into a geographical information system in the frame of agriculture 4.0 concept. *AgriEngineering*, *5*(3), 1280–1301.
13. Birrell, S., Hughes, J., Cai, J. Y., & Iida, F. (2020). A field-tested robotic harvesting system for iceberg lettuce. *Journal of Field Robotics*, *37*(2), 225–245.
14. Bjornlund, V., Bjornlund, H., & van Rooyen, A. (2022). Why food insecurity persists in sub-Saharan Africa: A review of existing evidence. *Food Security*, *14*(4), 845–864.
15. Bohman, B. J., Rosen, C. J., & Mulla, D. J. (2019). Evaluation of variable rate nitrogen and reduced irrigation management for potato production. *Agronomy Journal*, *111*(4), 2005–2017.
16. Buyya, R., & Dastjerdi, A. V. (Eds.). (2016). *Internet of Things: Principles and Paradigms*. Elsevier.
17. Cambra, C., Sendra, S., Lloret, J., & Garcia, L. (2017, May). An IoT service-oriented system for agriculture monitoring. In *2017 IEEE International Conference on Communications (ICC)* (pp. 1–6). IEEE.
18. Champ, J., Mora-Fallas, A., Goëau, H., Mata-Montero, E., Bonnet, P., & Joly, A. (2020). Instance segmentation for the fine detection of crop and weed plants by precision agricultural robots. *Applications in Plant Sciences*, *8*(7), e11373.
19. Daware, V., Mondal, K., & Khandagale, S. (2024). Precision agriculture and biotechnology.
20. Duhan, J. S., Kumar, R., Kumar, N., Kaur, P., Nehra, K., & Duhan, S. (2017). Nanotechnology: The new perspective in precision agriculture. *Biotechnology Reports*, *15*, 11–23.
21. Elferink, M., & Schierhorn, F. (2016). Global demand for food is rising. Can we meet it. *Harvard Business Review*, *7*(04), 2016.
22. Feng, Q., Zou, W., Fan, P., Zhang, C., & Wang, X. (2018). Design and test of robotic harvesting system for cherry tomato. *International Journal of Agricultural and Biological Engineering*, *11*(1), 96–100.
23. Fortino, G., Savaglio, C., Spezzano, G., & Zhou, M. (2020). Internet of things as system of systems: A review of methodologies, frameworks, platforms, and tools. *IEEE Transactions on Systems, Man, and Cybernetics: Systems*, *51*(1), 223–236.
24. Fue, K. G., Porter, W. M., Barnes, E. M., & Rains, G. C. (2020). An extensive review of mobile agricultural robotics for field operations: Focus on cotton harvesting. *AgriEngineering*, *2*(1), 150–174.
25. García, L., Parra, L., Jimenez, J. M., Lloret, J., Mauri, P. V., & Lorenz, P. (2020). DronAway: A proposal on the use of remote sensing drones as mobile gateway for WSN in precision agriculture. *Applied Sciences*, *10*(19), 6668.
26. Gawande, V., Saikanth, D. R. K., Sumithra, B. S., Aravind, S. A., Swamy, G. N., Chowdhury, M., & Singh, B. V. (2023). Potential of precision farming technologies for eco-friendly agriculture. *International Journal of Plant & Soil Science*, *35*(19), 101–112.
27. Gokool, S., Mahomed, M., Clulow, A., Sibanda, M., Kunz, R., Naiken, V., & Mabhaudhi, T. (2024). Exploring the potential of remote sensing to facilitate integrated weed management in smallholder farms: A scoping review. *Drones*, *8*(3), 81.
28. Guimarães, N., Fraga, H., Sousa, J. J., Pádua, L., Bento, A., & Couto, P. (2024). Comparative evaluation of remote sensing platforms for almond yield prediction. *AgriEngineering*, *6*(1), 240–258.

29. Gupta, N., & Gupta, P. K. (2024). Robotics and Artificial Intelligence (AI) in agriculture with major emphasis on food crops. In: *Digital Agriculture: A Solution for Sustainable Food and Nutritional Security* (pp. 577–605).
30. Gürsoy, S. (2024). A review of the factors affecting adoption of precision agriculture applications in cotton production.
31. He, L. (2022). Variable rate technologies for precision agriculture. In: *Encyclopedia of Smart Agriculture Technologies* (pp. 1–9). Springer International Publishing: Cham.
32. Heikal, N. H., Rady, M. H., Merdan, B. A., El-Abbassi, T. S., El-Genaidy, M. A., Azazy, A. M., Yones, M. S., & Essa, E. E. (2024). Early detection of Bactrocera zonata infestation in peach fruit using remote sensing technique and application of nematodes for its control. *Kuwait Journal of Science*, *51*(2), 100191.
33. Herraiz, Á. H., Marugán, A. P., & Márquez, F. P. G. (2020). Photovoltaic plant condition monitoring using thermal images analysis by convolutional neural network-based structure. *Renewable Energy*, *153*, 334–348.
34. Hespeler, S. C., Nemati, H., & Dehghan-Niri, E. (2021). Non-destructive thermal imaging for object detection via advanced deep learning for robotic inspection and harvesting of chili peppers. *Artificial Intelligence in Agriculture*, *5*, 102–117.
35. Hussain, N., Farooque, A. A., Schumann, A. W., McKenzie-Gopsill, A., Esau, T., Abbas, F., Acharya, B., & Zaman, Q. (2020). Design and development of a smart variable rate sprayer using deep learning. *Remote Sensing*, *12*(24), 4091.
36. Ireri, D., Belal, E., Okinda, C., Makange, N., & Ji, C. (2019). A computer vision system for defect discrimination and grading in tomatoes using machine learning and image processing. *Artificial Intelligence in Agriculture*, *2*, 28–37.
37. Jawhar, I., Mohamed, N., Al-Jaroodi, J., & Zhang, S. (2014). A framework for using unmanned aerial vehicles for data collection in linear wireless sensor networks. *Journal of Intelligent & Robotic Systems*, *74*, 437–453.
38. Joshi, A., Pradhan, B., Gite, S., & Chakraborty, S. (2023). Remote-sensing data and deep-learning techniques in crop mapping and yield prediction: A systematic review. *Remote Sensing*, *15*(8), 2014.
39. Karunathilake, E. M. B. M., Le, A. T., Heo, S., Chung, Y. S., & Mansoor, S. (2023). The path to smart farming: Innovations and opportunities in precision agriculture. *Agriculture*, *13*(8), 1593.
40. Kaswan, S., Chandratre, G. A., Upadhyay, D., Sharma, A., Sreekala, S. M., Badgujar, P. C., Panda, P., & Ruchay, A. (2024). Applications of sensors in livestock management. In: *Engineering Applications in Livestock Production* (pp. 63–92). Academic Press.
41. Kganyago, M., Adjorlolo, C., Mhangara, P., & Tsoeleng, L. (2024). Optical remote sensing of crop biophysical and biochemical parameters: An overview of advances in sensor technologies and machine learning algorithms for precision agriculture. *Computers and Electronics in Agriculture*, *218*, 108730.
42. Kultongkham, A., Kumnon, S., Thintawornkul, T., & Chanthasopeephan, T. (2021). The design of a force feedback soft gripper for tomato harvesting. *Journal of Agricultural Engineering*, *52*(1).
43. Kumar, L., Ahlawat, P., Rajput, P., Navsare, R. I., & Singh, P. K. (2021). Internet of things (IOT) for smart precision farming and agricultural systems productivity: A review. *IJEAST*, *5*, 141–146.
44. Kumar, P., Motia, S., & Reddy, S. R. N. (2023). Integrating wireless sensing and decision support technologies for real-time farmland monitoring and support for effective decision making: Designing and deployment of WSN and DSS for sustainable growth of Indian agriculture. *International Journal of Information Technology*, *15*(2), 1081–1099.
45. Li, Z., Yu, T., Paul, R., Fan, J., Yang, Y., & Wei, Q. (2020). Agricultural nanodiagnostics for plant diseases: Recent advances and challenges. *Nanoscale Advances*, *2*(8), 3083–3094.
46. Liang, C., & Shah, T. (2023). IoT in agriculture: The future of precision monitoring and data-driven farming. *Eigenpub Review of Science and Technology*, *7*(1), 85–104.
47. Longchamps, L., & Khosla, R. (2015). Improving N use efficiency by integrating soil and crop properties for variable rate N management. In: *Precision Agriculture 15* (pp. 249–256). Wageningen Academic.
48. Lytridis, C., Kaburlasos, V. G., Pachidis, T., Manios, M., Vrochidou, E., Kalampokas, T., & Chatzistamatis, S. (2021). An overview of cooperative robotics in agriculture. *Agronomy*, *11*(9), 1818.

49. Ma, Y., Chen, S., Ermon, S., & Lobell, D. B. (2024). Transfer learning in environmental remote sensing. *Remote Sensing of Environment, 301*, 113924.
50. Magalhães, S. A., Castro, L., Moreira, G., Dos Santos, F. N., Cunha, M., Dias, J., & Moreira, A. P. (2021). Evaluating the single-shot multibox detector and YOLO deep learning models for the detection of tomatoes in a greenhouse. *Sensors, 21*(10), 3569.
51. Marjani, M., Nasaruddin, F., Gani, A., Karim, A., Hashem, I. A. T., Siddiqa, A., & Yaqoob, I. (2017). Big IoT data analytics: Architecture, opportunities, and open research challenges. *IEEE Access, 5*, 5247–5261.
52. Masi, M., Di Pasquale, J., Vecchio, Y., & Capitanio, F. (2023). Precision farming: Barriers of variable rate technology adoption in Italy. *Land, 12*(5), 1084.
53. Mehta, S. S., & Burks, T. F. (2016). Adaptive visual servo control of robotic harvesting systems. *IFAC-PapersOnLine, 49*(16), 287–292.
54. Moeller, R., Deemyad, T., & Sebastian, A. (2020). Autonomous navigation of an agricultural robot using RTK GPS and Pixhawk. In *2020 Intermountain Engineering, Technology and Computing (IETC)* (pp. 1–6). IEEE.
55. Navas, E., Fernandez, R., Sepúlveda, D., Armada, M., & Gonzalez-de-Santos, P. (2020). A design criterion based on shear energy consumption for robotic harvesting tools. *Agronomy, 10*(5), 734.
56. Nijak, M., Skrzypczyński, P., Ćwian, K., Zawada, M., Szymczyk, S., & Wojciechowski, J. (2024). On the importance of precise positioning in robotised agriculture. *Remote Sensing, 16*(6), 985.
57. Nyamuryekung'e, S., Cox, A., & Perea, A. (2023b). Training beef cattle to use virtual fence systems. In: Zhao, Y., Berckmans, D., Gan, H., Ramirez, B., Siegford, J., & Wang-Li, L., (Eds.), *U.S. Precision Livestock Farming* (pp. 672–678).
58. Nyamuryekung'e, S., Duff, G., Utsumi, S., Estell, R., McIntosh, M. M., Funk, M., Cox, A., Cao, H., Spiegal, S., Perea, A. & Cibils, A. F. (2023a). Real-time monitoring of grazing cattle using LORA-WAN sensors to improve precision in detecting animal welfare implications via daily distance walked metrics. *Animals, 13*(16), 2641. https: //doi.org/10.3390/ ani13162641.
59. Nyamuryekung'e, S., Cibils, A. F., Estell, R. E., VanLeeuwen, D., Spiegal, S., Steele, C., González, A. L., McIntosh, M. M., Gong, Q., & Cao, H. (2022). Movement, activity, and landscape use patterns of heritage and commercial beef cows grazing Chihuahuan Desert rangeland. *Journal of Arid Environments, 199*, 104704. https: //doi.org/10.1016/j.jaridenv.2021.104704.
60. Oliveira, L. F., Moreira, A. P., & Silva, M. F. (2021). Advances in agriculture robotics: A state-of-the-art review and challenges ahead. *Robotics, 10*(2), 52.
61. Ongadi, P. A. (2024). A comprehensive examination of security and privacy in precision agriculture technologies. *GSC Advanced Research and Reviews, 18*(01), 336–363.
62. Paarlberg, R. (2022). *Resetting the Table: Straight Talk About the Food we Grow and Eat*. Vintage.
63. Pablo, N. H., Zarco-Tejada, J., & Loudjani, P. (2014). Precision agriculture: An opportunity for EU farmers-potential support with the cap 2014–2020.
64. Quan, X., Guo, Q., Ma, J., & Doluschitz, R. (2023). The economic effects of unmanned aerial vehicles in pesticide application: Evidence from Chinese grain farmers. *Precision Agriculture, 24*(5), 1965–1981.
65. Raihan, A. (2023a). An econometric evaluation of the effects of economic growth, energy use, and agricultural value added on carbon dioxide emissions in Vietnam. *Asia-Pacific Journal of Regional Science, 7*, 665–696.
66. Raihan, A. (2023b). The dynamic nexus between economic growth, renewable energy use, urbanization, industrialization, tourism, agricultural productivity, forest area, and carbon dioxide emissions in the Philippines. *Energy Nexus, 9*, 100180.
67. Rani, T. J. (2024). Nanotechnology in plant physiology: Revolutionizing agriculture for sustainable and resilient crop production. *Peer-Reviewed Refereed International Journal (UIJES), 5*(2), 62–73.
68. Raza, I., Zubair, M., Zaib, M., Khalil, M. H., Haidar, A., Sikandar, A., Abbas, M. Q., Javed, A., Liaqat, M., Ain, A., & Ashfaq, M. (2023). Precision nutrient application techniques to improve soil fertility and crop yield: A review with future prospect. *International Research Journal of Educational and Technology, 5*(8).
69. Roopaei, M., Rad, P., & Choo, K. K. R. (2017). Cloud of things in smart agriculture: Intelligent irrigation monitoring by thermal imaging. *IEEE Cloud Computing, 4*(1), 10–15.

70. Roshanianfard, A., Mengmeng, D., & Nematzadeh, S. (2021). A 4-DOF SCARA robotic arm for various farm applications: Designing, kinematic modelling, and parameterization. *Acta Technologica Agriculturae*, *24*(2), 61–66.
71. Rysz, M. W., & Mehta, S. S. (2021). A risk-averse optimization approach to human-robot collaboration in robotic fruit harvesting. *Computers and Electronics in Agriculture*, *182*, 106018.
72. Sachithra, V., & Subhashini, L. D. C. S. (2023). How artificial intelligence uses to achieve the agriculture sustainability: Systematic review. *Artificial Intelligence in Agriculture*. *8*, 46–59.
73. Saleem, S. R., Zaman, Q. U., Schumann, A. W., & Naqvi, S. M. Z. A. (2023). Variable rate technologies: Development, adaptation, and opportunities in agriculture. In: *Precision Agriculture* (pp. 103–122). Academic Press.
74. Sangeetha, C., Moond, V., Damor, J. S., Pandey, S. K., Kumar, P., & Singh, B. (2024). Remote sensing and geographic information systems for precision agriculture: A review. *International Journal of Environment and Climate Change*, *14*(2), 287–309.
75. Scarfone, A., Picchio, R., del Giudice, A., Latterini, F., Mattei, P., Santangelo, E., & Assirelli, A. (2021). Semi-automatic guidance vs. Manual guidance in agriculture: A comparison of work performance in wheat sowing. *Electronics*, *10*(7), 825.
76. Schimmelpfennig, D., & Ebel, R. (2016). Sequential adoption and cost savings from precision agriculture. *Journal of Agricultural and Resource Economics*, 97–115.
77. Si, J., Niu, Y., Lu, J., & Zhang, H. (2019). High-precision estimation of steering angle of agricultural tractors using GPS and low-accuracy MEMS. *IEEE Transactions on Vehicular Technology*, *68*(12), 11738–11745.
78. Suksawat, B., & Komkum, P. (2015, May). Pineapple quality grading using image processing and fuzzy logic based on Thai agriculture standards. In *2015 International Conference on Control, Automation and Robotics* (pp. 218–222). IEEE.
79. Tamta, S., Vimal, V., Verma, S., Gupta, D., Verma, D., & Nangan, S. (2024). Recent development of nanobiomaterials in sustainable agriculture and agrowaste management. *Biocatalysis and Agricultural Biotechnology*, *56*, 103050.
80. Tedesco, D., de Oliveira, M. F., dos Santos, A. F., Silva, E. H. C., de Souza Rolim, G., & da Silva, R. P. (2021). Use of remote sensing to characterize the phenological development and to predict sweet potato yield in two growing seasons. *European Journal of Agronomy*, *129*, 126337.
81. Thomopoulos, V., Bitas, D., Papastavros, K. N., Tsipianitis, D., & Kavga, A. (2021). Development of an integrated IoT-based greenhouse control three-device robotic system. *Agronomy*, *11*(2), 405.
82. Udomkun, P., Nagle, M., Argyropoulos, D., Mahayothee, B., & Müller, J. (2016). Multi-sensor approach to improve optical monitoring of papaya shrinkage during drying. *Journal of Food Engineering*, *189*, 82–89.
83. Viana, C. M., Freire, D., Abrantes, P., Rocha, J., & Pereira, P. (2022). Agricultural land systems importance for supporting food security and sustainable development goals: A systematic review. *Science of the Total Environment*, *806*, 150718.
84. Vimal, V. (2023, November). Integrating IoT-based environmental monitoring and data analytics for crop-specific smart agriculture management: A multivariate analysis. In *2023 3rd International Conference on Technological Advancements in Computational Sciences (ICTACS)* (pp. 368–373). IEEE.
85. Waterhouse, T. (2023). Virtual fencing systems: Balancing production and welfare outcomes. *Livestock*, *28*(5), 227–234. https: //doi.org/10. 12968/live.2023.28.5.
86. Williams, H. A., Jones, M. H., Nejati, M., Seabright, M. J., Bell, J., Penhall, N. D., Barnett, J. J., Duke, M. D., Scarfe, A. J., Ahn, H. S., & MacDonald, B. A. (2019). Robotic kiwifruit harvesting using machine vision, convolutional neural networks, and robotic arms. *Biosystems Engineering*, *181*, 140–156.
87. Xiao, X., Wang, Y., & Jiang, Y. (2024). Review of research advances in fruit and vegetable harvesting robots. *Journal of Electrical Engineering & Technology*, *19*(1), 773–789.
88. Yadav, A., Yadav, K., & Abd-Elsalam, K. A. (2023). Exploring the potential of nanofertilizers for a sustainable agriculture. *Plant Nano Biology*, *5*, 100044.
89. Yadav, N., & Sidana, N. (2023). Precision agriculture technologies: Analysing the use of advanced technologies, such as drones, sensors, and GPS, In precision agriculture for optimizing resource management, crop monitoring, and yield prediction. *Journal of Advanced Zoology*, *44*.

90. Yorozu, A., Ishigami, G., & Takahashi, M. (2021, June). Human-following control in furrow for agricultural support robot. In *International Conference on Intelligent Autonomous Systems* (pp. 155–164). Springer International Publishing: Cham.
91. Zaman, Q. U. (2023). Precision agriculture technology: A pathway toward sustainable agriculture. In: *Precision Agriculture* (pp. 1–17). Academic Press.
92. Zhang, C., & Kovacs, J. M. (2012). The application of small unmanned aerial systems for precision agriculture: A review. *Precision Agriculture*, *13*, 693–712.
93. Zhang, J., Huang, Y., Pu, R., Gonzalez-Moreno, P., Yuan, L., Wu, K., & Huang, W. (2019). Monitoring plant diseases and pests through remote sensing technology: A review. *Computers and Electronics in Agriculture*, *165*, 104943.
94. Zhang, L., Huang, Z., Liu, W., Guo, Z., & Zhang, Z. (2021). Weather radar echo prediction method based on convolution neural network and long short-term memory networks for sustainable e-agriculture. *Journal of Cleaner Production*, *298*, 126776.
95. Zhang, Q. (2016). *Precision Agriculture Technology for Crop Farming* (p. 374). Taylor & Francis.

CHAPTER 15

Augmented and Virtual Realities and Data-Driven Mobile Apps as Panacea for Sustainable Agriculture

M. S. SADIQ[1*], I. P. SINGH[2], M. M. AHMAD[3], and M. BABAWACHIKO[4]

[1]*Department of Agricultural Economics and Extension, FUD, Dutse, Nigeria*

[2]*Department of Agricultural Economics, SKRAU, Bikaner, Rajasthan, India*

[3]*Department of Agricultural Economics and Extension, BUK, Kano, Nigeria*

[4]*MCA-Google Apps Suresh Gyan Vihar University, Jaipur, Rajasthan, India*

*Corresponding author

ABSTRACT

Over the past 60 years, there have been many changes to the agriculture marketing system due to an increase in marketed surplus, a rise in rapid urbanization and levels of income, changes in the matrix of demand for marketing services, and a boost in connections with distant and international markets. Either e-Agriculture or e-Agribusiness refers to the use of information and communication technologies in the agricultural industry. Small-scale farmers can use ICT to find a variety of consumers for producers willing to pay a higher price. Market distortions can be reduced by using a mobile application that gives farmers price information. Because of fields of research including Precision Agriculture, Smart Agriculture, and Agriculture 4.0, which have boosted food supply in a sustainable way, the agricultural sector has profited from computer technology. Therefore, researchers and experts in virtual reality and augmented reality can decide to focus on the agriculture sector. For the benefit of farmers, the government and numerous businesses are investing in agricultural marketing solutions. In the upcoming years, digital agricultural marketing will be essential to doubling farmer output and tripling farmers' income.

15.1 INTRODUCTION

Since it meets the requirements of individuals and their livestock plus determines the survival of any society, its agriculture and food industry is an essential part of any economy [1].

Demystifying Sustainable Farming Using Smart Technologies.
Gunjan Chhabra, Keshav Kaushik, Priya Chugh, Susheela Dahiya, and Aryan Chaudhary (Eds.)

DOI: 10.1201/9781779640932-15

Any economy's contribution to gross domestic product (GDP) from agriculture is substantial, even exceeding 50% in the case of poor countries. Since the population employment rate in this sector is higher in nations like India, economic literature highly values its employment-creating potential [2]. Agriculture becomes less desirable as a sector of the economy as a result of a fall in its contribution to GDP and ability to create jobs [3].

In order to maintain its importance, the agriculture sector has undergone developments that have made it easier to implement new business methods that may result in the growth of the global market for agricultural products [4]. Agriculture's contribution can be raised through creating closer relationships within the industry and improving postharvest processes, storage facilities, logistics, and distribution, all of which are essential elements of agribusiness value chains. The agribusiness sector is significant in an economy like the US, which has a highly productive agriculture sector, and it consists of several levels of marketing, storage, and processing activities. The main factor influencing agribusiness growth is productivity. Agribusiness expansion and productivity are essential for the economic development of any nation, and technological innovation in all facets of company operations is one of the new drivers driving economic growth globally [5].

Given the significant increase in global food demand, the availability of natural resources such as fresh water and arable land is also a worry for agrifood industries. More efficient, functional, productive, sustainable, inclusive, open, and resilient food systems are needed in order to meet the UN Sustainable Development Goal of a "world free hunger" by 2030 [6]. The current agrifood system needs to be transformed in order to do this. Modern marketing strategy is one of the main tactics used by policy makers in the agriculture sector. Since none of us can avoid it as consumers or producers, not even those of us who prefer to lead simple lives, marketing has an impact on everyone [7]. Appropriate marketing strategies are necessary for accurate client demand research, creative product development, awareness raising of certain value-added products, creating an effective distribution system, and offering postsale services.

Since experts contend that marketing is everything, it is widely understood that the agriculture industry may perform better when using marketing tactics and methods [8]. Marketing strategies are rather prevalent in agriculture since academics hold believe marketing is marketing, irrespective of the product or market. However, this market differs from others in consumer goods due to a few unique features. Instead of concentrating on individual consumers, the agriculture industry frequently employs a business-to-business strategy that includes market segmentation, assessment, and the execution of a marketing plan [9]. Second, due to the vast quantity of agricultural products and their perishable nature, unique and proper delivery is required because delays could ruin the product, which would be bad for the customer. To get over the barriers, modern, efficient technology like digital innovation is required.

As part of the "Fourth Industrial Revolution," "disruptive" digital technologies like blockchain, the Internet of Things (IoT), artificial intelligence (AI), and virtual reality (VR) are rapidly revolutionizing a variety of industries [10, 11]. The proliferation of mobile devices, remote sensing services, plus distributed computing are similar advances in the agriculture sector. The "digital agricultural revolution," as it is being dubbed by experts, is now transforming the agriculture sector. This is the most recent change to guarantee that agriculture continues to satisfy the demands of the world's population.

Every step of the agrifood supply chain will shift as a result of the "digitalization" of the agriculture sector, based on the Food & Agriculture Organization of the United Nations [12]. This is because resource management will become more efficient, personalized, intelligent, and proactive across the system. It will be data-driven and function in real-time with high connectivity. Value chains will indeed be easier to trace and coordinate, making it possible to accurately manage different farms, crops, plus animals to achieve their own optimal circumstances. It is expected that digital agriculture will create systems that are incredibly productive, proactive, and changeable. As a result, sustainability, profitability, and food security may all increase. The functional online marketing of agricultural commodities, the expansion of the marketing channels of agricultural goods, as well as the online sale of agricultural products have all been significantly aided by the broad use of the IoT, e-commerce, and these technologies [13].

15.2 AGRICULTURAL MANAGEMENT

The task of maximizing plant cultivation is known as agricultural management. In order to positively impact agribusiness, agricultural management strives to improve plant quality and productivity. The procedure is carried out while taking into consideration factors such as the availability of water and nutrients, the physical and chemical composition of the soil, proper drainage and irrigation methods, the behavior of fruit maturity, ambient temperature, and plant stress levels. In order to functionally plan irrigation flows utilizing sensors based on the water's acidity (pH) and level, the IoT can also be used in plant water supply management.

15.3 AGRICULTURE MARKETING

The limited direct business-to-consumer (B2C) contact is what stimulated the widespread adoption of mobile agricultural apps. Mobile marketing solutions are thus anticipated to improve the potential for B2C interactions between farmers and consumers. The smartphone app can be utilized for management and marketing tasks like providing details on various seed varieties, fertilizer or pesticide types, and agricultural activities. The selling of agricultural products through mobile applications is one of the developments that maintain the trend as the agricultural business becomes more digitalized in the era of Industrial Revolution 4.0.

Marketing operations seek to establish lucrative business partnerships between purchasers and sellers by identifying consumer requirements, producing goods to meet those needs, and trading those items for money with prospective customers [14]. This helps to promote and ease the interchange of products and services. Since marketing plays a role in every economic sector, the agricultural industry is no exception and has undergone significant changes as a result. The manufacturers and customers will both reap reasonable benefits from a solid marketing plan. With the main goal of raising the worth of the agricultural goods and subsequently maximizing profit, agriculture marketing is focused on the employment of marketing strategies and tactics in the agricultural industry. By accurately identifying their needs and providing the benefits they want, marketing has contributed to a higher level of customer satisfaction among agricultural consumers. Agriculture marketing's most impressive accomplishment is the timely delivery of factual data to potential clients. As a result, both the buyer and the seller are able to organize their operations for a better result. As a result, agricultural marketing also includes the services required to move agricultural commodities from the farm to the end consumer. This involves a number of interrelated tasks, including planning, production, sorting, packing, shipping, storage, refining of agricultural and food products, distribution, advertising, and sales.

In the past two decades, there have been many changes to marketing itself, mostly as a result of technical advancements. Disruptive technologies have a clear impact on every element of human life [15]. Agriculture will soon experience the same technological revolution that has fundamentally altered the organization of industrial production in recent years. Markets, agricultural production systems, consumer patterns, and agro-innovation systems all undergo change. Therefore, those working in the agriculture industry are also adopting new technology to enhance their performance in order to stay economically viable and competitive in the market. Therefore, a healthy market is necessary for the agricultural sector to promote development, employment, and economic success. To improve the efficiency and efficacy of the agricultural marketing system, new marketing strategies and modes of exchange must be developed [16].

The marketing of agricultural harvests is fraught with difficulties, according to numerous studies in this field. These difficulties include farmers' lack of market knowledge about agricultural products, their low literacy rates, particularly in developing nations, and the fact that there are numerous distribution channels that reduce their profit margins. The number of middlemen who split the farmer's earnings is too great. As a result, the influence of technology in rural areas is minimal and out of the price range of low-income farmers. These factors all support the application of cutting-edge technology in the agricultural industry. The major goal is to find solutions to the issues that are keeping this sector from performing up to expectations, as well as the difficulties that the upcoming global agriculture industry will provide.

15.4 AGRICULTURAL MARKETING'S DIGITALIZATION

Digital marketing is a quasi-online channel, particularly on the internet, for recognizing and accepting clients' needs, advertising goods and services, and developing relationships with customers. It does this via the use of digital technology and gadgets [14, 17]. Given its importance to the economy, digitalization cannot be ignored in the agriculture industry [20]. Running a business nowadays without a presence on the internet is impossible, as statistics show that the shift in consumption is in the direction of online consumption [18, 19].

The development of agricultural brands or products via one or more kinds of electronic media is known as digital agriculture marketing, or just the application of the internet age in agriculture marketing. Utilizing a range of strategies, digital marketing's primary goal is to promote firms, develop preference, and boost sales [21]. A growing number of agribusinesses are employing the internet as a vehicle for marketing, governance, service, and coordination, according to Al-Hazmi [22]. Anyone enthusiastic about digital marketing ought to have a unique website or app through which they may link and use numerous promotions and resources.

A company may rapidly and cheaply reach its target audience by using digital marketing. Above-the-line marketing frequently refers to conventional forms of pricey advertising that raise the cost of marketing. These include print ads in periodicals and newspapers, billboards, and TV commercials. The goal of below-the-line advertising, on the other hand, is to reach more niche markets at a reasonable price. Digital technology makes below-the-line advertising easier.

E-marketing, commonly referred to as "internet marketing," "website marketing," "digital marketing," or "online marketing," is the name given to marketing strategies and tactics that employ online channels to interact with target consumers, collect pertinent data, and provide the necessary products [23]. Digital technology is used to run the entire process. Online distribution of agricultural products from producers to enterprises, consumers, or other groups is referred to as "e-marketing" [24]. E-marketing or digital marketing of agricultural products, is becoming more and more well-liked throughout the world as a result of its capacity to address some of the most significant difficulties relating to agriculture, from original production through consumption.

15.4.1 AGRICULTURE-RELATED PRODUCT DIGITAL MARKETING ENABLERS

Three crucial elements (enablers) that have driven the development of digital agriculture marketing and will continue to do so in the future have been recognized by the Food and Agriculture Organization (FAO) in its 2018 study on the subject. As follows:

1. Farmers and agricultural extension officers are using the internet, mobile devices, and social networks more frequently;
2. The rural population's digital literacy; and
3. A society that supports digital entrepreneurship and innovation in the agricultural sector.

For people involved in the production and distribution of agricultural products, high-speed internet connections, like 4G connections, smartphones, mobile apps, social media, plus digital engagement channels, have dramatically enhanced access to information and services [25]. People find it convenient to use digital technologies with newly learned skills [26]. Besides, the scope and constant accessibility have made a difference.

As educated people tend to accept new technology more quickly and effectively, education and income are two significant variables. The need for developing digital skills is growing as a result of digitization [27]. As a result, as literacy rates rise, so does the demand for digital skills, which accelerates adoption. The process of digitalization will quicken as potential consumers' levels of competence rise.

15.4.2 BENEFITS OF DIGITIZATION

Digitalization has the potential to greatly enhance the economic, social, and environmental situations, according to numerous research studies in this field. It can get through a number of obstacles that the conventional approaches cannot. It achieves a result that no other mode can because of its novel strategy, which optimizes reach among its target consumers. According to the Luckyardi et al. [28] study, targeted product positioning, which promotes product distinction, is positively influenced by the use of digital technology in agricultural product marketing, lending credence to this idea. A significant component driving the change of the entire agricultural sector is the growth of e-commerce, a new transaction method founded on technological innovation that has increased agricultural competitiveness and widened the global market for agricultural commodities.

As per Chatterjee [29], there are two key components to agricultural product marketing. The physical steps involved in getting goods from producers to consumers are covered in the first chapter. These steps include gathering, packaging, transporting, processing, storing, and selling agricultural goods at retail. The market price mechanism is the second factor. In addition to enhancing the efficiency of these processes, digitalization has the potential to greatly expand the options available to farmers and other agricultural stakeholders. The following categories can be used to classify the benefits:

1. **Market Growth**

   The market will grow as a result of agricultural marketing being digitalized [30]. Farmers can access the global market with a greater number of potential clients by using the internet. Eventually, the easy and trouble-free interchange of agricultural output will be the outcome of heightened awareness put on by the free and easy flow of information. The accessibility of data on all parts of agriculture and its marketing is helpful in the expansion of this business as Internet use has spread to rural areas. This industry is now operational 24 hours a day thanks to the internet. Farmers, who are the most significant stakeholders, can now use it whenever it is convenient for them, which has led to a growth in market size. In essence, the process of digitization is fostering a corporate climate where people are free to conduct transactions at their convenience because everything happens in real time. Because accessing another website requires so little effort, clients may find alternatives from other businesses much more easily online [31]. As a result, it may be claimed that digitalization enhances the effectiveness and efficacy of the entire system while also enabling people to engage at the time of their choosing.

2. **Cost-cutting**

   The primary goal of marketing is to carry out these tasks as cheaply as possible while simultaneously facilitating the exchange of products and services. The market's reach has risen due to the use of the internet, which has also helped to lower transaction costs. Many agricultural business owners have adopted internet marketing since Zhang and Du [32] discovered that the average return on online marketing investment is very substantial, making the entire process quite profitable.

Real-time prices for agricultural products are made available, intermediaries' services are less expensive, and developing transportation networks is made simpler thanks to easy information access. Through digital marketing, the marketing chain for agricultural products can be shortened, saving time and money [33]. Customer information is incredibly simple to obtain, inexpensive to retain, and profitable to use when using digital technology. Compared to traditional marketing strategies, an e-marketing campaign that is well thought out and strategically targeted can reach the right clients for a lot less money [34]. Businesses use the internet as a resource to reduce the cost of customer support, maintain client relationships, tailor marketing messages, and enable mass customization. The infrastructure needed for the reduction of marketing costs, which is essential for raising farmer income and lowering consumer costs, is created through digitalization.

3. **Reducing Agricultural Waste**

The transaction procedure has grown more efficient and quicker with the use of digital technologies. Agriculture produces an enormous amount of readily perishable goods. Time lost because of a convoluted transaction process actually increases the likelihood of waste and production degradation. Utilizing digital technology facilitates easy, quick, and accurate transaction processes, which actually reduce agricultural product waste. Because there is less waste, farmers' and other agricultural marketers' net income rises. The establishment of an online agricultural market provides a platform for farmers to market their goods while also simplifying the buying and selling of agricultural products.

4. **Promotes Positive Competition**

With thousands of suppliers and hundreds of millions of buyers, it goes without saying that there is intense rivalry for agricultural products. Building trust and understanding between customers and agricultural producers is essential when the gap between the two sides widens. The emergence of digital technology has made it easier than ever to obtain data about the availability, characteristics, sources, and costs of various products. It serves as a resource that buyers and sellers can access and use with ease. Due to the little effort required to access another website, buyers can find alternatives from rival companies much more easily online. Public platforms encourage users to publish these data points online and compare their product usage experiences with those of competitors' offerings. As a result, it encourages consumer involvement in the marketing process, which improves the message's credibility.

Increased customer and producer participation in the selling of agricultural production is encouraged by digitalization, which fosters healthy competition. The buyer may make their purchase without bias if there is a fair possibility of comparison, and the producer can see where they can make improvements to the caliber of their product. It is easier for many small producers to participate in the transaction process than it would have been with the conventional approach. The advantage of branding their product is available to producers, who can also understand the necessity of upgrading the packaging of their goods. Using digital technology, agribusiness marketing firms in underdeveloped countries were able to market to consumers in developed countries and sell their products for more money. Malaysia, Thailand, and Indonesia, three small Asian nations, have benefited greatly from this practice.

5. **Simple Access to Scarce Commodities**

   Agriculture has become more digitalized as a result. Producers are now more likely to market their products online, increasing the accessibility of goods like “Dates’ from Arabia, Darjeeling tea, Kiwi from New Zealand, etc., that were previously restricted by geography and scarcely sold in open markets. These farmers have been inspired to raise their produce and sell their goods online as a result. These products have been added to many companies’ product lines with appropriate branding that are engaged in online marketing.

Digital technology is actually applicable to every element of agriculture, from planting to consumer consumption. It is used by agriculturalists to predict pest infestation in the cultivated areas and the potential need for pest control agents. By correctly applying this technology, the growers can assess the anticipated yield from the existing crops and design a strategy for enhancing the return. With the least amount of work and price, this technology may help agricultural product marketers provide the right kind of commodity to the right customer at the right location. If implemented properly, digital technology is able to completely transform the agriculture and food industries.

Even if digital technology’s promise in the fields of agriculture and its marketing seems exceptional, much work needs to be done to boost its effectiveness in these fields. The FAO’s 2019 report on the digitalization of agriculture and its marketing emphasizes positive factors based on lowering smartphone prices, growing Internet access, and an aging population. This demonstrates the considerable potential for mobile phone adoption in rural areas and the enormous potential for performance enhancement. Aspects that could hinder achieving the ultimate objective were also mentioned in the same report.

15.5 AUGMENTED REALITY (AR) APPLICATIONS IN AGRICULTURE

A new technology, AR is already being used in the gaming, culture, health, transportation, and industry sectors [35, 36]. Because VR/AR technology is pricey, the agrisector mostly uses it in areas with heavy equipment. The United States, Great Britain, Germany, France, Japan, Arab countries (due to a lack of manpower), and Scandinavian countries are among the top nations producing serious games (virtual training and simulations). Agricultural producers and farmers could potentially benefit greatly from AR’s ability to help them with day-to-day work, training, and business planning [37]. The exponential increase in the world’s population has led to a sharp rise in the demand for resources such as food and agricultural goods. The pace of productivity for farmers must be doubled.

Farmers may benefit from new technologies in attaining their objectives. Farmers might utilize AR in addition to AI, big data, blockchains, and IoT to manage the rising food demand [38]. Utilizing AR in agriculture can assist farmers in reducing crop loss, increasing output, and mentoring new farmers. The following are some important ways that AR is used in food production:

1. ***Visually monitoring of farms:*** To determine the best crop to plant on their farms, farmers must assess the fertility of their soils. AR can improve fertility testing. Farmers can use AR to visualize their entire operation on a single dashboard, keep an eye on production quality overall, and spot pest and bug infestations.
2. ***Training new farmers:*** Through immersive e-learning environments, AR can help new farmers become comfortable with agricultural machinery. The usage of sophisticated agricultural tools will become easier for novice farmers to understand visually, reducing serious accidents at the same time. Additionally, it will make it easier for farmers who are masters at using cutting-edge precision farming technologies and procedures to instruct you remotely.
3. ***Assessing facilitation tools:*** There are numerous tools available in agricultural production to carry out specific duties. Based on the demands and requirements of the farmer and the task itself, AR devices can assist in choosing the best instruments for a given operation.

The use of AR applications in agriculture will undoubtedly increase in the coming years as the cost of AR devices continues to decline. It is expected that AR will play a big part there in the value chain of agricultural output.

15.5.1 THE GROWING USE OF AR IN AGRICULTURE

AR is a tool that could be utilized to raise overall productivity in the agriculture industry. Other technologies, such as automated machinery, the IoT, predictive analytics, deep learning (AI), and precision algorithms, can be coupled with it [39]. Other microeconomic reasons, besides the rising food demand, are anticipated to support the expansion of AR in the agriculture business [40]. VR and AR currently dominate the extended reality market. These technologies are based on the ideas of spatial computing, which are expected to be the next advancement in computing [41]. Technologies like AR and VR, digital twins, and computer vision are being developed as a result of the growing trend of digital transformation. The worldwide digitization efforts depend heavily on all of these technologies. The adoption of AR in many applications is also significantly influenced by the increased focus on spatial computing technologies and applications [42]. Basic agricultural techniques are being combined with digital technologies like AR, which is causing a significant upheaval in the agriculture sector. By 2026, the market for agricultural AR is anticipated to reach $4.43 million, growing at a compound annual growth rate of 37.4% from 2021 to 2026. Forecasts indicate that the industry will grow quickly as a result of the growing need for precision farming and cutting-edge agricultural methods. This, in turn, increases the need for digital technologies like AR.

15.5.2 FACTORS INFLUENCING AGRICULTURE'S USE OF AR

1. ***Growth Motivators:*** Currently, one of the largest societal issues facing the entire globe is food security. It is anticipated that technologies like AR, data analytics, autonomous machines, and AI would make agricultural methods more efficient, which will improve the industry's overall yield. The global food security may be improved as a result of this. The market for AR in agriculture will continue to expand due to the growing demand for spatial computing applications. When a person interacts with a machine in which actual items are superimposed or virtually stored, this is referred to as spatial computing. Spatial computing algorithms are the foundation of the complete interactive reality ecosystem, which includes AR technology, VR, and mixed reality. Users can interact with machines as if they were in the actual world and in real time thanks to spatial computing. Due to the need for a prototype or a live item for teaching, this capacity has addressed learning issues faced by a variety of sectors.
2. ***Growth Obstacles:*** Since 2008, immersive reality has been commercially produced, including AR. These technologies have grown over the previous five years and are anticipated to continue growing over the next five due to the ongoing trends of digitalization and Industry 4.0. However, just a few apps are seeing growth. Immersive technologies, such as AR, are anticipated to grow more quickly between 2021 and 2026 as a result of the large demand. However, despite the huge potential of some industries and applications, developers are not putting much emphasis on them. Despite the fact that immersive reality holds promise for a range of business and consumer areas, the early adoption of both AR and VR technologies has been slow due to some health concerns and hazards associated with utilizing AR and VR hardware.
3. ***Agricultural AR market segmentation:*** Numerous countries, notably North America, Europe, the United Kingdom, China, Asia-Pacific (APAC), and the rest of the globe have conducted substantial research on the markets for AR in agriculture (RoW). In 2020, North America was the market leader for AR in agriculture, and it is predicted that it will continue to do so for the duration of the forecast period. The market is expanding primarily as a result of increased research and development efforts as well as widespread use of AR and other digital technologies. Due to smaller agricultural holdings and overpopulation, farmers experience lower profit margins, which are ultimately reinvested back into the firm, limiting the opportunity for investments in developing countries. The market's growth is significantly impacted by the escalating trend toward digital transformation. Agriculture-related AR market segments consist of applications and solutions [43]. Indoor and outdoor farming are among the applications, and hardware, software, and service solutions are among the solutions. The two primary uses of AR in the agriculture sector are indoor and outdoor farming. Other subcategories of the outdoor farming portion of the AR in the agricultural industry include precision and smart farming, animal monitoring, simulated training, and weather tracking and forecasting. Some of the indoor applications that are dominating the AR in the agricultural market by

application include vertical farming, hydroponics, aquaponics, and cannabis cultivation. Although they generate more income than indoor farming, outdoor farming applications have better potential. Applications for outdoor farming are dominated by those for livestock monitoring. Currently, the market is one of the most lucrative and promising applications. Next is the segment for smart and precise agricultural applications, which is predicted to grow at the second-fastest rate [44].

15.5.3 AR REDUCES COSTS FOR PLANT PROTECTION EQUIPMENT

You can decide how to conserve the crop and reduce expenditures by being aware of the pests, diseases, and equipment to use (Figure 15.1). VR can be used to conduct a field survey. It is simpler for farmers to assess the soil's quality, choose the best crops, and maximize the area's capacity for agricultural production. How is it carried out? The crops are scanned by sensors that are placed in the field. Agronomists may easily acquire information about the weather, temperature, plant health, nutrient requirements, and harvesting date by pointing a tablet or smartphone camera at the plant. How can AR fit into this picture? AR can be used on smartphones and tablets or with the aid of specialist glasses. One of the cameras on some smartphones can transmit infrared images, and the other can capture volumetric photos. On the ground are sensors that communicate data to the camera. After reading the information, the camera shows pop-ups on the screen. The only tools required to use the technique are a smartphone and some sensors. These sensors are very reasonably priced.

FIGURE 15.1 Ekkasit Keatsirikul.

Source: https: //www.isabel.io/

15.5.4 AGRICULTURE USE CASES FOR AR

Plant vision is the initial application (previously known as Huxley) (Figure 15.2). With the help of AR, machine learning, and AI, this crop management pattern completely digitizes plant farming. A facility is installing RGB and infrared cameras as well as sensors to collect data. Crops are photographed every minute, and the AI scans the images to do a health check. Wearables like Google Glasses can be used by farmers to obtain AR data on the weather, plant health, and other topics.

FIGURE 15.2 Plant vision.

Source: https: //www.isabel.io/

Isabel is the second example of AR in agriculture (Figure 15.3). A 177-hectare hydroponic farm is managed using this standalone mobile app. San Francisco was the starting point for the project as a social endeavor. Isabel's triumph in San Francisco led to the program being transferred to Peru. It had developed into a fully functional management tool for greenhouses.

Lastly, the Amazone Group is a significant manufacturer of agricultural machinery. With the help of their SmartInstruction solution, Amazone technicians may use AR to develop maintenance and repair instructions.

15.6 MOBILE AGRICULTURE APPS

Utilizing mobile applications is one of the most efficient strategies since it may reach a large audience. With the inclusion of features like specific commodities delivery services, autonomous geolocation services, and data on various agricultural products, mobile apps for farm marketing activities have been developed. The restricted B2C contact may be

FIGURE 15.3 Isabel.

Source: https: //www.isabel.io/

the cause of the widespread adoption of mobile agriculture apps. Thus, it is projected that mobile marketing solutions would increase the possibility for B2C relationships between farmers and customers. The smartphone app can be utilized for management and marketing tasks like providing details on numerous seed varieties, fertilizer or pesticide types, and agricultural activities. The selling of agricultural goods through mobile applications is one of the developments that maintains the trend as the agricultural business becomes more fully digital in the epoch of Industrial Revolution 4.0.

Access to information on agriculture and related industries is now possible through an expanding variety of mobile apps [46]. On a mobile phone or tablet, an app is a software program that allows individuals to send messages, access particular information, and conduct financial transactions [47]. The application (app) is downloaded from a wireless network from an online shop (for free or for a fee), and it might need a live connection to run properly. Information is readily accessible on farmers' mobile devices, which is one of the main benefits of mobile apps for farmers. Data are maintained on the mobile device itself for convenience, including information about a series of procedures, specifics about pests and diseases, and scheme-related data, among other things [48]. The mobile app requires internet access to retrieve information from the back-end server databases when the information is continuous, such as weather information, market prices, or technical assistance. The rural community only receives one-way information from mobile services, especially SMS. For day-to-day farming, thc farmer requires dynamic information and two-way communication. In order to meet their unique needs, farmers require timely information. These apps aid in the dissemination of market data, the facilitation of market connections, the use of extension services, the gathering of farm information, etc.

For the advantage of farmers and other stakeholders, the Indian government has developed a range of web- and mobile-based apps for the free conveyance of data on agricultural-related activities. Both the Google Play Store and the official website, mkisan.gov.in, offer the ability to download these apps. Additionally, there are apps created by nongovernmental organizations (NGOs), the corporate sector, and agricultural organizations. Farmers and other stakeholders are receiving information from agricultural research and extension through these apps, which are also facilitating stakeholder communication. Mobile apps developed by various entities along the agricultural value chain enable access to agricultural and related information. The Google Play Store or the appropriate website mentioned with the information about the mobile app is where you may download these apps.

1. **Kisan Suvidha**

The comprehensive smartphone app Kisan Suvidha was created by the Ministry of Agriculture and Farmers Welfare, Government of India, with the goal of assisting farmers by offering helpful information. Farmers can get information on the soil health card (SHC), godowns, cold storage, dealers, plant protection, integrated pest management approaches, seeds, and the weather through the app (Figure 15.4). The information is currently available in English, Hindi, Tamil, Gujarati, Odia, and Marathi. The initial step for a farmer is to enroll their mobile number, language preference, state, district, and block information.

Characteristics

- ***Information on Weather:*** Information about the current day's humidity, temperature, wind, and rainfall, as well as the forecast for the following five days' worth of weather, is provided. Extreme weather warnings like hailstorms or unseasonal rains are examples of distinctive traits.
- ***Market Price:*** This information gives the most recent market price for a commodity in the closest mandi as well as the highest price in the district, state, and country as a whole. Prices for crops traded in a district's authorized farm market, or mandi, are provided.
- ***SHC:*** Farmers who have registered get access to information on soil health.
- Dealers' Information: Information about vendors is provided, including dealers of seeds, insecticides, fertilizer, and farm machinery.
- ***Plant Protection:*** From seedling/nursery to harvest, this part covers information on pests, weeds, and diseases as well as best practices for crop management. Additionally, one can upload a photo of the impacted crop and receive feedback.
- ***Call to Kisan Call Centre (KCC):*** Using the app, a farmer can also make a direct connection to the KCC, where technical graduates respond to their questions.
- Cold Stores and Godowns: Information is also added regarding cold stores and godowns.

- ***Agro-Advisory:*** The agro-advice area contains messages for farmers in the local tongue from state colleges and agriculture officials.

FIGURE 15.4 Kisan Suvidha Mobile App.

2. **Pusa Krishi**

The Ministry of Agriculture and Farmers Welfare, Government of India, created and disseminated this smartphone app for farmers in an effort to bring technology to the fields (Figure 15.5).

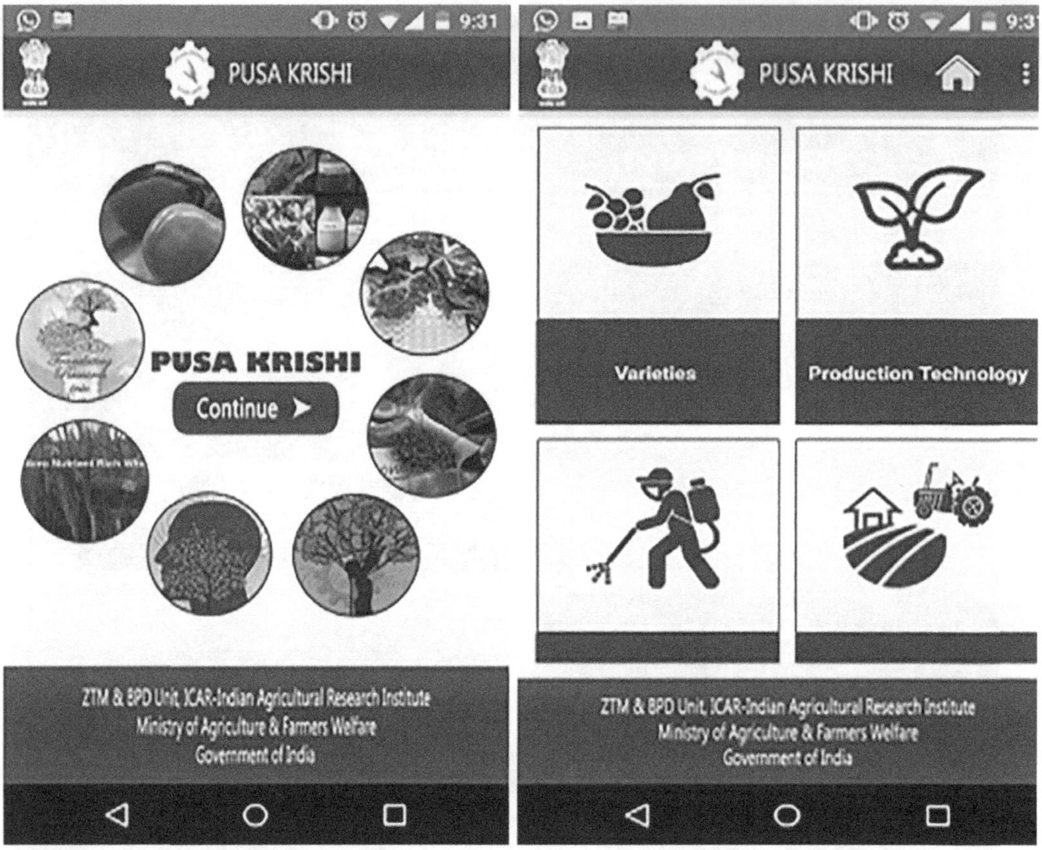

FIGURE 15.5 Pusa Krishi Mobile App.

- Information about novel crop varieties created by the Indian Council of Agricultural Research (ICAR), resource-conserving farming techniques, agricultural machinery, and their use and production technologies are made available to farmers.
- Through a feedback section, farmers can engage with stakeholders in real time.

3. **SHC**

The Department of Agriculture, Co-operation and Farmers Welfare, Ministry of Agriculture and Farmers Welfare and the States and Union Territories are implementing the Government of India's SHC Scheme (Figure 15.6). A farmer's soil nutrient level is indicated on a SHC along with recommendations for how much fertilizer and soil amendments are needed to keep the soil healthy over the long term. Every three years, all landowners will receive SHCs, allowing for the documentation of patterns in soil fertility changes brought on by plant nutrient uptake or other natural sources. This can also be used to treat the nutrient shortages in the soil identified by the SHCs.

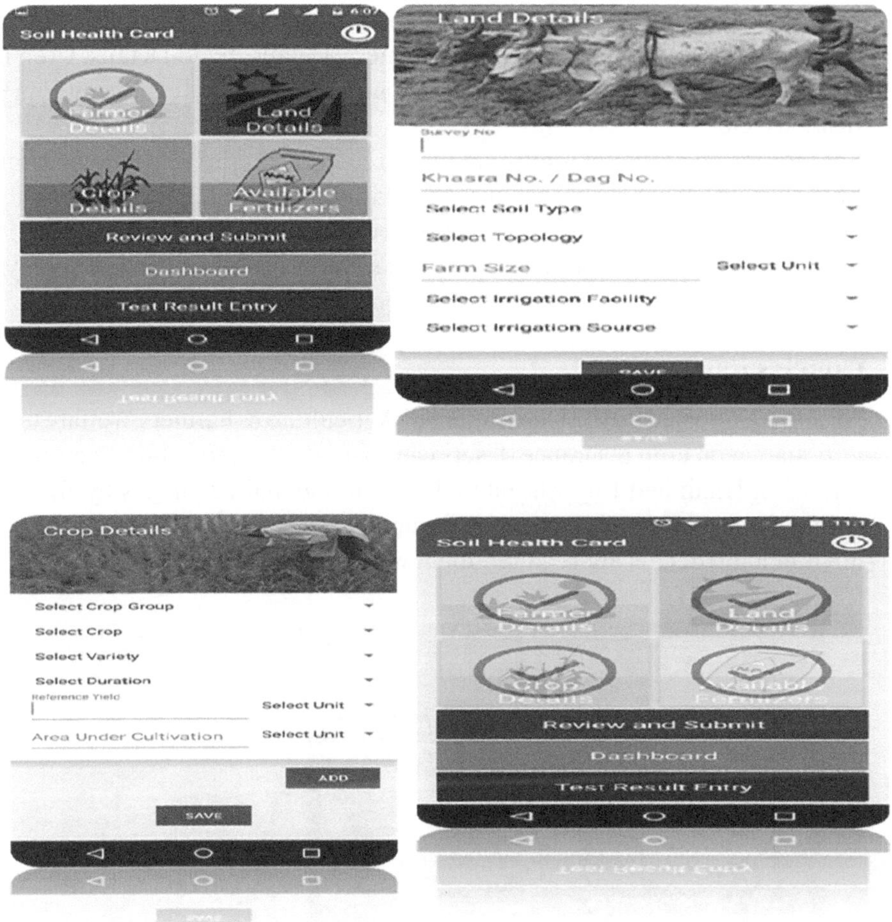

FIGURE 15.6 Soil health card.

Grid-based soil sampling is done on farmer fields, with a 10-ha grid for rain-fed areas and a 2.5-ha grid for irrigated areas. The concerned officials enter the sample information in the application portal along with their location information, and each soil sample is given a special Sample ID. The samples are examined in the laboratories, and the corresponding Soil Testing Lab personnel enter the test results. The Soil Testing Crop Response formulas created by ICAR or the general fertilizer recommendations provided by State Governments/State Agricultural Universities are used to determine the amount of nutrients necessary for a given crop based on the amount of nutrients that are nowadays present in the soil (state agricultural universities).

Based on the required nutrients, the application then determines the proper crop-wise fertilizer quantities and displays them on the SHC. When "Location" is turned on, this app logs latitude and longitude. You can enter information on the farmer, the property, the crops, and the fertilizer using this smartphone app.

- The icons on the dashboard for the sections: Each of the following sections includes a link to the relevant page—Farmers' Details, Land Details, Crop Details, Accessible Fertilizers, Assessment & Submit, Dashboard, Test Result Entry, Notifications, etc.
- For farmers, the following information is recorded: Address, phone number, clarification in the local language, and Aadhaar number. One needs to click the Save button after entering all the aforementioned settings.
- This covers the kind of soil, the irrigation system, and the irrigation supply.
- The crop group, crop, variety, and duration options are available here.

4. **Crop Insurance**

An app was created by the Department of Agriculture & Farmers Welfare to provide information on crop insurance. Users can translate the app, which is currently only offered in Hindi and English, into additional regional languages by filling out an online form at www.farmer.gov.in/insurance/appTrans.aspx (Figure 15.7). Your mKisan login details are required to view this form.

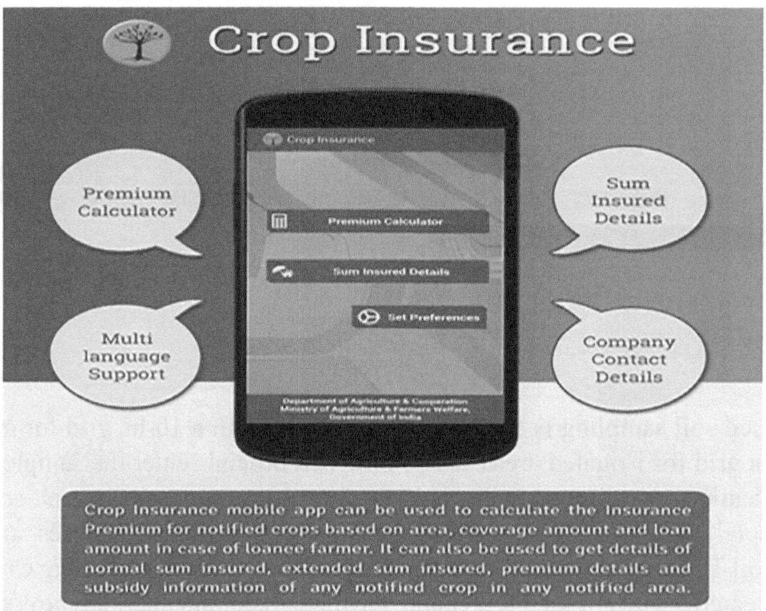

FIGURE 15.7 Crop insurance mobile app.

- The insurance premium for registered crops can be calculated using the Crop Insurance mobile app based on the region, the degree of coverage, and the loan amount in the case of a loanee farmer.
- The regular sum insured, expanded sum insured, premium information, and subsidy information for any notified crop in any name given can also be found using this tool.

5. **Bhuvan Hailstorm App**

This mobile app was created by the Ministry of Agriculture and Farmers Welfare, Government of India, to document agricultural loss brought on by hailstorms using images and geographic coordinates. To assess hailstorm damage, an agriculture inspector would approach the site with a smartphone or tablet executing this mobile program (Figure 15.8).

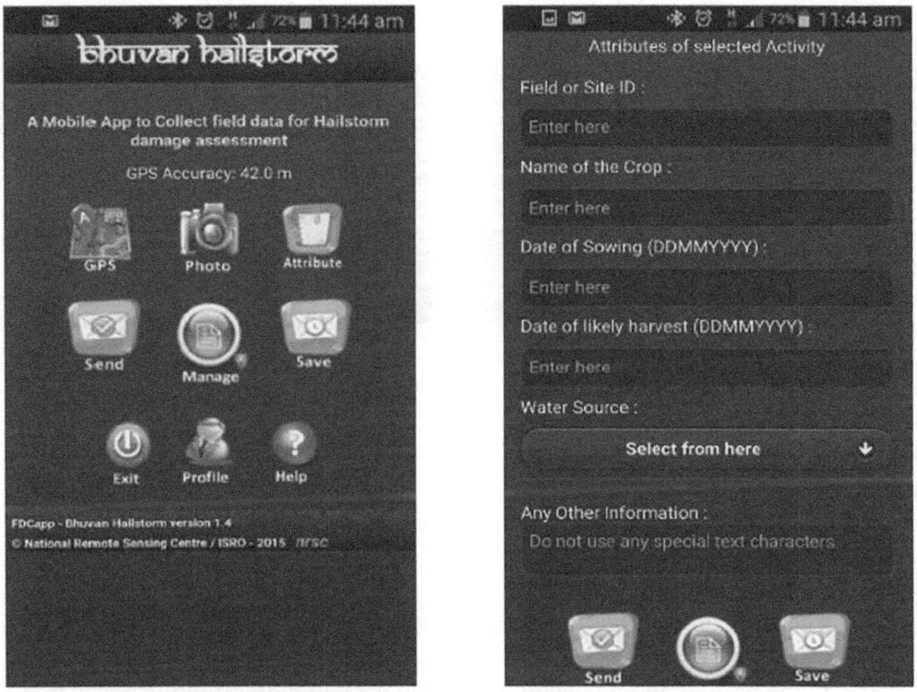

FIGURE 15.8 Bhuvan hailstorm app.

- An image of the field can be taken using this smartphone app, and the latitude, longitude, crop name; sowing date, anticipated harvest date, and irrigation source can all be recorded.
- Analysis is made easy by the data's automatic plotting on the Bhuvan Portal.

6. **The eNAM Mobile App**

The National Agriculture Market (NAM), a pan-Indian electronic trading portal, is being promoted by the Indian government (Figure 15.9). It was developed by the Small Farmers' Agribusiness Consortium, Ministry of Agriculture & Farmers Welfare, and merges the existing mandis to create a unified national market for agricultural goods. Farmers and other interested parties can use the mobile app to get arrivals and price-related information on their cellphones, while merchants can more easily place remote bids. Currently, only traders can see trade-related data and bid using the e-NAM mobile app version 1.0. The web version of the e-NAM Application is not an exact replica of this App. The mobile app has the following features:

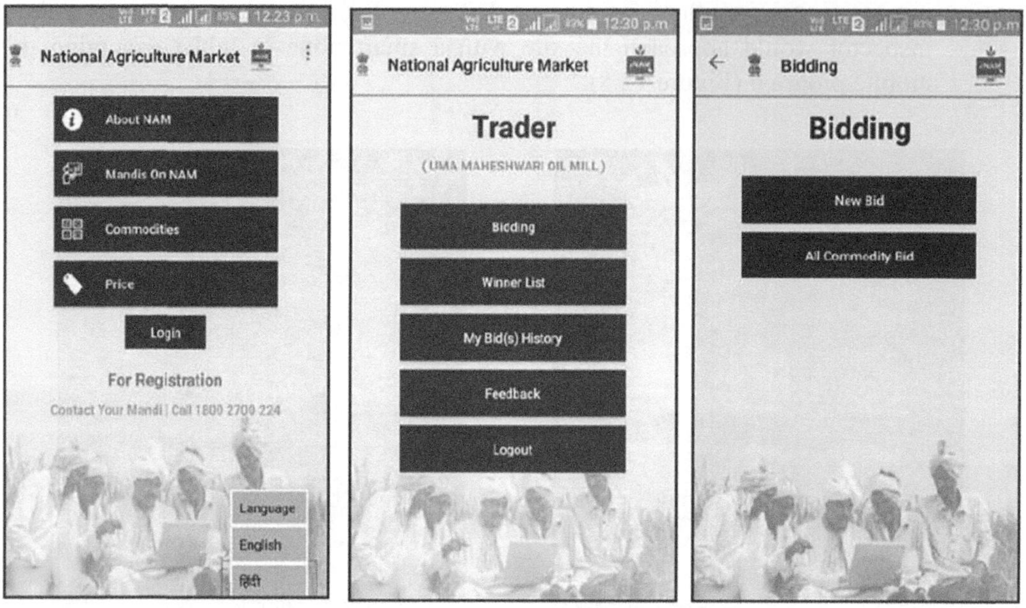

FIGURE 15.9 eNAM mobile app.

For Traders:

- There is a bid choice for the exchangeable lot.
- Can enter a new bid and/or alter the prices of the previous bid.
- In the case of an open auction, the lowest and maximum bid price are visible.
- View the winner list, view past bids, and give feedback.

For users, including farmers:

- View the e-NAM mandis list by state.
- Watching the arrivals at the mandi.
- Is able to see the minimum and maximum prices being charged in any mandi.

7. **Agrimarket Mobile App**

- With the help of the AgriMarket smartphone app, users can learn the cost of crops at markets 50 kilometers distant from their device (Figure 15.10).
- This app uses mobile global positioning system (GPS) to find users automatically and retrieve crop prices from markets within a 50-kilometer range.
- If a person prefers not to use GPS location, there is an alternate way to obtain the exact price of any market and any crop.

FIGURE 15.10 Agrimarket mobile app.

8. **The Digital Mandi Mobile App**

- With the aid of this app, one can easily view the most recent Mandi prices for various agricultural commodities in India's states and districts (Figure 15.11).
- Farmers, traders, and everyone else can access the app anytime to get the most recent Mandi price because it is simple to use and straightforward.
- Look through the different commodity categories.
- View the costs in various states.
- Simplified flow to arrive at the mandi price of the chosen commodity.
- Synchronizes information from the Agmarknet.nic.in government portal in India.

FIGURE 15.11 Digital mandi mobile app.

9. **riceXpert Mobile App**
   - Created by the National Rice Research Institute, Cuttack (Figure 15.12).
   - Farmers can access riceXpert for information on rice varieties for diverse ecologies, agricultural equipment for various locations, and harvesting methods. Additionally, it offers current data on weeds, nematodes, nutrients, insect pests, and disease-related problems.
   - This app allows farmers to construct personalized queries for quick problem-solving by transmitting text, images, and recorded voice in their rice fields as a diagnostic tool. Farmers can consult a team of specialists through the smartphone app, who can offer advice on growing rice.

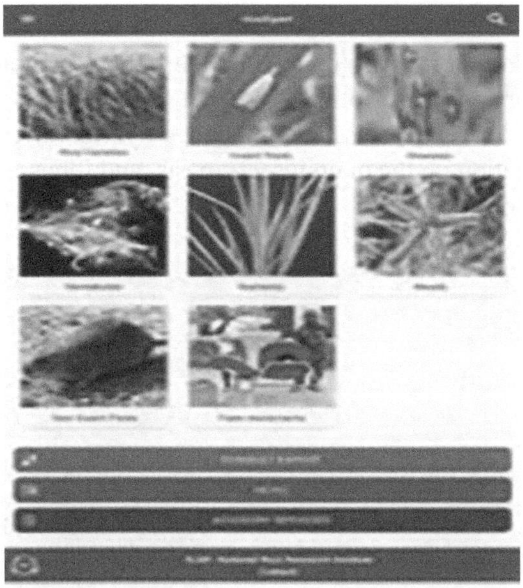

FIGURE 15.12 riceXpert mobile app.

10. Mobile app for Cane Adviser

- In Coimbatore, Tamil Nadu, India, the ICAR-Sugarcane Breeding Institute developed a smartphone application called Cane Adviser (Figure 15.13).
- It includes text and visuals with information on the entire growing cycle.
- It contains a scheduler app that is personalized for each registered user, and reminder messages appear in real-time mode.
- The query handler assists in submitting text or graphic questions for responses that are sent via SMS or email.

FIGURE 15.13 Cane adviser mobile app.

11. Castor Crop Mobile App

- An app for smartphones was created by the Indian Institute of Oilseeds Research on the castor crop (Figure 15.14).
- The castor farmers can access recommended hybrid varieties, intercropping techniques, information on the main insects, pests, and illnesses, as well as treatments, through this mobile app.

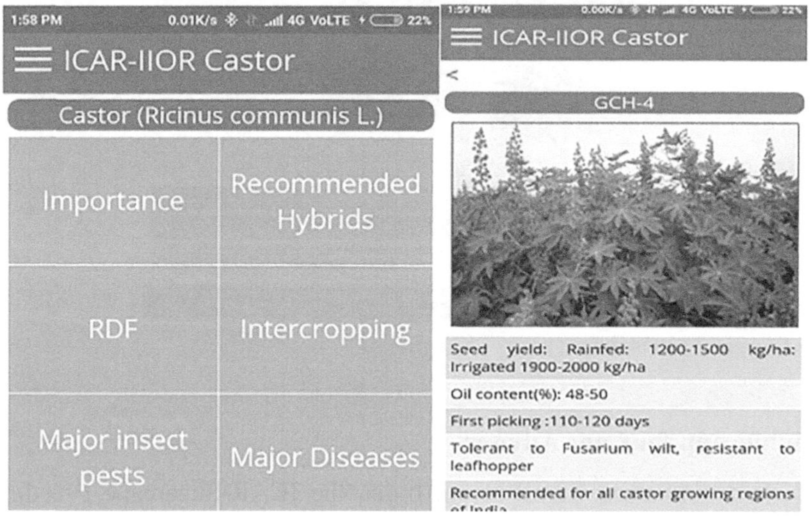

FIGURE 15.14 Castor crop mobile app.

12. Pomogranate Production Mobile App

- This mobile application was created by ICAR—National Research on Pomegranate (Figure 15.15).
- Local weather forecasts, current pomegranate market prices from all of India, announcement from the National Research Centre on Pomegranate (NRCP) about future events such as trainings, seminars, etc., and a single click to contact NRCP scientists for all issues relating to pomegranate.
- English, Hindi, and Marathi are all available on the smartphone app.

13. Crop Cutting Survey (CCE-Agri) mobile App

- The CCE-Agri was created by the Ministry of Agriculture and Farmer's Welfare of the Indian government (Figure 15.16).
- **The home screen offers the following four options:**

  1. The CCE Survey
  2. CCE Data Saved
  3. Check the Sent CCE Data
  4. Accept User

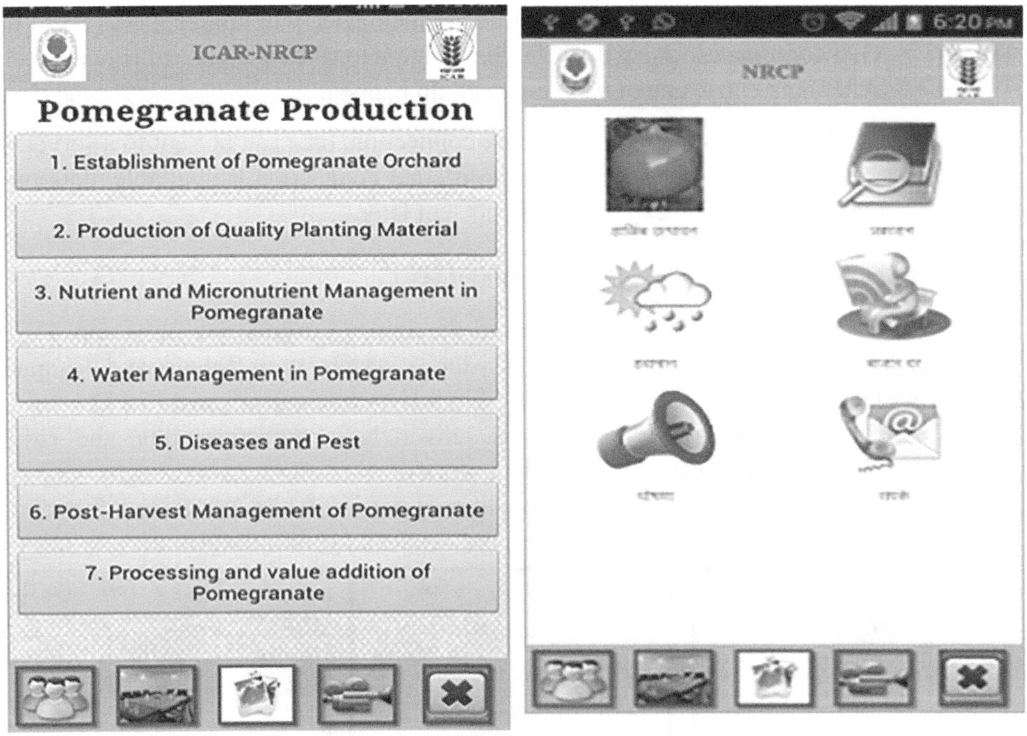

FIGURE 15.15 Pomegranate production mobile app.

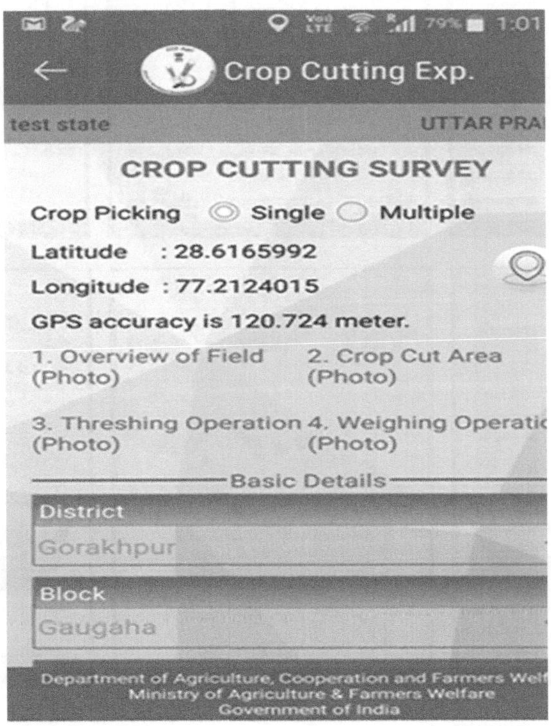

FIGURE 15.16 CCE-Agri mobile app.

- When filling out the CCE form, the mobile device will immediately record the GPS coordinates. Using the Add Photo button to take a picture (min1 and max4).
- After data entry, one of two things could happen:

  1. Send: When there is a network connection, data can be sent to a server.
  2. Save and send later: When an internet connection is available, data on a mobile device can be saved and sent later.

14. Dairy Mobile App (Telugu and Kannada)

- The Bangalore-based NGO created the animal husbandry mobile app. High-end analytics and a decision assistance system with Telugu and Kannada language support are both included in this mobile app (Figure 15.17).
- The smartphone app's content is delivered as interactive audio and video, which makes it easy to understand farmers.
- The software was created to dispel literacy barriers while delivering information.

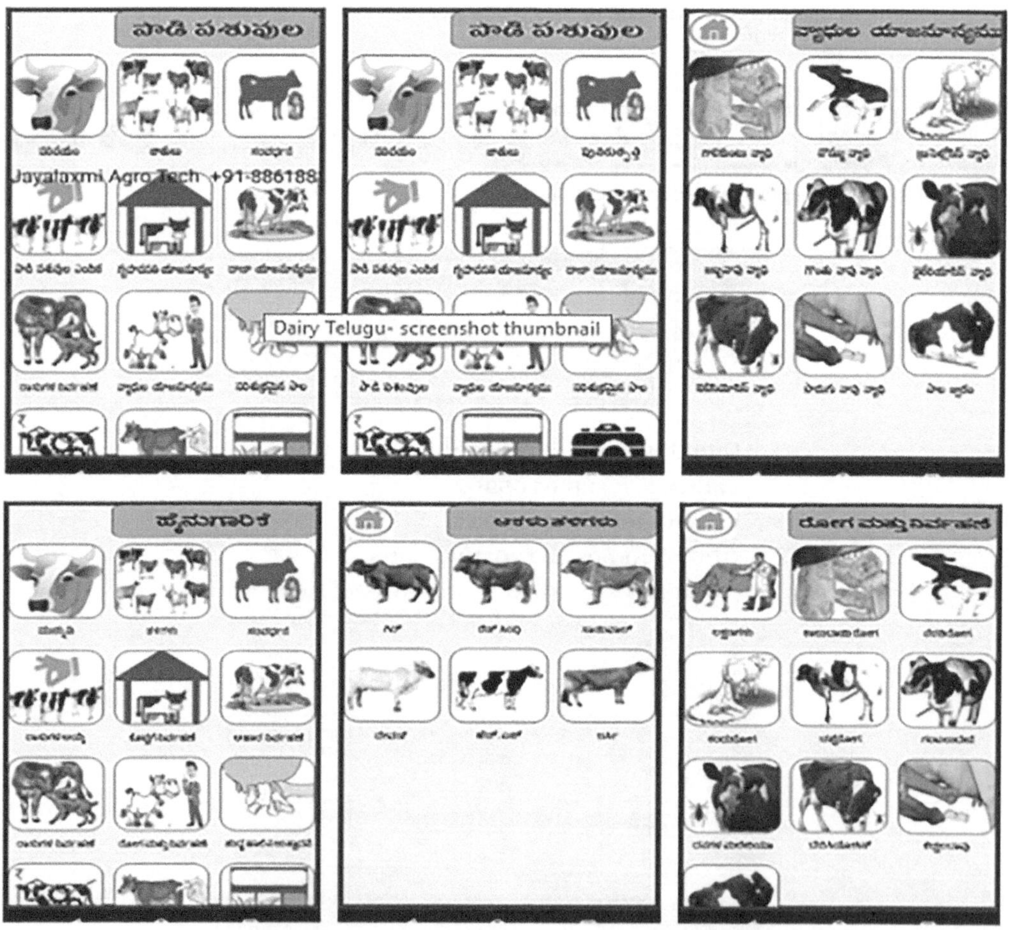

FIGURE 15.17 Dairy mobile app.

15. Indian Farmers Fertilizer Cooperative Limited (IFFCO) Kisan Mobile App

- The "IFFCO Kisan" smartphone app offers farmers specialized agricultural information catered to their unique needs, assisting them in making educated decisions.
- The smartphone app provides up-to-date mandi pricing, weather predictions, agricultural advisories, best practices guidelines for farming, animal husbandry, and horticulture, as well as a buyer and seller platform, news about agriculture, and government initiatives (Figure 15.18).
- For farmers who want to communicate in their native tongue, this smartphone app will also offer agriculture warnings and advisories in 11 Indian languages as well as agriculture audio snippets.

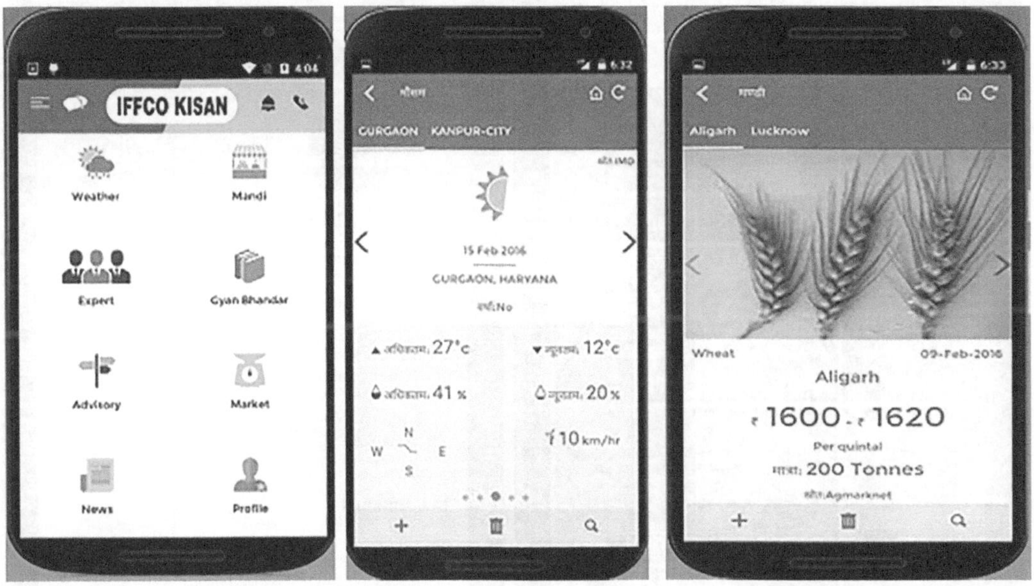

FIGURE 15.18 IFFCO Kisan Mobile App.

16. Plantix Mobile App

- Plantix is a mobile app that makes diagnosing and tracking plant diseases simple. PEAT, a German nonprofit, created this smartphone app to give farmers access to internet diagnostic tools (Figure 15.19).
- Plantix gives users the option to send photographs of afflicted plants directly from their smartphones and walks them through an easy identification process to identify the plant infection. Coordinates are attached to every photo shared through the mobile app.
- Real-time diagnosis: A database of 500,000 images covering 30 crops is used to analyze uploaded crop photos. This technology provides treatment recommendations for more than 120 crop illnesses.

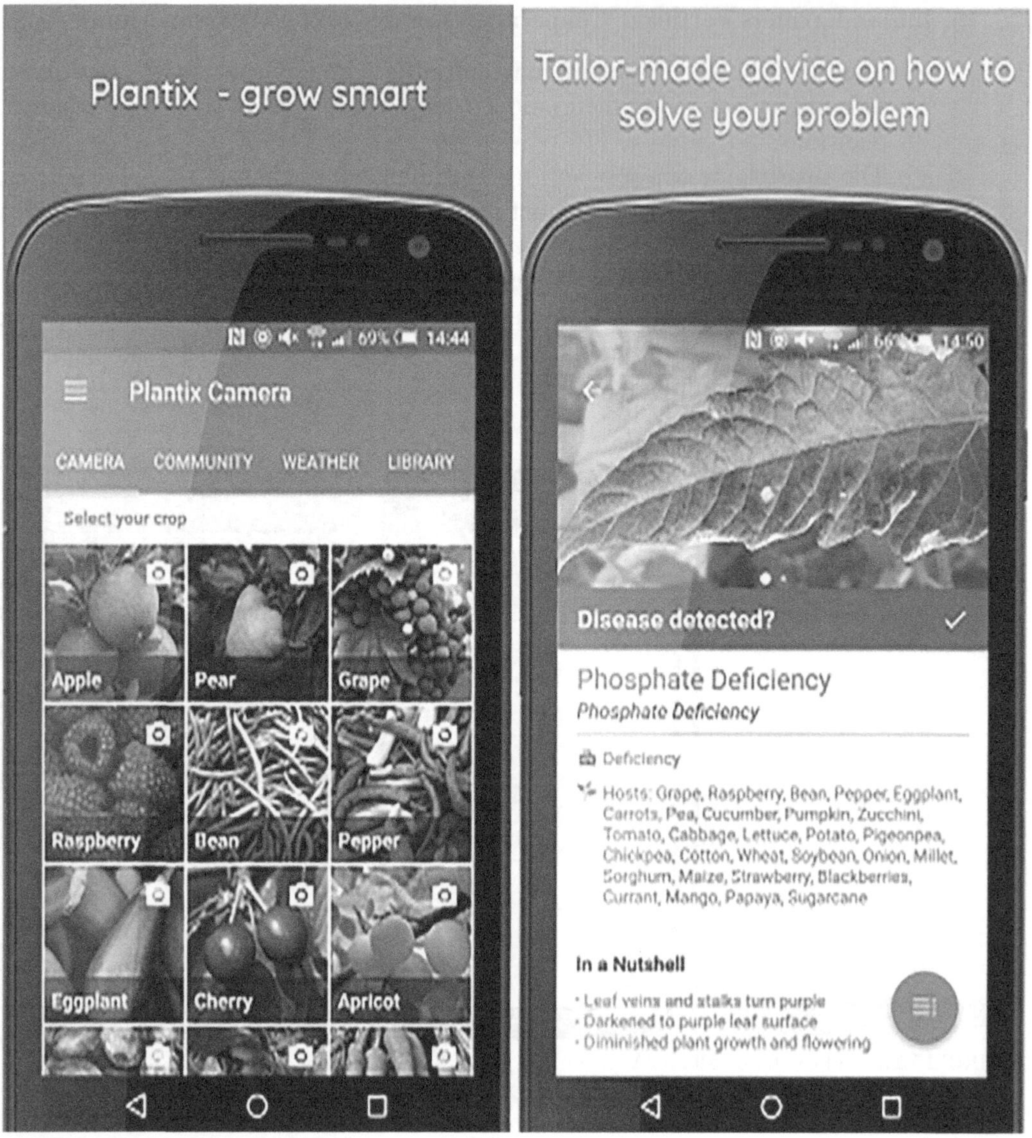

FIGURE 15.19 Plantix mobile app.

17. Pashu Poshan

The National Dairy Development Board created a program that enables the development of a balanced diet while minimizing costs and accounting for the animal's profile, which may include age, milk production, milk fat, and feeding schedule, among other things (Figure 15.20). The amount of locally accessible feed materials and mineral mixture that milk producers feed their animals should also be adjusted, according to this advice. A dairy farmer can use this software to determine the ideal feed and forage mixture to feed their dairy animals.

FIGURE 15.20 Pashu poshan.

18. mKrishi Fisheries App

This software was developed by the Indian National Centre for Ocean Information Services (INCOIS) and the Central Marine Fisheries Research Institute (CMFRI) of the ICAR in Hyderabad (Figure 15.21). Relying on sea surface temperature, the amount of phytoplankton—which is a source of food for a variety of species—and remote sensing information from National Oceanic and Atmospheric Administration satellites, INCOIS provides a forecast of fish shoals known as the potential fishing zone (PFZ). These data are compiled by the app, which also provides advisories in the regional tongue. This service was trialed in 56 fishermen cooperatives in Raigad, Maharashtra, by the Mumbai Research Center of the ICAR-CMFRI. Only logged-in users are able to access this service. Using this tool, fisherman may plan their fishing trip and decide whether to travel offshore only if a PFZ is nearby. This will help them avoid making pointless expeditions and save money on diesel, ice, and labor. The software tailors the information delivery based on user and device characteristics like location and language. According to a study by ICAR-CMFRI, it might have saved up to 30% on diesel expenses. The CMFRI-created mKrishi Fisheries mobile app took first place in the Ministry of External Affairs' 2016 National Social Innovation Contest, which was highly competitive.

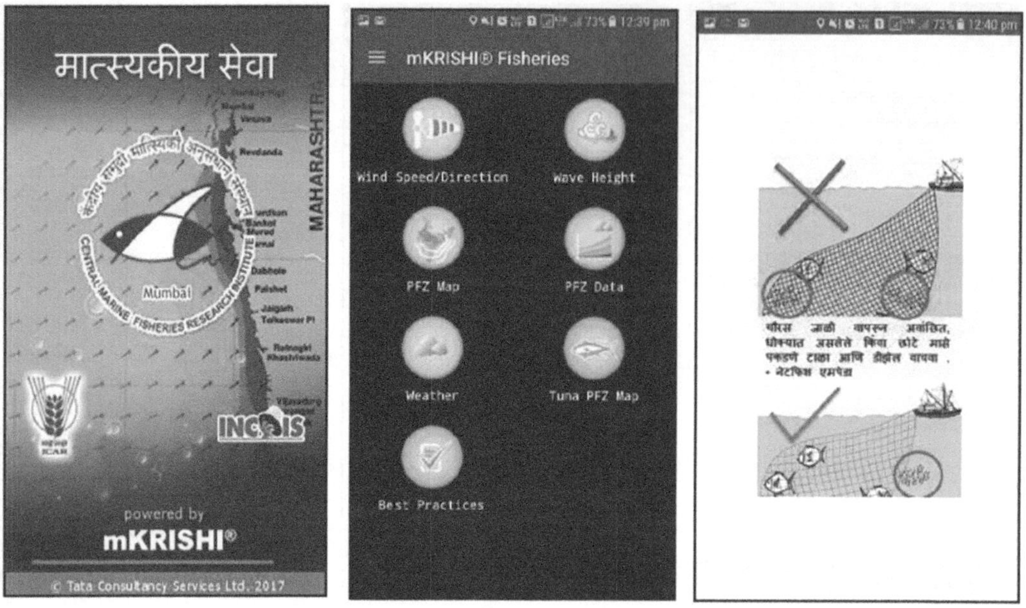

FIGURE 15.21 mKrishi fisheries app.

15.7 CRITIQUES AND PLAN OF ACTION FOR THE FUTURE

Marketing agricultural products is fraught with difficulties. The benefits that farmers are expected to receive are being eaten up by too many vultures. The chapter also highlights a number of problems with the current setup and demonstrates the absence of a coordinated, controlled marketing structure for the sale of agricultural goods. The largest obstacle preventing the complete value of digitization from becoming realized is the absence of cellphones and internet connection in developing nations generally, and in rural areas in particular [49]. Low income and ignorance are factors for such a disease. Farmers, on the other hand, take longer than expected to adapt new technologies for use in farming. Therefore, any plan to improve the agriculture sector's digitalization should start with creating the necessary infrastructure. Through mobile apps, online movies, and social media, this will be possible to access information and services.

Another factor contributing to the lower adoption rate in rural areas is the low literacy rate. According to Mamgain et al. [25], FAO expressed considerable worry about the lack of digital skills and knowledge in its yearly report for 2019 and claimed that it was a serious barrier to the use of new technologies. To improve digital literacy and skill, it is critical to increase youth literacy levels. For the benefit of instructors and pupils, school management should make an effort to integrate digital disciplines into their curricula. The appropriate agencies should seek to create opportunities for raising literacy rates and simultaneously developing digital abilities in order to encourage widespread usage of this technology. It is necessary to increase public awareness of the advantages of such technologies. Education and auxiliary services need to be improved if digital technology acceptability is to rise.

Government activities are crucial to the formation and maintenance of the digital environment, and they have a big impact on how successfully digital projects are carried out. Only with the active participation of the public and commercial sectors will the significant investment needed for this initiative be obtained. Only with significant government involvement can the necessary infrastructure and regulatory bodies be successfully established. The market has seen a significant infusion of start-ups and new businesses from the technology sector as a result of the expanding demand for information farming and related services. By encouraging start-ups to enter this market, the utilization of new technologies can be significantly expanded, enhancing the performance of this business. In actuality, any ground-breaking technology that can boost the production and efficiency of the agricultural industry is urgently needed in farm marketing. Digital technology is viewed by many academics and decision-makers as the sole cause of such radical transformation.

15.8 CONCLUSION AND RECOMMENDATIONS

Despite being a significant economic and social sector, policymakers continue to be most concerned about the agriculture sector's underwhelming performance. Technologies based on VR and AR can be used in many sectors, such science, services, education, medical, and the military. Despite the fact that technology has been heavily incorporated into numerous aspects of the agricultural industry in recent years to increase productivity, the desired outcome is not always as satisfying due to its unique qualities. People who work in farm marketing are motivated by this to use cutting-edge techniques like digital technology. Digitalization has been effectively implemented in numerous economic areas, leading to benefits like improved consumer contentment, reduced transaction costs, open competition, and greater information accessibility. Researchers, decision-makers, and agriculturalists have used digital technologies in this sector as a result. Despite only being employed in a few places, the results have been positive and have sparked more interest. The employment of digital technologies in agricultural marketing is improving the industry's efficiency. As potential clients become more conscious of digitization, this method becomes more important and desired. The importance of this industry can be further increased with the use of digital marketing in conjunction with technological integration. The method is beneficial to businesses engaged in digital agriculture marketing in that it broadens their client base, improves information accessibility, lowers transaction costs, raises customer satisfaction levels, and reduces waste. In comparison to businesses that have not yet embraced the new technology, these variables increase the efficiency and profitability of these businesses. It is anticipated that more agriculture-based businesses will be able to digitalize their marketing efforts once the obstacles have been removed.

KEYWORDS

- **mobile**
- **applications**
- **data**
- **sustainability**
- **agriculture**

REFERENCES

1. Sahoo, J. P., Samal, K. C., & Behera, D. (2020). A blueprint to boost Indian agriculture in the era of Covid-19 pandemic. *Biotica Research Today*, *2*(8), 780–782.
2 Kumar, A., Padhee, A. K., & Kumar, S. (2020). How Indian agriculture should change after COVID-19. *Food Security*, *12*(4), 837–840.
3. Priyadarshini, P., & Abhilash, P. C. (2021). Agri-food systems in India: Concerns and policy recommendations for building resilience in post COVID-19 pandemic times. *Global Food Security*, *29*, 100537.
4. Amjath-Babu, T. S., Krupnik, T. J., Thilsted, S. H., & McDonald, A. J. (2020). Key indicators for monitoring food system disruptions caused by the COVID-19 pandemic: Insights from Bangladesh towards effective response. *Food Security*, *12*(4), 761–768.
5. Nedumaran, D. G., Kumar, M. A., & Alaguraja, M. (2020). Effect of mobile applications on farming in Virudhunagar District—A study. *SSRN Electronic Journal*, *68*(1), 12718–12727.
6. Pemarathna, R. (2018). Beyond Agricultural tackle: An Analysis of e-Agriculture with paddy Sector in Sri Lanka. *IOSR Journal of Business and Management (IOSRJBM)*, *20*(7), 46–52.
7. Butt, T. M., Shahbaz, B., Hassan, M. Z. Y., & Khan, M. (2020). A critical analysis of e-learning agricultural technical capacity for rural development: Lessons from Pakistan. *GeoJournal*, *87*, 463–475.
8. Alkahtani, M., Khalid, Q. S., Jalees, M., Omair, M., Hussain, G., & Pruncu, C. I. (2021). E-agricultural supply chain management coupled with blockchain effect and cooperative strategies. *Sustainability*, *13*(2), 816.
9. Jadav, A., Sinha, A., & Swarnalatha, K. S. (2022). E-agricultural portal for farmers using decentralized ledger and machine learning tools. In: *Emerging Research in Computing, Information, Communication and Applications* (pp. 345–359). Springer: Singapore.
10. Lin, J., Shen, Z., Zhang, A., & Chai, Y. (2018). Blockchain and IoT based food traceability for smart agriculture. In *Proceedings of the 3rd International Conference on Crowd Science and Engineering* (pp. 1–6).
11. Sylvester, G.(2019). *E-Agriculture In Action: Blockchain for Agriculture, Opportunities and Challenges*. FAO.
12. Bankar, T., & Tamboli, M. A. (2020). Digital agricultural marketing: Step towards sustainable agricultural sector. *MS Journal*, *4227*.
13. Hu, T., Li, Y., & Le, C. (2022). Optimization of "internet+" agricultural product industry chain based on partial differential equation inversion layer by layer. *Mathematical Problems in Engineering*, *2022*, 3352927.
14. Bose, S. C., & Kiran, R. (2021). Digital marketing: A sustainable way to thrive in competition of agriculture marketing. In: *Bioinformatics for Agriculture: High-Throughput Approaches* (pp. 135–144). Springer: Singapore.

15. Dabbara, R., Chandrakumar, M., Anandhi, V., & Muruganandhi, D. (2020). Data driven marketing applications for agriculture services. *International Journal of Current Microbiology and Applied Sciences*, *9*(2), 2914–2920.
16. Novytska, I., Chychkalo-Kondratska, I. R. Y. N. A., Chyzhevska, M., Sydorenko-Melnyk, H., & Tytarenko, L. (2021). Digital marketing in the system of promotion of organic products. *WSEAS Transactions on Business and Economics*, *18*, 524–530.
17. Farooq, M. S., Riaz, S., Abid, A., Abid, K., & Naeem, M. A. (2019). A survey on the role of IoT in agriculture for the implementation of smart farming. *IEEE Access*, *7*, 156237–156271.
18. Noorbehbahani, F., Salehi, F., & Zadeh, R. J. (2019). A systematic mapping study on gamification applied to e-marketing. *Journal of Research in Interactive Marketing*, *13*(3), 392–410.
19. Liu, W., Shao, X. F., Wu, C. H., & Qiao, P. (2021). A systematic literature review on applications of information and communication technologies and blockchain technologies for precision agriculture development. *Journal of Cleaner Production*, *298*, 126763.
20. Oyedele, Y., Dlamini, P., van Greunen, D., & Chizema, T. (2021, October). Lessons learnt from deploying an IoT sensing system for e-Agriculture in South Africa. In *2021 IEEE Global Humanitarian Technology Conference (GHTC)* (pp. 208–212). IEEE.
21. Gopinath, R., & Kalpana, R. (2021). A study on adoption of ICT in Farming practices with special reference to E-commerce in agriculture.
22. Al-Hazmi, N. M. (2022). E-marketing by mobile apps for tourism products and its impact on the performance of travel and tourism organizations. *International Journal of Interactive Mobile Technologies*, *16*(12), 145–153.
23. Ashokkumar, K., Bairi, G. R., & Are, S. B. (2019). Agriculture E-commerce for increasing revenue of farmers using cloud and web technologies. *Journal of Computational and Theoretical Nanoscience*, *16*(8), 3187–3191.
24. Kusumawati, Y. A., Huda, M. N., Dirgantara, S., & Carvenoriega, O. A. (2021). Architecture design of agriculture marketing mobile apps during pandemic era. In *IOP Conference Series: Earth and Environmental Science* (Vol. 794, No. 1, p. 012126). IOP Publishing.
25. Mamgain, A., Joshi, U., & Chauhan, J. (2020). Impact of social media in enhancing Agriculture extension. *Agriculture & Food: E-Newsletter*, *2*(9), 367–370.
26. Azad, A. K., Mandi, K., & Singh, S. (2020). Farming revolution through e-evolution. *Informatics Studies*, *7*(2), 53–57.
27. Xiao, J., Wang, W., & Tsai, S. B. (2021). Coupling of agricultural product marketing and agricultural economic development based on big data analysis and "internet+." *Mobile Information Systems*, *2021*(8), 1–10.
28. Luckyardi, S., Hurriyati, R., & Disman, P. D. D. (2022). Mobile application design for digital marketing and management of horticultural crops. *Journal of Engineering Science and Technology*, *17*(1), 0064–0074.
29. Chatterjee, R. (2020). Indian agriculture and role of agricultural extension system to cope up with COVID-19 crisis. *Food and Scientific Reports*, *1*, 10–15.
30. Osmëni, T., & Ali, M. (2021, August). LoRa IoT WSN for E-agriculture. In: *International Conference for Emerging Technologies in Computing* (pp. 85–93). Springer: Cham.
31. Chilukuri, P., Anusuya, R., & Prabhu, M. R. (2022). Comprehensive design analysis of digital marketing in agriculture sector. *International Journal of Early Childhood Special Education (INT-JECSE)*, *14*(5), 814–821
32. Zhang, J., & Du, M. (2020). Utilization and effectiveness of social media message strategy: How B2B brands differ from B2C brands. *Journal of Business & Industrial Marketing*, *35*(4), 721–740.
33. Malesev, S., & Cherry, M. (2021). Digital and social media marketing-growing market share for construction SMEs. *Construction Economics and Building*, *21*(1), 65–82.
34. Zahay, D. (2021). Advancing research in digital and social media marketing. *Journal of Marketing Theory and Practice*, *29*(1), 125–139.

35. Biswas, P., Orero, P., Swaminathan, M., Krishnaswamy, K., & Robinson, P. (2021). Adaptive accessible AR/VR systems. In *Extended Abstracts of the 2021 CHI Conference on Human Factors in Computing Systems* (pp. 1–7).
36. Priya, J., Palanisamy, C., & Vinothini, C. (2021). Sensor informatics of IoT, AR/VR, and MR in healthcare applications. In: *Efficient Data Handling for Massive Internet of Medical Things* (pp. 103–127). Springer: Cham.
37. Hirankerd, K., & Kittisunthonphisarn, N. (2020). E-Learning management system based on reality technology with AI. *International Journal of Information and Education Technology*, *10*(4), 259–264.
38. Shanu, S., Narula, D., Pathak, L. K., & Mahato, S. (2022). AR/VR technology for autonomous vehicles and knowledge-based risk assessment. In: *Virtual and Augmented Reality for Automobile Industry: Innovation Vision and Applications* (pp. 87–109). Springer: Cham.
39. Klaas, E., & Roopaei, M. (2021). Immersive analytics application in smart agriculture and animal behavior. In *2021 IEEE 11th Annual Computing and Communication Workshop and Conference (CCWC)* (pp. 0290–0296). IEEE.
40. Mabiletsa, O., Viljoen, S. J., Farrell, J. A., Ngqwemla, L., & Isafiade, O. E. (2020). An immersive tractor application for sustainability: A South African land reform and learners' perspective. *International Journal of Virtual and Augmented Reality (IJVAR)*, *4*(1), 35–54.
41. Gaspard-Richards, D. (2021). Technology integration, e-learning and course design, and practical uses of social media apps, web portals and AI/AR/VR to transform education systems. *International Journal of Education and Development Using Information and Communication Technology*, *17*(3), 2–5.
42. Raval, M. S., Chaudhary, S., & Adinarayana, J. (2022). Computer vision and machine learning in agriculture. In: *Data Science in Agriculture and Natural Resource Management* (pp. 97–126). Springer: Singapore.
43. Rejeb, A., Rejeb, K., & Keogh, J. G. (2021). Enablers of augmented reality in the food supply chain: A systematic literature review. *Journal of Foodservice Business Research*, *24*(4), 415–444.
44. Hurst, W., Mendoza, F. R., & Tekinerdogan, B. (2021). Augmented reality in precision farming: Concepts and applications. *Smart Cities*, *4*(4), 1454–1468.
45. Anonymous (No Date). How virtual and augmented realities agriculture help agriculture. Accessed from https: //www.visartech.com/blog/how-virtual-and-augmented-realities-help-agriculture/
46. Dwivedi, Y. K., Ismagilova, E., Hughes, D. L., Carlson, J., Filieri, R., Jacobson, J., Jain, V., Karjaluoto, H., Kefi, H., Krishen, A. S., Kumar, V., Rahman, M. M., Raman, R., Rauschnabel, P. A., Rowley, J., Salo, J., Tran, G. A., & Wang, Y. (2021). Setting the future of digital and social media marketing research: Perspectives and research propositions. *International Journal of Information Management*, *59*, 102168.
47. Yao, B., Shanoyan, A., Peterson, H. H., Boyer, C., & Baker, L. (2019). The use of new-media marketing in the green industry: Analysis of social media use and impact on sales. *Agribusiness*, *35*(2), 281–297.
48. Shafinah, K., Sahari, N., Sulaiman, R., Saif, A. S., & Ikram, M. M. (2020, April). m-Naslada: A mobile application for black pepper crop disease information. *Journal of Physics: Conference Series, 1529*(2), 022041. IOP Publishing.
49. Cole, H. S., DeNardin, T., & Clow, K. E. (2017). Small service businesses: Advertising attitudes and the use of digital and social media marketing. *Services Marketing Quarterly*, *38*(4), 203–212.

CHAPTER 16

Cyber Biosecurity in Food and Agriculture

VRIKSHA SINGH[1*] and SACHI CHOUDHARY[2]

[1]*School of Agriculture, Graphic Era Hill University, Dehradun, Uttarakhand, India*

[2]*School of Computer Science, UPES, Dehradun, Uttarakhand, India*

*Corresponding author

ABSTRACT

The world is currently undergoing what has been called the fourth industrial revolution, a rapid digital transformation. In this wave of change, it is increasingly evident that many doctors are not prepared for the new era; agriculture is one such area. As a factory producing food for the world, this critical infrastructure requires exceptional research and analysis using highly improved technology. Despite its critical importance to modern society, many aspects of commercial farming have yet to change in the digital age. This is reflected in the multiple vulnerabilities that now exist in agricultural systems, as well as the absence of regulation and fragmentation in managing cybersecurity environments—a discipline that combines both and natural security to create multiple jobs fence in the life sciences. This emerging oversight threatens sound agricultural policy, putting companies, economies, and individuals at risk. While there are many ways to mitigate these risks, they ultimately rely on precautionary consideration of cybersecurity by all stakeholders, which affects device manufacturers, machine engineers, and most importantly end users. This study examines several weaknesses in advanced agricultural automation, explores alternative solutions to the increasing risks posed by them, and examines the future possibilities of robust continuous agricultural automation in cybersecurity programs.

16.1 INTRODUCTION

16.1.1 WHAT IS CYBER ECOSYSTEM SECURITY?

Cyberbiosecurity lies at the crossroads of cybersecurity, cyber-physical security, and biosecurity (Figure 16.1).

Demystifying Sustainable Farming Using Smart Technologies.
Gunjan Chhabra, Keshav Kaushik, Priya Chugh, Susheela Dahiya, and Aryan Chaudhary (Eds.)

DOI: 10.1201/9781779640932-16

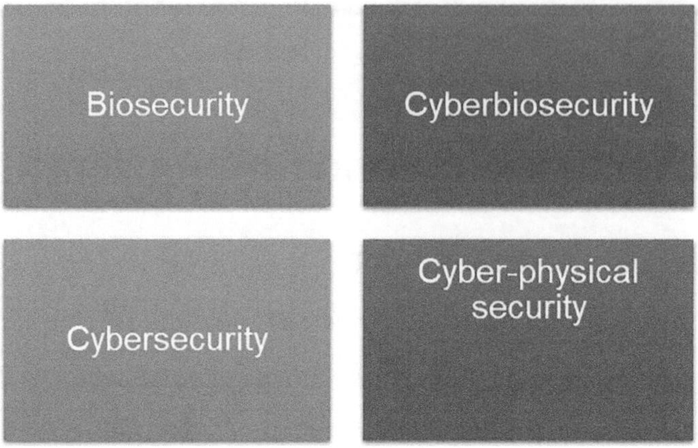

FIGURE 16.1 Evolving theme of cybersecurity, which combines biosecurity, cybersecurity, and cyberphysical security.

Source: Adapted from Ref. [3].

Cyberbiosecurity is unique among other fields as it deals with safeguarding biological data and information related to electronic or digital recording, editing, and accessing of life facts.

Because processing automation in the food industry involves the use of virtual records, data storage, and other technologies, these systems and processes become susceptible to attacks from cyberspace.

Errors caused by automation and technology consist of global positioning system devices, automated sensors, appliances, industrial or commercial software, operating systems, electronic data storage, mobile phones, and use of laptops, and most mechanical chains used in the food and agricultural industries or for records, or record keeping and supporting or production records. This knowledge is commonly utilized to make decisions, so data security and integrity are important components of equipment safety and a key issue in cybersecurity and food and agriculture.

Tens of millions of people rely on food and agriculture for their livelihood, just as the rest of the billions of people rely on healthcare, power, transportation, and economic development. With the shift of most business operations to digital infrastructures, the likelihood of large-scale cyberattacks keeps on increasing. Cybersecurity is no longer an option; it is a necessity to protect the food supply of the region. It also becomes imperative that all new operating systems are designed with cybersecurity built in from the ground up. The whole food delivery industry is very sensitive and interwoven; in the end, we all have to prioritize what is most important and has to be protected using the best and most complete tools and laws in place.

16.1.2 WHO SHOULD BE INVOLVED IN CYBERSECURITY?

Cyberbiosecurity is essential for everyone involved in food distribution, from the farm to the dog. Cyber threats can occur throughout the supply chain, causing vulnerabilities throughout the supply chain because employees are the most weak link in the machine's safety, personnel, and supervisors; factory team employees, accountants, and truck drivers should be aware of cyber biosecurity risks and warning signs reasons for integrating technology Because of its pervasiveness, it is impossible to provide cyberbiosecurity information technology departments or areas in the food and agricultural industries. Cyberbiosecurity involves several persons or functions. Employees and providers must collaborate to ensure successful cyberbiosecurity.

The combination of past, present, and future food and agriculture sectors' use of technology and the Internet for communication and data storage will reveal vulnerabilities and security breaches. If cybersecurity technology is owed to food agriculture, it can increase food distribution security. What are the principles of concern?

The main cybersecurity concerns can fall into four broad categories (Figure 16.2).

1. Data embodiment.
2. Automatic system capture.
3. A fake node.
4. Learning algorithm risk

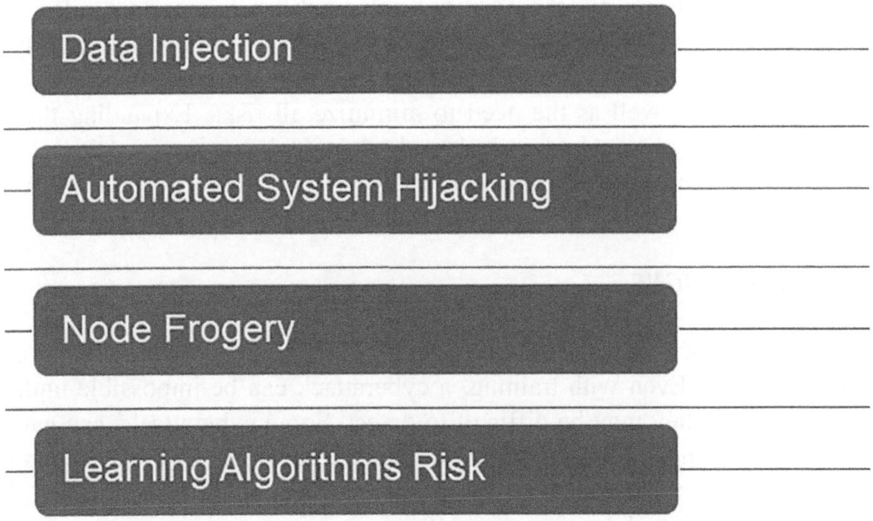

FIGURE 16.2 It identifies cyberbiosecurity hazards as falling into four categories: (1) data injection, (2) automated system hijacking, (3) node networking, and (4) learning algorithm dangers.

Source: Adapted from Ref. [14].

Data injection is the injection of incorrect information into a device or data set. In the food and agriculture industries, data units are used to monitor policy implementation and make decisions. Altering this data can cause systems and processes to continue to operate outside their optimal operating limits, or make decisions that harm the customer or the product.

As food and agricultural practices become increasingly digitized, the phenomenon of being captured by automation will severely hamper food delivery. The loss of control in a system can disrupt volumes associated with the distribution chain, and can result in increased food waste, decreased productivity, reduced supply, and drifting demand and costs top. The economic pressure created by such a move can be beneficial for smallholders in the supply chain. Node networking is the use of counterfeit detection sensors to increase efficiency or optimize food processing. Manufacturers continue to employ sensors to collect data on meteorological variables like temperature and humidity, as well as soil health factors, including temperature, humidity, and pH. There are concerns about the number of sensors used in the entire food supply chain, which will enhance detection sensors, safety, and security. Learning algorithm risk is defined as the sensitivity of system learning processes, which can pose a risk to records. Cyberattacks in this space will have a profound impact on the systems and techniques that enable them to develop and use algorithmic learning.

16.1.3 SUPPORTIVE ACTION

Certain methods of preserving and protecting food are certainly useful for cybersecurity. Good file security, traceability, chemical controls, and other important rules contribute not only to a good food safety and lifestyle but also to a strong cybersecurity culture, which may be beneficial. Food scientists and the food industry understand preventive and corrective strategies, as well as the need to minimize all risks. Expanding the scope of cyber threats to cover physical, chemical, and ecological dangers is a tiny step toward establishing a robust and secure food supply chain.

16.1.4 WARNING SIGNS

The vulnerability of our food supply highlights the importance of recognizing cyber and natural security risks. Even with training, a cyberattack can be impossible under certain circumstances. Also, they may be difficult to detect. Some cyberattacks can cause issues with accessing and using commonly used documents, packages, or devices, and although these issues may at first seem unfair, they are it is an example of possible red flags that would indicate a cyberattack has occurred [19].

- Unauthorized access to documents or systems.
- Unauthorized password changes.
- Unauthorized software installation or removal.
- Unauthorized file storage.
- Slow network speed.
- Unreported email interests.
- Lack of computer control.

Identifying red flags is the first and most critical step in averting a cyberattack. Next, disconnect the device from the Internet and unplug the network router [18]. Before starting your repairs, you should consult with a professional to ensure that you can resolve the issue quickly and avoid further risks.

16.1.5 HAS THE INTERNET OF THINGS (IOT) ADDED A SIMILAR LAYER OF COMPLEXITY?

In agriculture, the IoT poses specific challenges because of the variety of related devices and structures concerned. Unlike some other sectors, agricultural generation frequently lacks standardized cybersecurity measures, making it more prone to attack. The huge-scale capacity effect of any disruption to those structures underlines the significance of efficiently addressing these demanding situations. This is one of the reasons why we developed the Food and Agriculture Risk Modelling tool.

16.1.6 DUAL USE AND CYBERBIOSECURITY

Technology and research cause advances in scientific treatments, enhance agricultural yields, beautify food protection, and defend in opposition to organic threats. Genetic engineering can lead to breakthroughs in treating sicknesses, and artificial intelligence (AI) can help in predicting outbreaks. Technology and studies can also be misused to increase organic guns, conduct bioterrorism, or perpetrate bio-hacking attacks. Genetic enhancing gear can be used to create dangerous pathogens. Cyber assaults are employed to steal sensitive genetic information and sabotage bioinformatics systems. The convergence of biotechnology with different technologies like AI, robotics, and nanotechnology adds layers of complexity to the cybersecurity challenges. Each technological area brings its own personal set of cyber risks, and their convergence amplifies the potential for each revolutionary advantages and dangers of misuse.

According to Regulation (EU) 2021/821 (on export, trade, technical assistance, transport, and conversion of dual-use goods), a dual-use machine is a machine that has software generation that can be used by the citizen, which includes products that can be used in military operations and for the design, development, manufacture, or consumption of nuclear, chemical, or nonchemical weapons use or shipping, including all usable nonexplosive materials.

According to the law, various categories of people are concerned with twin-use objects, which include herbal people including carrier carriers, researchers, consultants, and people transmitting twin-use items electronically. It is crucial that every one, men and women, is aware of the dangers and the modus operandi of threat actors.

In particular, educational and research institutions face wonderful demanding situations because of, *inter alia*, their popular commitment to the free exchange of thoughts and scientific developments. Persons operating in these areas every so often accept as true with that understanding is releasing, and should be free and the property of all humanity. But while information is also a weapon in hybrid warfare, we must do not forget what Sun Tzu had stated: The preferred art of battle is to subdue the enemy without preventing.

16.1.6.1 WHO ELSE IS WORRIED?

State-backed organizations, foreign intelligence corporations, organized crime, and other cyber actors are exploiting vulnerabilities at the intersection of biological activities and facts protection. The human detail is sort of always the weakest link in cybersecurity.

According to the World Health Organization (WHO) Global Guidelines Framework for the Responsible Use of Life Sciences, biosafety, biosecurity, and applied research are increasingly recognized as teaching methods, is twice important in social science research. However, an increasing number of biosecurity programs include protection against external risks. Hospitals, medical research institutes, genetic databases, biotechnology institutes, and clinical vaccine manufacturing facilities all need to have a new perspective.

According to the WHO, the new threat goes beyond bacteria and organisms. For example, it is assumed that emerging neural characteristics should be manipulated (e.g., in order to improve or degrade overall human performance).

One of the breakthroughs in nanotechnology is the application of nanocarriers in improving therapeutic efficacy; however, there is worry that nanoparticles may be misused (e.g., delivered as aerosols which can cross blood–brain barrier).

However, today, attackers view the sector's dependence on an efficient food supply chain as an avenue for dropping malware like ransomware so as to achieve their criminal objectives.

16.1.6.2 CYBERSECURITY OF GENOMIC DATA

According to the National Center for Standards and Technology, a genomic record is created by reading the genome structure and characteristics of an organism, which consists of genes and other factors that affect the function of genes. The type of information varies depending on the application such as vaccine development and production, drug formulation and manufacturing, disease diagnosis, and agricultural advancements to boost food production, biofuel generation, basic and translational research, clinical testing, genetic tracing, regulatory processes, and more. As these fields evolve, new applications continue to emerge.

Sequencing technologies have advanced to the point where all genome sequencing is practical and cost-effective. Genome sequences for many species of microorganisms, plants, and animals are available in whole or in part from the National Institutes of Health, the Federal Bureau of Investigation (FBI), and referral genetic testing companies direct-to-consumer internally, to name a few. As time progresses so does the risk posed by genetic-focused cybersecurity attacks for the United States, and so does the growing awareness of national security, economics, biotechnology, and its citizens.

Cyberattacks on genetic information threaten the security, integrity, and availability of records. Cyberattacks on data privacy could threaten our economy through stealing valuable intellectual property controls by the US biotechnology industry, allowing adversaries to infiltrate and gain unfair economic advantage over holder's genetic facts. Attacks on accurate information can destroy biopharmaceutical production, agrifood manufacturing, and bioproduction as entertainment. Attacks to accounting systems include ransomware encryption, data compromise, and attacks on critical computing resources used in research and development, including those in mechanical engineering as well as construction.

The potential consequences of cyberattacks on genetic information threaten our national security, including biological weapons and brutality and theft against our civilian, military, and intelligence-based institutions. Cyberattacks on genomic data also allow for potentially harmful benefits, discrimination based on disease threats, and loss of privacy through congeners or hidden phenotypes such as by health, emotional strength, mental capacity, appearance, and expression of physical abilities.

In addition to privacy risks from cyberattacks, privacy risks related to cybersecurity can occur when processing genetic data. This threat arises when processing of genetic statistics is not predictable, cannot not handled, and shared if the people who have them do not know what is going on with the genetic information.

Insufficient processing power in record processing arises when genetic information does not have characteristics to allow precise granular control, for example, one may choose to remove some or all of their genetic information from a data set. Granting permission creates privacy problems due to poor disaggregation in statistics usage. Each of these privacy risks may hinder the ability to realize the benefits of using genetic data.

US research community, government, and private industry all need genetic data exchange to give the United States a competitive advantage in biotechnology and to improve clinical trials. Genetic data exchange and sharing is crucial to understand human resources, improve health, and access advanced scientific research The data represent tens of millions of recorded individuals. In all forms, such information is managed through a number of organizations (e.g., national, global, charitable, for-profit) that collect, access, exchange, and use genetic fitness data to address this issue. Research marketing needs the widespread use of this shared information and the allowance of technical and structural restrictions, as well as assessing the informed consent and privacy of data subjects, relying on protection preventing their rediscovery.

The loss of genetic information can threaten privacy, personal security, and national security, as genomic statistics can be used by adversaries for malicious purposes. Additional statistical losses in a genomic database breach could lead to intellectual property theft, costing the United States a significant amount of money in the biotechnology sector.

Security risks may also arise from the introduction of biological weapons or the arbitrary selection of national security brokers. Cyberattacks have been developed against genomic databases, commercial organizations storing genetic information, DNA sequencing devices, and genomic software program tools, and other attack scenarios and exploits focus on hardware, firmware, software, local networks, cloud systems, and physical ground security. Optimized cybersecurity and privacy risk management measures to be taken. We must bridge the gap in law and international cooperation. Understanding threats and barriers, not just opportunities, is the most important step.

16.1.7 FOOD AND AGRICULTURAL INDUSTRY OVERVIEW

Cybercriminals have not targeted the food and agriculture industries at all. However, today, threat operators are finding prey in the industry's overreliance on a well-managed food supply chain to exploit malware, including ransomware, to serve their nefarious purposes.

A cyberattack on a company dealing with food or agriculture could have its main aim being to disrupt the production and distribution of food, thereby causing a shortage and higher prices. It can also contaminate food, making it easier for consumers to contract food-borne diseases.

16.1.7.1 NOTABLE CORPORATE ATTACK

JFC International: The giant food distributor fell victim to ransomware on numerous IT systems in March 2021. JFC International is the biggest distributor and wholesaler of Asian kitchenware in Europe and the United States. The incident, by reports, hit JFC's Europe operations, but they returned to normality soon after they informed law enforcement, staff, and partners about the occurrence.

Breadfish: Breadfish is a nonprofit food distribution of balanced ingredients to persons and households that suffer from short-term disasters using mobile, "pressure-through" food delivery facilities. It was announced on August 2020, as part of a major Blackbaud attack in which sensitive customer data were compromised.

Blackbaud, which hosts websites in software clouds, repelled the ransomware's attack, which encrypted documents but nonetheless attempted to ransom it back to Blackbaud or their customers with threats of sharing critical information, including one of Loaves and Fishes from 2013. According to Blackbaud, to date, there is no indication that there was a sale over the internet, but it could occur at any time.

Home Chef: Home Chef is a service that provides customers with ingredients and recipes. In May 2020, security researchers found Home Chef usernames and passwords available on dark web marketplaces. The Chicago-based company said a few days later that a security incident had forced it to expose information about its customers, without saying how many. Measures of such protective nature do not pose threats to food delivery businesses, but rather increase risks for supply chain consumers.

Harvest Sherwood Food Distributors: In May 2020, information had been posted on the Happy Blog and Tor's hidden service showing that the Harvest Sherwood Food Distributors was under attack by hackers using the REvil ransomware. The attackers had stolen sensitive information from the company, then went ahead to threaten it by publicizing the same. The Revil ransomware is the same that was used against JBS Meats. More than 2600 documents were stolen from the food distributor. Among these were cash flow analysis records, references to commercial agencies, as well as supplier lists. People in the Harvest Sherwood distribution area were driving scanned license photos are also available.

Data Analysis: Over the 2021 Memorial Day weekend, the biggest in the industry, J.B.S, the largest meat processing company in the industry, gave in to a ransomware attack from an organized Russia-wide criminal group. The cyberattack destroyed much of the beef supply chain and reverberated shockwaves throughout the entire food industry. Confirming from the FBI, the cyberattack was carried out using the REvil ransomware. These targeted attacks had paralyzed the work for a short time at the 13 meat processing plants in Colorado, Iowa, Minnesota, Pennsylvania, Nebraska, and Texas.

This ransomware by GOLD SOUTHFIELD was reportedly behind it: an operation into the so-called money-loving group through Ransomware as a service. The organization sprayed ransomware by exploit kits, scan-and-make maximum, remote desktop protocol servers, and backdoored software installations. The breach of JBS had a monumental impact, inasmuch as it set off a chain reaction that quickly spread throughout the country. Wholesale meat prices, in quick order, sent a shock wave through the supply–demand balance.

This developed an oversupply in the market and lowered the prices of meat because the farmers and ranchers could not deliver their beef to the market. Retailers, such as restaurants, across the country could not accept packaged and processed meat. All these useless results gave rise to increasing consumer taxes. The effect of one attack on just one part of the food supply of the world had a catastrophic consequence to show how fragile food distribution is. Restaurant owners were having a tough time getting their hands on dependable meat then, as the industry was trying to recover from the outbreak. Trying to recover a reliable food source is big business, where JBS ultimately found itself bowing to the pressure of the hackers. They first negotiated with the hackers, then agreed to pay an $11 million ransom in Bitcoin.

As many people argue, such catering to the needs of ransomware hackers makes us all worse off in the long run. After all, though production was lost for less than a day in the end at the fast-moving JBS, and the company's costs for customer orders were only 3% of normal production, less than the damage done by an employer that some can see it completely from a strong storm.

16.1.7.2 COMMON CYBER THREATS

Organized crime uses malware and ransomware to implement malicious software into a computer system that threatens the data integrity and normal operations of the software in place. Such incidents can really hit a food or agricultural industry with damages. For instance, malware may be targeted at the control systems designed to control some key infrastructure, such as irrigation or food production plants, while ransomware has been known to encrypt sensitive information to be decrypted later for a ransom. For example, the Danish transport firm Maersk was attacked by NotPetya in 2017, with its systems falling apart, and Norsk Hydro, one of the largest aluminum producers globally, fell victim to LockerGoga ransomware in 2019. More recently, in 2021, JBS; in 2022, ransomware attacks on Hood Dairy; and in 2023, Dole reflect the increasing interest such criminal enterprises take in the operations of big corporations.

Often, employees become victims of fake e-mails or phone calls in which they provide their passwords or download malicious software by mistake. Other tactics may include the distribution of infected USB flash drives at trade shows or the use of QR codes, which would direct a visitor to a website with hosted malware. Other tactics can include infected USB flash drives given at trade shows, or QR codes that direct users to a website where malware is hosted. These can be used to gain access to sensitive resources, with sensitive information being stolen. Attackers will be able to disguise themselves as a provider supplies, superiors, or colleagues and so do various exploitations on the hapless. Very great proof of this lies in the 2019 phishing attack on the Food Standards Agency of the UK, which used fake emails in its attempt to penetrate the organization.

In the process, supply chain attacks violate weak points in the interlaced network of suppliers, distributors, and service providers. As such, such attacks can paralyze the whole supply chain, threaten food security, and stall travels. These are threats from adversaries that may damage third-party services by setting up malicious software or adjusting the products at the time of transmission. More than 18,000 nongovernmental organizations and governmental agencies across the industry were said to have been affected by the SolarWinds event by the end of 2020. The malicious code was delivered to users by way of adulterated SolarWinds software updates and went undetected for a span of over six months.

Denial of service (DoS) and distributed denial of service (DDoS) attacks make networks or systems inaccessible to legitimate customers. Online offerings—be it websites, e-mails, etc.—are especially susceptible to this former attack because such unavailability denies users access to visiting or buying sources. In these times, for every new technology that appears, a critical impact of DoS or DDoS attacks on operational technology systems, critical infrastructure, sensor networks, or control infrastructure of equipment, among other examples, is placed. Good examples of potential candidates for this type of attack might be agricultural precision technology-based sensors in the fields, connected by insecure wireless networks.

For example, insider threats might be malicious people within a company who will use their login credentials or location to do things such as break policy or steal valuable information. Food security might be at risk, intellectual property could be stolen, or key operations could be affected by malicious people. People who have been committing such incidents have been motivated financially or are dissatisfied and under pressure. Organizations will, therefore, have to put strict access restrictions, control measures, and personnel focus in an attempt to mitigate such risks.

This huge number of IoT devices can be taken together to crack passwords, attack other systems, or spread malware when assembled into "botnets" of compromised devices. IoT devices are at the forefront of stationary and mobile applications; also, drones in agriculture are an emerging trend.

These are typically advanced persistent threats (APTs) from well-financed and well-trained attackers. Most APTs indicate a hidden and long-term plan to retain access in order to steal valuable statistics or intellectual property from APTs in the food and agricultural sectors. Self-guided training development records sensitive information of customers or suppliers' operational data and target large databases, although destruction of critical services is not a usual goal of APT. The continuing conflict in Ukraine is evidence that Russia is prepared to target farmers and manufacturers through electronic communications.

Structured Query Language (SQL) injection is an attack where an attacker injects a web server or a vulnerable device that does not check user input before it is passed to an internal SQL server for processing. If carried out correctly, a SQL server vulnerability can be used by an attacker to restore confidential or sensitive information, such as credit card information or account passwords, from the database.

16.1.7.3 CONSEQUENCES OF CYBER THREATS IN FOOD AND AGRICULTURE

One of the biggest implications of cyberattacks on the food and agriculture sector is the pollution of food security. Food satisfaction can be changed by attackers, pathogen testing can be tampered, and traceability or related information can be altered, which may lead to the distribution of dangerous or tainted food, thus posing a great hazard to public health, product recalls, creation, and reputation damage for food producers and suppliers.

Cyberattacks have the capability to disrupt supply chains in this industry. Suppliers themselves could come under attack as well as distributors or logistics companies; such an incident would cause delays in delivery time frames; sometimes goods may not reach their destination at all due to a shortage created by these types of attacks, or they could end up being delivered as counterfeits. In general, any alteration made within the supply chain has wide-reaching effects because it not only affects company profits but also leads lack thereof plus increased consumer prices through reduced availability.

Organizations and businesses face huge financial impacts from cyberattacks too. Costs associated with recovering from incidents can run into millions if not billions of dollars, depending on severity levels reached during the attack period; enterprises would spend much money in trying to return back normalcy after being hit by a malware infection, for instance. News of data breaches, threats to food safety, or any kind of disturbances in the supply chain may potentially lead to the loss of trust in the minds of consumers and, hence, result in loss of consumers and prospects. Restoration of a damaged reputation may be a lengthy process with long-term consequences to the affected businesses.

16.1.7.4 THE BEST PRACTICES

Implementing a few best practices when possible can also help reduce the likelihood of a cyberattack. Below are principles and best practices that will assist in improving the cybersecurity posture of the region:

- Raise awareness among employees about cyber risks and security hygiene.
- Use a strong password or passphrase.
- Use multifactor authentication to access systems or databases.
- Maintain up-to-date software, packages, and equipment.
- Provide regular updates. Businesses need to implement a comprehensive cybersecurity strategy to effectively manage cyber risks.
- Control physical access to computers, phones, and other network devices by means of physical controls and personal passwords.

Promote a culture of cybersecurity that underscores the importance of security at all levels. This may involve increasing the knowledge of employees, providing training programs, as well as setting clear safety standards and procedures. Staff should be knowledgeable about potential risks and good practices regarding information security, alongside any reported suspicious activities. All workers need to have an easy-to-understand Information Security Policy, which they should review at least once every 12 months.

Secure access to critical systems and information must be ensured through robust authentication mechanisms and access controls, such as Multifactor Authentication Methods must be tested, access is monitored and changed on a regular basis, and rights should be granted based on need only; this will minimize chances for external attackers or insider threats who gain unauthorized entry into systems. System administrators plus other high-privilege users should have two accounts—one for normal use while another for administrative tasks that require extra privileges.

Continuous learning initiatives among employees are necessary in order to enlighten them about cyber risks, phishing methods, and securing online activities. Workers ought to be urged to immediately communicate any suspicious emails, phone calls, or incidents. Simulated game systems for phishing can also be used to test and improve the detection and response skills of employees towards phishing attacks.

Use strong monitoring capabilities for detecting and responding to direct cyberattacks. Intrusion Detection Systems and Security Information Event Management solutions should also be utilized. Continuously plan for incidents building, testing, and enhancing response capabilities to provide quick, strong responses against cybersecurity attacks. Step-by-step instructions should follow NIST (National Institute of Standards and Technology) Cybersecurity Framework 10 as well as NIST Incident Response Framework 11.

Software development companies or those that deploy custom applications must have a secure software program development lifecycle (SDLC). It is important to use consistent documentation techniques while developing software internally. SDLC processes should include risk modeling, regulatory analysis, and vulnerability analysis. Regular patching and updates should also be done on known vulnerabilities.

Joining ISACs (Information Sharing Analysis Centers) can facilitate collaborative discussions, which might help in identifying emerging cyber threats among other activities. In addition, government agencies and industry organizations together with industry stakeholders need to establish networks aimed at sharing threat intelligence, best practices, and lessons learnt from different cybersecurity incidents.

This coordinated endeavor has the potential to significantly enhance the global security situation. Follow professional norms and regulations for data security, food safety, and supply chain security. Adherence to principles such as ISO 27001 (Records Maintenance Management) and the implementation of recommendations such as the FDA's Food Safety Modernization Act can assist organizations in establishing a solid cybersecurity foundation.

16.2 CONCLUSION

The complex connected systems of a quarter are exposed to a variety of different types of cyber threats: malware, ransomware attacks, social engineering and phishing attacks, supply chain attacks, DoS and DDoS attacks, insider threats, IoT exploits, and many more. Such cyberattacks, from APTs to other stronger tendencies, have momentous consequences for the food and agriculture industries. The impacts range from compromise of food safety, disruption of the supply chain, economic losses, reputational damage, and punitive laws that such implications may carry with them. This threat, of course, extends its effect on the entire food supply and consumer confidence, but not so much just in a man's or woman's business. Cybersecurity should be one of the key priorities for businesses in the food and agriculture industries. Some of the important steps for process improvement are the implementation of strategies and best practices, which include lifestyle setup in cyber-security, implementation of strong authentication and access control, regular awareness training programs, inspections, and the response to incidents; adherence to good practices in software development; collaboration and sharing of data, and ensuring compliance with industry standards and regulation.

Cybersecurity is a dynamic and persistent challenge, a problem-solving effort requiring cooperation between companies, industry groups, and government agencies. As we pointed out in previous articles, the best method to try and stay a bit ahead of cyber risks is to participate in one or more shared-accountability research organizations—either in the form of a formal, regional ISAC or a group of peer workshops with similar tools and vulnerabilities.

The sharing of resources, the sharing of intelligence on threats, and the coordination of the activity make it possible for companies to considerably scale up their prevention, detection, and responses to attacks. Collaboration permits larger insight into strategies of attack that are arising, enables preventive action, and contributes to an effective response. Further, joint reactions will maintain cooperation against threats in cyberspace, creating a stable and robust cybersecurity environment. Third, the power of the collective is so much stronger than that of any single organization, making collaboration a cornerstone component of today's cybersecurity strategies.

KEYWORDS

- **agriculture**
- **cybersecurity**
- **biocybersecurity**
- **cyberbiosecurity**
- **biosecurity**
- **biomanufacturing**
- **COVID-19**
- **bioeconomy**

REFERENCES

1. Drape, T., Magerkorth, N., Sen, A., Simpson, J., Seibel, M., Murch, R. S., & Duncan, S. E. (2021). Assessing the role of cyberbiosecurity in agriculture: A case study. *Frontiers*, WorldCat.org. https: //doi.org/10.3389/fbioe.2021.737927

2. Duncan, S. E., Zhang, B., Thomason, W., Ellis, M., Meng, N., Stamper, M., Carneiro, R., & Drape, T. (2020). Securing data in life sciences—A plant food (Edamame) systems case study. *Frontiers in Sustainability*, https: //www.frontiersin.org/article/10.3389/frsus.2020.600394.
3. Duncan, S. E., Reinhard, R., Williams, R. C., Ramsey, F., Thomason, W. E., Lee, K., Dudek, N., Mostaghimi, S., Colbert, E., & Murch, R. S. (2019). Cyberbiosecurity: A new perspective on protecting US food and agricultural system. WorldCat.org. https: //doi.org/10.3389/fbioe.2019.00063.
4. Duncan, S., Carneiro, R., Bradley, J., Hersh, M., Ramsey, F., & Murch, R. (2021). Beyond ransomware: Securing the digital food chain. *Food Technology Magazine*. https: //www.ift.org/news-and-publications/~/link.aspx?_id=A0625C29CA904AFB8FA31CDF794CA3A9&_z=z
5. Federal Trade Commission. Children's Online Privacy Protection Rule ('COPPA'). https: //www.ftc.gov/legal-library/browse/rules/childrens-online-privacy-protection-rule-coppa.
6. Federal Trade Commission. Gramm-Leach-Bliley act. https: //www.ftc.gov/business-guidance/privacy-security/gramm-leach-bliley-act.
7. Food and Ag-ISAC. (2023). https: //www.foodandag-isac.org/.
8. Goodspeed, L. (2022). PCI DSS v4.0 resource hub. *Security Standards Council*. https: //blog.pcisecuritystandards.org/pci-dss-v4-0-resource-hub.
9. Intersoft Consulting. General Data Protection Regulation: GDPR. https: //gdpr-info.eu/.
10. Murch, R. S., So, W. K., Buchholz, W. G., Raman, S., & Peccoud, J. (2018). Cyberbiosecurity: An emerging new discipline to help safeguard the bioeconomy. *Frontiers in Bioengineering and Biotechnology, 6*. https: //www.frontiersin.org/article/10.3389/fbioe.2018.00039.
11. Sachs, M., & Norton, R. (February/March 2023). An information sharing and analysis center for the food and agriculture sector. *Food Safety Magazine*, https: //www.food-safety.com/articles/8325-an-information-sharing-and-analysis-center-for-the-food-and-agriculture-sector.
12. Sachs, M., & Norton, R. (April/May 2023). Cybersecurity and food defense: Establishing an ISAC for the food and agriculture sector. *Food Safety Magazine*. https: //www.food-safety.com/articles/8488-cybersecurity-and-food-defense-establishing-an-isac-for-the-food-and-agriculture-sector.
13. Sachs, M., & Norton, R. (June/July 2023). What exactly is 'information sharing?' *Food Safety Magazine*. https: //www.food-safety.com/articles/8670-what-exactly-is-information-sharing.
14. Schmale, D. G., Ault, A. P., Saad, W., Scott, D. T., & Westrick, J. A. (2019). Perspectives on Harmful Algal Blooms (HABs) and the cyberbiosecurity of freshwater systems. *Frontiers in Bioengineering and Biotechnology*, 7. https: //doi.org/10.3389/fbioe.2019.00128.
15. U.S. Department of Commerce. National Institute of Standards and Technology. *Cybersecurity Framework*. https: //www.nist.gov/cyberframework.
16. U.S. Department of Commerce. National Institute of Standards and Technology. *Computer Security Incident Handling Guide*. Special Publication 800-61, Revision 2. August 2012. https: //nvlpubs.nist.gov/nistpubs/specialpublications/nist.sp.800-61r2.pdf.
17. U.S. Department of Health and Human Services. Health information privacy. https: //www.hhs.gov/hipaa/index.html.
18. Additional Sources
19. ESDS. (2022). Signs of cyber attack and how to respond to them? https: //www.esds.co.in/blog/signs-of-cyber-attack-and-how-to-respond-to-them
20. Ministry of Health, Singapore. (2021). Common signs of a cyber-attack. https: //www.moh.gov.sg/licensing-and-regulation/cybersecurity-for-healthcare-providers/common-signs-of-a-cyber-attack.

CHAPTER 17

AI-Based Approaches for Smart Agriculture

SAYANTIKA BHATTACHARYA[1], SHIV VENDRA SINGH CHAUHAN[2], and RAJNEESH BHARDWAJ[1]

[1]School of Agriculture, Graphic Era Hill University, Dehradun, Uttarakhand, India

[2]Department of Agronomy, Rani Lakshmi Bai Central Agricultural University, Jhansi, Uttar Pradesh, India

*Corresponding author

ABSTRACT

Smart agriculture has emerged as a transformative paradigm, leveraging artificial intelligence (AI) to revolutionize traditional farming practices. The focus spans across key facets of AI applications in agriculture, encompassing precision farming, crop monitoring, disease detection, yield prediction, and resource optimization. The utilization of machine learning algorithms for data analysis and decision-making has revolutionized precision farming. AI-driven sensors and drones facilitate real-time monitoring of crop health, enabling early detection of diseases and optimizing resource utilization. Furthermore, the integration of computer vision techniques plays a pivotal role in crop monitoring by providing accurate and timely information on plant growth stages and identifying anomalies. In the context of pest detection, AI offers innovative solutions through image recognition and sensor data analysis. Deep learning algorithms enable intelligent systems to recognize and categorize pests, facilitating timely and focused intervention measures. With the introduction of AI, yield prediction—a crucial component of agricultural planning—has seen major improvements. By utilizing historical data, weather trends, and soil properties, machine learning models are able to predict crop yields more accurately. This empowers farmers with valuable insights for better decision-making, resource allocation, and overall farm management. In conclusion, this chapter provides a comprehensive overview of the manifold applications of AI in smart agriculture.

17.1 INTRODUCTION

17.1.1 ARTIFICIAL INTELLIGENCE

AI is an interdisciplinary field of research that attempts to imitate human intelligence in robots by modeling human actions and cognition, such as problem-solving and learning.

Demystifying Sustainable Farming Using Smart Technologies.
Gunjan Chhabra, Keshav Kaushik, Priya Chugh, Susheela Dahiya, and Aryan Chaudhary (Eds.)

DOI: 10.1201/9781779640932-17

This area employs a variety of methods, to learn and interpolate data in order to resemble human minds. These networks connect input and output variables dynamically to provide predictions. These forecasts can aid in the development of numerous solutions for both simple and complex problems that arise in our everyday lives; AI-powered technologies are currently widely used.

AI technology is currently being used by research scientists and extension workers to address issues related to agricultural production. By aiding in the selection of appropriate crop varieties, the adoption of better fertilizer and soil management techniques, the control of pests and diseases, the estimation of agricultural productivity, and the forecasting of commodity prices, AI can help farmers enhance yields.

17.1.2 SMART AGRICULTURE

The world has seen several technological developments throughout the years, which have greatly advanced many facets of human existence, including the agricultural industry. The digital age, which makes it possible for mobile devices and the internet to proliferate, has proven advantageous for farmers and owners of agribusinesses as well as huge corporations. Agriculture, which is currently experiencing its fourth revolution, has never been easier because to the growing application of technology in what is referred to as smart agriculture. Using modern technology to enhance a variety of farming and agricultural management techniques is known as precision agriculture, sometimes referred to as smart agriculture. Through the use of cutting-edge technology to optimize many elements of agricultural activities, it represents a paradigm shift in conventional farming approaches.

17.1.2.1 WHY SMART AGRICULTURE?

Agriculture is one of the prominent sectors that is affected by climate change. The dynamics of environment, such as heat stress, moisture stress, and floods, hamper productivity, thereby causing yield losses. In 2020, a study found that approximately 690 million people lacked sufficient food to eat, which was an increase of almost 60 million people in just five years. In the future, we will need to produce more food to feed 9 billion people. This will make the food security problem worse. So, how do we fix this problem? The only way is to adopt smart agriculture technologies. As per the data presented in 2022, about 15% of farmers around the world use smart agriculture. Although this percentage is very low compared to the risk of food insecurity, but the world needs to fight, farmers continue to understand and adapt positive perception and readiness for smart agriculture.

17.2 COMPONENTS OF SMART AGRICULTURE

17.2.1 SENSORS AND IOT DEVICES

To gather field data in real time, sensors are placed in fields. IoT devices make it possible to remotely monitor and manage agricultural activities, giving farmers important information on the health of their crops and the surrounding environment.

17.2.2 DATA ANALYTICS AND AI

Massive amounts of data gathered from sensors and other sources are processed by sophisticated analytics and machine learning algorithms. Farmers can make data-driven decisions about crop management, fertilization, irrigation, and pest control with the aid of AI-powered analytics.

17.2.3 PRECISION FARMING TECHNIQUES

Precision farming is the combination of various methodologies for improving crop yields and productivity. It is also assisting management decisions by using high-tech sensors and analysis tools. Farmers can make precise maps of their fields and monitor their crops to GPS technology, drones, and satellite imagery (Figure 17.1). By implementing interventions only where and when they are required, these strategies maximize yields, minimize input costs, and optimize resource utilization.

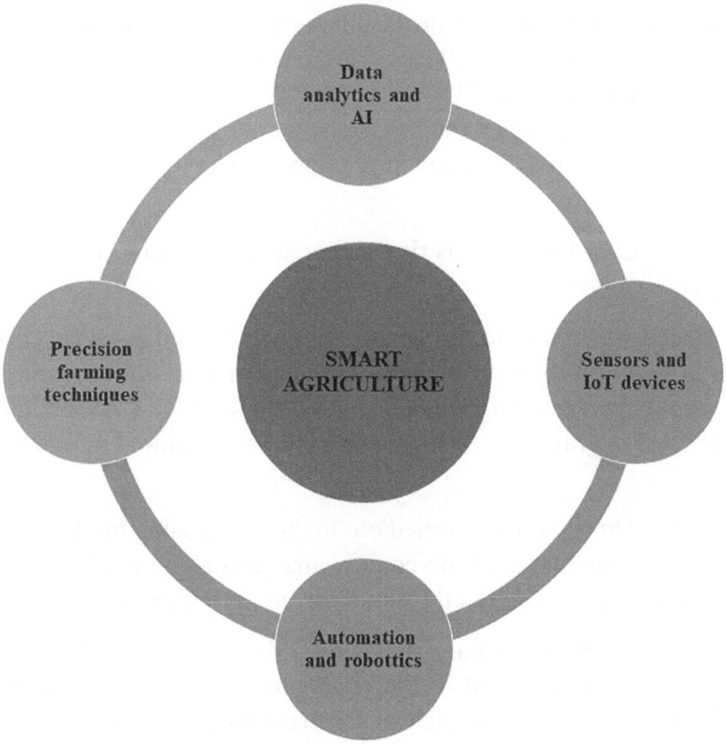

FIGURE 17.1 Components of smart agriculture.

17.2.4 AUTOMATION AND ROBOTICS

Numerous operations, including planting, harvesting, weeding, and spraying, are carried out by automated machinery and robotics. It guarantees accurate and constant farming operation execution, reduce the need for labor, and enhance operational efficiency.

17.2.5 REMOTE MONITORING AND CONTROL

Remote monitoring systems allow farmers to monitor and control agricultural equipment, irrigation systems, and environmental conditions from anywhere using mobile devices or computers. This capability enhances operational flexibility and enables proactive decision-making to address emerging challenges.

17.3 AGRICULTURAL CYBER-PHYSICAL SYSTEMS (A-CPS) BASED ON THE INTERNET OF AGRO-THINGS (IOAT)

The Internet of Things (IoT) is a network of physically connected items, devices, and things that have unique identifiers that allow them to connect to other systems and devices online and share data. Cyber-physical systems (CPS) are created when the IoT is implemented in physical systems. CPS combines software or computing power with physical components. It is a modern method of defining a sector. One or more CPS, such as smart citizens, renewable energy, smart transportation, smart health, smart agriculture, and smart citizens, are included in smart cities and smart villages. The foundation of smart agriculture is A-CPS. It changes the agriculture industry [1].

Data is the driving force behind the IoAT. The workflow is made extremely efficient by ongoing data processing, collecting, and improvement. The following five phases make up the cycle:

1. **Data Collection**: First, data is first gathered at the sensor or end level by a variety of objects.
2. **Data Processing:** Second, this phase involves any edge-level processing of data which is required to align the data with the model. For example, this is where things are done if the sensor data is out of range, if the UAV photographs need to be converted to grayscale, or if data encryption is required before being sent to the cloud.
3. **Prognostic**: This is mostly carried out in the cloud using modern technology. This is where data processed at the edge is analyzed using pre-built models or rules. Information is also stored in the cloud for future reference.
4. **Fix:** The fix is recommended as soon as the problem in the cloud platform is identified. Either the cloud or the edge can be used for this step. For instance, in order to irrigate a dry stretch of farmland as effectively as possible, this step suggests which irrigation system valves to open, at what pressure, and for how long.
5. **Measures Taken**: The application of the solution is carried out at this, the last, phase of the cycle. IoT device handles this. This is where the irrigation system valve is opened in the scenario that was previously stated. The cycle keeps on, serving the farming process as a whole as best it can. Figure 17.2(a) illustrates how decisions or solutions based on machine learning are transmitted to IoT devices so that

actions can be made there when machine learning based activities are completed in cloud and edge settings. However, TinyML as-a-Service is integrating embedded and machine learning. The ML-based work is carried out at the low resource IoT device exclusively, rather than outsourcing decisions to external systems. Figure 17.2(a) is becoming Figure 17.2(b) in this new period.

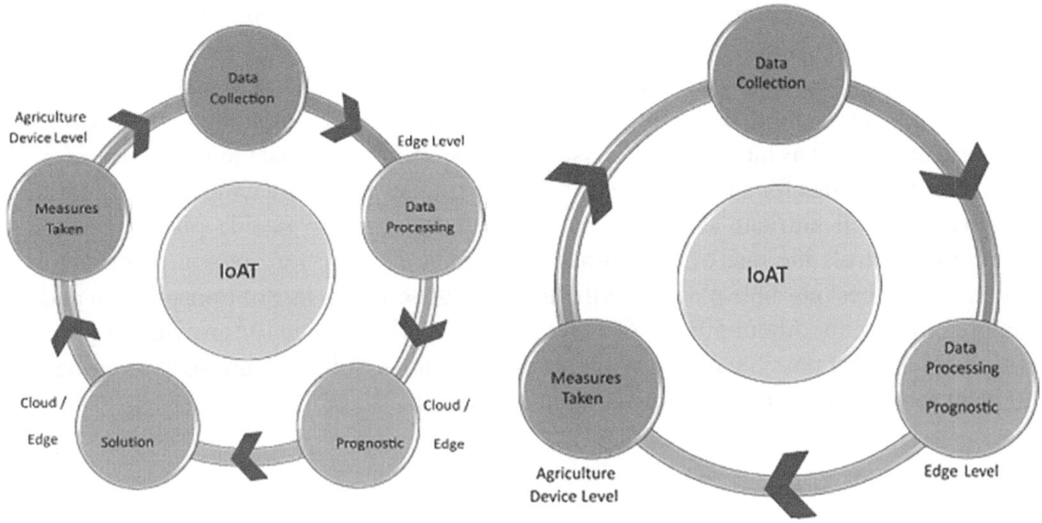

(a) TinyML Era (Before) (b) TinyML Era (After)

FIGURE 17.2 IoT-based smart agriculture cycle.

17.4 APPLICATIONS IN SMART AGRICULTURE

The various application areas of smart agriculture include crop management, autonomous tractor and UAV, hydroponic system, smart irrigation, animal monitoring, and pest control that are discussed below.

17.4.1 CROP MANAGEMENT

Crop management is the field of studying economics, ecological, and social aspects that are important in crop selection, growth, and marketing. Crop growth depends upon the availability of water resources, labor, and environmental factors that affect cropping patterns. Depending on ecological factors, cropping patterns may change. For example, areas with diminishing groundwater tables and water supplies are unable to maintain traditional crops like paddy cultivation, which requires substantial amounts of water. The market for the agricultural product, export and import laws, and international policy all have an impact on crop choice. Following crop selection, crop cultivation is a crucial next stage.

Farmers can monitor plant growth with the use of sensors installed in the field and the newest technology thanks to the IoT. For example, in the field, ultrasonic sensors are used to track the presence of insects and pests that hinder plant development. High frequency sound waves are produced to eradicate pests once they are identified, and the farmer is also informed of the pests' presence for further assistance.

17.4.2 SOIL MONITORING

Soil moisture is essential to the farming process as a whole. It oversees transpiration, photosynthesis, respiration, and mineral transport in plants during growth. Soil monitoring is essential for decision-making on farms. Cropping patterns are influenced by a wide range of factors, such as pests, soil salinity, moisture content, pH, and humidity. All these components aid in the assessment of soil health. Field sensors monitor variations in soil temperature as well as humidity, processing and transmitting the data to the cloud. Farmers will obtain warnings for a range of reasons, and variables including temperature, humidity, salinity, and soil nutrient level will be assessed and used to decide planting patterns. Humidity controls the rate of transpiration and food delivery for optimal plant development. For the best possible plant growth, humidity governs the rate of transpiration and the supply of nutrients. About 50% to 60% humidity is the best range for growing vegetables. To enable the best possible use of available water sources, soil moisture sensors embedded in plant roots measure the moisture content of the soil.

17.4.3 SMART IRRIGATION

Smart irrigation means using new technology to grow more and better crops with less water. It saves water by watering the plants with just the right amount. There are two kinds of irrigation systems: one depends on sensors that measure soil moisture, and the other depends on weather conditions. Weather-based irrigation systems use a machine to water plants after getting information about the temperature and rain from a nearby weather station. Soil moisture sensor irrigation systems use sensors placed near plants to measure how much water is in the soil. In this method of watering plants, you need to know how damp the soil is and how hot the air is. You also need to keep an eye on the weather and how the plants are growing to track the water requirement of the field. Actuators such as sprinklers, transfer their data to the cloud. The irrigation schedule for each and every farm area is determined by the obtained values of the soil moisture sensor. Irrigation scheduling, efficient actuation and micro-level analysis ensure 100% water utilization efficiency and maximum crop growth (CR). Farmers can control the irrigation system using the mobile application on their smartphone. The proposed irrigation system would run on the information derived from a set of sensors in the field, including temperature, humidity, soil moisture, and ultrasonic sensors. Farmers can activate and control the irrigation pumps for their farm through an intuitive mobile application, which connects the smartphone-based automatic irrigation system to the cloud for analysis.

17.4.4 LIVESTOCK MONITORING

The management of livestock is an important aspect of smart agriculture. Farmers are able to monitor cow herds, grazing animals, and best practices for cattle raising with the aid of an IoT-enabled health monitoring device. The animal's blood pressure, heart rate, and respiration rate can all be automatically monitored with the use of an RFID tag or collar. This has two benefits: it saves labor costs and treats the animal promptly to stop the spread of illness. GPS tracking is used in this regard. Additionally, this technique can shield the animal from harm. RFID tags are further utilized for tracking and animal identification. The use of AI in monitoring livestock adds to increased efficiency, health management, and enhancing general productivity in agriculture. Specifically, AI is used in the monitoring of livestock through:

Health Monitoring: AI-powered systems can monitor individual health based on various indicators, like body temperature, heart rate, level of activities, or even facial recognition. They can be able to point out changes in individual animals' behaviors and physiological parameters in time to ensure timely interventions and disease prevention.

Image Recognition: Applied within the framework of AI algorithms, image recognition allows the facility to manage all the imagery and video material obtained through cameras mounted for these purposes. Every single animal can be identified; movements can be traced, and feeding behaviors can be monitored in such a way that even anomalies might be distinguished, such as injuries or signs of stress.

Predictive Analytics: Analyzing historical data related to, say, weather, feed quality, and animal behavior, the AI algorithms could predict potential health-related issues or trends in performance to assist farmers in informed allocation and management of resources. **Reproduction Management:** To optimize the breeding programs, AI can be extremely helpful in monitoring the reproductive cycles of animals. This can also help in analyzing data on the configurations of fertility, estrus behavior, and genetic information, which enable farmers to optimize their breeding efficiencies and selecting the best pairs.

Feed Management: The composition of feed with respect to animal growth, availability of feed and nutrition required can also be optimized by the use of AI.

Environmental Monitoring: AI sensors can also be used to monitor environmental conditions in a livestock facility, such as temperature, humidity, and air quality. Such information can be double-checked in real time so that the farm animals are kept in the best living conditions and any health disorders arising out of stress are averted. **Precision Livestock Farming:** Precision agriculture can be applied at the individual or group level, it can apply itself well to the particularities and characteristics of the animals. It uses a combination of data sources, such as sensors, RFID tags, and satellite images, in general information that can advise the farmers on the best ways to use their resources in achieving high productivity with minimal environmental impact.

Decision Support Systems: AI-based decision support systems generate actionable insights and recommendations to farmers based on real time data. By this method, farmers can make good decisions regarding the best way of managing their herds, controlling the diseases, and allocating resources to reach maximum productivity and profitability levels.

In conclusion, AI-assisted livestock monitoring is an activity in livestock management that is highly beneficial, as it results in better health management, improved productivity, and enhanced decision-making capabilities for the farmer. With growing technology, further innovations in this area are anticipated, and this will continue to transform farming practices.

17.4.5 REMOTE SENSING

Remote sensing is the science of recording and detecting the radiation emitted and reflected from the surface of the Earth without having direct physical contact with the object. Targets of interest interact with the incoming radiation during the process. The most common types of electromagnetic radiation relevant to remote sensing are visible light (VIS), near infrared (NIR), shortwave infrared (SWIR), thermal infrared (TIR), and microwave. Active remote sensors transmit their own radiation to the target under study; on contact, the radiation is returned to the measuring device. On the other hand, passive sensors simply measure the incoming radiation that is reflected or emitted from the objects.

Mapping of farm fields for remote sensing in agriculture could hence be done with the help of drones, capturing quality pictures of farm fields and providing real-time data on the state of crop development to the farmers. The use of drones can also support remote sensing of field conditions and crop health. The correct type of soil that is conducive for a crop is thus provided to the farmer through data on soil conditions and remote sensing. It also helps identify pests and weeds and adapt appropriate pest management practices. Weather monitoring and forecasting are among the most significant use of remote sensing. It can be used to monitor rainfall, drought conditions, and water resources, informing farmers in advance about water availability and weather so that crop and capital planning can be completed in advance. Values from the Normalized Difference Vegetation Index (NDVI) are used to forecast yield and track plant growth. Using remote sensing equipment, farmers may monitor abiotic stressors with the highest possible spatial precision. Several applications of remote sensing in agriculture in details are mentioned below:

17.4.5.1 CROP OUTPUT FORECASTING

Using remote sensing, one may predict how much crop will be harvested under particular circumstances as well as estimate predicted crop output and yield across a given area. Scientists are able to forecast how much crop will be grown on a specific farm over a specific period of time.

17.4.5.2 CROP DAMAGE AND CROP PROGRESS ASSESSMENT

To assess field conditions in terms of crop damage or progress, remote sensing technology can be used, allowing for the precise measurement of crop damage and the advancement of the residual crop on the farm.

17.4.5.3 CROP IDENTIFICATION

In situations where the crop being observed has certain peculiar qualities, remote sensing has proven to be quite helpful in identifying crops. The acquired agricultural data is brought to laboratories for evaluation of other aspects.

17.4.5.4 CROP ACREAGE ESTIMATION

Assessing the amount of farmland that has been planted with a crop has also benefited greatly from remote sensing. Because the areas being estimated are so large, doing this manually is typically a labor-intensive process.

17.4.5.5 CROP OUTPUT MODELING AND ESTIMATION

By measuring the crop's quality and the size of the farmland, remote sensing also enables farmers and specialists to forecast the expected crop output from a specific plot of land.

17.4.5.6 IDENTIFICATION OF PESTS AND DISEASE INFESTATION

Remote sensing technology greatly helps to recognize pests in an agricultural field and provides information about the best strategies for their control and eradication of pests and diseases from the farm.

17.4.5.7 ESTIMATING SOIL MOISTURE

Without the use of remote sensing equipment, measuring soil moisture is very difficult. Remote sensing data on soil moisture help determine the extent of moisture in the soil and consequently the type of crop to be grown on that particular piece of land.

17.4.5.8 SOIL MAPPING

The use of soil mapping helps in classifying which soils need irrigation and which do not, thus enabling farmers to establish which soils are best for each crop. This information is useful in precision farming.

17.4.5.9 MONITORING OF DROUGHTS

A specific area's weather pattern is observed using remote sensing equipment. Additionally, the device tracks the area's drought trends. The data can be utilized to forecast an area's patterns of precipitation and to determine the interval between the present and the next rainfall, which aids in monitoring the drought.

17.4.5.10 MAPPING OF WATER RESOURCES

Over a specific farmland, the mapping of water resources that can be exploited for agriculture is made possible in large part by remote sensing. Farmers can utilize remote sensing to determine the locations and availability of water resources on a specific plot of land.

17.4.6 SMART GREENHOUSE

Technologically assisted farming methods are welcomed by the agriculture sector in light of the changing global environment and depletion of natural resources. One example of this is the smart greenhouse. It is a regulated indoor environment designed specifically for plants. It's an ecosystem for self-isolated agricultural monitoring that incorporates AI, machine learning (ML), and IoT technology. The farm is shielded from wind, storms, and flooding by it. It boosts productivity without requiring physical assistance.

Vegetables, fruits, and other horticultural crops are kept under observation in the greenhouse using IoT sensors that run on solar power. Tree roots equipped with soil moisture sensors can be used for automatic drip irrigation. The in-field actuator waters the farm according to the threshold value once it is achieved. LED lighting is a more effective way to meet the needs of plants. An exact wavelength and intensity of controlled lighting can revitalize plant development and yield throughout the year.

It is possible to apply the right quantity of minerals, such as potassium, phosphorus, and other minerals needed for healthy and optimal plant growth, by using drip fertigation techniques. The use of smart greenhouse techniques to cultivate organic fruits and vegetables in greenhouses is becoming more popular as farmers have access to more advanced technologies [2]. In order to boost rose plant productivity, a decision-support (DS)-based Internet of Things-friendly smart greenhouse system has been proposed in [3].

17.4.7 URBAN FARMING

In densely populated cities, the rising rate of urbanization presents a concerning issue. In those locations, a new method of farming has emerged to provide a sustainable farming option. As a result, urban or vertical farming has become more popular among those living in cities. It requires three-dimensional area for farming with artificial lighting, minimal herbicides, controlled water, and nutrients. Due to the high expenses and need for artificial light sources for plant growth, the vertical farming system is somewhat limited practically[7].

Hydroponics is a water-based method, as the name implies, in which plants receive all of their nutrition from a nutrient-rich water solution, [8] Continuous nutrient input is required in hydroponic systems. Mobile apps can be used to control these systems. In [9], an Arduino controller is managed via a smartphone app to water the plants and maintain the hydroponic system.

Similar techniques are used in aeroponics, however with this method, the roots are misted rather than immersed in water. Studies reveal that plants grown aeroponically receive more nutrients than those grown hydroponically [10].

This method is utilized on the International Space Station to produce plants.

An additional contemporary farming method is aquaponics, which is basically a hydroponic system with no external fertilizer mixing. The fish in the same tank produce those nutrients.

17.4.8 AUTONOMOUS TRACTOR

Modern technology is transforming the agricultural sector. From crop management, soil monitoring, and smart irrigation to pest control, animal management, and agro-marketing, the Industrial Internet of Things, or IIoT, has rapidly advanced. In the near future, farming with intelligent, autonomous machinery is something we can anticipate. The most crucial component of this equipment is an autonomous tractor. It is an autonomous vehicle that can be programmed. It is capable of fertilizer spraying and tillage. Since they have cameras, lasers, and GPS, farmers do not need to keep an eye on them because they can operate independently. To sustain agriculture, autonomous drones are used together with smart tractors for weed identification, pesticides and fungicides applications, field monitoring, and surveillance. Autonomous tractors used in mowing and spraying in an orchard have a perception system, which is able to detect obstacles and navigate through remote assistance in the execution of agricultural activities. The perception system is based on cameras identifying a path and obstacles based on geometry.

17.5 BENEFITS OF SMART AGRICULTURE

17.5.1 INCREASED EFFICIENCY AND PRODUCTIVITY

Smart agriculture techniques increase efficiency by optimizing resource usage, reducing wastage, and enhancing output by ensuring that crops receive the right amount of water, nutrients, and care.

17.5.2 SUSTAINABILITY AND ENVIRONMENTAL CONSERVATION

Smart agriculture supports the achievement of conservation and reduction in the use of environmental resources, basically performed by the use of reduced amounts of water, fertilizers, pesticides, and energy in agriculture.

17.5.3 IMPROVED CROP HEALTH AND QUALITY

Crop monitoring from remote locations helps in early detection of pests, diseases, and nutrient deficiencies allow timely interventions of crop management practices to maintain crop health and quality.

17.5.4 COST REDUCTION AND RISK MITIGATION

Farmers can minimize production losses, cut input costs, and lessen the risk of unfavorable weather, pests, and market fluctuations with the help of precision farming.

17.5.5 DATA-DRIVEN DECISION MAKING

It is in this regard that farmers can reduce losses in production, cut input costs, and lower the chances of adverse weather, pests, and market fluctuations by practicing precision farming. This enable farmers to respond to actionable insights stemming from historical trends as well as real-time data using analytics and artificial intelligence. In brief, smart agriculture is an innovative approach to farming that makes use of technology to address the challenges of agriculture and tap into opportunities that exist for the sustainable and efficient production of food.

As technology continues to evolve, the adoption of smart agriculture practices is expected to grow, driving innovation and positive change in the agricultural industry.

17.6 CHALLENGES IN SMART AGRICULTURE

Smart agriculture practices have simplified and modernized traditional agriculture. However, a lot of issues still need to be resolved before technologies can be widely adopted. These problems are related to several different factors that are covered in the current section. A few of the main hindrances to smart agriculture are depicted in Figure 17.3.

17.6.1 POWER ISSUES

The majority of smart agriculture initiatives make use of heavily automated machinery, which consumes a significant amount of energy to run. Farms can have quite high-power requirements because there are large areas that prerequisite a lot of electronic equipment. Large farms have found it difficult to widely implement these automated methods because of this. Using clean/green energy from renewable resources like solar, wind, and hydro to power the machines continuously and without interruption is one of the alternatives suggested [11]. Numerous academics have expressed interest in this field, and work is still being done to develop and apply such renewable energy sources for smart farming [12, 13]. A few problems with these alternative power sources are power transmission and storage, as well as inconsistent energy needs at various farm locations. To address these problems, an effective microgrid architecture is required [14, 15].

17.6.2 POWER CONSUMPTION

The efficient, dependable, and sustainable operation of the smart agricultural farm depends on the deployed models using less electricity. This is a result of the need for alternative power sources to power IoT devices. They ought to be able to work in a setting with limited resources.

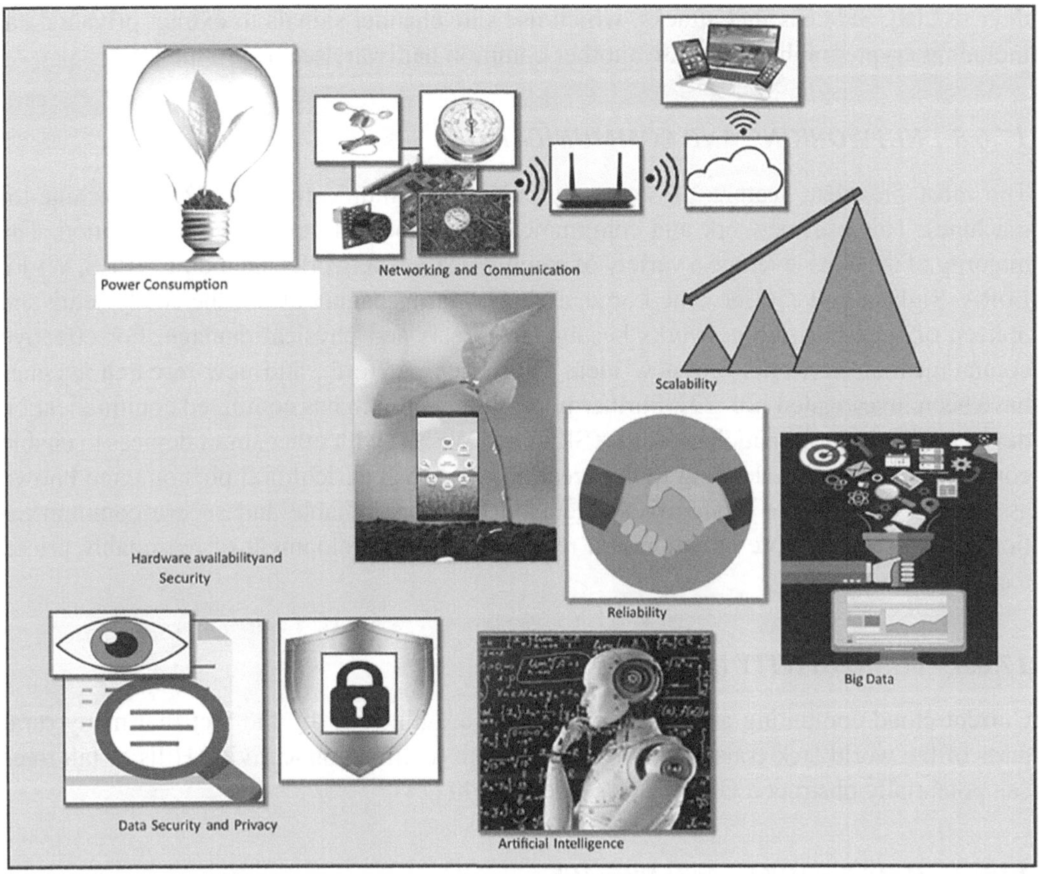

FIGURE 17.3 Challenges of smart agriculture.

17.6.3 HARDWARE AVAILABILITY

Smart agriculture requires a variety of sensors and devices to monitor different environmental and system conditions. The devices make uses of the signals derived from the data to deliver a better and more reliable yield. The availability of certain hardware in this instance acts as a bottleneck.

17.6.4 HARDWARE SECURITY

By 2020, it was predicted that 50 billion devices will be linked to the IoT [16]. These IoT devices need to be robust and able to withstand many kinds of attacks. However, hardware security is compromised by the necessity for low-cost, basic technologies. The most common security threats for IoT devices are Hardware Trojan and Side Channel Attacks which limits the extensive usage of IoT networks. Hardware Trojans use malicious hardware components, and they are extremely difficult to detect. A number of methods include using an electronic microscope to scan de-metalized chips [17], checking the circuit's power and delays, and searching via the PUF, which acts as a signature for these electrical devices [18]. Side channel attacks, which use side channel signals to extract private data, including cryptographic keys, are another common hardware security issue [19].

17.6.5 NETWORKING AND COMMUNICATION

The most prevalent feature of smart agriculture communication is M2M (machine-to-machine). This uses network and communication protocols to exchange information. The majority of the apps leverage a variety of communication networks, including GPRS, Wi-Fi, LoRA, SigFox, and ZigBee. The Large, and open farms cannot afford the costly setup and upkeep of such massive networks because of threats and physical damage. For effective communication networks few new ideas have been put forth, and new research avenues have been investigated [20–22]. Furthermore, some research has combined communication devices—like solar insecticidal lamps (SILs) and WSN—with other smart devices to enable continuous connection, leading to the creation of a novel agricultural phenomenon known as SIL-IoT [23]. There is significant demand for more reliable and secure communication, which require more attention and research on the development of reasonably priced technologies

17.6.6 CONNECTIVITY ISSUES

Current cloud computing and smart agriculture are hindered by the fact that many rural parts of the world lack consistent, high-bandwidth Internet connectivity. Hills or tall trees can potentially obstruct a GPS signal's line of sight [24].

17.6.7 DATA SECURITY AND PRIVACY

Data security and privacy during data transport need to be protected, robust encryption techniques and security measures are needed. However, the low resource demanding nature of IoT sensor nodes and their associated protocols stems from their basic architecture. Considering the current state of technology, practicing security measures on a low-resource device is difficult. As a result, data security and privacy are becoming major issues in smart agriculture. Since the majority of smart agriculture's processes are automated, a malicious party might control them to destabilize the network. This could have extremely negatively effects on the yield and general caliber of agricultural output.

17.6.8 SCALABILITY AND RELIABILITY

The sizes of agricultural farms range from smaller, private farms to bigger, commercial farms. The amount of field sensors required varies. The quantity of data produced by these sensors varies. As a result, scalable technology is required for any usage in agriculture. The devices must be trustworthy in order to minimize the quantity of redundant devices required to enable fault tolerance. The price will reduce significantly.

17.6.9 BIG DATA CHALLENGE

In smart agriculture, cameras or sensor nodes gather enormous volumes of diverse data. Big Data (BD) analysis is used since traditional methods of processing this massive amount of data are insufficient. Large datasets can be explored with the help of big data. It reduces concerns related to food security [25], enhances the effectiveness of the entire supply chain in smart agricultural systems, offers real-time decision-making and predictive analysis [26, 27], and introduces new business models. Big data platforms have been integrated for milk supply chain security using support vector machines and artificial neural networks [28]. The big data process in smart agriculture systems is depicted in Figure 17.4 and is based on [25, 27]. It starts with data collection at various sensor nodes and ends with the various data analysis methods including both traditional and big data analysis.

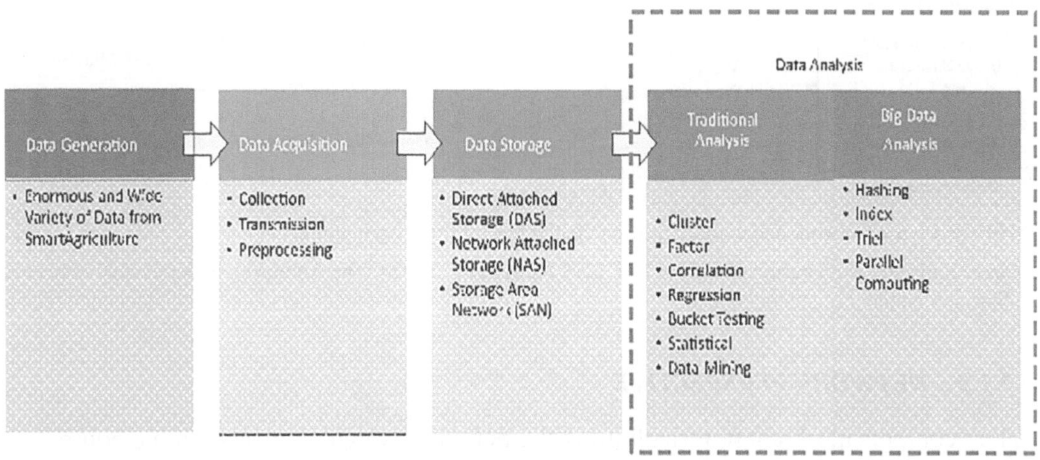

FIGURE 17.4 Big data work flow in context of smart agriculture.

17.7 SMART AGRICULTURE OPEN RESEARCH PROBLEMS

17.7.1 TECHNOLOGY PERSPECTIVE

As was previously indicated, smart agriculture faces a number of difficulties. Technologies, both old and new, must be adapted to meet these challenges. Up until this point, the majority of AI models for smart agriculture have been cloud-based, cloud-edge-based, or cloud-fog-edge-based. Developments in hardware have accelerated the paradigm change in computing. The newest trend in IoT devices is the incorporation of intelligence [29]. Network accessibility, latency, and bandwidth are no longer obstacles to the smooth, effective operation of agricultural systems. This provides a fresh line of inquiry for study. The use of edge AI in smart agriculture is a broad field that is expected to gain popularity soon. Various open research problems in a technological environment are displayed in Figure 17.5.

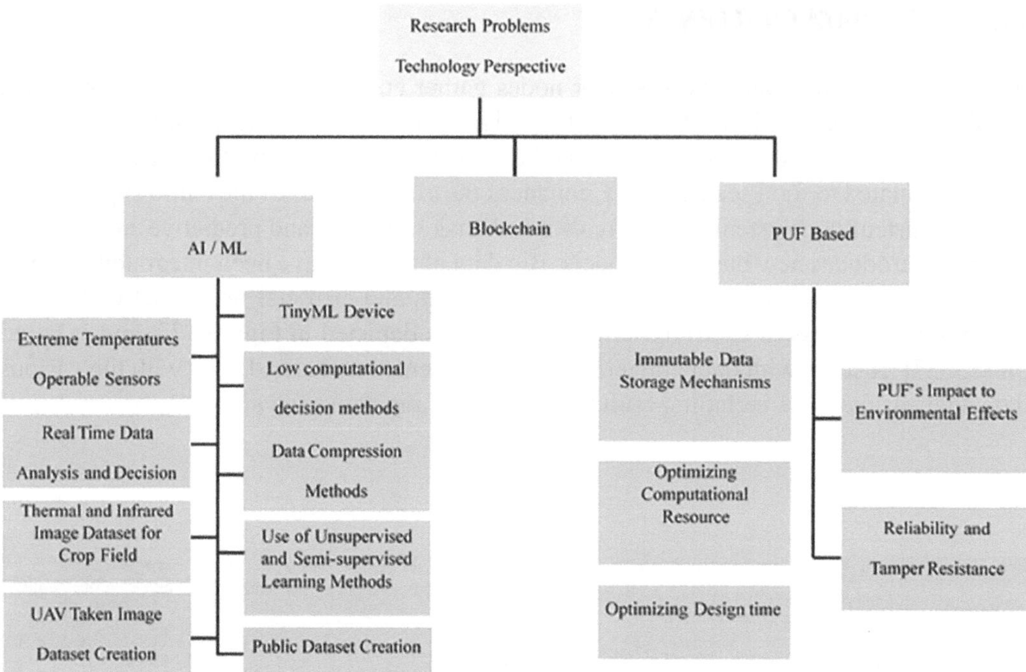

FIGURE 17.5 Network and communication challenges on smart agriculture.

Source: Reprinted with permission from Ref. [29]. Copyright © 2024, The Author(s), under exclusive licence to Springer Nature Singapore Pte Ltd.

17.7.2 NETWORK PERSPECTIVE

A key component of smart agriculture is the network, which connects remote equipment and facilitates data transfer via a variety of Information and Communication Technologies (ICT). Many security issues have emerged as a result of newly developed insecure network layer protocols for IoT devices with constrained resources. Figure 17.6 provides a classification of research problems that require attention.

17.8 CONCLUSIONS AND FUTURE DIRECTIONS

Given that a high-quality diet strengthens immunity, there is now greater emphasis on research in agriculture, food security, and the food supply chain. This discussion offers a thorough overview of current research trends in smart agriculture, discussing everything from current technological advancements to unresolved research issues. The authors hope that this work will provide readers with a general understanding of the technologies, issues, and research problems in smart agriculture.

Modern agriculture is intelligent, sophisticated, and automated thanks to technological advancements and the explosive rise of ICT. By using sustainable, green farming practices, cutting back on fertilizer and pesticide use, and making the most use of available natural resources, smart agriculture lowers its carbon footprint. Agriculture 5.0 is about to arrive in the agricultural sector [32]. By doing this, yields will increase and the system will remain environmentally sustainable. Developing nations will likewise pursue the same course as industrialized nations. More than ever, humanity will accept the efficient production and distribution of food from an ecological and economic standpoint [33].

KEYWORDS

- **smart agriculture**
- **artificial intelligence**
- **machine learning**
- **sustainability**
- **crop health**

REFERENCES

1. Saraju, P. M. (2021). Internet-of-agro-things (IoAT) makes smart agriculture. *IEEE Consumer Electronics Magazine*, *10*(4), 4–5.
2. Kodali, R. K., Jain, V., & Karagwal, S. (2016). IoT based smart greenhouse. In *Proceedings of IEEE Region 10 Humanitarian Technology Conference (R10-HTC)* (pp. 1–6). https: //doi.org/10.1109/R10-HTC.2016.7906846.
3. Tripathy, P. K., Tripathy, A. K., Agarwal, A., & Saraju, P. M. (2021). Mygreen: An IoT-enabled smart greenhouse for sustainable agriculture. *IEEE Consumer Electronics Magazine*, *10*(4), 57–62.
4. Isakovic, H., Fasching, A., Punzenberger, L., & Grosu, R. (2019). CPS/IoT ecosystem: Indoor vertical farming system. In *2019 IEEE 23rd International Symposium on Consumer Technologies (ISCT)*. IEEE. https: //doi.org/10.1109/isce.2019.8900974.
5. Mbaga, M. (2015). The prospects of sustainable desert agriculture to improve food security in oman. *Consilience: The Journal of Sustainable Development*, *13*, 44–59.
6. Vidhya, R., & Valarmathi, K. (2018). Survey on automatic monitoring of hydroponics farms using IoT. In *Proceedings of 3rd International Conference on Communication and Electronics Systems (ICCES)*. IEEE. https: //doi.org/10.1109/cesys.2018.8724103.
7. Vertical Farming. https: //mbc.studentlife.umich.edu/2021/04/23/urban-farming-the-introduction-of-vertical-farming/.
8. Liu, J., Chai, Y., Xiang, Y., Zhang, X., Gou, S., & Liu, Y. (2018). Clean energy consumption of power systems towards smart agriculture: Roadmap, bottlenecks and technologies. *CSEE Journal of Power and Energy Systems*, *4*(3), 273–282. https: //doi.org/10.17775/cseejpes.2017.01290.
9. Ram, S. K., Sahoo, S. R., Das, B. B., Mahapatra, K., & Mohanty, S. P. (2021) Eternal-thing: A secure aging-aware solar-energy harvester thing for sustainable IoT. *IEEE Transactions on Sustainable Computing*, *6*(2), 320–333. https: //doi.org/10.1109/tsusc.2020.2987616.
10. Huang, K., Shu, L., Li, K., Yang, F., Han, G., Wang, X., & Pearson, S. (2020). Photovoltaic agricultural internet of things towards realizing the next generation of smart farming. *IEEE Access*, *8*, 76300–76312. https: //doi.org/10.1109/access.2020.2988663.
11. Desai, A., Mukhopadhyay, I., & Ray, A. (2021). Techno-economic-environment analysis of solar PV smart microgrid for sustainable rural electrification in agriculture community. In *Proceedings IEEE 48th Photovoltaic Specialists Conference (PVSC)*. https: //doi.org/10.1109/pvsc43889.2021.9518454.
12. Ebrahimi, M., Kelati, A., Nkonoki, E., Kondoro, A., Rwegasira, D., Dhaou, I. B., Taajamaa, V., & Tenhunen, H. (2019). Creation of CERID: Challenge, education, research, innovation, and deployment

"in the context of smart MicroGrid". In *Proceedings IST-Africa Week Conference (IST-Africa)*. https: // doi.org/10.23919/istafrica.2019.8764845.
13. Evans, D. (2011). How the next evolution of the internet is changing everything.
14. Courbon, F., Loubet-Moundi, P., Fournier, J. J., & Tria, A. (2015). A high efficiency hardware trojan detection technique based on fast SEM imaging. In *Proceedings Design, Automation & Test in Europe Conference & Exhibition (DATE)*. IEEE Conference Publications. https: //doi.org/10.7873/date.2015.1104.
15. Sankaran, S., Shivshankar, S., & Nimmy, K. (2018). LHPUF: Lightweight hybrid PUF for enhanced security in internet of things. In *Proceedings IEEE International Symposium on Smart Electronic Systems (iSES) (Formerly iNiS)*. https: //doi.org/10.1109/ises.2018.00066.
16. Tang, M., Luo, M., Zhou, J., Yang, Z., Guo, Z., Yan, F., & Liu, L. (2018). Side-channel attacks in a real scenario. *Tsinghua Science and Technology, 23*(5), 586–598. https: //doi.org/10.26599/tst.2018.9010047.
17. De La Concepcion, A. R., Stefanelli, R., & Trinchero, D. (2014). Adaptive wireless sensor networks for high-definition monitoring in sustainable agriculture. In *Proceedings IEEE Topical Conference on Wireless Sensors and Sensor Networks (WiSNet)*. https: //doi.org/10.1109/wisnet.2014.6825511.
18. Zhang, Y. (2011). Design of the node system of wireless sensor network and its application in digital agriculture. In *Proceedings International Conference on Computer Distributed Control and Intelligent Environmental Monitoring*. https: //doi.org/10.1109/cdciem.2011.371.
19. Sahitya, G., Balaji, N., & Naidu, C. D. (2016) Wireless sensor network for smart agriculture. In *Proceedings 2nd International Conference on Applied and Theoretical Computing and Communication Technology (iCATccT)*. https: //doi.org/10.1109/icatcct.2016.7912049.
20. Yang, F., Shu, L., Huang, K., Li, K., Han, G., & Liu, Y. (2020). A partition-based node deployment strategy in solar insecticidal lamps internet of things. *IEEE Internet of Things Journal, 7*(11), 11223–11237. https: //doi.org/10.1109/jiot.2020.2996514.
21. R. Bejgam and T. K. G. V, "Integrating Machine to Machine Communication (M2M) and MQTT Protocol Techniques for Conversion of Water Motor Pump into a Smart System," 2021 Second International Conference on Electronics and Sustainable Communication Systems (ICESC), Coimbatore, India, 2021, pp. 982-987, doi: 10.1109/ICESC51422.2021.9532848.
22. Chen, M., Mao, S., & Liu, Y. (2014). Big data: A survey. *Mobile Networks and Applications, 19*(2), 171–209.
23. Poppe, K., Wolfert, J., Verdouw, C., & Renwick, A. (2015). A European perspective on the economics of big data. *Farm Policy Journal, 12*(1), 11–19.
24. Wolfert, S., Ge, L., Verdouw, C., & Bogaardt, M. J. (2017). Big data in smart farming—A review. *Agricultural Systems, 153*, 69–80, 2017.
25. Kempenaar, C., Lokhorst, C., Bleumer, E. J. B., Veerkamp, R. F., Been, T., van Evert, F. K., Boogaardt, M. J., Ge, L., Wolfert, J., Verdouw, C. N., et al. (2016). *Big data analysis for Smart Farming: Results of to2 Project in Theme Food Security*. Technical Report, Wageningen University & Research.
26. Signoretti, G., Silva, M., Andrade, P., Silva, I., Sisinni, E., & Ferrari, P. (2021). An evolving tinyml compression algorithm for IoT environments based on data eccentricity. *Sensors, 21*(12), 4153.
27. Saiz-Rubio, V., & Rovira-Más, F. (2020). From smart farming towards agriculture 5.0: A review on crop data management. *Agronomy, 10*(2), 207, 2020.
28. Fraser, E. D. G., & Campbell, M. (2019). Agriculture 5.0: Reconciling production with planetary health. *One Earth, 1*(3), 278–280.
29. Mitra, A., Vangipuram, S.L.T., Bapatla, A.K. et al. (2024). Smart Agriculture: A Comprehensive Overview. SN COMPUT. SCI. 5, 969 . https: //doi.org/10.1007/s42979-024-03319-w

CHAPTER 18

Blockchain for Smart Farming and Agriculture

SHREEJAYA SIVADAS[1], ANNU[1*], PRIYA GOND[1], DEEPAK GUPTA[2,3], MADHVENDRA SAXENA[4], and BHAVANA MATHUR[5]

[1]*Department of Post-Harvest Process and Food Engineering, Govind Ballabh Pant University of Agriculture and Technology, Pantnagar, Uttarakhand, India*

[2]*Mechanical Engineering Department, Graphic Era Hill University, Dehradun, Uttarakhand, India*

[3]*Mechanical Engineering Department, Graphic Era (deemed to be) University, Dehradun, Uttarakhand, India*

[4]*Department of Mechanical Engineering, Roorkee Institute of Technology, Roorkee, Uttarakhand, India*

[5]*Department of Mechanical Engineering, Anand International College of Engineering, Jaipur, Rajasthan, India*

*Corresponding author

ABSTRACT

Blockchain technology holds significant potential to transform agriculture through secure, transparent, and traceable solutions. As demand grows for efficient peer-to-peer agricultural data systems, blockchain-based smart farming initiatives are gaining traction. This technology replaces conventional methods, driving innovation via immutable and reliable data management. This examination explores blockchain core principles and its practical, future-oriented applications in addressing current and future agricultural challenges. It demonstrates blockchain ability to provide innovative solutions, highlighting the role of smart contracts and different blockchain types for farming advantages. The discussion extends to blockchain capabilities in data science and management, including decision-making, supply chain efficiency, food safety, and environmental sustainability. Ethical and regulatory considerations are addressed. Real-life examples showcase successful deployments and blockchain transformative influence on farming practices. Blockchain emerges as a disruptive force in agriculture, capable of overcoming obstacles and reshaping the industry. This investigation provides a comprehensive understanding of its role, setting the stage for seamless integration

Demystifying Sustainable Farming Using Smart Technologies.
Gunjan Chhabra, Keshav Kaushik, Priya Chugh, Susheela Dahiya, and Aryan Chaudhary (Eds.)

DOI: 10.1201/9781779640932-18

18.1 INTRODUCTION

The agricultural sector significantly contributes to the economic development of respective countries. Continuous improvement has been seen from the civilization era to the present era of technology in agriculture practices, but when it comes to embracing the latest technologies, this sector always lags, which results in affecting the productivity. Farmers invest most of their time watching and monitoring the crop and land conditions continuously. It is very difficult for them to keep an eye on every aspect related to pre- and postfarming processes. This problem was most evident in micro-farming, where farmers cultivate different crops in a large number of isolated plots, each of which requires specific soil, water, and weather conditions. To resolve these issues, smart agriculture farming was introduced. Smart farming is a management concept with the idea of helping farmers through the use of high-tech tools like big data, cloud computing, and the Internet of Things (IoT). This helps the farmers in monitoring, automating the tasks, and to analysis the data for sustainable farming practices. It helps the farmers in resource management. However, smart farming management systems have less transparency and often centralized results in various alterations and inaccuracies in the data collection process. There is also a high risk of cyberattacks. Consequently, sometimes it leads to poor communication, misleading consumers, unfair remuneration for farmers, and middlemen profiting from inflated retail pricing. However, blockchain technology provides a better solution for all these problems.

Blockchain technology allows users to create a digital record of transactions and share that record with everyone involved. It helps to maintain a secure, single, trustworthy, transparent, and traceable source for information transfer and exchanges. Blockchain has completely changed the way of handling data. All stakeholders involved in agricultural activities, from planting of seeds to selling of its products, can easily monitor the related data in real-time with this technology. This means no one needs to worry about whether the information transferred in the chain is correct or not, since this technology makes everything clear and certain.

Integration of blockchain technology in agricultural activities has showed promising results to bring revolutionary change in the agricultural work style. Blockchain technology's fundamental character of immutability, transparency, and decentralized consensus, agricultural stakeholders may effectively tackle persistent issues as well as open up novel prospects along with the entire value chain. Blockchain has the potential capabilities for transforming the agricultural systems all around the world completely. It originated from supply chain optimization, including fair practices for trading to traceability, and provenance verification. But for achieving this, it is required to have a deeper knowledge of agricultural operations and harvested products, goods, as well as the inherent understanding of blockchain technology complexity.

While it is very important to be concerned about the difficulties and barriers present in both fields, including working to overcome them as well. This chapter, therefore, aims to clarify the importance of incorporating blockchain technology into agriculture, investigate its applications and ramifications, and outline the existing issues in agriculture that blockchain technology can assist in resolving. Readers will obtain important insights into the cooperative link between blockchain technology and agriculture from this investigation, opening the door to a more open, effective, and sustainable agricultural ecosystem.

18.2 OVERVIEW OF BLOCKCHAIN

The inception of blockchain was initiated by the individual known as "Satoshi Nakamoto" in 2008 to act as a communal register of bitcoin dealings [12]. Blockchain technology stands as a decentralized, disseminated ledger that keeps track of time-stamped transactions in a peer-to-peer network. This arrangement guarantees that no single document can be modified retroactively. Within blockchain, all participants can review and validate transactions independently and openly. Consequently, blockchain is devised as an expanding pile of archives, referred to as "blocks," which are intertwined using cryptographic approaches. Every block includes the hash code of the preceding block, a chronology, and a collection of all verified transactions [30].

In simple terms, blockchain is akin to a manuscript where a ledger is structured in a decentralized way. It simulates an endless number of pages, and every page (block) accommodates new material termed transitions. The database within the blockchain is self-monitored through a peer-to-peer network and a communal time-stamping server [31].

18.3 FEATURES AND FUNCTIONALITY

The significant problem of double spending is resolved by the unique features of a secure and encrypted transaction procedure and the open, decentralized structure of blockchain [4]. Every block has a distinct hash value that acts as its unique identifier. The main block is referred to as the "genesis block" and is devoid of antecedents, while a sequence of blocks with hash code in the blockchain system has been discussed in their investigation by [30].

The block's general layout is a complex layout, where the block header and block body together make up a single block unit [30]. Specifically, the block header consists of the following six segments as delineated below:

- **Block version:** signifies a software identifier that dictates which consensus protocol to comply with.
- **Markle tree root hash:** It is employed for validating the hash code that determines all transactions stored in a block. It can be characterized as the recursive binary tree of hash codes.
- **Timestamp (since January 1, 1970):** It is expressed in seconds. It is employed to permanently track the establishment and revision time of the block to guarantee the reliability and authenticity of the blockchain system.
- **N-Bits:** It is set by a certain threshold, which defines the degree of difficulty for miners to find a valid block hash code. Higher difficulty is implied by lower N-Bits, which means more processing power is needed to generate a hash below the threshold.
- **Nonce:** Nonce denotes a unique number used once in an encryption interaction. It typically starts with "0s" and escalates with each hash calculation.
- **Parent block hash:** It is a hash code referencing a preceding block. It constructs a link between the chronological sequences of blocks in the blockchain.

According to [33], in a blockchain system, block transaction counter (BTC) counts the number of confirmed and approved transactions that are kept within the block. The block provides details about who received the data at what timestamp and from whom. Only if the transaction between the two nodes is put in a block, which is subsequently validated and tallied on the blockchain, can it be uniquely identified. It is imperative that this ledger be publicly accessible to demonstrate the critical role that peer-to-peer networks play in blockchain technology. Using the peer-to-peer network in a blockchain program provides different advantages like: (1) The user can easily browse and check the transaction details of the blocks with a blockchain explorer without the interference of a third party. (2) To compromise the blocks, a malicious attacker faced the difficulties of attacking the millions of nodes simultaneously. (3) Since blockchain systems all the nodes are in peer-to-peer networks, blockchain can never be removed.

18.4 BLOCKCHAIN FEATURES

Blockchain system provides different key characteristics, as can be seen in Figure 18.1, which makes it set apart from other traditional centralized database systems.

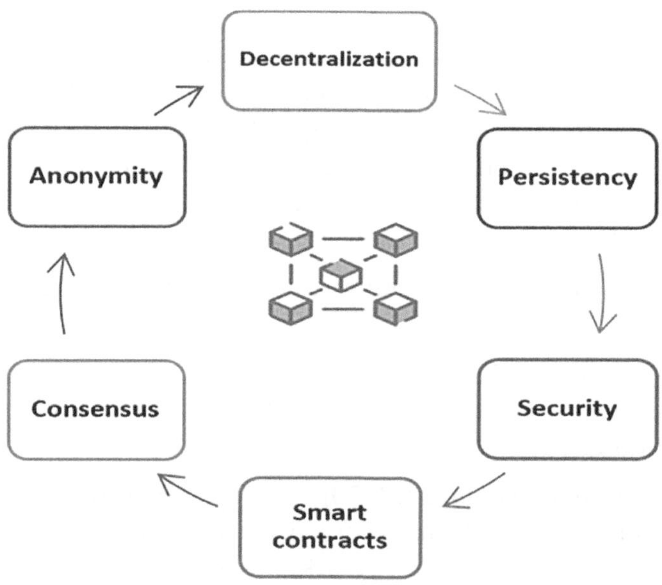

FIGURE 18.1 Features of blockchain system.

1. **Decentralization:** It means that there is no single authorized party that governs the system. All the nodes in the block are in peer-to-peer network system that works together to verify and validate the transactions. Having a decentralized nature, blockchain systems are less prone to failure. It provides the facility of creating a more transparent profile for every user. Thus, every manipulation is traceable and more concrete.

2. **Persistency:** Every node in the peer-to-peer network has a digital replica of the database. A new transaction will be added to the network only when every node verifies the transaction and the majority approves it as a valid transaction. In the blockchain system, once the transaction is added and validated, it is not possible to roll back. Furthermore, the alteration in the transaction could be found out immediately. This feature makes it a tamper-proof system.
3. **Anonymity:** In the blockchain, every user has a virtual identity code through which they can communicate and does not require any original identity card. However, this aspect creates the challenge of privacy and security in blockchain transactions.
4. **Security:** All the information in the blockchain is encrypted individually. It provides overall security to the system. Every block in the blockchain has a different hash code of its own as well hash value of the previous block which makes every piece of information hashed cryptographically.
5. **Consensus:** All kind of blockchain systems has a framework of agreements to facilitate the system to decide quickly and objectively. Consensus is a decision-making algorithm that helps in efficient decision-making. For example, nodes do not have trust in each other but have faith in algorithms to validate the transaction. The consensus algorithm plays an important role in blockchain systems.
6. **Smart contracts:** Smart contracts are self-executing computer programs that run on a blockchain network. When specific criteria are satisfied, they will automatically carry out preestablished rules and agreements, enabling transparent and trustless transaction execution without the need for middlemen. These decentralized contracts can support asset management, programmable logic across several domains, and safe multiparty agreements because they are immutable and guarantee the integrity of the agreed terms.

18.5 TYPE OF BLOCKCHAIN

Concerning what are mines, and who validates the blocks, who authenticates the access of the blockchain system, the design, privacy, and efficiency of blockchain are mainly divided into three parts: public, private, and consortium blockchain [29], which are contrasted in Table 18.1.

TABLE 18.1 Different Blockchain Types

Property	Public Blockchain	Private Blockchain	Consortium Blockchain
Blockchain authentication	Every node	Single organization	Particular nodes
Access authority	All	Public or private public	Public or private public
Persistency and safety	Very tough to manipulate	Easy manipulated	Easy manipulated
Performance effectiveness	Low	High	High
Centralization	No	Yes	Partial
Authorization	Unrestricted	Restricted	Restricted

18.5.1 PUBLIC BLOCKCHAIN

As the name suggests, the public blockchain system is for the public and of the public. In this type of blockchain, no one has the authority, and open to everybody. As no one has the charge, all the additions and approvals of transactions are done by the decentralized consensus mechanism.

18.5.2 PRIVATE BLOCKCHAIN

Unlike the public blockchain, a private blockchain system functions as a person's or organization sector's private asset. In this type of blockchain, some particular entity has the authority to manage the important tasks. As it restricts who can be a participant in transactions and validations is termed a permissioned blockchain. The participants in this type of blockchain are selected by the respective authority and the developers while creating the application of blockchain. The company does not want it to be accessible to the public as it handles the internal ledger of the company.

18.5.3 CONSORTIUM BLOCKCHAIN

A collective blockchain represents a kind of blockchain technology where multiple organizations manage the platform. Also known as a federated blockchain, this type is a permissioned system and not public. In a collective blockchain, a predetermined node group can oversee the consensus process. Many financial institutions utilize this type, each operating its own individual blockchain.

18.6 BLOCKCHAIN APPLICATIONS FOR FARMING OPERATIONS

Blockchain technology is holding an important place in technological development, including exploring the new opportunities for the transformation and reshaping of many industries, including agriculture. The blockchain presents an interesting and innovative solution for supply chain management and farming operations problems since it provides an immutable and secure means of recording and tracking transactions as a decentralized and transparent digital ledger. This section will discuss the blockchain technology methods and procedures that may be used in farming operations and how this disruptive technology is able to transform the agriculture sector completely.

18.6.1 SUPPLY CHAIN TRACKABILITY

Supply chain management and traceability are the most promising applications in agriculture regarding the incorporation of blockchain. The lack of honesty and corruption involvement in traditional supply chains makes it challenging to follow agricultural products from the farm to the corresponding fork. But now with the incorporation of blockchain technology in supply chain tracing, the related consumers, merchants, and regulatory agencies are now able to obtain information regarding the origin, handling, processing, and shipping of the goods they bought [31].

Farmers can have information related to weather and its moving patterns, soil conditions, farming and harvesting operations, and other pertinent topics. They can further record this information as well on a ledger to examine the pattern and forecasting. Downstream stakeholders can have access to this data, which also promotes transparency and trust in the supply chain [9]. Additionally, food safety, quality, and sustainability concerns can also be observed and resolved with the effective use of blockchain-based traceability systems. This will lead to boost consumer confidence and lower the possibility of associated fraud [25].

18.6.2 FARM INPUTS AND SUPPLIES MANAGEMENT

Blockchain in tracking the provenance and flow of agricultural supplies, including herbicides, fertilizers, and seeds, is another application. Major dangers to crop productivity and quality, as well as the environment, are modelled by fake and subpar inputs. It makes sure that the farmers and stakeholders are getting real, high-quality products from reliable sources by tracking the provenance and distribution of agricultural inputs on the blockchain [31].

Additionally, a smart bond can be implemented easily with blockchain incorporation because they automate the fulfillment of predetermined and recognized terms and conditions. For example, the use of disburse payments to suppliers upon confirmation of input delivery for tracing the payment, so reducing the unnecessary need for middlemen and maximizing efficiency as a result [14].

18.6.3 FOOD SAFETY AND QUALITY ASSURANCE

Food producers and consumers both have very serious concerns about ensuring food safety and quality. Weather patterns and other associated elements can affect food safety and quality. Blockchain technology is very helpful in this regard with facilitating the recording and exchange of data about agricultural techniques, related conditions of soil [6].

Farmers may have gained more confidence and respect of consumers, merchants, and regulatory bodies by presenting substantiated evidence of their goods compliance, including sustainable farming practices as well as food safety regulations. The blockchain makes it very easier accesses to, and examination of this data, promoting accountability to authorized one and full transparency throughout the supply chain [9].

18.6.4 FINANCE AND INSURANCE FOR AGRICULTURE

Continue operating and reducing risks, including access to financing and insurance, is essential for farmers. Blockchain technology can make traceable and secure transactions between lenders, insurance companies, and farmers. According to [31], smart bonding has the ability to make it automatic for the cash distribution, related loan repayments, and claiming of insurance under recognized criteria. This eliminates any extra need for third parties and also reduces the related expenses. An automatic payout of the insurance upon the fulfillment of specific weather or harvested crop yield conditions gives farmers and stakeholders timely financial support during hard times [14]. These ability makes it very supportive and sustainable to the agriculture practices and increase the flexibility of agriculture industries as well.

18.6.5 OWNERSHIP OF LAND AND IDENTIFYING FARMERS

Blockchain effectively offers a safe and transparent register of land titles as well as farmer identities in the areas where land ownership and farmer identity are complicated concerns. This also supports equitable distribution of the agricultural resources, access to subsidies, and associated land rights. Governments and authorized regulatory agencies can lower the possibility of fraud and conflicts with the implementation of blockchain technology to preserve an unchangeable record of land ownership and farmer information. Additionally, it also makes it very easy to involve associated policies and smallholder farmer support programs into action, raising inclusive and sustainability in agricultural growth [6].

18.6.6 TRADING IN AGRICULTURAL COMMODITIES

Together, the above solution technologies can bring about the possibility of conducting the safest and transparent movement of transactions among all stakeholders, farmers, traders, and buyers, simplifying trade in agricultural products. Smart contracts can automate trade order execution, eliminating the requirement for intermediaries, thereby enhancing productivity. In the agricultural commodity markets, the blockchain can enhance accountability and transparency by maintaining a transparent and unmodifiable record of all transactions. Impersonation will be less common; there will be less fluctuation in prices, and farmers will be compensated fairly [14].

18.6.7 COOPERATIVE FARMING AND INFORMATION EXCHANGE

As noted by Kamilaris et al. [16], one of the major health challenges to sustainable agriculture is the fear of data tampering or unauthorized access among stakeholders sharing data and insights. With blockchain technology, this concern is mitigated since the decentralized and secure feature is likely to abolish such worries, thereby encouraging cooperation among stakeholders and fast-tracking innovation and information exchange, queue reusable [6].

18.7 CHALLENGES AND FUTURE PERSPECTIVES

While blockchain has numerous benefits to offer farming operations, the technology may face some roadblocks in this sector due to infrastructure, network scaling and development, and adoption challenges. Access to the internet, processing power, and technological knowledge are all crucial requirements for implementing blockchain solutions. According to Tripoli and Schmidhuber [31], these resources will be frequently scarce in numerous locations due to a lack of these resources among smallholder farmers. Additionally, blockchain networks need to expand, and the industry is working on solutions to handle the vast quantities of data and transactions generated by agricultural supply chains. As Kamilaris et al. [16] point out, to meet the requirements of the agriculture industry, current research and development are concerned with enhancing blockchain systems' scalability and performance. Nonetheless, even with the emerging challenges, several players and stakeholders such as governments, commercial enterprises, and other international organizations have increasingly been interested in and committed to the blockchain technology given the many expected benefits on the farming activities and their supply chain in terms of overall transparency, traceability, and efficiency [9]. In conclusion, the use of the blockchain methodology in farming activities has enlightened various innovative methods that can effectively be utilized to address the agriculture-related issues. The technology is powered by the decentralized, open, and transparent ledgers that provide a new opportunity to reinvent the entire supply chain management, food safety and quality monitoring, finance, as well as a particularly cooperative farming practice. However, such an effort will attract extensive investments and the need to create an effective stakeholder coordination. Nevertheless, the potential of the technology in revolutionizing the entire agricultural sector continues to dominate with the increasing technological advancements.

18.8 BLOCKCHAIN USED FOR DATA MANAGEMENT AND DECISION-MAKING

In this topic, we are discussing the use of blockchains for data management and decision-making. As we know that it is used to enhance data protection and privacy in each step of the decision-making process. It also provides a decentralized and secure infrastructure for processing and storing data, while making it suitable for applications such as democratic on-chain decision-making. They give a chance for dispersion of power and information, and a great esoteric opportunity for broad collaboration in the digital era [32]. By leveraging the decentralized and immutable nature of blockchain, traditional centralized identity management systems can be transformed into more secure and privacy-enhancing solutions [21]. There are several components to the agricultural system, supply chain management, crop insurance, healthcare, energy, and transshipment. Thus, it can relate with plenty of dishonest parties like users (framers, vendors, clients, distributors, etc.), departments, and bureaucrats. Therefore, the direct participation of the middle man and agencies run by people has shown some concerns that are escalated and are inhibited by clarity, promptness, traceability, defense, and immutability, resulting in money loss, crop soilage, and spoiling [24]. The cryptographic ledger has gained traction in its distributed form and extremely safe approach for making data manipulation. Incorporating blockchain into agriculture, one could eradicate double-dealing dubious partners. Here are some advantages that are available to enhance sustainability in the agriculture sector.

18.8.1 TRANSPARENT SUPPLY CHAIN

Sometimes in every deal that is saved in the chain, which cannot be improved. Thus, every transaction can be viewed by the stakeholder's system. The data consist of harvesting, cargo time, cost, and so on, are recorded which cannot be amended because it is stored in the ledger. This helps in the transparency of the tracks of locations and the quality of the crop.

18.8.2 FAIR PRICE AND PAYMENTS FOR CROPS AS WELL AS STAKEHOLDERS

The original price of crops is transparently viewed by the stakeholders. Therefore, the alteration in prices is impossible as it is written in the blockchain system. An increase or decrease of price in an uncertain manner is not possible; thus there is a rare chance of fraud occurring.

18.8.3 TRACING AND AUDITING

From farm to fork or farmer to consumer, all the information about crops and transportation is digitally connected to the crop through technology. The representatives, including farmers, to end users/ consumers can track the whole supply chain. The blockchain-based food supply chain can provide the stakeholders with access to all information on food quality and its contamination at every stage. As a result, food supply chain networks become more open with one another, and it helps in to figure out how and when the food is contaminated.

18.8.4 TRACKING AND PAYMENT SUBSIDIES

In the instance of weather-related loss, the growers can file for the crop coverage and recover the funds via Ethereum. Because of the information's accessibility and consistency behavior helps all the insurance firms and other legal parties can simply view the data given by the farmers. The smart contracts support the direct querying of the blockchain for the required information. Once the insurance claim is granted to the farmers, they will receive the required amount in their e-account. As a result, farmers could track the whole procedure or observe the entire subsidy approval through the blockchain. Hence, it gives compensation seamlessly and quickly.

18.8.5 DATA IMMUTABILITY

It ensures that once data is fed or recorded in the system, it cannot be deleted or tampered with by any means. This helps in maintaining the integrity and trust among people or parties.

In a blockchain system, the content includes the total number of transactions verified and authenticated within the block. The BTC counts the actions that occur in trading, since each block includes data transferred in bitcoin for a specific date. The recognized transaction between the servers takes place after all blocks are involved, validated, and uploaded to the blockchain. To obtain the database, it should be publicly available, which also understands the value of peer-to-peer network in blockchain [33]. It is also explained by [30] in their investigation that how does the blockchain system operate, as user "A" transfers the transactional data to user "B" [30].

18.9 ADVANCEMENTS IN SMART AGRICULTURE: USING BLOCKCHAIN FOR DATA SECURITY AND PRIVACY

In the agriculture field, the blockchain system ensures that data security and privacy are predominant because of its sensitive nature to collect the information regarding farming operations such as temperature, moisture and humidity, soil nutrition, irrigation, and livestock/poultry monitoring. The growing torrent of agricultural-related information, together with the greater ability for interpreting data, accompanied by rising concerns about information's privacy and security which has triggered the development of numerous digital agriculture platforms [27]. Smart agriculture framing is based on IoT technology, which helps farmers to improve their crop yield, gives optimize irrigation efficiency, and also reduces the farming cost. Combining both IoT and agriculture gives three aspects [8].

18.9.1 SATELLITE AGRICULTURE

It is a farming technique based upon monitoring, measuring, and responding to inter- and intrafield variation in agricultural output. It also has the technology which can be used for enhancing the crop yield within them, with a wireless sensor network employed as a main source of growth. It helps farmers in the production process with the least risk, also helps to make accurate and controlled framing practice by developing an enormous number of multifunctional, insufficient power, wireless communication devices in environments such as poultry and livestock breeding and collection of vital data in agricultural production.

18.9.2 PROVISION AGRICULTURE

It is an advanced agricultural yield form that seeks high quality and quantity, which belongs to the high-input, high-output, capital, technology, and labor-costly firm. Which supplies crop production protection facilities designed by engineers to meet the purpose of ensuring that the environmental elements do not affect the agricultural production. Furthermore, it makes easier for the growth of crops free from locational constraints, which also shortens the production cycle of the produce, which also improves the quality and harvest of the crop.

18.9.3 CONTRACT FARMING

It is an agricultural production carried out according to an agreement between the framer and the buyer. It is a reformed harvesting and handling of crops, with its headway gap between rural and urban areas is expanding. To solve this problem, contract framing requires an outsourced production demand such as decrease sowing and breeding hazards of growth, and to prevent sightless production as an economical market-based output. It also involves supply chain management of farm goods, traceability of product safety, logistics, and trading system. As it helps in tracking the food supply chain which is traceability of farm-to-fork system.

18.10 SUPPLY CHAIN MANAGEMENT AND FOOD SAFETY-BLOCKCHAIN TECHNOLOGY

In today's era, the food supply chain ensures the quality and safety of food which we consume or purchase; thus, it is a most important area. As Ethereum has the capability to record the products data, cultivation location, its origin, and its handling procedural information. There is a revolutionary shift in this domain as it has drawn the attention of developing countries. The food supply chain has enthralled several salient and advanced technologies for processes like artificial intelligence, analytics, IoT, and so on. In the food supply chain, there is a network of stakeholders (input suppliers, food producers, processors, distributors, and consumers) which gives significant information about the transaction with each other in the form of growing, processing, and selling food to the customer. They also make sure that the consumer gets the right quantity and quality of products that reach at right time in the right destination as shown in Figure 18.2 [2].

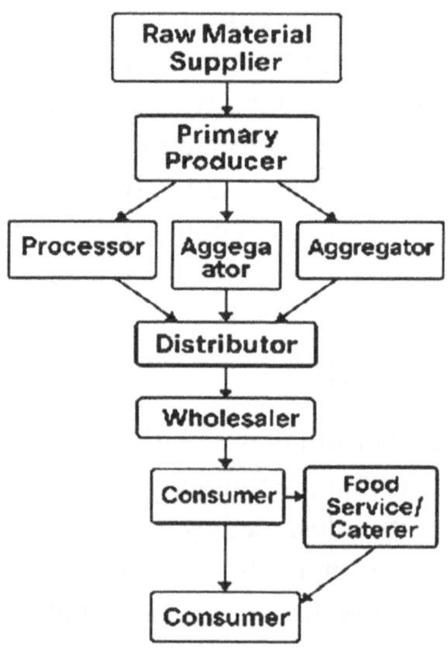

FIGURE 18.2 Agriculture food supply chain [2].

Agriculture the largest and important sector of the world economy, which gives a solid backbone to human life and the growth of the country's economy, too. Unluckily, most of the farmers till the date rely on traditional farming and are far away from modern techniques and technology. Thus, results of low productivity and yields are much below the meeting endpoint or are below their potential. Therefore, to overcome this situation, agriculture sector requires modern technologies such as blockchain technology, which can resolve its problems. Some of the technologies used in agricultural field are shown in Figure 18.3.

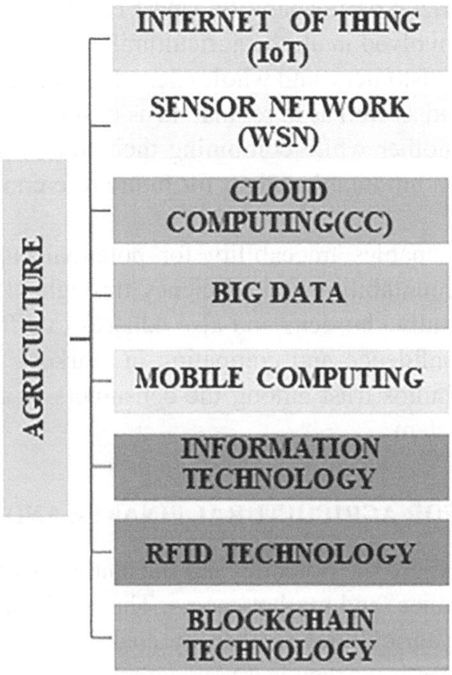

FIGURE 18.3 Technology presently used in agriculture.

Through this section, we are explaining the blockchain technology, which has reduced the paperwork and improved the product traceability with the help of a controlled food supply chain. The studies have proven that the blockchain technology has improved the products documentation, food quality, food safety, wrapping, and different fields developments. The documentation of products consists of barcodes, RFID tags, and electronic information about the product. Moreover, the adoption of all the technology has given the strength to build overall all the logistics and to improve the supply planning, and also enables the food tracking system. By decentralizing, blockchain has gained trust in the food and management sector. It has eliminated the function of mediators and third parties and also increased the trusted transaction between the different stakeholders through the ensured security of information with the use of cryptography. In agribusiness, data loss is a serious issue that is resolved by a blockchain platform. As it provides a secure solution for the problems by decentralizing all the data, automation, immutation, and transport management [24].

The entities are broadly classified in three categories that is

1. Data
2. Process, and
3. Distributors

Data is a commodity that collects all the information from different sources, as farmers, crop production, seed quality, quantity, soil moisture, weather and atmosphere, demand and selling price of different produce and harvest. These various processes assimilate the biome, which are categorized as traceability, smart contract, and cryptocurrency. Thus, several stakeholders are involved in all the agricultural areas and resolve among the users, such as retailers, farmers, customers, and wholesalers; they also serve the agencies as banks and insurance company and as well as in regulators as in surveyors. Different members can communicate with one another while combining their transaction data into a chain. All these entities are kept in an immutable ledger for future referencing in physical and digital movement.

However, blockchain enables traceability for potential supply and food safety by proving transparency, accountability, and efficiency throughout the food supply. Thus, it also reduces the administrative burdens and also adheres to different industrial norms by enhancing consumer's confidence and competing in markets, which ensures the safety and quality of products, builds trust among the consumers in a leveraging complex and interlinked global food system.

18.11 BLOCKCHAIN FOR AGRICULTURAL FINANCE AND MARKET ACCESS

In this section, we will explore the potential and the challenges faced by blockchain technology for agriculture finance, and market access. The blockchain technology has craved a huge transformation in agricultural finance by enabling the farmers to reach out and access markets more efficiently, as they were facing prolonged challenges as limited access to credit, lack of transparency in markets, and several barriers to fair trading. There are various difficulties in farming supply chains that cause farmers and consumers to lose money or face adverse scenarios. The agriculture sector faces diverse issues in transparency and efficiency of supply chain as it is a risky and complex linkage that relies on a number of intermediaries. The consumers with poor transparency get no conscious of how food comes and is produced. So, to reduce and eliminate these challenges, blockchain in the agriculture sector relies on strengthening the linkage between the farms, markets, and consumers, which can generate the growth of income and also generate job opportunities. Thus, to elevate the efficiency and to enhance the agricultural supply chain, the blockchain has integrated food safety, which reduces the risk of trade finance and generates smarter agriculture market information on the legal certainty of the land tenure system [31]. Lack of access to finances causes huge impact on the agricultural chain, where farmers cannot improve their yields, and as well as decreases the value of buyers, which has a great impact on their lives. Therefore, to bring transparency to the procedure, blockchain is used throughout the steps of agricultural finances. Each step consists of records of the transaction in the blockchain and also provides easy access to all the parties in the system. On the other hand, it also has the ability to record all the information permanently so that there are no issues in the verification of transactions during audits. So that audits can be conducted easily through its ledger from its adequate parties [2].

Due to the lack of trust in financial industry, the blockchain has used cryptocurrency to develop the decentralized database for storing valuable knowledge, which can be encrypted and kept in an environment [22]. For easier and efficient, which provide low costs than the current system, blockchain should discard the middleman system and central authority. This will give blockchain a new and profound base for finances [3]. To make payments easily accessible for farmers, there are several apps for real-time payment. Thus, they receive wages through a blockchain mobile app. Moreover, the use of mobile apps in real-time transaction of money has made several ongoing issues an effortless task. As farmers can transact faster with the transparency of farmers information, which is anyhow protected in the blockchain. There are some of the currencies used by the farmers which benefits them are [30]:

- **Cryptocurrency:** It helps cultivators to make remittance deals with dealers and collect their wages on time.
- **Electronic wallet/e-wallet:** This app helps farmers to reserve the money and helps in managing their assets in e-form.
- **Digital harvest tracker:** With the help of this app, farmers are updated about information on trade rates and market dynamics for various agricultural produce, so that the transaction can be based on cryptocurrency based.
- **Retail app:** By using digital currency, it helps farmers to conduct payments and thus empowers them.
- **Smart contract:** In this app, it allows farmers to identify any type of fraud or illegal operation about digital assets.
- **AgriDigital:** It is a virtual currency transfer platform of cloud based blockchain which helps in a simple, easy, and secure supply chain in agriculture field which is held between farmers and the stakeholders. It also helps in real-time payment from one place for all contracts, deliveries, orders, and payments. Its primary and important feature is to form digital assets in the form of tokens. This shows the physical numbers of goods and produce.
- **Foodcoin:** They consist of 1000 Eco farms as a new blockchain ecosystem. It was developed to bring the agricultural products and the global marketplace together. Foodcoin works on cryptocurrency which uses smart contracts and verified food transactions. They also used colored tokens for buying and selling of goods.
- **IBM blockchain:** This app gives high security, flexibility for multicloud-based transactions and a trusted expertise which helps in logistical transactions. It also improves the global trade relationship by bringing the trader, stakeholders, suppliers, and regulators together.

Thus, this concludes that blockchain technology gives a possible transformation in the agricultural field of finances and market access to transparency, efficiency, and creates trust in the supply chain. Blockchain also helps farmers to easily access their credits, gives a fair market, and helps to directly connect with consumers, which creates a comprehensive and legitimate ecosystem in agricultural field.

18.12 ENVIRONMENTAL SUSTAINABILITY AND CLIMATE RESILIENCE

The agricultural industry is experiencing tremendous pressure to supply the rising demand for food in the world while going through issues like resource depletion, climate change, and environmental degradation. Promising options to improve climate resilience, traceability, and sustainability in smart farming techniques are offered by blockchain technology.

18.12.1 TRACEABILITY AND TRANSPARENCY IN FOOD SUPPLY CHAINS

The ability of blockchain to enable end-to-end traceability and transparency in food supply chains is one of the primary applications in agriculture. Stakeholders can monitor agricultural products from farm to fork using the blockchain's distributed and determinate ledger to ensure the products' sustainability, authenticity, and quality [19]. Using blockchain-based traceability systems, important data points, such as farm location, growing conditions, use of fertilizers and pesticides, shipping routes, and storage conditions, can all be recorded [26]. To this transparency, customers have more knowledge to make decisions about the ethical and environmental impacts of their food and are better informed to prioritize products that do not exact a high social and environmental cost.

18.12.2 PRECISION AGRICULTURE AND SUSTAINABLE RESOURCE MANAGEMENT

Blockchain and IoT-powered or IoT-supported smart farming can significantly impact the indicated areas to ensure that they are climate-resilient and sustainably managed for competent resources. Once deployed in farms, IoT sensors and devices deployed in the agricultural lands gather real-time data on the status of crop health, weather, soil moisture, and nutrients [16]. The adoption of blockchain technology will enable farmers to optimize their water usage or minimize chemical application in their farms more efficiently through the safe storage and sharing of gathered data across a blockchain platform. Blocks of information about farm activity can be supported by smart contracts to reduce soil erosion, water pollution, and greenhouse gas emissions [13].

18.12.3 CLIMATE-RESILIENT AGRICULTURAL FINANCING AND INSURANCE

Finally, blockchain technology can significantly reduce the risks of the impact of climate change on agriculture with the help of innovative insurance and finance systems. For example, platforms recorded in a blockchain that allow transparent and secure transactions between lenders, insurers, and farmers. The smart contract mechanisms of the blockchain allow making insurance payments to farmers based on unjustified conditions, for example, on yield, climatic cataclysms, or market prices. Therefore, all payments for farmers' insurance are reduced, and the compensation for a climate-game maker and a cryptocurrency price will be simplified and made faster. Furthermore, thanks to blockchain, farmers will collaborate with peer-to-peer lenders and crowd funders. It will give the opportunity to open credit lines on the most convenient terms for the implementation of investment projects in more resistant infrastructure, friendly farming methods, and adaptation mechanisms [17].

18.12.4 INCENTIVIZING SUSTAINABLE AGRICULTURAL PRACTICES

Those could include farmers preserving water in the form of regenerative farming; moreover, those related to carbon sequesters stored in soil, those undertakings could be rewarded through tokenization and other crypto processes [16, 20]. It is possible to identify that the creation of incentive programs might facilitate the development of a sustainable and climate-resilient sector of agriculture due to the realization of the potential of the variance provided by the blockchain [17].

18.12.5 CHALLENGES AND FUTURE DIRECTIONS

Although blockchain technology has a long way to go before its potential in agriculture is realized, it holds great promise. Research into low-energy consensus algorithms is in progress, as are the methods to merge them into current systems and the clearing of technical and legal entanglements that prevent their extensive use in the agriculture industry [1]. Consequently, this is a sturdy tool for achieving autonomous climate-resilient food systems and sustainable agriculture by fostering novel ways of working, accountability, and transparency [20]. It is for this reason that it is critical to SDG13 and the Paris Agreement.

18.13 REGULATORY AND ETHICAL CONSIDERATIONS

Assess and analyze the major legal and ethical issues presented by the use of blockchain in smart farming and agriculture. Despite the numerous benefits that come along with the use of blockchain technology, supply chain traceability and safe data sharing are among the potential benefits but the two have risks.

18.13.1 DATA PRIVACY AND SECURITY

One of the major regulatory worries is ensuring adequate data security and privacy safeguards for the use of blockchain for agricultural data purposes. Blockchain is a permanent, nonchangeable, distributed database, and satisfying the right to erasure, for example, following data protection rules like General Data Protection Regulation can be difficult for end-users [9]. There are also risks to cybersecurity if blockchain networks are not properly encrypted and protected [15]. They need to be clear about the guidelines, regulatory oversight, and the establishment of governance structures that strike an appropriate balance between the benefits of data tracking and transparency and sufficient data privacy protections. Strong access controls, data encryption, and data minimization should be implemented. Data rights and security best practices should also be effectively communicated to farmers and other participants in the agriculture supply chain.

18.13.2 INTEROPERABILITY AND STANDARDS

The compatibility of several blockchain platforms and traditional systems will be critical to the success of blockchain agricultural solutions at scale. Common standards, protocols, and governance will potentially have to be developed to facilitate seamless integration and data sharing [16]. At present, the lack of convergence between common frameworks risks market fragmentation and siloed systems. Industry coalitions, public–private partnerships, as well as user engagement from international standards bodies will be essential for focusing the alignment.

18.13.3 FAIR MARKET COMPETITION

The use of blockchain technology in agriculture may further commercialize the authority over market power, allowing a few major technology providers to dominate [5]. According to [30], there are worries that this could provide such corporations with an advantage over smaller farmers by granting them control over vital infrastructure, fees, and vendor lock-in. Regulatory oversight may be required to promote fair competition, prevent monopolistic practices, and ensure a level playing field with reasonable costs and negotiating benefits for all participants. Decentralized and open-source blockchain platforms could reduce these risks compared to more centralized, exclusive solutions.

18.13.4 ENVIRONMENTAL IMPACT

The high energy requirements of some consensus algorithms, such as proof-of-work, have led to concerns about blockchain's environmental impact. However, better protocols with considerably lower energy consumption are starting to emerge [18]. Nevertheless, greater resource consumption and electronic waste could result from the scalability of blockchain and related technologies like IoT sensor networks [23].

Careful life-cycle assessments will be important to understand and mitigate environmental impacts. Circular economy principles, like improving recycling and repairability of devices, should be priorities. Alternatives to energy-intensive consensus algorithms should be a priority in sustainable agricultural projects.

18.13.5 ETHICS AND HUMAN RIGHTS

In addition to regulatory aspects, blockchain agriculture initiatives should also be concerned with wider ethical issues. Systems should be meticulously designed to incorporate basic ethical notions, such as transparency, fairness, resilience, and human rights protection from the inception. Various abuses that might result from carelessly constructed systems with inextricable records include but are not limited to human rights concerns, including exploitative labor arrangements, unfair working situations, discrimination, and privacy invasions [7]. There is a need to analyze uneven power dynamics across agricultural chains to ensure that it is impossible for corporations to exploit marginalized small producers, local populations, or hired labor. Ethical standards and accountability actions promoting the principles of privacy by design and periodic algorithmic independent assessment is critical. In addition, an inclusive governance notion that allows the process of creating affected stakeholders is necessary. In general, incorporating ethical systems in its conceptualization phase may assist blockchain agriculture projects in gaining public support and legitimacy while minimizing unintended consequences to society.

18.14 REAL-WORLD EXAMPLES OF SUCCESSFUL BLOCKCHAIN IMPLEMENTATIONS IN AGRICULTURE SECTOR

Overall, the possible application of blockchain technology in agriculture is already widely recognized, and many pilot projects and large-scale programs are being initiated to discover the potential and develop its value. At the same time, the real examples of blockchain implementation and success stories are based on the unique cases and experience, which means that the actual advantages, challenges, and lessons learned are also observed.

18.14.1 AGRIDIGITAL: OPTIMIZATION OF GRAIN SUPPLY CHAIN

AgriDigital is an Australian company that successfully implemented a blockchain platform aimed at optimizing the grain supply chain. The key objective of the solution is to increase transparency, cut costs, and boost performance by digitalizing and automating various processes, including forming contracts, making payments, and tracking inventory. AgriDigital's platform, based on smart contracts and a distributed ledger, allows for real-time monitoring of grain flows, automatic payment processing, and data exchange among all actors in the supply chain, such as farmers, buyers, and carriers. This blockchain platform has been adopted by major agribusinesses and agricultural institutions, showcasing how the supply chain can be shortened and mutual trust between the stakeholders reinforced. Finally, all the aforementioned categories of supply chain actors should be engaged and catered to, which is also pivotal in terms of implementation success. Besides, the merger of blockchain and current systems and practices should be carefully planned and executed, and the transition must be gradual.

18.14.2 IBM FOOD TRUST: ENHANCING FOOD TRACEABILITY

IBM Food Trust is an initiative that exemplifies a blockchain-based platform designed to boost the transparency and traceability of the global food supply chain. Created with some of the leading food producers and retailers, this endeavor helps safeguard consumers, minimize waste, and bolster confidence by providing a safe, unchangeable history of how a product such as a piece of meat or fish is sourced and processed on the tables [10, 11]. IBM Food Trust focuses on the secure exchange of information and data to and from the supply chain and delivers users a secure record of food production, storage, and processing circumstances, from pesticide use to product success beyond the expiration date.

18.14.3 RIPE.IO: EMPOWERING SMALL-SCALE FARMERS

With the use of blockchain, Ripe.io captures farm data such as activities, crop yields, and market transactions that are secure and transparent. This data can be used by agricultural bodies and financial institutions to render services and support to smallholder farmers by offering, for instance, micro-financing crop insurance, and ensuring fair payment.

18.14.4 TERRACERTA: ENHANCING LAND OWNERSHIP RECORDS

TerraCerta ensures the integrity of land ownership records. TerraCerta, a blockchain-based system that addresses the multifaceted challenge of land ownership and property rights, primarily in developing countries. TerraCerta uses blockchain tech to create an immutable and transparent record for land tenure, title transfers, and property boundaries [28]. The possibilities of commercially licensing this prove the worth of the pilot in multiple countries, which appears to help in boosting land governance and securing the rights of small farmers and communities. Although these examples demonstrate how blockchain might help solve various issues in agriculture, some major challenges and potential lessons learned emerge:

1. **Scalability and performance:** The primary concern as blockchain networks grow and incorporate more members is to guarantee scalability and performance. Matters including transaction speed, data storage, and network congestion will have to be addressed in real-world situations.
2. **Interoperability and integration:** The key to universal blockchain adoption is the seamless interface of blockchain-based products with existing systems and processes. Establishing industry-wide protocols and standards for data exchange and cross-blockchain coordination is a challenge.
3. **User adoption and education:** In full acknowledgment of the technological and cultural obstacles to user adoption, education and training initiatives should be undertaken to overcome them. This is especially valid for smallholder farmers outside the digital technologies' traditional user base. The potential of the software to generate user-friendly interfaces is realized, and existing concerns about data privacy and security are addressed.
4. **Regulatory frameworks:** Since the blockchain eliminates a considerable number of traditional processes and business models, the best possible regulatory frameworks should be implemented and regularly amended. And participants are not only farmers and their cooperatives, but–depending on the volume and scale–producers, physically not connected consumers, and even "subject to market regulation" states. Consequently, this inclusive and completely well-aligned effort is required to enforce the best possible implementation. Summarizing the points raised so

far, the potential of blockchain technology to enhance transparency, traceability, and efficiency in the agricultural market is worth more than the chances that these implementations are impossible. Working examples create potential applicability and expansions of blockchain in the agricultural sector as this market grows, effectively generating a safer, more reliable, and more equitable food system.

18.15 CONCLUSION

Integrating blockchain with smart farming technology is very supportive to tackle the problems and issues associated with the agricultural sector. Blockchain not only simplifies operations by improving traceability, transparency, and resource optimization along the supply chain but also attributes new approaches to increase efficiency through the cutting-edge technology. On the other hand, blockchain systems feature smart contracts that automate a variety of processes, secure transactions, and strengthen trust relationships between participants in the agricultural industry. Additionally, the technology facilitates the decision-making process based on data through the secure exchange of information, interoperability, and predictive analyses arising from real-time data from IoT devices and sensor exploitation. In other words, incorporating blockchain technology in agricultural practices translates into increased efficiency and affordability, pointing out concrete examples and success stories that demonstrate the technology's potential. The trials and full deployments life have taught many lessons by demonstrating the real effect of blockchain efforts on the practices of farming. The future of blockchain-based intelligent farming looks optimistic. Evolving patterns and tech advancements involving IoT, AI, and 5G connections will further unlock blockchain power in agriculture. However, it cannot be achieved without concerted efforts from policy influencers, industry leaders, researchers, but most importantly, the farmers. Addressing the regulatory and ethical concerns associated with utilizing the blockchain, including legislative issues, data privacy, and intellectual property rights, is vital. Cooperation among players is critical for fostering a stronger, transparent, and more secure food system for generations. This chapter is a call to action for ongoing initiatives that are to realize blockchain's full potential in agriculture. By innovation, creating partnerships, and confronting industry obstacles together, we can make blockchain-driven farming more common in the future and redefine how we grow and distribute agricultural products involved.

KEYWORDS

- **blockchain technology**
- **smart agriculture**
- **sustainability**
- **agriculture**
- **supply chain management**

REFERENCES

1. Akram, S. V., Malik, P. K., Singh, R., Anita, G., & Tanwar, S. (2020). Adoption of blockchain technology in various realms: Opportunities and challenges. *Security and Privacy*, *3*(5), e109. https: //doi.org/10.1002/spy2.109
2. Awan, S. H., Ahmed, S., Safwan, N., Najam, Z., Hashim, M. Z., & Safdar, T. (2019). Role of internet of things (IoT) with blockchain technology for the development of smart farming. *Journal of Mechanics of Continua and Mathematical Sciences*, *14*(5), 170–188. http: //doi.org/10.26782/jmcms.2019.10.00014
3. Chinaka, M. (2016). *Blockchain Technology--applications in Improving Financial Inclusion in Developing Economies: Case Study for Small Scale Agriculture in Africa* (pp. 44–45) (Doctoral dissertation, Massachusetts Institute of Technology). http: //dspace.mit.edu/handle/1721.1/7582
4. Chohan, U. W. (2017). The double spending problem and cryptocurrencies. Available at SSRN 3090174. https: //dx.doi.org/10.2139/ssrn.3090174
5. Clapp, J. (2021). The problem with growing corporate concentration and power in the global food system. *Nature Food*, *2*(6), 404–408. https: //doi.org/10.1038/s43016-021-00297-7
6. Cohn, J., Curry, E., & El Ama, R. (2019). *Blockchain for Agriculture: Opportunities and Challenges*. White Paper, Deloitte Development LLC.
7. Dierksmeier, C., & Seele, P. (2019). Blockchain and business ethics. *Business Ethics: A European Review, 29*, 348–359 https: //doi.org/10.1111/beer.12259
8. Ferrag, M. A., Shu, L. X., Yang, A., & Maglaras, L. (2020). Security and privacy for green IoT-based agriculture: Review, blockchain solutions, and challenges. *IEEE Access*, *8*, 32031–32053, https: //doi.org/10.1109/ACCESS.2020.2973178
9. Ge, L., Brewster, C., Spek, J., Smeenk, A., & Top, J. (2017). *Blockchain for Agriculture and Food: Findings from the Pilot Study*. Wageningen Economic Research. ISBN 9789463438179-33 https: //doi.org/10.18174/426747
10. Gupta, D., Chaudhary, A. K., Singh, V. K., Verma, D., Goh, K. L., & Sharma, M. (2023). Thermo-mechanical analysis of bhimal fiber (Grewia Optiva)-$CaCO_3$/Flyash/TiO_2 reinforced epoxy bio-composites. *Industrial Crops and Products*, *204*, 117341. https: //doi.org/10.1016/j.indcrop.2023.117341.
11. Gupta, D., Chaudhary, A. K., Verma, D., & Sharma, M. (2024). Chapter 16 - bioresorbable polymers/HNT blend composite wound dressings. In: Verma, D., Okhawilai, M., Goh, K. L., Ramakrishna, S., Pasbakhsh, P., & Mohit B T - Bioresorbable Polymers and their Composites Sharma (Eds.), *Woodhead Publishing Series in Biomaterials* (pp. 385–419). Woodhead Publishing. doi: https: //doi.org/10.1016/B978-0-443-18915-9.00008-2.
12. Jaroodi, J., & Mohamed, N. (2019). Blockchain in industries: A survey. *IEEE Access*, *7*, 36500–36515. https: //doi.org/10.1109/ACCESS.2019.2936094
13. Javaid, M., Haleem, A., Singh, R. P., & Suman, R. (2022). Enhancing smart farming through the applications of Agriculture 4.0 technologies. *International Journal of Intelligent Networks*, *3*, 150–164. https: //doi.org/10.1016/j.ijin.2022.09.004
14. Kamble, S. S., Gunasekaran, A., & Arha, H. (2019). Understanding the Blockchain technology adoption in supply chains—Indian context. *International Journal of Production Research*, *57*(7), 2009–2033. https: //doi.org/10.1080/00207543.2018.1518610
15. Kamble, S. S., Gunasekaran, A., & Sharma, R. (2020). Modeling the blockchain enabled traceability in agriculture supply chain. *International Journal of Information Management*, *52*, 101967. https: //doi.org/10.1016/j.ijinfomgt.2019.05.023
16. Kamilaris, A., Fonts, A., & Prenafeta-Boldv, F. X. (2019). The rise of blockchain technology in agriculture and food supply chains. *Trends in Food Science & Technology*, *91*, 640–652. https: //doi.org/10.1016/j.tifs.2019.07.034
17. Kshetri, N. (2018). 1 Blockchain's roles in meeting key supply chain management objectives. *International Journal of Information Management*, *39*, 80–89. https: //doi.org/10.1016/j.ijinfomgt.2017.12.005
18. Lasla, N., Al-Sahan, L., Abdallah, M., & Younis, M. (2022). Green-PoW: An energy-efficient blockchain Proof-of-Work consensus algorithm. *Computer Networks*, *214*, 109118. https: //doi.org/10.1016/j.comnet.2022.109118

19. Lezoche, M., Panetto, H., Kacprzyk, J., Hernández, J. E., & Alemany Díaz, M. M. (2020). Agri-food 4.0: A survey of the supply chains and technologies for the future agriculture. *Computers in Industry*, *117*, 103187. https: //doi.org/10.1016/j.compind.2020.103187
20. Lin, W., Huang, X., Fang, H., Wang, V., Hua, Y., Wang, J., Yin, H., Yi, D., & Yau, L. (2020). Blockchain technology in current agricultural systems: From techniques to applications. *IEEE Access*, *8*, 143920–143937.http: //dx.doi.org/10.1109/ACCESS.2020.3014522
21. Mahmood, Z., & Jusas, V. (2022). Blockchain-enabled: Multi-layered security federated learning platform for preserving data privacy. *Electronics*, *11*(10), 1624.https: //dx.doi.org/10.3390/electronics11101624
22. Mzuku, K. (2019). A decentralised asset registry to expand access to finance for the agricultural sector in South Africa. Faculty of Commerce, African Institute of Financial Markets and Risk Management. Retrieved from http: //hdl.handle.net/11427/31069
23. Nizetic, S., Solic, P., Gonzalez-De, D. L. D. I., & Patrono, L. (2020). Internet of Things (IoT): Opportunities, issues and challenges towards a smart and sustainable future. *Journal of Cleaner Production*, *274*, 122877. https: //doi.org/10.1016/j.jclepro.2020.122877
24. Patel, H., & Shrimali, B. (2023). *AgriOnBlock: Secured Data Harvesting for Agriculture Sector using Blockchain Technology*. ICT Express. https: //doi.org/10.1016/j.icte.2021.07.003
25. Pearson, S., May, D., Leontidis, G., Swainson, M., Brewer, S., Bidaut, L., & Zisman, A. (2019). Are distributed ledger technologies the panacea for food traceability. *Global Food Security*, *20*, 145–149. https: //doi.org/10.1016/j.gfs.2019.02.002
26. Prashar, D., Jha, N., Jha, S., Lee, Y., & Joshi, G. P. (2020). Blockchain-based traceability and visibility for agricultural products: A decentralized way of ensuring food safety in India. *Sustainability*, *12*(8), 3497. https: //doi.org/10.3390/su12083497
27. Runck, B. C., Joglekar, A., Silverstein, K. A., Chan-Kang, C., Pardey, P. G., & Wilgenbusch, J. C. (2022). Digital agriculture platforms: Driving data-enabled agricultural innovation in a world fraught with privacy and security concerns. *Agronomy Journal*, *114*(5), 2635–2643. https: //doi.org/10.1002/agj2.20873
28. Shen, C., & Pena-Mora, F. (2018). Blockchain for cities—A systematic literature review. *IEEE Access*, *6*, 76787–76819. https: //doi.org/10.1109/ACCESS.2018.2880744
29. Sylvester, G. (2019). *E-agriculture in Action: Blockchain for Agriculture Opportunities and Challenges*. Food and Agriculture Organization of the United Nations. http: //www.fao.org/3/ca2906en/ca2906en.pdf
30. Torky, M., & Hassanein, A., E. (2020). Integrating blockchain and the internet of things in precision agriculture: Analysis, opportunities, and challenges. *Computers and Electronics in Agriculture*, *178*, 105476. https: //doi.org/10.1016/j.compag.2020.105476
31. Tripoli, M., & Schmidhuber, J. (2018). *Emerging Opportunities for the Application of Blockchain in the Agri-food Industry*. FAO and ICTSD: Rome and Geneva. Licence: CC BY-NC-SA. 3. IGO, ISSN 1817-356X
32. Zhang, B., Zhang, A., Nastenko, H., & Oliynykov, R. (2023). Privacy-preserving decision-making over blockchain. *IEEE Transactions on Dependable and Secure Computing*, *20*(6), 4648–4663, "Privacy-Preserving Decision-Making Over Blockchain." https: //dl.acm.org/doi/10.1109/TDSC.2022.3231237.
33. Zyskind, G., & Nathan, O. (2015). Decentralizing privacy: Using blockchain to protect personal data. *IEEE Security and Privacy Workshops (SPW)* (pp. 180–184). IEEE. https: //doi.org/10.1109/SPW.2015.27

CHAPTER 19

Cultivating a Greener Future: Agriculture 5.0 and the Dawn of Sustainable Farming

DEEPAK GUPTA[1,2], ALKA VERMA[3*], ARUN KUMAR CHAUDHARY[4], DIPESH KUMAR MISHRA[5], BHAVANA MATHUR[6], and ROHIT SAHU[7]

[1]*Mechanical Engineering Department, Graphic Era Hill University, Dehradun, Uttarakhand, India*

[2]*Mechanical Engineering Department, Graphic Era (deemed to be) University, Dehradun, Uttarakhand, India*

[3]*College of Agriculture, G. B. Pant University of Agriculture and Technology, Pantnagar, Uttarakhand, India*

[4]*Industrial and Production Engineering Department, G B Pant University of Agriculture and Technology Pantnagar, Uttarakhand, India*

[5]*Mechanical Engineering Department, Madan Mohan Malaviya University of Technology, Gorakhpur, Uttar Pradesh, India*

[6]*Mechanical Engineering Department, Anand International College of Engineering, Jaipur, Rajasthan, India*

[7]*Department of Mechanical Engineering, G L Bajaj Institute of Technology and Management, Greater Noida, Uttar Pradesh, India*

*Corresponding author

ABSTRACT

This chapter discovers the new revolution in the area of agriculture via Agriculture 5.0. It signifies a fundamental shift in farming activities by the incorporation of smart technologies. The integration of artificial intelligence (AI), Internet of Things (IoT), and robotics, including big data statistical analysis, has played a major role in developing the upcoming future of sustainable agriculture. Although it has potential innovations, Agriculture 5.0 has barriers that include the digital divide, data privacy challenges, a skills deficit, and legal restrictions. Stakeholders can overcome these challenges and completely achieve the Agriculture 5.0 promises to a resilient and sustainable agricultural future through promoting collaboration and strategic actions. Agriculture's future relies on being able to feed the rising population, including environmental protection. By integrating automated machinery, AI, and the IoT to promote higher productivity, optimal resource use, and increased sustainability,

Demystifying Sustainable Farming Using Smart Technologies.
Gunjan Chhabra, Keshav Kaushik, Priya Chugh, Susheela Dahiya, and Aryan Chaudhary (Eds.)

DOI: 10.1201/9781779640932-19

Agriculture 5.0 provides a glimmer of hope. This chapter explores the basic ideas, enabling technology, and upcoming problems of Agriculture 5.0, stressing its innovative capability. Through the deep understanding of its novel approach, this chapter emphasizes the foundation for agriculture to thrive harmoniously with the environment.

19.1 INTRODUCTION

Sustainable farming practices are important for optimizing the rate of food production and conserving the environment [60]. The implementation of innovative new technologies, such as precision agriculture activity and genetically modified crops, can contribute to sustainable agriculture [6]. Agroecological practices like crop rotation and organic farming also hold an important place in enhancing soil productivity as well as reducing pesticide use [6]. The utilization of local resources and conventional wisdom holds an important place in sustaining the equilibrium of ecosystems within agricultural practices. Drawing upon local resources and traditional knowledge not only enhances the ecological harmony but also contributes to the long-term viability of farming systems [51]. The future of farming, including space agriculture, can benefit from the interventions of Industrial Revolution (IR) 5.0 technologies [51]. Decision support systems that utilize digital technologies and artificial intelligence (AI) algorithms can provide actionable insights for sustainable farming. These technologies can optimize crop yields, improve resource management, and open up new opportunities for sustainable agriculture. Overall, the concept of Agriculture 5.0, which encompasses the utilization and incorporation of cutting-edge technologies, presents optimistic and encouraging prospects for development to a more environmentally friendly future by implementing and following sustainable agricultural methodologies that support to minimize negative impacts on the ecosystem and promote ecologically responsible farming practices.

The agricultural sector is confronted with a significant challenge. The nourishing and expanding global population amidst the diminishing resources as well as the shifting of climate. While old Agriculture 4.0 is characterized by automation and data-driven decision-making techniques, but Agriculture 5.0 envisions a future of heightened efficiency, including best resource utilization, and better sustainability through the incorporation of advanced skills such as AI, robotics, and IoT [84]. This chapter discussed the potential possibilities of Agriculture 5.0, also investigating its capability to work for sustainable farming.

During the constantly evolving scenario of the agriculture sector, the intelligent technologies integration has brought forth a new era known as Agriculture 5.0. This chapter delves into the forthcoming trajectory of farming, emphasizing the rise of Agriculture 5.0 and the obstacles it brings. Through a deeper exploration of this subject, we will demystify sustainable farming practices and clarify how intelligent technologies are reshaping the agricultural landscape [61]. Agriculture 5.0, also referred to as the fifth phase of agriculture, signifies a transition toward more sustainable and environmentally friendly farming methods. This new agricultural era prioritizes leveraging technology's capabilities to reduce environmental impact while amplifying productivity.

One fundamental aspect of Agriculture 5.0 involves incorporating precision farming methods (Figure 19.1), utilizing advanced tools like drones, sensors, and GPS systems to oversee and enhance crop production. By precisely gauging elements such as moisture in soil, nutrient availability, and pest numbers, farmers may have more information to make choices on when and where to apply resources, ultimately leading to greater yields and reduced wastage [80]. Another key element of Agriculture 5.0 is the adoption of regenerative farming techniques, focusing on the restoration and improvement of soil health. It is possible through the implementation of agricultural techniques such as cover cropping, which involves planting crops primarily to maintain soil health and fertility during periods when the land would otherwise be left bare, and crop rotation. It is a method of alternating the variation of crops grown in a particular agricultural land from season to season for preventing the depletion of soil nutrients as well as minimizing pest outbreaks. It is also a practice regarding dropping the frequency and amount of soil disturbance while farming. Thus, it is practicable to enhance the sustainable production rate in farming systems. Farmers can boost their resistance to climate change, lessen reliance on synthetic inputs, and sequester carbon in the soil through improving the soil quality and structure.

FIGURE 19.1 Precision crop protection.

Source: Reprinted from Ref. [46]. https: //creativecommons.org/licenses/by/4.0/

Agriculture 5.0 highlights the importance of biodiversity conservation and enriching natural habitats and supplements to the farms, simultaneously aiding the support of pollinators, beneficial insects calling, and related wildlife essential for maintaining the ecosystem. Maintaining a productive and balanced environment, it also contributes to better crop pollination and pest management [63]. Agriculture 5.0 embodies a fresh approach to farming that gives priority to sustainability, resilience, and environmental stewardship. By approving its ideologies and techniques, farmers will be able to enhance their own livelihoods, including helping to promote a cleaner and more sustainable future for all.

The beginning of Agriculture 5.0 signifies a novel model in smart agriculture, capitalizing on cutting-edge technologies like robotics, AI, quantum sensing, and 4D-communication, which collectively aim to reform the agricultural sector for smart farming to increase the productivity, sustainability, as well as efficiency through advanced digital solutions and interconnected systems [53]. The main objective is to make it advance in automation for agricultural processes in order to enhance productivity by minimizing the manual labor requirement simultaneously improving proficiency. All the while striving to minimize the water, fertilizers, and pesticides utilization, as well as the release of harmful pollutants into the environment. Agriculture 5.0 has the potential to take advantage of the developing 5G network infrastructure; however, the complete realization of Agriculture 5.0 will necessitate the utilization of the technological innovations provided by 6G-IoT networks. The incorporation of these innovative networks can significantly increase the efficiency as well as the productivity of agricultural practices. It is also preparing the path for a higher sustainable and technologically advanced, smart agricultural farming [46]. The utilization of AI, machine learning (ML), and other advanced innovative technologies holds an important place in effectively tackling the present and upcoming obstacles related to agricultural farming protection in Agriculture 5.0 context. These advanced technologies are essential for upgrading the proficiency and sustainable approach of agricultural practices against the evolving challenges [77]. Digitalization based on Society 5.0 principles can enhance the elasticity and sustainability in agricultural value chains as per the weather variation [31]. Industry 5.0 offers various advantages to the agricultural domain, such as enhanced flexibility, efficacy, output, and cost-efficiency [17]. Urban Agriculture 5.0 is a multifaceted concept that integrates systemic agriculture into smart cities, promoting sustainability and a resilient urban food system.

19.2 THE EVOLUTION OF AGRICULTURE FROM MANUAL LABOR TO AI-POWERED SYSTEMS

Throughout history, agriculture has heavily depended on manual labor and basic tools. The 20th century helped in recent decades with the introduction of mechanization and chemical fertilizers, resulting in improved yields.

However, these advancements brought about environmental degradation and soil depletion as significant issues. In our current phase of Agricultural 4.0, the focus has shifted toward precision farming through the utilization of sensors, data analysis, and automation to enhance resource efficiency [74].

The evolution of agriculture from manual labor to AI-powered systems has been driven by the need to fulfill the increasing demands of food for the continuously increasing population and improve productivity. Traditional farming methods were insufficient to fulfill these requirements, leading to the adoption of automation in agriculture [2]. AI technologies have played a crucial role in this evolution, enabling agriculturalists to achieve more with fewer, as well as improving the crops quality. AI-powered systems have helped protect yields from climate change, food security issues, and employment problems. These systems utilize AI sensors to detect and target weeds, identify diseases in plants, and optimize the use of herbicides [49]. The induction of AI in farming has seen significant evolution recently. While, with several AI techniques, such as ML, convolutional neural networks, IoT, big data, robotics, and computer vision, are being used [11]. Despite challenges such as cost, lack of expertise, and data requirements, AI has much potential to revolutionize agricultural activities and address the needs for a growing world population (Figure 19.2).

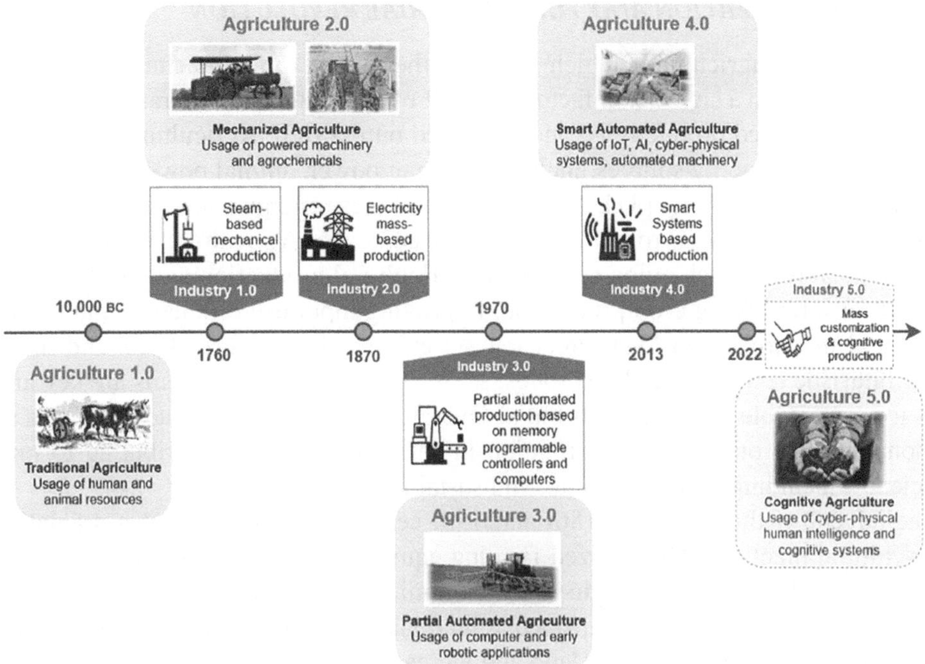

FIGURE 19.2 Agricultural revolutions.

Source: Reprinted from Ref. [71]. Copyright © 2024 by the authors. Licensee MDPI, Basel, Switzerland. https://creativecommons.org/licenses/by/4.0/

19.2.1 EARLY AGRICULTURAL PRACTICES

Early agricultural practices varied across different regions and time periods. In the Haojiatai site in China, foxtail millet was the main staple crop in periods of the Longshan and Xinzhai, along with common millet, rice, and soybean. Cultural factors such as agricultural civilizations, and food inclinations likely influenced the crop pattern [12]. In southern countries, intercropping systems, such as maize/cowpea row intercropping, have been found to improve crop growth and address fertile soil pressure concerns [12]. The wood-made tools found at La Draga in the Iberian Peninsula provide insights into the agricultural practices of early Neolithic communities, particularly the use of digging sticks and sickle handles [75]. In Wales, the early Neolithic period saw the adoption of cereal cultivation alongside continued wild plant exploitation, with wheat being the leading produce agri-product. The cultivation conditions in that particular area may be less intensive in comparison to other regions [78]. Isotopic analysis of barley grains through unconventional harvesting plots in Morocco at present and old agricultural land in the east of the Mediterranean helps to recognize past agricultural practices and their environmental impact [70].

In the earliest stages of human civilization, agriculture emerged as a means of sustenance and survival. Hunter-gatherer societies transitioned to settled agricultural communities, cultivating crops and domesticating animals to meet their food needs. These early farmers relied on manual labor and traditional farming techniques, employing rudimentary tools such as hand-held implements and animal-drawn plows [76].

19.2.2 MECHANIZATION AND THE INDUSTRIAL REVOLUTION

Mechanization in agriculture, which involves the use of tools, machines, and powered equipment, has been a significant factor in the IR of agriculture. It has transformed farming techniques, increased productivity, and improved output [86]. Agricultural mechanization relies on different energy sources, including human power, animal power, and mechanical power fueled by technology [37]. The utilization of these energy sources varies by country and depends on regional surroundings [7]. In the context of Agriculture 5.0, mechanization plays a crucial role in achieving sustainable agricultural production and productivity. It is essential for intensifying crop production, improving input use efficiency, and promoting agricultural modernization [57]. Precision agriculture, digital agriculture, and the adoption of futuristic technologies like drones, sensors, and computer systems are key areas of focus for sustainable agriculture in India. Overall, mechanization in agriculture is a vital component of the ongoing transformation in smart agriculture, contributing to improve proficiency, including production rate, and sustainable practices.

The advent of the IR in the 18th and 19th centuries brought about a dramatic shift in agricultural practices. Mechanized farming equipment, powered by steam engines and later by internal combustion engines, revolutionized agricultural productivity. Tractors, combines, and other machinery replaced animal labor and human muscle power, enabling farmers to cultivate larger areas of land and increase yields [58].

19.2.3 PRECISION AGRICULTURE AND THE RISE OF TECHNOLOGY

Precision agriculture is also called site-specific farming. It is a management methodology and aims to increase the production of crops by professionally matching resource inputs with crop needs. It utilizes a forward-thinking innovative approach by integrating satellite imagery, GPS, sensors, and ML for detecting, measuring, and responding to the needs of crops. Precision agriculture practices include variable application of nitrogen fertilizer, precise irrigation and fertilizer management, accurate pest and disease management, and precise livestock farming [34, 48]. The agricultural evolution is having rise of Agriculture 4.0 and Agriculture 5.0, which leverage technologies like the IoT, robotics, AI, and 5G networks to further automate and increase agricultural productivity [53]. These technologies, along with AI and ML, hold an important place in addressing current and future challenges in crop protection, such as detecting and controlling diseases related to crop harvesting, including weeds, and pests [46]. Precision agriculture has become a key solution for improving the productivity, proficiency, and sustainability of agriculture. It is having future lies in technological innovation, sustainable farming practices, data analytics, and policy interventions [40]

The late 20th and early 21st centuries witnessed the emergence of precision agriculture, characterized by the use of advanced technologies to optimize farm management practices. Early innovations such as GPS-guided tractors and yield monitoring systems laid the groundwork for more sophisticated approaches to farming [44]. Precision agriculture enables farmers to precisely utilize fertilizers and pesticides, resulting in enhanced proficiency, including environmental sustainability [30]

19.3 UNDERSTANDING AGRICULTURE 5.0

Agriculture 5.0 embodies an incorporation of advanced innovative unconventional methodologies, including AI, IoT, robotics, and big data analytics into agricultural practices. In contrast to its predecessors, it also emphasizes sustainable practices implementation, proficiency, including resilience. Through utilizing data-driven insights and automation, farmers can optimize resource utilization, minimize environmental impact, and increase productivity.

Let us delve into the smart technologies that will drive Agriculture 5.0.

19.3.1 ARTIFICIAL INTELLIGENCE (AI)

AI holds an important place in Agriculture 5.0 by revolutionizing the sector through various applications. AI, ML, robotics, and computer applications [49], have been utilized for agricultural activities management, plant disease forecasting, as well as pest management [4]. These technologies enhance productivity, optimize resource utilization, and improve decision-making processes [36]. Additionally, AI enables precision farming, crop-monitoring, forecasting analysis, and supply chain management, including optimization [45], contributing to increased efficiency, reduced leftovers, and enhanced safety and quality of food products [73]. Despite the benefits, challenges like unpopularity and skill development among farmers and lesser investments exist. Collaborative efforts among students, researchers, industrialists, educators, and farmers are essential for removing these barriers to develop new business models in AI-driven agriculture. AI has been poised to examine extensive sets of data for forecasting the growth trends, detecting diseases, and enhancing resource allocation. Robots powered by AI will undertake intricate tasks such as weed recognition and precise spraying. Envision a setup where drone-captured images of a field are scrutinized to pinpoint specific weeds, guiding robots to apply precise herbicides solely to those areas, thereby reducing unintended harm to beneficial insects and the ecosystem. AI-driven systems analyze extensive data to offer practical insights for farmers. From using predictive analytics for crop management to detecting diseases in livestock, AI transforms decision-making processes within the agricultural sector.

19.3.2 INTERNET OF THINGS (IOT)

This is also an important tool in the evolution of Agriculture 5.0, offering innovative solutions to enhance agricultural practices [3, 14, 28, 50, 56]. These technologies enable the advancement of the smart irrigation systems that utilize accurate sensors used for moisture detection in soil, as well as ultrasonic sensors to optimize water usage and enhance soil moisture management. IoT applications in agriculture range from climate monitoring to crop prediction, contributing to increased productivity and resource efficiency. Challenges include real-time operational data requirements and understanding plant responses to environmental changes for effective system design. Overall, incorporating IoT in agriculture holds promise for meeting food demand, improving food quality, and ensuring sustainable agricultural practices against the climate variation challenges. An integrated group of sensors on the ground, plants, and machinery will gather up-to-the-minute information on soil moisture, nutrient levels, health status of livestock, and weather patterns in real-time. This information will then be inputted into AI systems to make informed decisions. Picture an arrangement of ground moisture sensors consistently transmitting data to a central hub [15]. This information can then be examined by AI to identify irrigation requirements in real time. This gathered information also helps farmers to have well-informed choices, improve watering practices, which result in boosting their crop production yields.

19.3.3 ROBOTICS AND AUTOMATION

Agricultural robots streamline labor-intensive activities such as planting, harvesting, and weeding. Implementing self-governing machinery enables farmers to enhance productivity, cut down on labor expenses, and decrease chemical usage.

This new paradigm leverages technologies like robotics, AI, and IoT to revolutionize farming practices (Figure 19.3). Agriculture 5.0 integrates advanced innovative methodologies for enhancing productivity, minimizing inputs, and ensuring sustainability. Robotics systems, such as the agricultural robotic system, are designed to automate tasks like planting, harvesting, and detection [85]. Additionally, the use of AI and ML for crop protection is a key aspect of Agriculture 5.0, enabling real-time decision-making and actuation for efficient crop management [53]. Furthermore, the transition to more sustainable agrifood production, as seen in Agriculture 5.0, emphasizes an important place in addressing resource depletion and the impact of the environment on agriculture [46].

FIGURE 19.3 Drones application.

Source: Iberdrola. https: //www.iberdrola.com/about-us/our-innovation-model/smart-farming-precision-agriculture

19.3.4 BIG DATA ANALYTICS

The incorporation of advanced data analysis in agriculture enhances precision farming by offering insights into the performance of crops, the health of soil, and market trends. By utilizing data-driven strategies, farmers can adjust resource allocation, reduce wastage, and maximize earnings effectively. Big data analytics plays a crucial role in the evolution of agriculture toward Agriculture 5.0 by leveraging vast amounts of data to enhance productivity and sustainability [5, 8, 26, 39, 41]. It enables the development of smart solutions for agricultural challenges, such as climate change impacts, crop selection based on location-specific parameters, and long- and short-term predictions for production and marketing. Farmers will be able to make correct decisions to optimize resource allocation as well as agricultural practices through big data examination. The integration of advanced tools like data fusion further enhances the efficiency and accuracy of agricultural processes, ensuring a smooth transition toward Agriculture 5.0.

19.4 AGRICULTURE 5.0 PROMISES

It builds for advancements in integrating human expertise with the power of AI and robotics. These next-generation techniques are categorized as:

19.4.1 HYPER-PERSONALIZATION

Hyper-personalization in agriculture, within the context of Agriculture 5.0, involves tailoring agricultural practices to individual needs using these AI-based technologies. This approach provides a customized solution for farmers with the help of real-time data collection and examination to support enhancing the efficiency and effectiveness of agricultural management. Agriculture 5.0 represents the evolution of smart precision farming, integrating digitization to optimize agricultural processes [64]. By leveraging personalized sustainable agriculture through techniques like conditional average treatment effect estimation and causal machine learning, farmers can enhance their practices, transform agriculture into a carbon sink, and contribute to combating climate change [1, 13]. This personalized approach is crucial for achieving sustainable intensification and maximizing the global carbon sink potential. Sensors and AI will assess up-to-the-minute data concerning soil quality, weather trends, and plant development, allowing for precise, customized actions for each individual plant. Picture an AI system adjusting nutrient distribution with extreme accuracy to cater to the specific requirements of each plant, enhancing growth while reducing wastage.

19.4.2 PRECISION RESOURCE MANAGEMENT

Irrigation systems are poised to shift toward automation, channeling water precisely where and when it is most required. The optimization of fertilizer application will be tailored to address real-time nutrient deficiencies. Picture agricultural fields equipped with embedded sensors continuously observing moisture levels, initiating automated irrigation only when deemed essential [52]. Such advancements hold the potential to significantly diminish water wastage, marking a crucial advancement toward sustainable water management [35].

19.4.3 AUTONOMOUS OPERATIONS

Robotic technology will be responsible for repetitive actions related to planting, weeding, and crop harvesting, etc., which will enable human workers to focus on more complex decision-making processes. Robots integrated with AI and advanced visual systems have the capability to precisely maneuver through fields, identify specific crops with accuracy, and carry out tasks with minimal need for human involvement [52]. This advancement improves overall proficiency while simultaneously minimizing the physical requirements on farmers, empowering them to concentrate on strategic farm management.

19.4.4 ENHANCED TRACEABILITY

Blockchain technology aims to trace the journey of crops produced from the agricultural land site to the end consumer's table, guaranteeing transparency and building trust in sustainable farming methods. Picture a scenario where buyers can scan a QR code on their produce and promptly access a comprehensive breakdown of the agricultural techniques applied, starting from seed selection through to harvest. This process cultivates confidence among consumers, encouraging the adoption of sustainable food options.

19.4.5 CLIMATE-SMART AGRICULTURE

AI is poised to forecast weather patterns and recommend proactive steps for controlling the climate variation impact. Modern AI systems have the capability to examine historical data and upcoming weather predictions to anticipate occurrences like droughts, floods, or pest infestations. This helps farmers for taking proactive measures like adjusting planting timelines, adopting water-saving techniques, or cultivating pest-resistant crops [43].

19.5 THE PILLARS OF SUSTAINABLE FARMING IN AGRICULTURE 5.0

Agriculture 5.0 goes beyond mere efficiency. It places sustainability at the core, with the following key pillars:

19.5.1 ENVIRONMENTAL SUSTAINABILITY

By optimizing the utilization of resources, Agriculture 5.0 has the capability to significantly decrease water usage, reduce the overflow of fertilizers, and enhance the health of the soil [59]. Precision in irrigation and the application of fertilizers work toward minimizing wastage, while monitoring systems powered by AI can identify and tackle potential problems such as soil erosion or disease outbreaks before they become serious [54].

19.5.2 ECONOMICALLY SUSTAINABLE

Efficient resource utilization and automation-integrated advancements are able to achieve cost reductions for farmers, which results in improving their financial sustainability as well [81]. The adoption of more precise water usage, targeted fertilizer application, and decreased labor expenses will all play a role in bolstering the economic success of farms [32].

19.5.3 SOCIAL SUSTAINABILITY

Agriculture 5.0 has the potential to generate fresh employment opportunities in technology innovation, data analysis, and agricultural management, rejuvenating rural areas. Although some traditional farming roles may become automated, novel pathways will emerge in fields such as system development, data organization, and AI incorporation [79]. This transition could help close the gap between urban and rural regions, enticing experienced professionals to return to countryside communities [55].

19.6 BENEFITS OF AI IN AGRICULTURE

Agriculture 5.0 offers several benefits to the farmers as well as agricultural industries, including fast and better quick responsiveness, proficiency, production rate, and accurate decision-making ability in an economical manner [31]. It can contribute to the prosperity of farmers by enabling the steady conversion to efficient energy-smart farms, providing economical processes, climate information in real time, remote access online monitoring, and sustainable energy solutions [55]. In combination with greener energy resources, it will lead to sustainable and energy-efficient farms, through cost reduction and promoting sustainability [77]. Additionally, by integrating technologies 4.0 and Society 5.0 [17], Agriculture 5.0 is able to contribute to building an extra robust urban food system in smart cities, promoting social, economic, and environmental gains [53].

It also represents the culmination of centuries of agricultural innovation, combining advanced AI-based technologies to revolutionize farming practices [72]. AI-powered systems examine data collected through sensors, drones, and different sources to provide actionable insights for farmers. From forecasting yields of crop for optimizing irrigation schedules, AI enables farmers to make the right decision based on collected data that will also enhance productivity and a sustainable approach. The incorporation of AI into agriculture offers a plethora of benefits.

Precision farming: AI-based methodologies can examine the data related to soil composition, climate variation patterns, and harvested product's health for controlling the farming practices to specific conditions, maximizing yields while minimizing resource usage.

Predictive analytics: AI-powered predictive system can easily predict detailed crop yields, pest control, market trends, and help farmers take proactive action against the associated risks to achieve optimized production.

Crop monitoring and management: AI-based robotics drones, satellites can monitor agricultural activities and requirements from above, identifying areas of stress or disease outbreaks that are difficult to see via open human eyes. This capability of detecting diseases in the early stage helps farmers take timely reaction to mitigate losses by maintaining crop health [21].

Optimized resource allocation: By analyzing data on moisture quantity present in soil, nutrient availability, and climate variation prediction, AI can recommend accurate irrigation and fertilization schedules, reducing waste and environmental impact.

19.7 CHALLENGES OF AGRICULTURE 5.0

Agriculture 5.0 faces several challenges. These challenges include the weakening of agricultural land, farming activities, a higher pests utilization due to climate change and an increasing birth rate in correspondence to population that requires ensuring food security [46]. Climate change also threatens the production capability, which results in affecting the sustainable agricultural systems as well as food availability globally [77]. For example, the agriculture sector in Georgia faces challenges such as infrastructure weaknesses, limited technology involvement, and small land sizes [62]. Furthermore, agriculture is confronted with various challenges such as disappearing biodiversity, closing of deserts, weather variations, labor unavailability, and animal plague outbreaks [19]. To address these challenges, innovation is crucial, and the new era of agriculture is described through the newly induction crop harvesting and cultivation concepts, robots interference, fossil-free energy, smart connected systems, and animal welfare technology systems [47]. The application of Agriculture 5.0 is still in the early stage, and farmers need to adapt and develop their skills to meet the demands of the IR of Agriculture 5.0.

While Agriculture 5.0 offers promising solutions to modern agricultural challenges, it also presents several hurdles that must be addressed:

19.7 1 DIGITAL DIVIDE

Access to advanced technologies and digital infrastructure remains uneven, particularly in rural regions. Narrowing the digital gap is crucial to guarantee the fair integration of Agriculture 5.0 methods among farmers globally.

19.7 2 DATA PRIVACY AND SECURITY

As agricultural systems grow more interconnected, the concerns regarding data privacy and cybersecurity are heightened. Ensuring the protection of sensitive agricultural data against cyber threats and unauthorized access is crucial for upholding trust in smart farming technologies.

19.7 3 SKILLS GAP

To realize the complete potential of Agriculture 5.0, it is imperative to have a proficient workforce skilled in technology integration and data analysis. Essential steps include investing in educational programs and training for farmers to bridge the skills gap and enable the smooth adoption of technology (Table 19.1).

TABLE 19.1 Era of Digital Farming Application

Application	Year	References
GPS-based animal tracking system	2008	[27]
Thermal and narrowband multispectral remote sensing unmanned aerial vehicle	2009	[25]
Development of a web-based weather station for irrigation	2014	[18]
Wireless robot for moisture sensing	2017	[33]
Weather forecasting and image analysis	2017	[69]
Robotics in digital farming: AI-driven yield management and widespread adoption in agriculture	2018	[82]
Impact of ML in precision agriculture to increase productivity and maximize yield	2018	[10]
Using drones for precision agriculture	2019	[38]
Agriculture 5.0	2019	[20]
Use of drones for precision agriculture	2019	[38]
Employing IoT technology for automated irrigation and frost protection systems	2019	[23]
Smart sustainable agriculture harnesses IoT and AI: Monitoring crop and soil health and enabling predictive analysis	2019	[66]
Smart irrigation support system integrates fuzzy decision support and GIS-based decision support for optimized water management	2020	[87]
Utilizing sensor-based crop monitoring systems for disease identification	2020	[16]
Co-operative smart farming by using the cyber-physical system and AI	2020	[67]
IoT-based smart farming system using AI & ML	2020	[9]
Unmanned farming: Autonomous operations driven by AI, ML, and IoT technologies	2021	[83]
Smart modular farming encompasses organic, indoor vertical, and livestock farming technologies for sustainable agriculture	2021	[42]
IoT and sensors revolutionize farming: Enhancing production and bridging old and new methods	2021	[24]
Crop growth analysis system integrates weather forecasting and image analysis for comprehensive crop monitoring	2021	[68]
Study of coconut oil extraction using robots	2022	[22]
Smart farming includes remote sensing for crops and smart solutions tailored for the animal industry	2022	[29]
By using AI techniques to explore the susceptibility of smart farming	2022	[29]

19.7 4 REGULATORY FRAMEWORKS

The swift progress of technology surpasses existing regulatory frameworks, resulting in legal and moral uncertainties within smart agriculture. Crafting sturdy guidelines that harmonize innovation with safety and sustainability is essential to steer through the changing landscape of Agriculture 5.0.

19.8 CONCLUSION

Agriculture 5.0 has much potential for transforming food cultivation by focusing on sustainability, proficiency, and flexibility. Integrating AI-based smart methodologies enables farmers to tackle and work with global problems in the context of hunger issues, food supply chain security, climate variations, with limited availability of resources. Maximizing more benefits to farmers of Agriculture 5.0 for smart agriculture requires removing many correlated obstacles like digital access limitation, data examination and management, skill enhancement of farmers, and associated regulatory benefits and structures. A collective effort is required among all the stakeholders to address the associated challenges as well as to develop the advanced and sustainable futuristic farming approach. Agriculture 5.0 also supports to revolutionize of new smart agriculture practices, promoting an innovative future in agriculture for achieving efficiency and sustainability together. Although challenges like building infrastructure, protecting data, the incapability of farmers, lack of popularity of AI-based methodologies are having issues but implementing smart technologies is always significant. By implementing innovation, developing partnerships among farmers, industrialists, researchers, tutors, tech experts, and emphasizing ethics, we can navigate these obstacles and fully unleash the potential of Agriculture 5.0. This approach will lead us toward a smart future where farming not only nourishes a growing population but also helps to serve as guardians of a healthy environment, ensuring a secure and sustainable future for generations to come.

KEYWORDS

- **Agriculture 5.0**
- **artificial intelligence**
- **Internet of Things**
- **robotics**
- **sustainable agriculture**

REFERENCES

1. Ahmed, L., & Nabi, F. (2021). Agriculture 5.0—The future. In: *Agriculture 5.0: Artificial Intelligence, IoT, and Machine Learning* (pp. 187–203). CRC Press. https: //doi.org/10.1201/9781003125433-9.
2. Suresh, A., Archana, K. S., Kumar, A. S., Joseph, A., & Koshy, S. (2022). Overview of artificial intelligence in agriculture for enhancement of irrigation, application of pesticides and insecticide. *International Journal of Engineering Technology and Management Sciences*, *6*(5), 618–624. https: //doi.org/10.46647/IJETMS.2022.V06I05.098.
3. Babakhouya, A., Naji, A., Hnini, A., & Daaif, A. (2023). Agricultural IoT technology: An overview of usages, technologies, and challenges. In *2023 3rd International Conference on Innovative Research in Applied Science, Engineering and Technology (IRASET)*. Hassan II University,2IACS Laboratory, ENSET, Mohammedia, Morocco; Hassan II University, Signals, Distributed Systems and Artificial Intelligence (SSDIA) Laboratory ENSET, Mohammedia, Morocco; LAVETE Laboratory, Hassan First

University, Department of Mathematics and Informatics FST Settat, Settat, Morocco: IEEE. https: //doi.org/10.1109/iraset57153.2023.10153038.

4. Cavazza, A., Mas, F. D., Paoloni, P., & Manzo, M. (2023). Artificial intelligence and new business models in agriculture: A structured literature review and future research agenda. *British Food Journal, 125*(13). 436–461. https: //doi.org/10.1108/BFJ-02-2023-0132.
5. Cheng, J. (2022). Application of big data analysis to agricultural production, agricultural product marketing, and influencing factors in intelligent agriculture. *Journal of Computer Science and Technology, 29*(3), 151–165. https: //doi.org/10.20532/cit.2021.1005404.
6. Concepcion, R., Ramirez, T. J., Alejandrino, J., Janairo, A. G., Baun, J. J., Francisco, K, Relano, R.-J. et al. (2022). A look at the near future: Industry 5.0 boosts the potential of sustainable space agriculture. In *2022 IEEE 14th International Conference on Humanoid, Nanotechnology, Information Technology, Communication and Control, Environment, and Management (HNICEM)*, 1–6. https: //doi.org/10.1109/HNICEM57413.2022.10109559.
7. Cr, M., & Bhopal Director ICAR—Central Institute of Agricultural Engineering Madhya Pradesh. (2022). Trend in sustainable mechanization of Indian Agriculture. *Journal of Rice Researcg, 15*(Special Issue). https: //doi.org/10.58297/eqre1792.
8. Cravero, A., Bustamante, A., Negrier, M., & Galeas, P. (2022). Agricultural big data architectures in the context of climate change: A systematic literature review. *Sustainability, 14*(13), 7855. https: //doi.org/10.3390/su14137855.
9. Dahane, A., Benameur, R., Bouabdellah, K., & Benyamina, A. (2020). An IoT based smart farming system using machine learning. https: //doi.org/10.1109/ISNCC49221.2020.9297341.
10. Das, S., Ghosh, I., Banerjee, G., & Sarkar, U. (2018). Artificial intelligence in agriculture: A literature survey May.
11. Dawn, N, Ghosh, T., Ghosh, S., Saha, A., Mukherjee, P., Sarkar, S., Guha, S., & Sanyal, T. (2023). Implementation of artificial intelligence, machine learning, and Internet Of Things (IoT) in revolutionizing agriculture: A review on recent trends and challenges. *International Journal of Experimental Research and Review, 30*. https: //doi.org/10.52756/IJERR.2023.V30.018.
12. Deng, Z., Zhang, H., Li, W., Liang, F., & Cao, Y. (2021). A preliminary study of early agriculture practices at the Haojiatai Site in Luohe City, Henan Province. *Science China Earth Sciences, 64*(2). 307–317. https: //doi.org/10.1007/s11430-020-9694-3.
13. Desaid, D. (2022). Hyper-personalization: An AI-enabled personalization for customer-centric marketing. (pp. 40–53). https: //doi.org/10.4018/978-1-7998-7959-6.ch003.
14. Dhal, S., Wyatt, B.M., Mahanta, S., Bhattarai, N., Sharma, S., Rout, T., Saud, P., & Acharya, B. S. (2023). Internet of things (IoT) in digital agriculture: An overview. *Agronomy Journal, 116*(3). https: //doi.org/https: //doi.org/10.1002/agj2.21385.
15. Dhanaraju, M., Chenniappan, P., Ramalingam, K., Pazhanivelan, S., & Kaliaperumal, R. (2022). smart farming: Internet of Things (IoT)-based sustainable agriculture. *Agriculture, 12*(10), 1745. https: //doi.org/10.3390/agriculture12101745.
16. Farooq, M. S., Riaz, S., Abid, A., Umer, T., & Zikria, Y. B . (2020). Role of IoT technology in agriculture: A systematic literature review. *Electronics, 9*(2), 319. https: //doi.org/10.3390/electronics9020319.
17. Ferreira, A. C. D., Titotto, S. L. M. C., & Akkari, A. C. S. (2022). Urban agriculture 5.0: an exploratory approach to the food system in a super smart society. *International Journal of Mathematical Engineering and Management Sciences, 7*(4), 455–475. https: //doi.org/10.33889/ijmems.2022.7.4.030.
18. Fourati, M. A., Chebbi, W., & Kamoun, A. (2014). Development of a web-based weather station for irrigation scheduling. In *2014 Third IEEE International Colloquium in Information Science and Technology (CIST)* (pp. 37–42). https: //api.semanticscholar.org/CorpusID: 2210644.
19. Frankelius, P., & von Rosen, M. (2023). Towards the fifth innovation era in agricultural innovation. In: Keswani, C., Possas, C., Koukios, E., & B. T. Davide (Eds.), *Agricultural Bioeconomy Viaggi* (pp. 205 229). Academic Press. https: //doi.org/https: //doi.org/10.1016/B978-0-323-90569-5.00003-2.
20. Fraser, E., & Campbell, M. (2019). Agriculture 5.0: Reconciling production with planetary health. *One Earth, 1*(November), 278–280. https: //doi.org/10.1016/j.oneear.2019.10.022.

21. Friedl, M. A. (2017). Remote sensing of croplands. *Comprehensive Remote Sensing, 1–9*(January), 78–95. https: //doi.org/10.1016/B978-0-12-409548-9.10379-3.
22. Gagliardi, G., Cosma, A., & Marasco, F. (2022). A decision support system for sustainable agriculture: The case study of coconut oil extraction process. *Agronomy, 12*(January), 177. https: //doi.org/10.3390/agronomy12010177.
23. Gómez, R., Real, K., Morán, C., Grijalva, P., & Recalde, T. (2019). IoT applications in agriculture: A systematic literature review. In *Proceedings of the Second International Conference*, CITAMA 2019, (pp. 68–76), Guayaquil, Ecuador, January 22–25, 2019. https: //doi.org/10.1007/978-3-030-10728-4_8.
24. Hassan, S., Alam, M., Illahi, U., Ghamdi, M. A., Almotiri, S., & Mazliham, M. (2021). A systematic review on monitoring and advanced control strategies in smart agriculture. *IEEE Access, 9*(February), 1. https: //doi.org/10.1109/ACCESS.2021.3057865.
25. Berni, J., Jose, A., Zarco-Tejada, P., Suárez, L., & Fereres, E. (2009). Thermal and narrowband multispectral remote sensing for vegetation monitoring from an unmanned aerial vehicle. *IEEE Transactions On Geoscience and Remote Sensing, 47*(April), 722–738. https: //doi.org/10.1109/TGRS.2008.2010457.
26. Madhuri, J., & Indiramma, M. (2023). Big data analytics-based agro advisory system for crop recommendation using Spark platform. In: Hossain, M. S., et al. (Eds.), *Handbook of Research on AI and Machine Learning Applications in Customer Support and Analytics* (pp. 227–247). IGI Global Scientific Publishing. https: //doi.org/10.4018/978-1-6684-7105-0.ch012.
27. Jain, V., Bagree, R., Kumar, Aman ., & Ranjan, P. (2009). WildCENSE: GPS based animal tracking system. https: //doi.org/10.1109/ISSNIP.2008.4762058.
28. Jaliyagoda, N., Lokuge, S., Gunathilake, P. M. P. C., Amaratunga, K. S. P., Weerakkody, W. A. P. , Bandaranayake, P. C. G., & Bandaranayake, A. U . (2023). Internet of Things (IoT) for smart agriculture: assembling and assessment of a low-cost IoT system for polytunnels. *PLoS One, 18*(5), e0278440. https: //doi.org/10.1371/journal.pone.0278440.
29. Jerhamre, E., Carlberg, C., & van Zoest, V. (2022). Exploring the susceptibility of smart farming: Identified opportunities and challenges. *Smart Agricultural Technology, 2*(November), 100026. https: // doi.org/10.1016/j.atech.2021.100026.
30. Johansen, K., Maltese, A., & McCabe, M. F. (2023). Monitoring agricultural ecosystems. *Unmanned Aerial Systems for Monitoring Soil, Vegetation, and Riverine Environments*, 125–151. https: //doi.org/10.1016/B978-0-323-85283-8.00013-8.
31. Kankekar, S. S., Dadwal, S. S., & Ali, A. (2023). Investigating the factors, challenges, and role of stakeholders in implementing industry 5.0 and its impact on supply chain operations. In: Dadwal, S., et al. (Eds.), *Opportunities and Challenges of Business 5.0 in Emerging Markets* (pp. 124–150). IGI Global Scientific Publishing. https: //doi.org/10.4018/978-1-6684-6403-8.ch007.
32. Konfo, T. R. C., Chabi, A. B. P., Gero, A. A., Lagnika, C., Avlessi, F., Biaou, G., & Sohounhloue, C. K. Dominique. (2024). Recent climate-smart innovations in agrifood to enhance producer incomes through sustainable solutions. *Journal of Agriculture and Food Research, 15*, 100985. https: //doi.org/https: //doi.org/10.1016/j.jafr.2024.100985.
33. Krishna, K., Omayo, S., Malende, W., & Anuradha, K. (2017). Internet of things application for implementation of smart agriculture system. In *2017 International Conference on I-SMAC (IoT in Social, Mobile, Analytics and Cloud) (I-SMAC)* (pp. 54–59*)*, Palladam, India. https: //doi.org/10.1109/I-SMAC.2017.8058236.
34. Kumari, A. India Department of Computer Science R.N.S. Institute of Technology Bangalore, Bhavanishankar, and Associate Professor Department of Computer Science R.N.S. Institute of Technology. 2023. "Precision Agriculture." *International Journal of Computing Algorithm, 12*(1). https: //doi.org/10.20894/ijcoa.101.012.001.004.
35. Kushwaha, N.L., Elbeltagi, A., Patel, A., Zakwan, M., Rajput, J., & Sharma, P. (2022). Assessment of water resources using remote sensing and GIS techniques. pp. 85–98. https: //doi.org/10.1016/B978-0-323-91910-4.00006-6.
36. Lebid, O. V., Kiporenko, S. S., & Vovk, V. Yu. (2023). Use of artificial intelligence technologies in agriculture: European experience and application in ukraine. *Èlektronnoe Modelirovanie*, *45*(June), 57–71. https: //doi.org/10.15407/emodel.45.03.057.

37. Liu, X., & Li, X. (2023). The influence of agricultural production mechanization on grain production capacity and efficiency. *Processes (Basel), 11*(2), 487. https: //doi.org/10.3390/pr11020487.
38. Liuzza, D., Silano, G., Picariello, F., Iannelli, L., Glielmo, L., De Vito, L., & Daponte, P. (2019). A review on the use of drones for precision agriculture. In *IOP Conference Series: Earth and Environmental Science* (vol. 275). https: //doi.org/10.1088/1755-1315/275/1/012022.
39. Lohit, V. S., & Mujahid, M. M. (2022). Big data analytics in developing smart and sustainable solutions for the agricultural industry1. *Technoarete Transactions on Advances in Data Science and Analytics, 1*(2), 1–7. https: //doi.org/10.36647/ttadsa/01.02.a001.
40. Loveleen, L, & Pillai, S. (2023). Precision agriculture innovation in agriculture. *Cardiometry. 25*, 678–684. https: //doi.org/10.18137/cardiometry.2022.25.678684.
41. Lv, T., & Lv, H. (2023). Analysis of the application of agricultural big data in modern agriculture. Zhang, K. (Ed.), *International Conference on Mathematics, Modeling, and Computer Science (MMCS2022).* SPIE. https: //doi.org/10.1117/12.2671150.
42. Manida, M., & Ganeshan, M. K. (2021). *New Agriculture Technology in Modern Farming.*
43. Manfreda, S., & Dor, E. B. (2023). Remote sensing of the environment using unmanned aerial systems. In: *Unmanned Aerial Systems for Monitoring Soil, Vegetation, and Riverine Environments* (pp. 3–36), Elsevier. https: //doi.org/10.1016/B978-0-323-85283-8.00009-6.
44. Manfreda, S., McCabe, M. F., Miller, P. E., Lucas, R., Madrigal, V. P., Mallinis, G., Dor, E. B., et al. (2018). On the use of unmanned aerial systems for environmental monitoring. *Remote Sensing, 10*(4). https: //doi.org/10.3390/RS10040641.
45. Mathur, R. (2023). Artificial intelligence in sustainable agriculture. *International Journal for Research in Applied Science and Engineering Technology, 11*(6). 4047–4052. https: //doi.org/10.22214/ijraset.2023.54360.
46. Mesías-Ruiz, G. A., Pérez-Ortiz, M., Dorado, J., de Castro, A. I., & Peña, J. M. (2023). Boosting precision crop protection towards agriculture 5.0 via machine learning and emerging technologies: A contextual review. *Frontiers in Plant Science, 14*(March), 1143326. https: //doi.org/10.3389/FPLS.2023.1143326/BIBTEX.
47. Mirajiani, M. (2022). Farmers' rationality and skill preparedness to answer the challenges of industrial revolution 4.0 towards society 5.0: A case study in serang. *Komunitas, 14*(2), 193–206. https: //doi.org/10.15294/komunitas.v14i2.35801.
48. Nath, S. (2023). A vision of precision agriculture: Balance between agricultural sustainability and environmental stewardship. *Agronomy Journal, 116,* 1126–1143. https: //doi.org/10.1002/agj2.21405.
49. .Cavalcante de Oliveira, R., & Diogne de Souza e. Silva, R. (2023). Artificial intelligence in agriculture: Benefits, challenges, and trends. *Applied Sciences (Switzerland), 13*(13). 7405. https: //doi.org/10.3390/APP13137405.
50. Pandrea, V.-A., Ciocoiu, A.-O., & Machedon-Pisu, M. (2023). IoT-based irrigation system for agriculture 5.0. In *2023 17th International Conference on Engineering of Modern Electric Systems (EMES).* Transilvania University of Brasov, IEEE: Brasov, Romania. https: //doi.org/10.1109/emes58375.2023.10171631.
51. Pechlivani, E. M., Gkogkos, G., Giakoumoglou, N., Hadjigeorgiou, I., & Tzovaras, D. (2023). Towards sustainable farming: A robust decision support system's architecture for agriculture 4.0. In *2023 24th International Conference on Digital Signal Processing (DSP)* (pp. 1–5). https: //doi.org/10.1109/DSP58604.2023.10167922.
52. Perez-Ruiz, M., Martínez-Guanter, J., & Upadhyaya, S. K. (2021). High-precision GNSS for agricultural operations. In *GPS and GNSS Technology in Geosciences* (pp. 299–335). Elsevier. https: //doi.org/10.1016/B978-0-12-818617-6.00017-2.
53. Polymeni, S., Plastras, S., Skoutas, D. N., Kormentzas, G., & Skianis, C. (2023). The impact of 6G-IoT technologies on the development of agriculture 5.0: A review. *Electronics (Basel), 12*(12), 2651. https: //doi.org/10.3390/electronics12122651.
54. Qazi, S., Khawaja, B., & Farooq, Q. (2022). IoT-equipped and AI-enabled next generation smart agriculture: A critical review, current challenges and future trends. *IEEE Access, 10*(January), 1. https: //doi.org/10.1109/ACCESS.2022.3152544.

55. Ragazou, K., Garefalakis, A., Zafeiriou, E., & Passas, I. (2022). Agriculture 5.0: A new strategic management mode for a cut cost and an energy efficient agriculture sector. *Energies, 15*(9), 3113. https: //doi.org/10.3390/en15093113.
56. Raj, C. S. M., Kumar, A. V. S., Musirin, I., Khalid, S., Latip, R., Mishra, N., & Kaur, G. 2023. Role of IoT technologies in agricultural ecosystems. In: *Handbook of Research on Deep Learning Techniques for Cloud-Based Industrial IoT* (pp. 134–154). IGI Global.https: //doi.org/10.4018/978-1-6684-8098-4.CH009.
57. Rakhra, M., Deb, P., Dahiya, O., Chandel, S. S., Bhutta, B., Badotra, S., Kumar, S., Shaukat, A., & Singh, D. (2022). An analytical study of the types of implements used by farmers in mechanized agriculture. In *2022 International Mobile and Embedded Technology Conference (MECON)*. Lovely Professional University, School of Computer Science and Engineering, IEEE: Phagwara, Punjab, India. https: //doi.org/10.1109/mecon53876.2022.9751983.
58. Rasmussen, W. D. (1982). The mechanization of agriculture. *Scientific American, 247*(3), 76–89. http: //www.jstor.org/stable/24966681.
59. Rufí-Salís, M., Petit-Boix, A., Villalba, G., Sanjuan-Delmás, D., Parada, F., Montserrat, M. E., Arcas, V., Liesa, J. M., Rieradevall, J., & Durany, X. G. (2020). Recirculating water and nutrients in urban agriculture: An opportunity towards environmental sustainability and water use efficiency? *Journal of Cleaner Production, 261*(March), 121213. https: //doi.org/10.1016/j.jclepro.2020.121213.
60. Saikanth, D. R. K., Supriya, Singh, B. V., Rai, A. K., Bana, S. R., Sachan, D. S., & Singh, B. 2023. Advancing sustainable agriculture: A comprehensive review for optimizing food production and environmental conservation. *International Journal of Plant & Soil Science, 35*(16), 417–425. https: //doi.org/10.9734/ijpss/2023/v35i163169.
61. Saiz-Rubio, V., & Rovira-Más, F. (2020). From smart farming towards agriculture 5.0: A review on crop data management. *Agronomy, 10*(2), 207. https: //doi.org/10.3390/agronomy10020207.
62. Silagadze, D. (2022). Major challenges of agriculture. *Gs*. Association For Science (Publications). https: //doi.org/10.52340/gs.2022.04.02.11.
63. Singh, P., Pandey, P. C., Petropoulos, G. P., Pavlides, A., Srivastava, P. K., Koutsias, N., Deng, K. A. K., & Bao, Y. (2020). Hyperspectral remote sensing in precision agriculture: Present status, challenges, and future trends. In: *Hyperspectral Remote Sensing: Theory and Applications*, (pp. 121–146). https: //doi.org/10.1016/B978-0-08-102894-0.00009-7.
64. Sitokonstantinou, V., Giannarakis, G., Lorilla, R. S., & Kontoes, C. (2023). Personalizing sustainable agriculture with causal machine learning. EGU General Assembly 2023, Vienna, Austria. https: //doi.org/10.5194/egusphere-egu23-2028.
65. Smart Farming and Precision Agriculture—Iberdrola [Internet]. 2024 [cited March 26]. Available from: https: //www.iberdrola.com/innovation/smart-farming-precision-agriculture.
66. Smart Farming Data Innovation.Edited.Docx—Surname 1 Student's Name Professor's Name Course Due Date Smart Farming Data Innovation Data Innovation | Course Hero [Internet]. 2024 [cited March 26]. Available from: https: //www.coursehero.com/file/149620075/Smart-Farming-Data-Innovationediteddocx/.
67. Sree, S., Chukkapalli, L., Gupta, M., Abdelsalam, M., Joshi, A., Sandhu, R., Joshi, K., & Mittal, S. (2020). Ontologies and artificial intelligence systems for the cooperative smart farming ecosystem. *IEEE Access, 8*(September), 164045–164064. https: //doi.org/10.1109/ACCESS.2020.3022763.
68. Srinivas Kumar, B., Santhi, S. G., & Kumar, K. K. (2021). SAMS: Smart agriculture management system using emerging technologies IoT, AI—A study. In *IOP Conference Series: Materials Science and Engineering 1074*(1), 012017. https: //doi.org/10.1088/1757-899X/1074/1/012017.
69. Stévant, P., Rebours, C., & Chapman, A. (2017). Seaweed aquaculture in Norway: Recent industrial developments and future perspectives. *Aquaculture International, 25*(4), 1373–1390. https: //doi.org/10.1007/s10499-017-0120-7.
70. Styring, A. K., Ater, M., Hmimsa, Y., Fraser, R., Miller, H., Neef, R., Pearson, J. A., & Bogaard, A. (2016). Disentangling the effect of farming practice from aridity on crop stable isotope values: A present-day model from morocco and its application to early farming sites in the Eastern Mediterranean. Reviews in Anthropology, *3*(1), 2–22. https: //doi.org/10.1177/2053019616630762.

71. Symeonaki, E., Maraveas, C., & Arvanitis, K. G . (2024). Recent advances in digital twins for agriculture 5.0: Applications and open issues in livestock production systems. *Applied Sciences, 14*(2), 686. https: //doi.org/10.3390/app14020686.
72. Talaviya, T., Shah, D., Patel, N., Yagnik, H., & Shah, M. (2020). Implementation of artificial intelligence in agriculture for optimisation of irrigation and application of pesticides and herbicides. *Artificial Intelligence in Agriculture, 4*, 58–73. https: //doi.org/https: //doi.org/10.1016/j.aiia.2020.04.002.
73. Taneja, A., Nair, G., Joshi, M., Sharma, S., Sharma, S., Jambrak, A. R., Roselló-Soto, E., et al. 2023. Artificial intelligence: Implications for the agri-food sector. *Agronomy, 13*(5), 1397. https: //doi.org/10.3390/agronomy13051397.
74. Taylor, J. A. (2023). Precision agriculture. *Encyclopedia of Soils in the Environment* (pp. 710–725). Elsevier. https: //doi.org/10.1016/B978-0-12-822974-3.00261-5.
75. Terradas, X., Piqué, R., Palomo, A., Antolín, F., López, O., Revelles, J., & Buxó, R. (2017). Farming practices in the early neolithic according to agricultural tools: Evidence from La Draga Site (Northeastern Iberia) BT—Times of neolithic transition along the Western Mediterranean. In: García-Puchol, O., & Salazar-García, D. C. (Eds), (pp. 199–220). Springer International Publishing: Cham. https: //doi.org/10.1007/978-3-319-52939-4_8.
76. The Development of Agriculture [Internet]. 2024 [cited March 26]. Available from: https: //education.nationalgeographic.org/resource/development-agriculture/.
77. Tombe, R., & Smuts, H. (2023). Society 5.0-inspired digitalization framework for resilient and sustainable agriculture. *EPiC Series in Computing*. EasyChair. https: //doi.org/10.29007/xc5q.
78. Treasure, E. R., Gröcke, D. R , Caseldine, A. E., & Church, M. J. (2019). Neolithic farming and wild plant exploitation in Western Britain: Archaeobotanical and crop stable isotope evidence from wales (c. 4000–2200 Calbc). *Proc. Prehist. Soc.* 85. Cambridge University Press (CUP): 193–222. https: //doi.org/10.1017/ppr.2019.12.
79. Tsolakis, N., Harrington, T.S., & Srai, J. S. (2023). Leveraging automation and data-driven logistics for sustainable farming of high-value crops in emerging economies. *Smart Agricultural Technology, 4*, 100139. https: //doi.org/https: //doi.org/10.1016/j.atech.2022.100139.
80. Victor, N., Reddy, P., Raj, D., Murugan, R., Chengoden, R., Gadekallu, T., Rakesh, N., Zhu, Y., & Paek, J. (2024). Remote sensing for agriculture in the era of industry 5.0—A survey. *IEEE Journal of Selected Topics in Applied Earth Observations and Remote Sensing, 17*, 5920–5945. https: //doi.org/10.1109/JSTARS.2024.3370508.
81. Vijayakumar, S., Kumar, R. M., Choudhary, A., Murugesan, D., Sreedevi, B., & Sundaram, R. (2022). Artificial intelligence (AI) and its application in Agriculture *6*(March), 25–31.
82. Vijayanand, C. (2018). Artificial intelligence (AI) in agriculture. *International Journal of Current Microbiology and Applied Sciences,* 7(January), 2122–2128. https: //doi.org/10.20546/ijcmas.2018.712.241.
83. Wang, T., Xu, X., Wang, C., Li, Z., & Li, D. (2021). From smart farming towards unmanned farms: A new mode of agricultural production. *Agriculture, 11*(2), 145. https: //doi.org/10.3390/agriculture11020145.
84. Yahya, N. (2018). Agricultural 4.0: Its implementation toward future sustainability. In: *Green Energy and Technology* (pp. 125–145). https: //doi.org/10.1007/978-981-10-7578-0_5.
85. Yang, W. (2023). *Agricultural Robotic System: The Automation of Detection and Speech Control.*
86. Yedk, P. T. (2022). Farm mechanization in india and perspectives—A review paper. In: *International Journal of Advanced Research in Science, Communication and Technology* (pp. 847–852). School of Commerce and Management Sciences, SRTM University, Nanded: Naksh Solutions. https: //doi.org/10.48175/ijarsct-9305.
87. Zhai, Z, Martínez, J., Beltran, V., & Martínez, N.L. (2020). Decision support systems for agriculture 4.0: Survey and challenges. *Computers and Electronics in Agriculture, 170*(March), 105256. https: //doi.org/10.1016/j.compag.2020.105256.

[illegible]

CHAPTER 20

Future and Challenges in Agriculture

NITIN ARORA[1*], MAMTA MARTOLIA[2], DEEPAK GUPTA[3,4], SUBHASH C. SHARMA[1], ARUN KUMAR CHAUDHARY[5], and WARSHI SINGH[6]

[1]*Department of Computer Science, Thapar Institute of Engineering and Technology, Patiala, Punjab, India*

[2]*Department of Cyber Security, COER University, Roorkee, Uttarakhand, India*

[3]*Mechanical Engineering Department, Graphic Era Hill University, Dehradun, Uttarakhand, India*

[4]*Mechanical Engineering Department, Graphic Era (deemed to be) University, Dehradun, Uttarakhand, India*

[5]*Industrial and Production Engineering Department, G B Pant University of Agriculture and Technology, Pantnagar, Uttarakhand, India*

[6]*School of Business Management, CSJM University, Kanpur, Uttar Pradesh, India*

*Corresponding author

ABSTRACT

Agriculture 5.0 presents itself as the next revolutionary stage in farming as humanity approaches the start of the Fourth Industrial Revolution. This chapter explores how cutting-edge technologies and creative approaches will shape agriculture in the future. Building on the achievements of earlier agricultural revolutions, Agriculture 5.0 uses biotechnology, artificial intelligence, and the Internet of Things to transform crop cultivation and livestock husbandry. The study delves into the amalgamation of biotechnology, precision agriculture, smart farming, and data analytics to tackle the worldwide predicaments of environmental preservation, sustainability, and food security.

20.1 INTRODUCTION

The agricultural industry in the 21st century is about to enter a new age known as "Agriculture 5.0." Agriculture 5.0 is a defining juncture in the future of the sector and its prospects are both complex and hopeful [12].

Demystifying Sustainable Farming Using Smart Technologies.
Gunjan Chhabra, Keshav Kaushik, Priya Chugh, Susheela Dahiya, and Aryan Chaudhary (Eds.)

DOI: 10.1201/9781779640932-20

Such perfect farming practice are covered under this new concept labeled as Agriculture 5.0, which is a paradigm shift in agricultural system, incorporating semi-automation, and other technological innovations, as artificial intelligence (AI), Internet of Things (IoT), and precision agriculture that help in enhancing the productivity, save wastage, and reduce environmental impact associated with agriculture activities. One of the most capable tools to leverage this next generation of agriculture—which should lead to an improved monument of resource management, crop management, and decision-making, is big data and predictive analytics [36].

A few of the major barriers to the success of Agriculture 5.0, meanwhile, includes the push for sustainable agriculture due to the changing climate, concerns over data security and privacy, and challenges with the assimilation of new technologies. This chapter discusses the challenges in this and provides an in-depth view on how Agriculture 5.0 is going to address issues of rural development and global food security [9].

In this context, it is important to understand that the world is moving toward Agriculture 5.0 but this knowledge is essential for making the right policy hereafter because the future of agriculture is vital for our planet and our people.

20.1.1 THE EVOLUTION OF AGRICULTURE: FROM 1.0 TO 5.0

A journey in itself, agriculture has travelled through different phases with each marked by an epoch-making evolution. Agriculture 1.0 (characterized by manual labor and using beasts of burden to plow fields), set the tone for cultivated forms of agriculture. Agriculture 2.0 brought machinery, followed by the Industrial Revolution to farming, to increase efficiency and output (Figure 20.1). With Agriculture 3.0 came the introduction of synthetic fertilizers and pesticides, increasing yields to phenomenal heights [33].

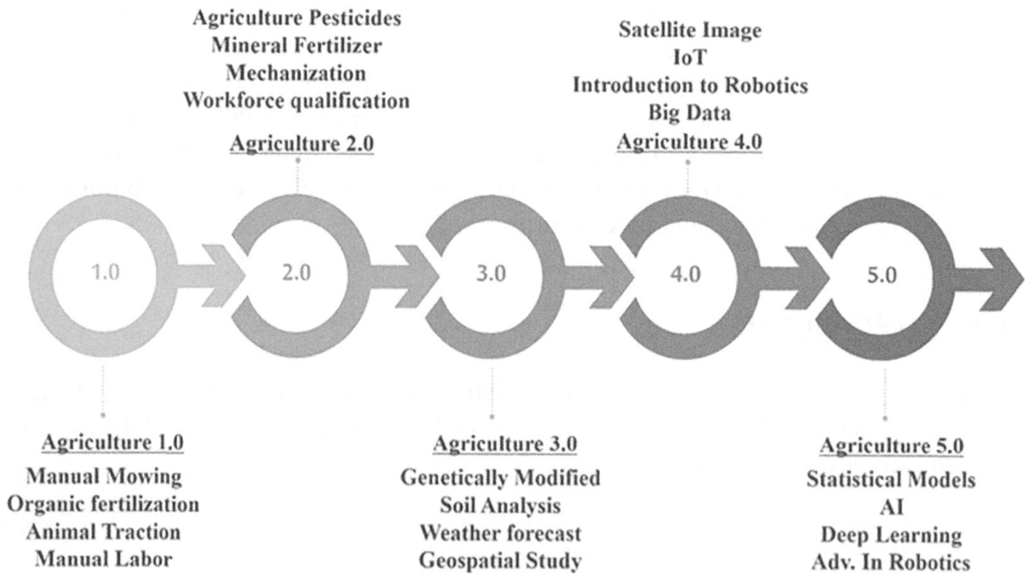

FIGURE 20.1 Sequential development in agricultural.

The era before this one, which could probably be characterized as Agriculture 4.0, was primarily about electrification and the infusion of computers and automation. These were the early days of precision agriculture, which unlocked agricultural practice optimization by being data driven. Now, the Agriculture 5.0 is leading to the Fourth Industrial Revolution. In this phase, use of technologies such as biotechnology, digitization/IoT, minimally invasive agriculture, accelerated through 5G networks, biologicals, AI, will lead to a smart yet connected agricultural ecosystem. Agriculture has undergone a series of advances over the course of time, with each phase in the process of crop improvement working in addition to, as well as on the platform created by its predecessor [1].

20.1.2 AGRICULTURE 5.0: DEFINITION AND SIGNIFICANCE

The most disruptive new model of agriculture, Agriculture 5.0 is where agriculture and technology have completely fused, and intelligent systems are working in partnership with farmers. Agriculture 5.0 is often described as the merger of biotech, AI, and IoT into the world of agriculture, but a way to think about this in more simplistic terms is that at its heart, Agriculture 5.0. Genetically modified crops, along with advanced breeding techniques, which are primarily developed using biotechnology, help farmers produce more food, while AI ensures that agricultural decision-making processes are as optimized as possible, and the IoT connects the tractors and combines on the farm with a network of other smart devices [37].

Agriculture 5.0 is important because it can meet some of the most urgent problems in the world today. This approach is achieved by using state-of-the-art technologies to improve in-crop cultivation, livestock management, and farm efficiency (Figure 20.2).

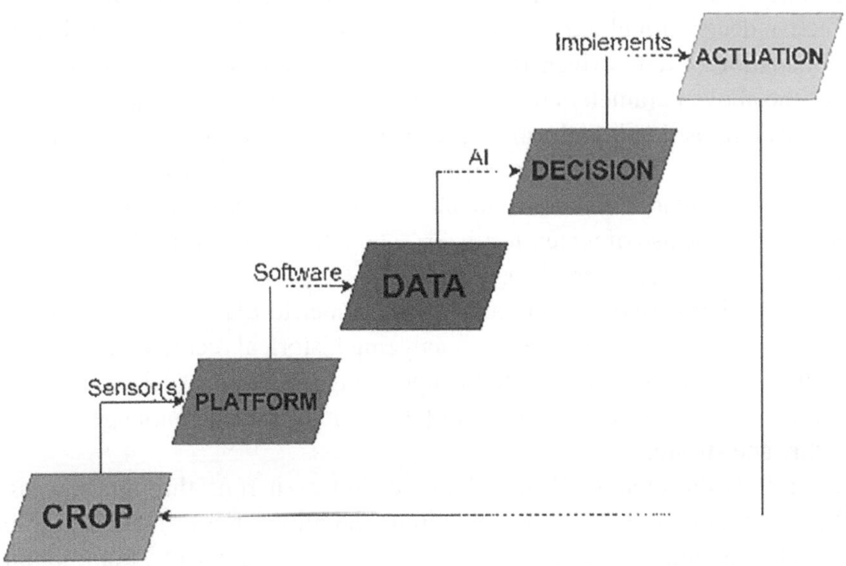

FIGURE 20.2 Integration of technology with a targeted focus on precision agriculture practices and advancements.

But by doing so through sustainable farming practices and the 5.0 Agricultural, Agriculture 5.0 also aims to solve some of the problems we have had in the food security and thus seek to preserve the environment and have a positive impact on society [55]. Here, in this chapter, we discover the complexity and necessity of Agriculture 5.0 and peek into its deep dimensions, which are critical in building the future of agriculture.

20.2 PRECISION AGRICULTURE

As the adoption of Agriculture 5.0 grows, data emerges as a key asset, reshaping the agriculture space.

20.2.1 HARNESSING THE POWER OF DATA IN FARMING

Data capture within agriculture runs the full spectrum from soil health to crop germination rates, all the way through to growth. While tools such as these have been available to farmers for several years, today's data analytics software provides even more detailed information about the farm, which aids in the making of very important and potentially profitable decisions. This part is about how data-driven methods enable farmers to increase productivity while wasting fewer resources and being more accurate in making interventions for crop management [22].

In the rapidly changing environment of contemporary farming data is paramount for farmers looking at getting the most out of their farming practice. Agriculture today is more precise, more efficient than it was before and this is largely in part due to data-driven approaches used to provide farmers with real-time and actionable information to improve their overall productivity. Farmers benefit from a deeper-level understanding of their fields, made possible by a comprehensive data analytics that furnishes them with information to make effective decision-making according to each stage of the farming cycle [28].

The foundation of data-driven farming has to do with the process of monitoring and analysis of the above parameters that are vital to crop health. Using sensors to gather data about the soil composition into the moisture level and nutrient content, farmers could make interventions that were quite targeted towards the needs of their crops. [19]. Since the level of insight is so granular, farmers are much less likely to waste resources as they can be more optimized in the use of water, fertilizers, and other inputs, this also leads generally to better financial and environmental sustainability.

Furthermore, data-based farming allows the farmer to plan the best possible management of the land. This can be resolved by analyzing historical record and predictive modelling that can help farmers to anticipate the upcoming challenges such as pest and diseases outbreak [46]. This vision allows them to take action, reduce challenges and keep their crops healthy and strong.

Indeed, this is the cream of this whole section as it runs through how data-driven approaches are changing the face of agriculture and allows the farmers to use these tools to navigate all that modern agriculture throws at them very precisely. For optimal farmers, this results in increased productivity and contributes toward a sustainable and resilient agricultural ecosystem overall by using data analytics.

20.2.2 DRONES, SENSORS, AND IOT IN PRECISION AGRICULTURE

At such a large scale, precision agriculture, which is the backbone of Agriculture 5.0, uses a sophisticated web of drones, sensors, and the IoT to transform the way farmers monitor and manage their crops. In addition, drones that carry high-resolution cameras provide birds-eye views to perform quick and thorough field assessment. Real-time sensors in the soil, the crop, and in the equipment itself are used to measure moisture, nutrient levels, and the condition of the equipment [53]. The IoT links these devices together, allowing farmers to get real-time alerts and analytics. This part pulls back the curtain on the interplay of drones, sensors, and IoT in uniting to take precision agriculture to new heights.

This section reveals how drones, sensors, and the IoT work together to demonstrate how they have enabled precision agriculture to soar to new heights. With the incorporation of these revolutionary technologies, farmers will be able to monitor crops, manage weeds, and utilize resources more effectively.

20.2.2.1 DRONES: AERIAL PRECISION AT SCALE

Today in modern farming, drones have started to replace tractors because their cameras and sensors make them very useful tools. With the help of these UAVs, high-resolution images and multispectral data of the whole farmland are captured. Such an abundance of data allow farmers to track the health of their crops, detect anomalies, and observe field conditions at a level never before achievable [27, 39]. Drones then provide the scalable, efficient way to survey these vast expanses of farmland in an overview that would be otherwise impractical.

20.2.2.2 SENSORS: GROUND-LEVEL PRECISION

Ground-level accuracy is validated through sensors located across the farmland. These sensors added to the model will continuously provide real-time measurements of soil moisture, temperature, nutrient levels, and so on [8]. Having sensor networks integrated in the agricultural environment results in accurate, up-to-date information for farmers so they can act quickly when conditions change. The sensors are the data generators, the backbone of informed decision-making that help farmers to optimize irrigation schedules, fertilization plans, and crop management in general [7].

20.2.2.3 IOT: CONNECTING THE AGRICULTURAL ECOSYSTEM

The various components of precision agriculture need to be in touch with each other, which is facilitated by the connective tissue in this discussion: the IoT. IoT operates drones, sensors, and automated machinery together, allowing such devices to work as a single unit and making the agricultural system one homogenous and responsive system [6]. This level of integration improves efficiency for farmers beyond any single piece of equipment, allowing them to monitor and control equipment from anywhere within range, receive live alerts based on sensor data, and analyze a single stream of data across a single portal [21].

Essentially, this section describes the manner in which drones, sensors, and IoT work together to usher in the era of precision in agriculture. This intelligent integration of these techs have enabled farmers to take data first decisions, optimizing resource usage, and enhancing crop yield to drive the modernization of the agriculture [17].

20.2.3 OPTIMIZING RESOURCE USE: REAL-TIME DECISION-MAKING

The most important need in Agriculture 5.0 the real-time decision-making that will help in managing the same and optimizing the resource being used. Thanks to all the data available to them, farmers can instantly take a call on where to put in resources most effectively. This could include the adjustment of irrigation levels based on soil moisture readings, or more specifically tailoring the application of fertilizer to the health of an individual crop, through leveraging better access to the related data—this real-time fine-tuning of resource use is a defining feature of this vision of agriculture [31].

This section explores the game-changing nature of real-time decision-making, explaining how it boosts sustainability, minimizes waste, and adds to the resilience of contemporary farming techniques. This development represents a discrete change in the field of agriculture that allows farmers to make decisions in real time [51] enabling farmers to respond to changes in the environment and to use their resources more efficiently.

20.2.3.1 HARNESSING TIMELY INSIGHTS FOR RESOURCE ALLOCATION

It uses streams of data such as data from sensors, drones, or weather forecasts, to enable continuous farmer decision-making in real time, providing real-time advice on what is currently going on the fields of a grower. With this real-time information in hand, the farmers can in turn dynamically adjust resource allocations, that is, to soil water, nitrogen fertilization, and pest control according to the current demand of the crops [50]. This agility means that resources are applied exactly when and where they are of most benefit, reducing inputs and therefore the environmental impacts of farming.

20.2.3.2 ENHANCING SUSTAINABILITY THROUGH PRECISION

Real-time decision-making—Adoption of real-time decision-making has a profound effect on maintaining the sustainability of farming operations. With more precision in using the resources, farmers could make agriculture to develop a more sustaining ecological footprint [35]. So it is not just watering and fertilizing judiciously—it is also about energy use and good stewardship of the land. Regenerative or sustainable farming practices, contributing to soil health, biodiversity, and ecosystem resilience, meet the broader goals of care for the planet.

20.2.3.3 MINIMIZING WASTE AND MAXIMIZING YIELD

Real-time decision-making is also beneficial that it helps in reducing waste in various agricultural processing so that the only remaining waste can be turn out into by-products [14, 54]. Farmers can prevent over-application of inputs, preventing excess runoff and lessening environmental impact by being able to precisely measure the needs of the crops as the crop is growing, before the crop sells. This simultaneously maximizes yield as crops are provided with precisely what they need for optimal growth [30]. These techniques can help improve the efficiency with which we farm our land, leading to higher crop yields and an overall lighter environmental footprint for agriculture.

In conclusion, we can see that this is a section that presents how real-time decisions have a central theme in resource optimization within new farming nowadays. In this way, embracing this dynamic approach can provide farmers with the tools to grow sustainably, reduce waste, and adapt with resilience in a rapidly changing sector.

20.3 SMART FARMING

This part walks us through the IoT and smart device landscape that has sprung up in the realm of agriculture, and the way in which it has been revolutionized farm lives.

20.3.1 IOT AND SMART DEVICES ON THE FARM

Integration of IoT and smart devices [10] as a game-changer precision based, efficient, data-centric decision-making in the agriculture has been the latest trend in the agriculture new era of technology advancement (Figure 20.3).

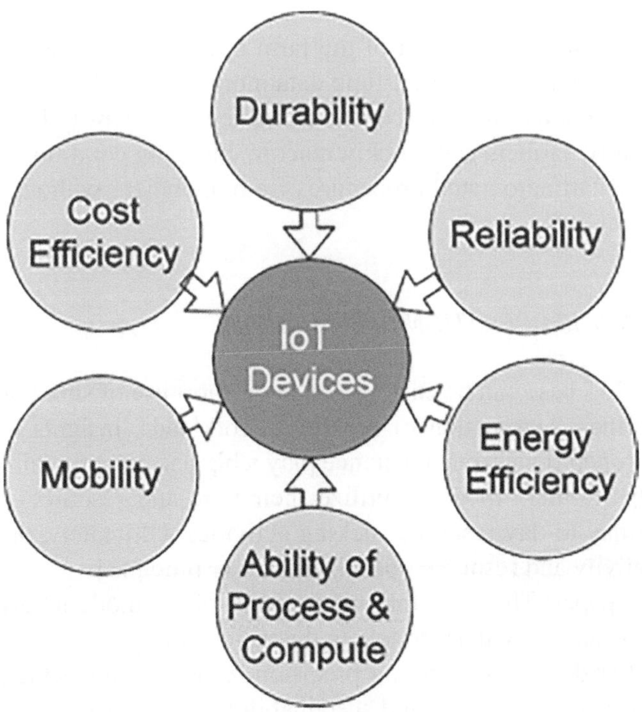

FIGURE 20.3 Smart farming with IoT integration to smart devices.

20.3.1.1 CONNECTING THE FARM: THE POWER OF IOT

IoT in agriculture has given new dimensions to how farmers track and control their daily farm operations. Whether it is soil sensors or weather stations or even coupled machinery, IoT devices are interlocked devices producing real-time data. These data include important metrics such as soil moisture, temperature, humidity, and device operational status [40]. Farmers assimilate this information to develop a full picture of the current status of the farm, empowering them to make better business decisions and maximize productivity [13].

20.3.1.2 SMART DEVICES FOR PRECISION AGRICULTURE

Agriculture-centric smart devices provide an advanced level of intelligence to the traditional way of farming. This includes devices used in precision agriculture like automated irrigation systems, drones with imaging capabilities, and GPS-guided tractors. Soil sensors, meanwhile, provide on-the-swivel, real-time data enabling farmers utilizing automated irrigation systems to calibrate those systems correctly to deliver just the right amount of water to their crops [58]. To this end, drones with imaging capability give farmers close up high-level aerial views of their own plot enabling them to spot crop stress, pest infestations, and nutrient deficiency.

20.3.1.3 REMOTE MONITORING AND CONTROL: A PARADIGM SHIFT

Remote monitoring and management of the farm operations is a signature feature of IoT and smart devices in agriculture. Real-time data monitoring and controlling machines and devices through their smartphone or computer are few clicks away for farmers [59]. With the remote capability, farmers can quickly react to changing conditions, tweak equipment settings, and even start automated procedures from anywhere without the need to physically be on the farm.

20.3.1.4 ENHANCING OPERATIONAL EFFICIENCY

Incorporating the IoT base can be customized in on-farm use of smart devices thus raising the efficiency further. Automation of repetitive workloads, insights based on data, and control over your operations from a distance play a big part in optimizing your workflows. This in turn helps farmers to better utilize their time and resources on strategic decisions rather than day-to-day resource sucking activities. Ultimately, this synergy enables improved productivity and resource optimization over time [43].

To sum up the paper: The most important pieces of the modern agriscape may be our things, and the IoT that we will use to talk to them. Technologies that work together to help farmers make better decisions—driving precision resource management so that they can implement new levels of efficiency and sustainability on the farm [23].

20.3.2 REMOTE MONITORING AND CONTROL

In this series, we explore how the IoT can take smart farming capabilities to the next level with various transformative applications in remote monitoring and control. Advancements in modern technology have provided farmers with tools that ensure they can run their Ag operations from just about anywhere with unprecedented precision and authority [47].

20.3.2.1 EMPOWERING FARMERS AT THEIR FINGERTIPS

The introduction of remote monitoring technology to agriculture has given farmers more control over their agricultural operations than ever before. Using connected sensors, cameras that allow farmers to get a birds-eye-view of the performance of their fields and facilities in real-time [5]. Such real-time data can show a number of critically important in-field information including environmental conditions, along with crop health signs, and the condition of any automated equipment that you might have.

20.3.2.2 REAL-TIME INSIGHTS FOR INFORMED DECISIONS

Farmers can access vital information and take quick decisions: This method helps farmers to take quick action and make faster decisions based on the current scenario. This will impact farmers, as they will have real-time contextual data regarding anything from soil-moisture levels, to irrigation-system efficiency to livestock conditions [56]. It is especially good at helping to respond to time-sensitive problems and improving resource usage.

20.3.2.3 AUTOMATED CONTROL FOR PRECISION AGRICULTURE

By including integration of remote control software, farmers are able to automate many functions of their operations. The features include the remote control of irrigation systems, fertilizers, and pesticide applications and management of autonomous machines [15]. The automation control process achieves better accuracy in agricultural practices and, at the same time, the machining waste is reduced and the automation is increased to streamline the process.

20.3.2.4 EFFICIENT RESOURCE ALLOCATION

Farmers can react immediately to changing conditions thanks to remote monitoring and control, which in turn improves resource allocation. This way, soil moisture level variations can be detected with the sensors in the field, and instances of irrigation adjustments can be remotely controlled by the farmers to allow the crops to grow in the best conditions possible [16]. Taking an upstream-first strategy for resource management by addressing nutrients at their source instead of downstream into a watershed addresses agricultural efficiency as a whole.

20.3.2.5 ENHANCING OPERATIONAL FLEXIBILITY

The flexibilities that farmers will realize due to the ability to monitor and control farm operations remotely are a paradigm shift. This type of flexibility is very useful in adverse weather conditions, where to make changes is possible without being present on the farm physically [34]. It also allows farmers to streamline operations and better adapt to real-time problems.

20.3.2.6 CHALLENGES AND CONSIDERATIONS

Despite the benefits, as with any human use case, boat monitoring and control remotely is not without its challenges and public concerns around cybersecurity and rural connectivity need to be addressed. The safety of data transmitted from smart devices and stable communication channels are essential to allow remote monitoring and control to proliferate in agriculture.

As this section detailed, remote monitoring and control are making history in farming practices today. Farmers who embrace these tools will gain a suite of powerful metrics through which to practice precision agriculture, optimize resources, and make quicker, wiser decisions, adding up to more sustainable and resilient agricultural systems.

20.3.3 ENHANCING OPERATIONAL EFFICIENCY

IoT: Get smarter with the on field with your new set of farming assistant. This section will provide more insights into how IoT and smart devices transition to the next level of operational efficiency at a farm level. Through the integration of these tools, we are now working in the realm of Precision Ag, a with a combination of data/sensors/image analysis providing farmers more control, and insight, than ever before when it comes to planning their crop management functions.

20.3.3.1 CONNECTED DEVICES FOR SEAMLESS OPERATIONS

These IoT devices and smart technologies are virtually uniting different parts of the agricultural processes resulting in streamlining of communication and coordination. A smart farm is a network of connected machinery and equipment and environmental conditions sensors monitoring the ambiance [20], and each and every one of them contribute to a more efficient operation.

20.3.3.2 REAL-TIME DATA FOR INFORMED DECISION-MAKING

The real-time data produced by smart devices in turn, vastly facilitates operational efficiency. Sensors from time to time collects data of soil health, weathering conditions, crops status, and machineries performances [45]. This treasure trove of knowledge enables farmers to act quickly, make data-driven choices, fine-tune resource allocation, and adapt to change.

20.3.3.3 PRECISION AGRICULTURE TECHNIQUES

Precision agriculture can be practiced with high accuracy due to the deployment of smart devices. This can include calibrating automated irrigation systems to deliver only the volume of water crops actually need by using soil moisture data in real time [49]. This fine-tuned nature of approach underlying in using models would help in reducing the wastage of resources while increasing efficiency in utilizing agricultural inputs.

20.3.3.4 OPTIMIZING RESOURCE USE

Optimized resource utilization is one of the significant contributors to operational efficiency. These include smart devices, for example, IoT-enabled sensors that provide an insight into resource consumption which empowers farmers to make data-driven decisions. This also manages to trickle down water usage, fertilizer application, and energy consumption, hereby promoting sustainability.

20.3.3.5 REMOTE MONITORING AND CONTROL

IoT devices enable remote monitoring and control, vastly improving operational efficiency. This will enable farmers to manage a host of different types of technology from livestock conditions to automated equipment from the same locale, saving valuable time for busy growers. Besides, this feature lets managers to respond quickly to urgent situations such as diseases affecting the growth of crops thereby building more agile and efficient farming practices.

20.3.3.6 CHALLENGES AND CONSIDERATIONS

With a lot of benefits that IoT and smart devices can bring in, but also challenges like data security, the standardization of protocols as well as the investment for the implementation of them are still an issue that has to be well-thought and prepared for. Overcoming these challenges is essential for the technology to be more widely used and adopted on the farm in the long term to help support operational efficiency [4].

The objective of this section is specifically to highlight how the IoT revolutionizes and improves the current state of IoT and smart devices in the realm of agriculture in terms of operational maturity. Through real-time data and precision techniques, combined with remote capabilities, farmers can operate their farms more efficiently specific to crop management, resulting in sustainable agriculture and sustainable resilience.

20.4 BIOTECHNOLOGY IN AGRICULTURE

This portion enters the foray of GMO (genetically modified organisms) crops, the science behind it, its implications, and effect on farming practices.

20.4.1 GENETICALLY MODIFIED CROPS

Genetically modified crops have enabled strides in agricultural biotechnology; the genetic construction of plants has been altered so that they carry particular traits (Figure 20.4).

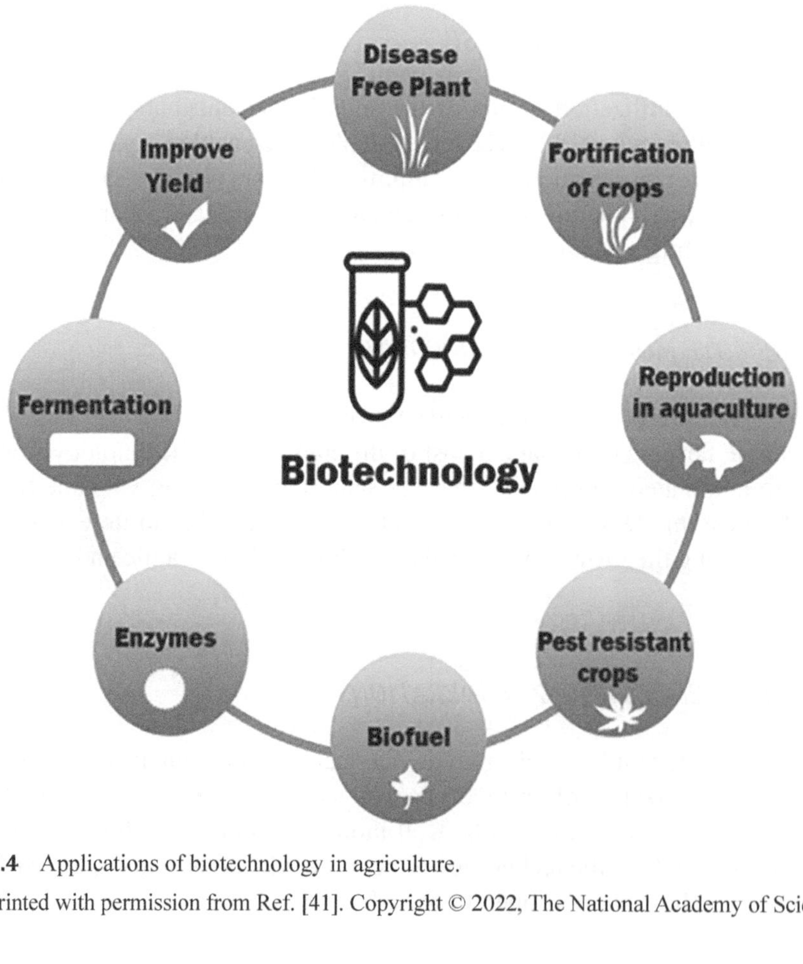

FIGURE 20.4 Applications of biotechnology in agriculture.

Source: Reprinted with permission from Ref. [41]. Copyright © 2022, The National Academy of Sciences, India

20.4.1.1 UNDERSTANDING GENETIC MODIFICATION

Simply put, genetic modification is changing the DNA of any organism to produce a particular trait or characteristic. It is mainly performed in the case of crops to maximize useful characteristics such as pest resistance, tolerance to environmental pressures, or nutritional values [52]. Using specific genetic engineering techniques, scientists can make precise modifications to the plant's genome.

20.4.1.2 BENEFITS OF GENETICALLY MODIFIED CROPS

A key benefit of GM crops is their improved pest and disease resistance. These genetically modified crops show added resilience, because of genes that help the plant produce natural pesticides, or provide resistance versus particular pests, considering that a bunch of these can be the harbinger of malady and decrease the need for chemical pests [25]. GM crops can also be engineered to be more tolerant of environmental stress like drought or harsh climates, to produce more consistent yields.

20.4.1.3 IMPROVED NUTRITIONAL CONTENT

Genetic modification allows us to boost the nutritional content of crops as well. This approach enables scientists to add genes that increase the levels of the most needed nutrients where areas suffer from widespread nutritional deficiencies. For example, crops may be engineered to be more nutritious, such as higher levels of vitamins or minerals, or proteins, improving food security and public health.

20.4.1.4 ENVIRONMENTAL IMPACT

Clever breeding practices and innovative technologies can correct this, as the use of genetically modified crops can have positive environmental effects. This would reduce the dependency on chemical pesticides/herbicides, which otherwise would have increased environment pollution and would have been detrimental to the biodiversity. Also, GM crops are less vulnerable and this is increasingly leading to good farming practices, which are sustainable respecting resources of cultivation.

20.4.1.5 CONTROVERSIES AND ETHICAL CONSIDERATIONS

Although GM crops have the potential for great benefits, they have also generated controversy and ethical debate. Questions include possible unforeseen consequences on ecosystems, seed patent consolidation in the hands of a few companies, and the degradation of biodiversity over time. Ethical concerns also extend to matters of informed consent, labeling transparency, and potential agribusiness monopolies by biotechnology companies.

20.4.1.6 REGULATORY FRAMEWORK

Many countries have put restrictions on production and marketing of GMO crops due to tight regulatory frameworks. Before GM crops are released to the market, these Safety, Environmental Impact, and Efficacy Global Regulatory bodies review and evaluate the USYN and EU-PLNs. Establishing rigorous regulatory safeguards is an attempt to reconcile the promise and perils of a new energy source. In summary, this chapter fully examines the body of genetically modified crops covering some scientific principles, possible benefits, environmental concerns, and ethical considerations of their application in farming practice of the present day. The narrated discussion is intended to provide a comprehensive perspective of this emerging part of crop management.

20.4.2 ADVANCED BREEDING TECHNIQUES

The section on Exotic Breeding Techniques moves to the cutting edge with new strategies that have revolutionized conventional breeding throughout Africa. They use the most recent technologies in helping out the development of crops with favorable characteristics, which results in greater efficiency and accuracy in the breeding process. Advancing breeding techniques are radically different from the conventional breeding tool, and represent new interpretations to fine tune and expedite future crop improvement. These include techniques that involve genetic recombination—methods that do not actually work for protein crops—while others use biotech, genomics, and other tools for more precise manipulation of plant genomes.

20.4.2.1 GENOMIC SELECTION AND MARKER-ASSISTED BREEDING

The best, and the newest, is genomic selection, a ground-breaking technique that delves into an organism's whole genome to seek out markers that are linked to traits it is considering. Then breeders use this information on the genetic potential of plants to identify, which ones to select at the earlier stages of development. Another related approach is marker-assisted breeding which uses genetic markers that are perched on a sequence of DNA tightly linked to a trait of interest to guide the breeding

20.4.2.2 CRISPR-CAS9 GENE EDITING

Breeding: The breakthrough in our ability to make precise changes to the plant genome has been through the development of CRISPR-Cas9 technology. CRISPR is a gene-editing tool that can be used to add, delete, or change specific DNA sequences and as a result, can enable revolutionized to the control of the expression of traits. Thanks to CRISPR-Cas9, the development of crops with resistant to diseases, enhanced nutritional content, and tolerance to environmental stresses has become much easier [2].

20.4.2.3 SPEED BREEDING

A novel technique known as speed breeding has been devised in order to reduce the length of the more traditional breeding cycle. Speed-breeding makes use of optimized growth conditions (e.g., light warmth) under continuous light regimes, thus decreasing the time for a plant to develop to adulthood and seed set [38]. This speeds up the breeding cycle, permits multiple generations a year, and enables faster selection for specific traits [48].

20.4.2.4 HYBRIDIZATION AND HETEROSIS

Traditional hybridization is the combination of the genetics from two plants to make progeny with some of the favored characteristics of both parents. Heterosis, also known as hybrid Vigor, means the enhanced performance of hybrids when compared to their parent lines. New technologies in genomics brought the advent of advanced breeding techniques to exploit hybridization and heterosis—and ultimately improve crop yield, tolerance, and traits of interest [24].

Though advanced breeding techniques provide unparalleled advantages, they also pose numerous ethical, regulatory, and environmental concerns. More questions concern whether gene-edited crops are safe, potential side effects, and who will benefit from and be at risk from them. This requires humanization within the bounds of innovation and responsibility as we navigate the intricacies of contemporary agricultural breeding landscapes.

The section ends with a note on the continuous changes in high breeding associated with the establishment of emerging technologies in agriculture. The mapping and sequencing of the cotton genome and development of clustering and genetic mapping techniques, gene expression and quantitative trait loci (QTL) mapping, gene sequencing, Recombinant Inbred Lines, DNA marker technology, gene tagging, marker-assisted selection and transformation technology will not only enhance our understanding of cotton genetics but will also lead to increasing the productivity and sustainability of cotton crops by introgressing novel genes and QTLs.

20.4.3 IMPROVING CROP YIELDS AND FOOD SECURITY

In this section, some pathways are elucidated from the perspective of the increasing role of advanced breeding techniques to the increase of crop productivity and global food security. These techniques are a key part of the answer to meeting the challenges of a growing population, changing climates, and evolving agricultural landscapes by focusing precision, resilience, and efficiency.

20.4.3.1 PRECISION AGRICULTURE FOR YIELD OPTIMIZATION

Modern breeding practices enhance the breeding of crops and increase the production of crops which have different types of characteristics to increases the yield.

Characteristics such as insensitivity to drought and pests, nutrient utilization efficiency, together provide an overall robustness in crops to enable them to survive in various environmental conditions. Today, these advanced traits guide the way for precision agriculture, helping farmers get closer to yield optimization by providing insights that direct the right intervention at the right time with the least possible use of resources.

20.4.3.2 RESILIENCE TO ENVIRONMENTAL STRESSORS

Global food security faces threats from climate change and uncertain environmental conditions. These stressors can weaken crops, but advanced breeding techniques impart stress-resistance to make them ready for any extreme; be it drought, floods, or temperature fluctuations. Enhancing crop resilience against adverse conditions by selectively incorporating stress tolerance-related genes through breeding, basically shoring up food production.

20.4.3.3 BIOTECHNOLOGY AND ENHANCED NUTRITIONAL CONTENT

Advanced breeding techniques also help to develop the nutritional quality of the crops along with the improvement in yield. Advances in biotechnological avenues by endogenous enrichment of vital nutrients through crops overcome nutrient deficiency from main food staple This strategy includes another important aspect, which is Bio fortification, a process that is used to increase the content of vitamins, minerals, and other beneficial compounds in a crop, thereby improving the nutritional value of the food supply [26].

20.4.3.4 ADDRESSING GLOBAL HUNGER AND MALNUTRITION

With the global population on the rise, and the demand for food more nutritious than ever, delivering an abundant food supply has perhaps never been more essential. This is where advanced breeding techniques, like genome editing, hold a key to nurturing future generations by optimizing food production. These techniques allow us to grow crops that produce more per acre and have better nutritional profiles, and are a net benefit for a more sustainable and resilient food system.

20.4.3.5 EMPOWERING SMALLHOLDER FARMERS

For millions of smallholder farmers who are an essential part of global agriculture, this is particularly important. When these farmers have access to advanced breeding techniques, this empowers them to do more with less land. This empowerment benefits farmers at a farmer level and also plays role in improving local and regional food security goals.

20.4.3.6 CHALLENGES AND CONSIDERATIONS

The scale of the potential benefits of increasing grain productivity with advanced breeding techniques is vast, but so too are the difficulties and caveats. Bioethics, regulatory framework, and equitable benefit sharing are crucial. The full benefits of these technologies for international food security will only be realized by finding the right balance of technological innovation and ethical obligations.

The section closes by discussing current and future directions in the development of advanced breeding techniques in efforts to increase crop yield and improve food security. The future, in terms of a more robust, sustainable, and supported food secure world for our growing global population, lies in further innovation, collaboration, and responsible deployment of these techniques.

20.5 VERTICAL FARMING

Vertical farming has arisen as a radical new way of growing food faster, creating options to tackle problems associated with urbanization, land scarcity, and environmental challenges [32].

This part examines the different dimensions of vertical farming from who it may especially benefit to answering to the issues posed by urbanization and using controlled environment agriculture (CEA) to alleviate the pressure from vast amounts of arable land (Figure 20.5).

FIGURE 20.5 A vertical farming model.

Source: Discovery. https: //www.discovery.com/science/vertical-farming--raising-agriculture-s-potential-and-lowering-i

20.5.1 MEETING URBANIZATION CHALLENGES

Urbanization means that there is less land being dedicated to agriculture and agricultural land needs to be closer to the consumer community. A solution may lie in vertical farming, which is positioning cultivation close to where people are spending their time—in urban areas.

Farm x (FtC): This method of farming allows for the integration of farms in vertical structures at the urban and metropolitan scale for increasing the distance in harvest and consumer. It targets issues around urban food supply chains, including the carbon emissions associated with moving produce from farm to urban areas.

20.5.2 CONTROLLED ENVIRONMENT AGRICULTURE

CEA is the term used to describe a technology-based approach to produce crops under light, temperature, humidity, and nutrient-controlled environment that do not rely excessively on the main traditional climate prediction [18]. Such precise control is what makes vertical farming possible, optimizing plant growth and enabling crops to grow in vertical constructions where nature conditions can be less than ideal. With CEA, you are sure of round the year production, no dependency on weather and the type of crop.

20.5.3 REDUCING THE NEED FOR ARABLE LAND

Traditional agriculture also necessitates wide swaths of arable land, which can supplement deforestation and habitat loss. This relatively new concept of vertical farming, is introducing us to an innovative way of utilizing vertical real estate and not only relying on the horizontal land. The difference in this method is that it allows the same productivity per square meter as conventional farming by cultivating crops in one or more layers on top of another, or on a vertical, inclined surface. The lower demand for arable land compliments sustainable land-use practices, which support both biodiversity and ecosystem preservation [29].

Vertical farming may present some great prospects but comes with its own set of challenges and overall frame of mind. Investing in vertical farming infrastructure, energy use, and developing cultivation systems are all expensive from the start. In addition, it is essential to resolve questions around the scalability, resource efficiency, and economic potential of vertical farming in order to finally bring it out of obscurity and make it market-ready.

The section ends with the prospects for the future of vertical farming as a natural component of the world of agriculture. Advances in technology along with an increased level of education are starting to make space for vertical farming as a crucial part of sustainable food production which might potentially meet the demands of current urbanization and reduced areas of arable land. Vertical farming fosters synergy between the realms of technology, agriculture, and urban planning that may have crucial and transformative utility in building food systems capable of adapting to radical increases in urban population pressures.

20.6 CHALLENGES IN AGRICULTURE 5.0

The advent of Agriculture 5.0, enabled primarily by new technology adoption, presents across numerous fronts, with challenges that must be thoughtfully measured and approached strategically for equitable solutions. In this section, we untangle the intricate web of hurdles of Agriculture 5.0, touching upon matters of data privacy, ethical issues of biotechnology [42].

20.6.1 DATA PRIVACY AND SECURITY CONCERNS

With the growing data-centric nature of farming, the need to collect and store large amounts of sensitive data surrounding every aspect of the operation raises potential concerns with regards to data privacy and security. Farmers, agribusinesses, and technology providers must navigate a very fine line that protects the rights of farmers to their data and the technology providers to benefit from aggregation of that data to produce valuable services and features for both the farmers and industry. Robust encryption, secure cloud storage, and strict access controls are all critical safeguards against unauthorized access and data breaches.

20.6.2 BRIDGING THE DIGITAL DIVIDE

The main problem for farmers, e-level or digital divide, it means, every farmer should have the right to access the advanced technology but not only few some. The challenge: Addressing these divides will necessitate a concerted effort to give people access to digital tools, training, and infrastructure. It would be important for benefits of Agriculture 5.0 to percolate and be used by farmers irrespective of the socio-economic background through collaborative initiatives by government, technologists, and agricultural stakeholders [44].

20.6.3 ETHICAL DILEMMAS IN BIOTECHNOLOGY

The inclusion of biotechnological infrastructure in Agriculture 5.0 like genetically modified crops and improved breeding or breeding prompts an examination that should be given careful consideration. The potential benefits of improving crop yield and pest resistance must be weighed against the ethical issues of GMOs and since this will strongly depend on the population under study and the proposed technology to be implemented, communication, good ethical framework as well as continuous dialogue between the scientists, policy makers and the public are needed.

20.6.4 ENVIRONMENTAL SUSTAINABILITY AND TECHNOLOGY

Technology has the potential to drive productivity, however, there exists a delicate balance between the level of technological innovation and the environmental sustainability. A challenge of Ag 5.0 will be reducing its ecological footprint, preserving the land, and sustainable agricultural practices. Innovations have to be in line with wider environmental goals—think regenerative agriculture and less dependence on harmful agrochemicals.

20.6.5 ECONOMIC DISPARITIES IN THE ADOPTION OF AGRICULTURAL TECHNOLOGIES

This also brings with it the challenge of inequality of farmers in economic terms as better advanced agriculture technologies introduce. Smallholder farmers and farmers who already have no access to land in the medium or long term could face high entry costs for the often not inexpensive acquisition and operation of the high-tech. To tackle this challenge, inclusive policies need to be developed, strengthened public–private partnership need to be fostered, and innovative financial incentives should be in place to ensure equitable access to app technology.

The last part around this concept explains that working on this requires a combined and multidisciplinary approach. Governments, research organizations, technology platform developers and local communities must work closely together to steer through the complex maze of challenges and ensure a sustainable and inclusive Agriculture 5.0 landscape. These challenges must be confronted head on and successfully if Agriculture 5.0 is to deliver to its full potential for global food security, environmental sustainability, and economic advancement as the agricultural sector navigates this through this transformative.

20.7 STRIKING A BALANCE

While Agriculture 5.0 goes on to change the face of how farming is done, the mandate to harmonize the scale of technological evolution with practiced land stewardship becomes the central mission. This section elaborates on the critical considerations fundamental to the successful junction of the state-of-the-art technologies and sustainable agricultural practices.

20.7.1 RESPONSIBLE STEWARDSHIP OF THE LAND

Agriculture 5.0 is founded on responsible land stewardship. And thinking in alignment with this restorative ethos means developing agriculture that maintains and enhances healthy soil, biodiversity, and resilience of ecosystems. The major methods such as targeted nutrient management, very precise use of fertilizers, and soil conservation measures like reducing soil erosion, plays an important role in prescribing and controlling environmental impact of farming activities. As well, this section looks at blending regenerative agriculture principles, those that seek to restore and improve ecosystem functions, to promote a comprehensive and sustainable strategy for land management.

20.7.2 ENSURING SUSTAINABLE AND PROSPEROUS AGRICULTURE IN THE FUTURE

What is needed to achieve sustainable, thriving agriculture in the future is more of a compound solution. Some of these measures go far beyond the simple incorporation of a range of modern technologies (in effect, they imply certain appropriate policy frameworks to favor sustainability, empower farmers, and ensure more equitable access to resources.) It further searches for agro-ecological principles like organic farming and agroforestry, which are vital for sustainable agriculture. The subtext states we must seek holistic solutions to global agriculture taking into account ecologic, social, and economic sustainability [11].

The section concludes by underscoring the link between technological innovation and environmental stewardship as we traverse through the uncharted waters of Agriculture 5.0. This balanced view necessitates a more nuanced relationship between agricultural practices and the health of the planet. Agriculture 5.0 does have the potential of meeting growing demands for food production not only through a transformative nature of agriculture but also a transformed future that is sustainable and prosperous.

20.8 CONCLUSION

A full circle: All in all, Agriculture 5.0 is OM's big vision about the future of farming: A new generation of farming that simply transcends past, present and future, and achieves something that has never been there before. A synthesis of possible transformative impacts along with inherent challenges enmeshed in the evolution towards Agriculture 5.0. One of the reasons why there is so much potential in Agriculture 5.0 is because it can disrupt and bring a revolution in every aspect of agricultural practices. Biotechnology, AI, IoT can achieve ultra-precision, hyper-efficiency, sustainability on a scale never before imagined by farmers. The combination of data-driven decision-making, smart farming technologies, and biotechnological innovations offer huge potential for boosting crop yields, resource efficiency, and environmental stewardship. But the path to Agriculture 5.0 is easier said than done. Providing a thoughtful survey of the ethical, social, and environmental issues tied to biotechnology, this section highlights the necessity for responsible and openly informed implementation. For our part, it provides new solutions to the so-called issues of data privacy, and security in a world of interconnected devices, stressing the need for adequate regulatory frameworks, and technological safeguards. The section also discusses the potential for economic inequities to develop as a result of the implementation of new agricultural technologies, thereby emphasizing the need for inclusive policies and access for all. This, in essence, Agriculture 5.0, it is the transformational force that requires innovation but with a sense of responsibility. Finding our way to be truly ethical, environmentally sustainable and inclusive requires that we are all in this together, together being the key word. But, at the same time, since the agriculture terrain is everchanging, policymakers and other partners should actively craft policies and practices that not only take advantage of the potential of Agriculture 5.0, but also mitigate against its perils. By working together with the same vision for a sustainable future, Agriculture 5.0 could revolutionize the status quo of global food systems, and the result will be a future of agriculture that will be more resilient and efficient.

KEYWORDS

- **Agriculture 5.0**
- **precision agriculture**
- **data-driven farming**
- **sustainability**
- **technology**
- **challenges**
- **future prospects**

REFERENCES

1. Anil Chougule, M., & Mashalkar, A. S. (2022). A comprehensive review of agriculture irrigation using artificial intelligence for crop production. *Computational Intelligence in Manufacturing*, 187–200. https: //doi.org/10.1016/B978-0-323-91854-1.00002-9
2. Arora, L., & Narula, A. (2017). Gene editing and crop improvement using CRISPR-Cas9 system. *Frontiers in Plant Science*, *8*, 1932. https: //www.frontiersin.org/journals/plant-science/articles/10.3389/fpls.2017.01932
3. Barui, P., Ghosh, P., & Debangshi, U. (2022). Vertical farming -an overview. *Plant Archives*, *22*, 223–228. https: //doi.org/10.51470/PLANTARCHIVES.2022.v22.no2.038
4. Cavazza, A., Dal Mas, F., Paoloni, P., & Manzo, M. (2023). Artificial intelligence and new business models in agriculture: A structured literature review and future research agenda. *British Food Journal*, *125*(13), 436–461. https: //doi.org/10.1108/BFJ-02-2023-0132
5. Cheng, J. (2022). Application of big data analysis to agricultural production, agricultural product marketing, and influencing factors in intelligent agriculture. *Journal of Computer and Information Technology*, *29*(3), 151–165. https: //doi.org/10.20532/cit.2021.1005404
6. Concepcion, R., Ramirez, T. J., Alejandrino, J., Janairo, A. G., Baun, J. J., Francisco, K., Relano, R.-J., Enriquez, M. L., Bautista, M. G., Vicerra, R. R., Bandala, A., & Izzo, L. G. (2022). A look at the near future: Industry 5.0 boosts the potential of sustainable space agriculture. In *2022 IEEE 14th International Conference on Humanoid, Nanotechnology, Information Technology, Communication and Control, Environment, and Management (HNICEM)* (pp. 1–6). https: //doi.org/10.1109/HNICEM57413.2022.10109559
7. Cr, M., & Director ICAR—Central Institute of Agricultural Engineering Madhya Pradesh, B. (2022). Trend in sustainable mechanization of Indian agriculture. *Journal of Rice Research*, *15*(Special Issue). https: //doi.org/10.58297/eqre1792
8. Dawn, N., Ghosh, T., Ghosh, S., Saha, A., Mukherjee, P., Sarkar, S., Guha, S., & Sanyal, T. (2023). Implementation of Artificial Intelligence, Machine Learning, and Internet of Things (IoT) in revolutionizing agriculture: A review on recent trends and challenges. *International Journal of Experimental Research and Review*, *30*, 190–218. https: //doi.org/10.52756/IJERR.2023.V30.018
9. Dhal, S., Wyatt, B. M., Mahanta, S., Bhattarai, N., Sharma, S., Rout, T., Saud, P., & Acharya, B. S. (2023). Internet of Things (IoT) in digital agriculture: An overview. *Agronomy Journal*, *116*(3). https: //doi.org/https: //doi.org/10.1002/agj2.21385
10. Dhanaraju, M., Chenniappan, P., Ramalingam, K., Pazhanivelan, S., & Kaliaperumal, R. (2022). Smart farming: Internet of Things (IoT)-Based sustainable agriculture. In: *Agriculture* (Vol. 12, Issue 10). https: //doi.org/10.3390/agriculture12101745

11. Farooq, M. (2023). Conservation agriculture and sustainable development goals. *Pakistan Journal of Agricultural Research, 60*, 291–298. https: //doi.org/10.21162/PAKJAS/23.170
12. Frankelius, P., & von Rosen, M. (2023). *Towards the Fifth Innovation Era in Agricultural Innovation* (pp. 205–229). In: Keswani, C., Possas, C., Koukios, E., & Viaggi, D. B. T.-A. B., (Eds.), Academic Press. https: //doi.org/https: //doi.org/10.1016/B978-0-323-90569-5.00003-2
13. Gowthaman, D. (2024). IoT: An internet of things-based intelligent city, agriculture, and healthcare system. *IJIREEICE, 12*. https: //doi.org/10.17148/IJIREEICE.2024.12113
14. Gupta, D., Chaudhary, A. K., Singh, V. K., Verma, D., Goh, K. L., & Sharma, M. (2023). Thermo-mechanical analysis of bhimal fiber (Grewia optiva)-CaCO3/flyash/TiO2 reinforced epoxy bio-composites. *Industrial Crops and Products, 204*, 117341. https: //doi.org/https: //doi.org/10.1016/j.indcrop.2023.117341
15. Jaliyagoda, N., Lokuge, S., Gunathilake, P. M. P. C., Amaratunga, K. S. P., Weerakkody, W. A. P., Bandaranayake, P. C. G., & Bandaranayake, A. U. (2023). Internet of things (IoT) for smart agriculture: Assembling and assessment of a low-cost IoT system for polytunnels. *PLoS One, 18*(5), e0278440. https: //doi.org/10.1371/journal.pone.0278440
16. Janmohammadi, M., Amanzadeh, T., Sabaghnia, N., & Dashti, S. (2016). Impact of foliar application of nano micronutrient fertilizers and titanium dioxide nanoparticles on the growth and yield components of barley under supplemental irrigation. *Acta Agriculturae Slovenica, 107*(2), 265–276.
17. Johansen, K., Maltese, A., & McCabe, M. F. (2023). Monitoring agricultural ecosystems. *Unmanned Aerial Systems for Monitoring Soil, Vegetation, and Riverine Environments*, 125–151. https: //doi.org/10.1016/B978-0-323-85283-8.00013-8
18. Kalantari, F., Mohd tahir, O., Akbari Joni, R., & Fatemi, E. (2017). Opportunities and challenges in sustainability of vertical farming: A review. *Journal of Landscape Ecology, 11*. https: //doi.org/10.1515/jlecol-2017-0016
19. Konfo, T. R. C., Chabi, A. B. P., Amoussouga Gero, A., Lagnika, C., Avlessi, F., Biaou, G., & Sohounhloue, C. K. D. (2024). Recent climate-smart innovations in agrifood to enhance producer incomes through sustainable solutions. *Journal of Agriculture and Food Research, 15*, 100985. https: //doi.org/https: //doi.org/10.1016/j.jafr.2024.100985
20. Korres, N. E., Norsworthy, J. K., Burgos, N. R., & Oosterhuis, D. M. (2017). Temperature and drought impacts on rice production: An agronomic perspective regarding short- and long-term adaptation measures. *Water Resources and Rural Development, 9*, 12–27. https: //doi.org/https: //doi.org/10.1016/j.wrr.2016.10.001
21. Kumar, P., Srivastava, P. K., Tiwari, P., & Mall, R. K. (2021). Application of GPS and GNSS technology in geosciences. *GPS and GNSS Technology in Geosciences*, 415–427. https: //doi.org/10.1016/B978-0-12-818617-6.00018-4
22. Kumari, A., Department of Computer Science R. N. S. Institute of Technology Bangalore, I., Bhavanishankar, & Technology, A. P. o. C. S. N. S. I. of. (2023). Precision agriculture. *International Journal of Computing Algorithm, 12*(1). https: //doi.org/10.20894/ijcoa.101.012.001.004
23. Kushwaha, N. L., Elbeltagi, A., Patel, A., Zakwan, M., Rajput, J., & Sharma, P. (2022). *Assessment of Water Resources Using Remote Sensing and GIS Techniques*. 85–98. https: //doi.org/10.1016/B978-0-323-91910-4.00006-6
24. Labroo, M. R., Studer, A. J., & Rutkoski, J. E. (2021). Heterosis and hybrid crop breeding: A multidisciplinary review. *Frontiers in Genetics, 12*, 643761. https: //doi.org/10.3389/fgene.2021.643761
25. Ma, J. K. C., Drake, P. M. W., & Christou, P. (2003). The production of recombinant pharmaceutical proteins in plants. *Nature Reviews Genetics, 4*(10), 794–805. https: //doi.org/10.1038/NRG1177
26. Mandal, O., Krishnamoorthi, A., Kaur, R., Kalaiselvi, P., Singh, N., Tiwari, U., & Mitra, M. (2024). Biofortification: Enhancing nutritional content in crops through biotechnology and fighting climate change. *Journal of Advances in Biology & Biotechnology, 27*, 186–210. https: //doi.org/10.9734/jabb/2024/v27i2710
27. Manfreda, S., McCabe, M. F., Miller, P. E., Lucas, R., Madrigal, V. P., Mallinis, G., Dor, E. Ben, Helman, D., Estes, L., Ciraolo, G., Müllerová, J., Tauro, F., de Lima, M. I., de Lima, J. L. M. P., Maltese, A., Frances, F., Caylor, K., Kohv, M., Perks, M., Ruiz-Pérez, G., Su, Z., Vico, G., & Toth, B. (2018). On the

use of unmanned aerial systems for environmental monitoring. *Remote Sensing, 10*(4), 641. https: //doi.org/10.3390/RS10040641

28. Mesías-Ruiz, G. A., Pérez-Ortiz, M., Dorado, J., de Castro, A. I., & Peña, J. M. (2023). Boosting precision crop protection towards agriculture 5.0 via machine learning and emerging technologies: A contextual review. *Frontiers in Plant Science, 14*, 1143326. https: //doi.org/10.3389/fpls.2023.1143326
29. Mir, M., Naikoo, N., Kanth, R., Bahar, F., Bhat, M., Nazir, D., Amin, Z., Singh, L., Raja, W., Saad, A., Bhat, T., Palmo, T., & Ahngar, T. (2022). Vertical farming: The future of agriculture: A review. *The Pharma Innovation International Journal, 11*(2),
30. Mishra, P. N., Gerala, P., & Maitra, S. (2022). Study on artificial intelligence applications uses in agriculture. *International Journal of Health Sciences, 6*(S2), 9162–9173. https: //doi.org/10.53730/IJHS.V6NS2.7391
31. Nath, S. (2023). A vision of precision agriculture: Balance between agricultural sustainability and environmental stewardship. *Agronomy Journal, 116*(3), 1126–1143 https: //doi.org/10.1002/agj2.21405
32. Oh, S., & Lu, C. (2022). Vertical farming—Smart urban agriculture for enhancing resilience and sustainability in food security. *The Journal of Horticultural Science and Biotechnology, 98*, 1–8. https: //doi.org/10.1080/14620316.2022.2141666
33. de Oliveira, R. C., & de S. e. Silva, R. D. (2023). Artificial intelligence in agriculture: Benefits, challenges, and trends. *Applied Sciences (Switzerland), 13*(13). https: //doi.org/10.3390/APP13137405
34. Patel, K. G., & Patil, M. S. (2022). Artificial Intelligence in Agriculture. *International Journal for Research in Applied Science and Engineering Technology, 10*(2), 624–627. https: //doi.org/10.22214/IJRASET.2022.40308
35. Pechlivani, E. M., Gkogkos, G., Giakoumoglou, N., Hadjigeorgiou, I., & Tzovaras, D. (2023). Towards sustainable farming: A robust decision support system's architecture for agriculture 4.0. In *2023 24th International Conference on Digital Signal Processing (DSP)* (pp. 1–5). https: //doi.org/10.1109/DSP58604.2023.10167922
36. Perez-Ruiz, M., Martínez-Guanter, J., & Upadhyaya, S. K. (2021). High-precision GNSS for agricultural operations. *GPS and GNSS Technology in Geosciences*, 299–335. https: //doi.org/10.1016/B978-0-12-818617-6.00017-2
37. Polymeni, S., Plastras, S., Skoutas, D. N., Kormentzas, G., & Skianis, C. (2023). The impact of 6G-IoT technologies on the development of agriculture 5.0: A review. *Electronics (Basel), 12*(12), 2651. https: //doi.org/10.3390/electronics12122651
38. Potts, J., Jangra, S., Michael, V. N., & Wu, X. (2023). Speed breeding for crop improvement and food security. In: *Crops* (Vol. 3, Issue 4, pp. 276–291). https: //doi.org/10.3390/crops3040025
39. Ragazou, K., Garefalakis, A., Zafeiriou, E., & Passas, I. (2022). Agriculture 5.0: A new strategic management mode for a cut cost and an energy efficient agriculture sector. In: *Energies* (Vol. 15, Issue 9). https: //doi.org/10.3390/en15093113
40. Randall, W. C., Lopez, R. G., Currey, C., Gerovac, J., Mahan, C., Buschkoetter, J., Hilligoss, A., & Patz, B. (2014). Comparison of supplemental lighting from high-pressure sodium lamps and light-emitting diodes during bedding plant seedling production. *HortScience, 49*(5), 589–595.
41. Ranjha, M. M. A. N., Shafique, B., Khalid, W., Nadeem, H. R., Mueen-ud-Din, G., & Khalid, M. Z. (2022). Applications of biotechnology in food and agriculture: A mini-review. *Proceedings of the National Academy of Sciences, India Section B: Biological Sciences, 92*(1), 11–15. https: //doi.org/10.1007/s40011-021-01320-4
42. Saikanth, D. R. K., Modi, R., Gagan, T., Rajesh, K., Anmol, G., Pandey, S. K., & Kumar, V. L. (2024). The impact of emerging technologies on sustainable agriculture and rural development. *International Journal of Environment and Climate Change, 14*(1), 253–263. https: //doi.org/10.9734/ijecc/2024/v14i13830
43. Saikanth, D. R. K., Supriya, Singh, B. V., Rai, A. K., Bana, S. R., Sachan, D. S., & Singh, B. (2023). Advancing sustainable agriculture: A comprehensive review for optimizing food production and environmental conservation. *International Journal of Plant & Soil Science, 35*(16 SE-Review Article), 417–425. https: //doi.org/10.9734/ijpss/2023/v35i163169
44. Samadder, S., Pandya, S., & Lal, S. (2023). Bridging the digital divide in agriculture: An investigation to ICT adoption for sustainable farming practices in Banaskantha district of Gujarat, India. *International*

Journal of Environment and Climate Change, *13*, 1376–1384. https: //doi.org/10.9734/ijecc/2023/v13i92367

45. Sharma, B., Kumari, P., Kumari, R., Tripathi, S. K., Kumar, A., & Dhara, P. K. (2020). Impact of multistorey system on yield and soil quality of mustard in alfisols of Eastern India. *ADRRI Journal of Agriculture and Food Sciences*, *4*(6(4)), 27–46. https: //doi.org/10.55058/ADRRIJAFS.V4I6(4).600
46. Shivaanivarsha, N., Vijayendiran, A. G., Saravanan, S., & Sriram, A. (2022). Farming using AI technology. In *2022 IEEE International Power and Renewable Energy Conference, IPRECON 2022*. https: //doi.org/10.1109/IPRECON55716.2022.10059618
47. Singh, P., Pandey, P. C., Petropoulos, G. P., Pavlides, A., Srivastava, P. K., Koutsias, N., Deng, K. A. K., & Bao, Y. (2020). Hyperspectral remote sensing in precision agriculture: Present status, challenges, and future trends. *Hyperspectral Remote Sensing: Theory and Applications*, 121–146. https: //doi.org/10.1016/B978-0-08-102894-0.00009-7
48. Singh, S., Singh, A., Singh, A., & Singh, S. P. (2021). *Speed Breeding: An Innovative Method for Crop Improvement. 7.*
49. *Smart Farming and Precision Agriculture—Iberdrola*. (n.d.). [cited 2024 March 26]; Available from: https: //www.iberdrola.com/innovation/smart-farming-precision-agriculture
50. Sommer, N. F., Fortlage, R. J., & Edwards, D. C. (1992). Postharvest diseases of selected commodities. In: Kader, A. A., (Ed.), *Postharvest Technology of Horticultural Crops* (*Vol.* 3311). University of California Davis Division of Agriculture and Natural Resources.
51. Srinivas Kumar, B., Santhi, S. G., & Kranthi Kumar, K. (2021). SAMS: Smart agriculture management system using emerging technologies IoT, AI—A study. *IOP Conference Series: Materials Science and Engineering*, *1074*(1), 12017. https: //doi.org/10.1088/1757-899X/1074/1/012017
52. Takeda, S., & Matsuoka, M. (2008). Genetic approaches to crop improvement: Responding to environmental and population changes. *Nature Reviews Genetics*, *9*(6), 444–457. https: //doi.org/10.1038/NRG2342
53. Talaviya, T., Shah, D., Patel, N., Yagnik, H., & Shah, M. (2020). Implementation of artificial intelligence in agriculture for optimisation of irrigation and application of pesticides and herbicides. *Artificial Intelligence in Agriculture*, *4*, 58–73. https: //doi.org/https: //doi.org/10.1016/j.aiia.2020.04.002
54. Tamta, S., Vimal, V., Verma, S., Gupta, D., Verma, D., & Nangan, S. (2024). Recent development of nanobiomaterials in sustainable agriculture and agrowaste management. *Biocatalysis and Agricultural Biotechnology*, *56*, 103050. https: //doi.org/https: //doi.org/10.1016/j.bcab.2024.103050
55. Tombe, R., & Smuts, H. (2023). Society 5.0-inspired digitalization framework for resilient and sustainable agriculture. In *EPiC Series in Computing*. EasyChair. https: //doi.org/10.29007/xc5q
56. Tsolakis, N., Harrington, T. S., & Srai, J. S. (2023). Leveraging automation and data-driven logistics for sustainable farming of high-value crops in emerging economies. *Smart Agricultural Technology*, *4*, 100139. https: //doi.org/https: //doi.org/10.1016/j.atech.2022.100139
57. *Vertical Farming: Raising Agriculture's Potential and Lowering Its Environmental Impact | Latest Science News and Articles | Discovery*. (n.d.). [cited 2024 April 8]; Available from: https: //www.discovery.com/science/vertical-farming--raising-agriculture-s-potential-and-lowering-i
58. Yedk, P. T. (2022). Farm mechanization in India and perspectives—A review paper. *International Journal of Advanced Research in Science, Communication and Technology*, 847–852. https: //doi.org/10.48175/ijarsct-9305
59. Zvyagin, L. S. (2022). Questions of the use of artificial intelligence and measurements in the modern agricultural sector. *Soft Measurements and Computing*, *1*(50), 63–74. https: //doi.org/10.36871/2618-9976.2022.01.007

[illegible]

CHAPTER 21

Advances in Precision Disease Identification: Cutting-Edge Diagnostic Tools and Techniques

SUPRIYA GUPTA[1*], ANUPAMA RAWAT[1], MANPREET SINGH PREET[1], VIVEK KUMAR PATHAK[1], NIKITA CHAUHAN[1], PANKAJ RAUTELA[2], and K. P. SINGH[2]

[1]*Graphic Era Hill University, Dehradun, Uttarakhand, India*

[2]*G.B. Pant University of Agriculture and Technology, Pantnagar, Uttarakhand, India*

*Corresponding author

ABSTRACT

The diagnosis of phytological diseases is important for implementing the required control measures, which finally leads to the better growth and development of the crop. The outbreak of new pathogens increases the prevalence of many plant diseases, and their adoption to environmental conditions requires advancement in developing the diagnostic tools. Conventional methods, such as macroscopic examinations and petri-plate culturing, have been aided by molecular biology, genomics techniques, imaging techniques, and remote sensing. Techniques such as polymerase chain reaction and DNA sequencing provide very rapid and precise pathogen diagnosis, while high-throughput screening, microarray, and next-generation sequencing is paying a new ray of hope in pathogen profiling. Advances in fluorescence microscopy, confocal laser scanning microscopy, and spectral imaging have led to detect pathogen visualization and also analysis of disease symptom simultaneously. Recent advances made through developing quick diagnostic kits, plant pathogen biosensors, and nanodiagnostic tools facilitate precise disease detection at the right time, although many technologies require high cost at the time of installation. Therefore, innovation and research demonstrations are essential for advancing affordable precision disease identification, which will lead to the ultimate safeguarding of global nutritional as well as food security finally leading to sustainable agriculture.

21.1 INTRODUCTION

Diagnosis of the diseases in plants is very vital when giving control measures on the diseases as well as on the health of the crop. Modern plant diagnostic technologies have brought new dimension to plant pathology through precise identification of diseases and pathogens by both the scientists and growers, thereby enabling correct and proper

Demystifying Sustainable Farming Using Smart Technologies.
Gunjan Chhabra, Keshav Kaushik, Priya Chugh, Susheela Dahiya, and Aryan Chaudhary (Eds.)

DOI: 10.1201/9781779640932-21

disease management. The review will highlight the latest relevant innovations in disease precision identification with special focus on identifying new diagnostic instruments and methods that enhance the identification and diagnosis of plant diseases to a significantly greater extent. In the recent past, problems associated with plant diseases have increased progressively in agriculture due to emergence of new diseases and pathogens or due to change in behavior of existing diseases to adaption of new environment. Diseases have always remained a significant menace to plant produce across the world, a factor that has greatly affected food production and wealth of farmers. Tools used for the diagnosis of the specific pathogen are crucial for efficient disease control and contribute to the development of management interventions that alleviate the impact. Of course, even in plant pathology, diagnostic tools and techniques have been developed and improved in the recent past, bringing about a lot of revolutionizing in the extension. As it has been deduced, previous approaches to achieve precision, sensitivity, and effectiveness are generally constrained to macroscopic examinations and plate culturing. Despite this concern, recent advancements in areas such as molecular biology, genomics, imaging techniques, and remote sensing are clearly helping in presenting precision identification of diseases that were difficult to identify. The various molecular diagnostic techniques that can be employed are polymerase chain reaction (PCR) and DNA sequencing among others; they are very useful in giving rapid diagnosis of the plant pathogen. PCR of specific gene sequences that are peculiar to the intend pathogen has proved to be very effective even at high dilution levels. Further, advances in DNA sequencing techniques positively inform on the genetic variation and pathogen evolution of plants; hence instructing measures of managing and eradicating them [3].

The philosophical neutrally of high-throughput screening techniques and technological evolution of microarray and next-generation sequencing (NGS) platforms has conferred tremendous boost to research diagnosing disease. These technologies enable processing of thousands of samples in a single run, making it possible to rapidly detect pathogens and analyze their genetic structure. The processes of fluorescence microscopy and confocal laser scanning microscopy offer researchers the opportunity to inspect the structure of the pathogen and disease symptom in plant tissues with the help of time and spatial resolution and image quality [6]. These detailed techniques help in direct visual estimation of pathogens in the plant tissues besides helping significantly in the disease diagnosis. Therefore, there has been a rise in quick diagnostic kits especially where they are used together with point of care devices regarding on-site diagnosis of diseases. These are user-friendly kits that afford correct results in the shortest time, and the grower can obtain info from them to begin combating a disease illness at once [8]. In terms of plant pathogen biosensors and nano-diagnostics tool development, early and sensitive and selective detection was also adapted [4]. These miniaturized and integrated devices, levering unique material properties at the nanoscale levels, can accurately and rapidly identify the pathogen presence at molecular levels, thus facilitating disease diagnosis. With reference to sufficiently develop spectral imaging and spectroscopy as a part of technologies used in remote sensing, it is possible to conclude that the last can be applied in disease analysis and identification. Evaluation of the physiology, structural, and molecular characteristics of plants allow for the understanding of disease-related stress and disease development. This is evidenced within the engagement of enhanced tools in plant disease management programs, which transforms disease control approaches by [11]. The correct identification of pathogens and understanding their genetics allows a grower to tailor the best control measure, thus maximizing the reduction of crop losses while minimizing the environmental impact. Despite the remarkable tool development in diagnosis, there are still challenges. These include the provision of a more affordable and user-friendly diagnostic technology, incorporation of diagnostic data into a decision support system for real-time disease management. Continued innovation and collaboration in the fields are mandatory to overcome these challenges for the advancing precision identification of the disease. Recent advancements in diagnostic tools and techniques have transformed our ability to detect and diagnose plant diseases with unprecedented accuracy and efficiency. From molecular techniques to remote sensing technologies, these cutting-edge tools represent the forefront of precision disease identification. By harnessing these advancements, researchers and growers can better understand disease dynamics, develop targeted control measures, and safeguard global food security.

21.2 MOLECULAR DIAGNOSTIC TECHNIQUE

The impact of plant diseases to world food security considering plant and agricultural sustainability. This chapter highlights the importance of the identification of plant pathogens since such information plays a vital role when mitigating diseases affecting plants. Molecular diagnostic technology has become a highly effective method to for the quick identification of the plant pathogen and plant pathology has been enriched with such technologies. This section overviews the major MDs essential to plant disease diagnosis in connection with the principles that drive them and their relevance to sustainable agriculture.

21.3 PCR

PCR is one of the versatile molecular techniques that finds application in the diagnosis of plant diseases [9]. PCR is used in enhancing the reproduction of precise DNA sequences that are particular to plant pathogens, which means that they can be easily identified irrespective of the concentration levels. PCR works on the basis of sequential process, which includes denaturation, annealing, and extension, which makes the process of amplification exponential. The PCR assays are sensitive, specific, and facile for different plant samples, which makes it a very helpful tool for early disease diagnosis and monitoring DNA sequencing technologies. Molecular diagnostics of plant pathogens have shifted with the availability of DNA. Real-time PCR (qPCR) and loop-mediated isothermal amplification (LAMP) are rapid, sensitive diagnostic techniques that are broadly applicable in the field of plant disease management [3]. These tests are described as simple, rapid, and performable in the field. Real-time PCR makes it possible to quantify the target DNA in real time. LAMP amplifies the DNA under isothermal conditions and does not need a complex thermal cycler. Both qPCR and LAMP assays are important disease surveillance and outbreak management tools for quarantine screens. Molecular diagnostic techniques for plant diseases are important because they detect plant pathogens quickly, sensitively, and accurately. It helps researchers and farmers in evolving focused control measures so that the losses owing to diseases should be minimized for overall productivity and food security. In conclusion, the developing molecular diagnostics promise an advancing future for sustainable agriculture in the innovative solutions for disease management (Figure 21.1).

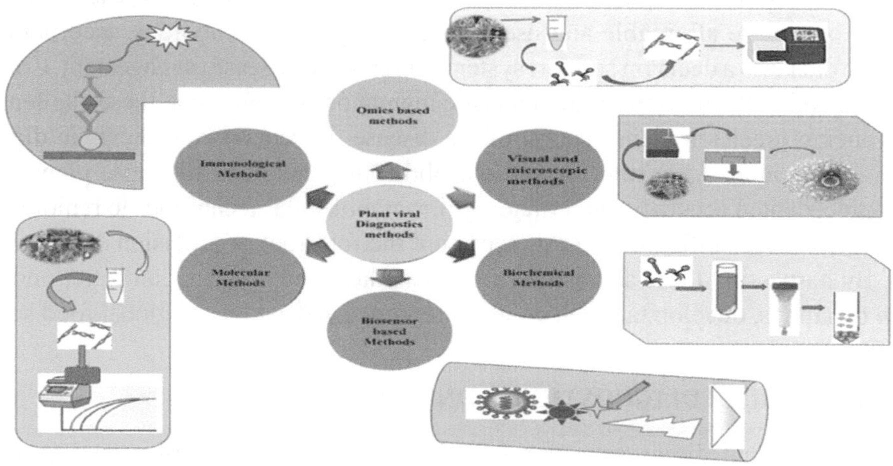

FIGURE 21.1 Illustration depicting use of molecular techniques for molecular diagnosis of plant viruses in various crop plants.

Source: Reprinted from Ref. [13]. http: //creativecommons.org/licenses/by/4.0/

21.3.1 HIGH-THROUGHPUT SCREENING METHODS

High-throughput screening techniques constitute an important set of tools for plant disease management and deliver rapid and in-depth analyses of plant samples toward the identification of pathogens. Such processes include an aspect of advanced technologies in performing the analysis of multiple analyses simultaneously in order to make the detection, characterization, and monitoring of plant diseases at the earliest. The subsequent section focuses on some of the important high-throughput screening methods, their applications, and importance in plant disease management for sustainable agriculture. Microarray technology: Microarray technology is a new high-throughput technique developed recently that can analyze thousands of genetic markers in one go [12]. Microarrays, in fact, are nothing but the solid substrate with thousands of immobilized DNA probes that hybridize with complementary DNA or RNA sequences present in the test sample. Microarray-based assays offer an approach to the detection and quantification of pathogen-specific nucleic acids, a result that is very informative in terms of the composition and abundances of the pathogens in the sample. The technology of microarray is more relevant in the identification of known pathogens and monitoring changes in the populations of pathogens over time.

21.4 NGS

The NGS techniques newly developed and commercialized revolutionized the high-throughput sequencing of nucleic acids at high speed. Illumina and PacBio can now run millions of DNA fragments in parallel, giving the opportunity to analyze multiple samples at the same time. NGS metagenomic sequencing-based methods provide a detailed view of genetic diversity and composition within microbial communities associated with plant disease. Pathogen genome sequencing also enables the identification of molecular markers associated with virulence, pathogenicity, and resistance to drugs in a direct manner, hence guiding the development of focused strategies for their control.

21.4.1 HIGH-THROUGHPUT PHENOTYPING

High-throughput phenotyping techniques, in turn, enable the fast and noninvasive estimation of plant traits related to disease resistance [1]. The technologies make use of advanced imaging technologies such as spectral imaging and infrared thermography to quantify various physiological and morphological plant traits. As a result, high-throughput phenotyping platforms allow for the screening of large populations of plants under controlled conditions in a bid to facilitate the identification of resistant germplasm and the characterization of plant–pathogen interactions. High-throughput phenotyping becomes very important for the acceleration of breeding programs aimed at developing disease-resistant crop varieties (Figure 21.2).

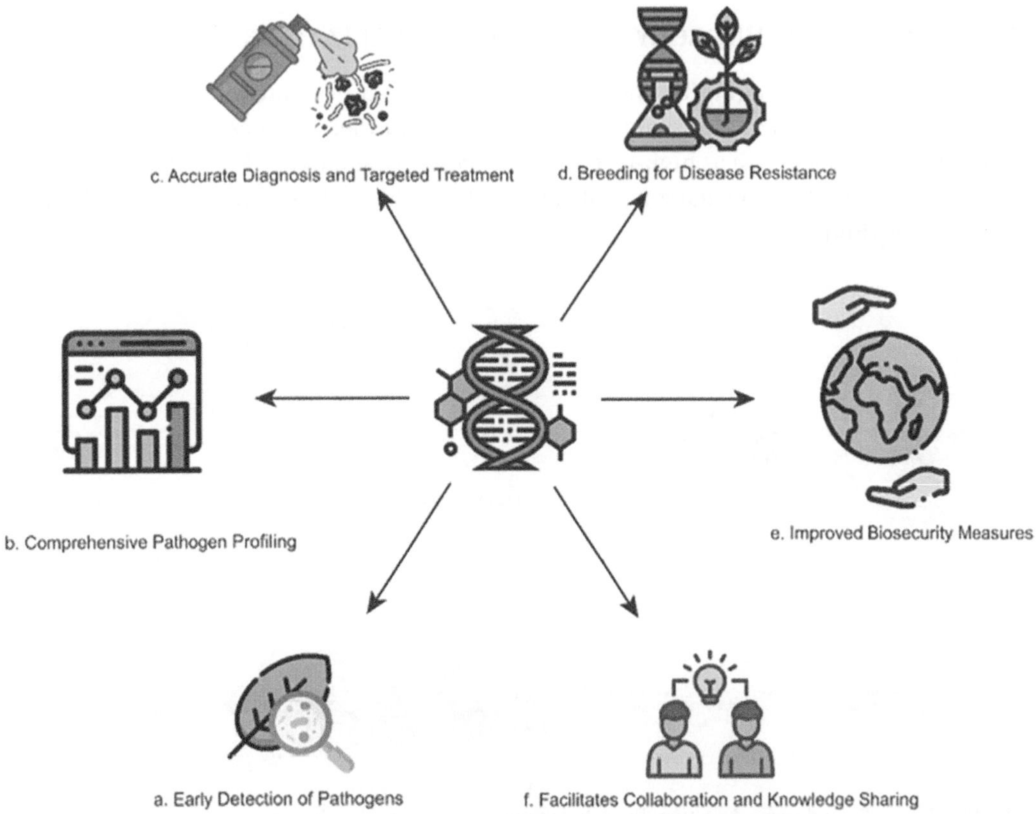

FIGURE 21.2 Illustration depicting management of plant diseases through high-throughput sequencing.

Source: Reprinted from Ref. [14]. http: //creativecommons.org/licenses/by/4.0/

21.5 MULTIPLEX IMMUNOASSAYS

The multiplex immunoassays techniques make it possible to identify multiple plant pathogens at the same time were identified by using antibody-based detection tools [12]. These assays employ the microbead-based techniques for the identification and quantification of the pathogen-specific antibodies in order to detect their presence in plant samples. There are benefits associated with multiplex immunoassays including high sensitivity, proven specificity, as well as high throughput performance, which can allow the analysis of large volumes of plant samples in diagnostic laboratories or in the field. The immunological techniques are more effective when it comes to the identification of the plant pathogens that include virus, bacteria, and fungi from different plants.

One very important role of high-throughput screening methods is that such system can take samples of plants and perform quick and accurate results in identifying the presence of pathogens. It highlights the methodologies involved in recognition of pathogen richness, density, and dispersion aiming at early identification and successful management of plant diseases. This information can allow researchers and growers to develop efficient disease management approaches for each crop based on the results of high-throughput screening methods which would reduce crop yield losses and promote sustainable agriculture.

21.6 IMAGING TECHNOLOGY FOR DISEASE VISUALIZATION

Recent advances in imaging technologies have significantly improved our understanding of plant–pathogen interactions and hold promise for further advances in plant pathology research. Techniques such as high-resolution microscopy and live cell imaging have proven to be powerful tools for visualizing subcellular dynamics and dynamic processes in living plant tissues (Figure 21.3).

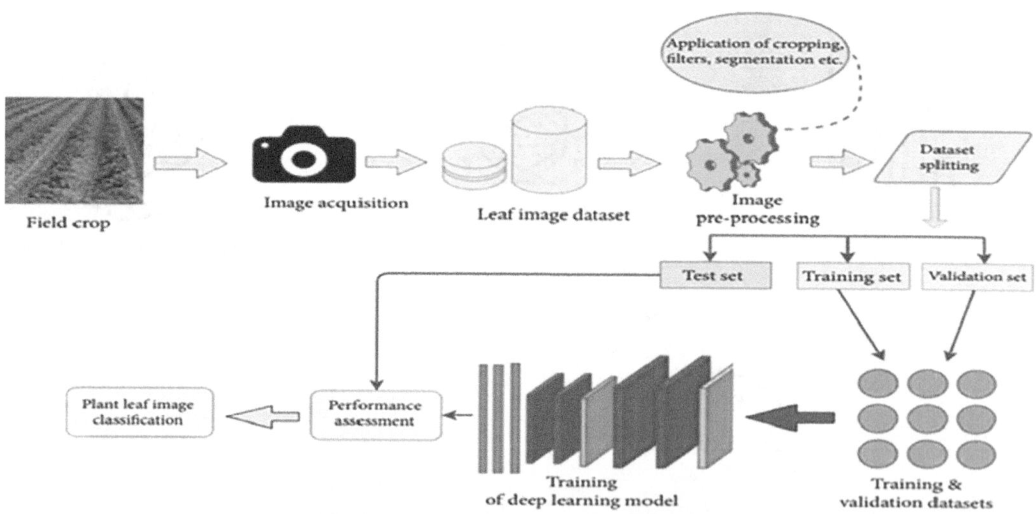

FIGURE 21.3 Plant disease detection through computer vision-based techniques.

Source: Reprinted from Ref. [15]. http: //creativecommons.org/licenses/by/4.0/.

21.7 SUPER-RESOLUTION MICROSCOPY

Super-resolution microscopy techniques, which include structured illumination microscopy and stochastic optical reconstruction microscopy (typhoon), have surpassed the diffraction restriction of traditional light microscopy, allowing researchers to visualize subcellular structures with extraordinary resolution [7]. These techniques make use of advanced optical standards and computational algorithms to reap resolutions at the order of tens of nanometers, taking into account distinct visualization of organelles, protein complexes, and molecular interactions inside plant cells. exceptional-decision microscopy has the potential to revolutionize our expertise of plant–pathogen interactions by means of revealing formerly unseen details of pathogen invasion strategies, host defense mechanisms, and mobile responses to infection [7].

21.8 LIVE-CELL IMAGING

In vivo imaging allows the real-time visualization of plant tissues by exploiting the natural metabolism of the sample, with many applications for investigating the ontogenesis of diseases and for analyzing the relationship between a plant and a pathogen at different stages of development [10]. Some of the techniques that are employed include the use of light and video of cells to determine how cells react to infections and illnesses and even get diseases. According to Ref. [10], live-cell imaging techniques help researchers monitor plant–pathogen interactions and document the changes in real time, which in turn helps them understand how plants develop resistance or vulnerability to diseases. This is because newer applications like high-resolution microscopy and live cell imaging that enable higher resolution visual imaging of cells and molecules in the plant diseases are available now. These methods can be used to clearly and precisely analyze variations of things and their relationship to each other. Imaging has provided significant inputs into understanding the dynamics of pathogen invasion, plant defense, and pathogen development, thereby becoming an extended framework in the fight against disease management strategies (Figure 21.4).

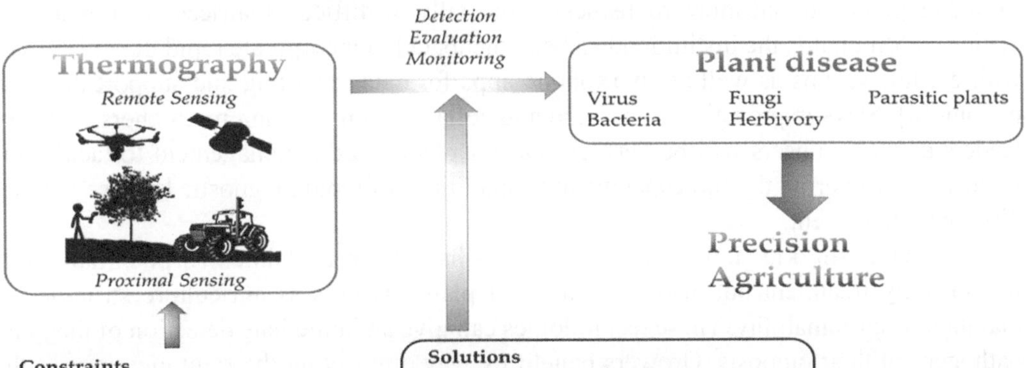

FIGURE 21.4 Illustration depicting plant imaging technologies for disease visualization in various crop plants.

Source: Reprinted from Ref. [16]. Copyright © 2020 by the authors. Licensee MDPI, Basel, Switzerland. http://creativecommons.org/licenses/by/4.0/

21.8.1 APPLICATIONS OF RAPID DIAGNOSTIC KITS AND POINT-OF-CARE DEVICES

A diverse range of diagnostic kits for rapid diagnostics and point-of-care devices are increasingly being used for disease management in plants such as plant disease surveillance, early detection, and disease outbreak management. These techniques are miracles of scientific development and as they aid in cases of remote places or resource restrictions which can make lab accessibility very difficult. Growers will be able to conduct testing on site by using the rapid diagnostic kits and thus can make decisions nearly immediately and adopt mitigation measures very fast [8].

21.9 CONTRIBUTIONS TO SUSTAINABLE AGRICULTURE

RID kits and handheld devices turn into valuable tools which support both the detection of diseases and the chemical-free methods of fight the pests. The pathogen detection on the early stage is helping the growers to take timely actions, such as crop rotation, quarantine measure, and targeted biocontrol agents appalling which will finally minimize the crop loss and have little environmental footprint [8]. It continues to be a very important part of quarantine checks as rapid diagnostic kits can also help prevent the epidemic outbreak of quarantine pests and diseases by enabling timely detection and confinement of outbreaks [8].

21.10 CHALLENGES AND FUTURE DIRECTIONS

Although the rapid diagnostic kits and the point-of-care devices have an ingrained potential, they are still met with problems concerning, limited sensitivity and specificity as well as shelf life. However, more research should be performed to update these technologies to higher precision and more robustness, especially in difficult matrices such as a field setting. Additionally, the institution has been incubated to incorporate rapid diagnostic kits with digital systems as well as smartphone apps for data recording and support decision making [8]. Stays in proactive innovation and collaboration, through researchers, industry leaders as well as users will be very critical for plant disease management to tackle this issue and discovering the full capacity and capabilities of rapid diagnostic kits use in plant disease management.

Fast diagnostic kits and point-of-care tools have been acclaimed as potential assets for not only treatment but also prevention of plant diseases as agriculture struggles to maintain its sustainability. These technologies can give an immediate detection of the plant pathogens at their outposts. Growers benefit by way of enacting the right measures at the right time and by doing so, informed decisions are made. While achievements are made in the development and improvement of (diagnostic kits[entities]) and point-of-care devices, these tools have the ability to disrupt plant disease management systems as they play a role in global food security (Figure 21.5).

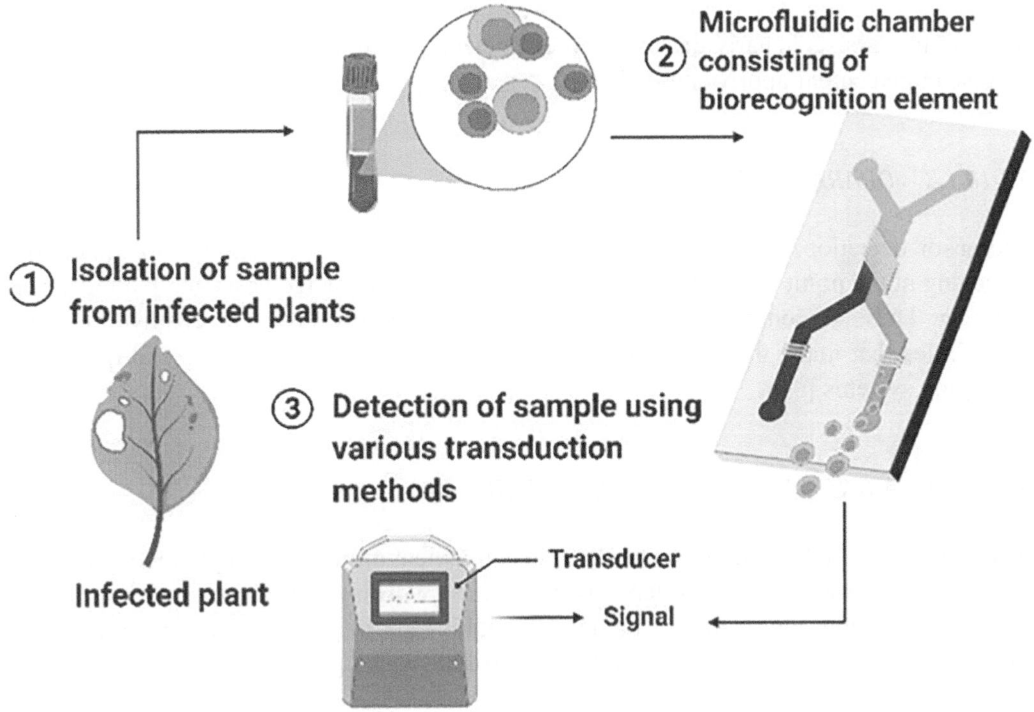

FIGURE 21.5 Illustration depicting Rapid diagnostics of plant diseases.

Source: Reprinted from Ref. [17]. http: //creativecommons.org/licenses/by/4.0/

21.10.1 BIOSENSORS AND NANOTECHNOLOGY IN DISEASE DETECTION

Sensing devices ranging from encapsulated biosensors to the nanoscale sensors permit quick, sensitive identification, and specific identification of pathogens in plant material that is tested. This part investigates the manifold avenues of utilizing biosensors and nanotechnology in the realm of plant disease detection, assorted examples take from research.

21.10.1.1 RAPID ON-SITE TESTING

Detection of plant pathogens via biosensors and nanotechnology-based detection platforms provides a rapid on-site facial for testing plant samples. These tools display the information in real-time for the farmers and experts to make immediate decisions and to carry out good disease management plan following the results [4].

21.10.1.2 DISEASE SURVEILLANCE

Sensors and nanotech enable the surveillance of pathogen populations in agro-watersheds that is made possible by tracking environmental situations, thus leading to hereditary detection. The monitoring of biosensor systems in a continuous mode is a timely measure that assists in early disease outbreak detection and, thus, introduces proactive and disease management approaches [4].

21.10.1.3 QUARANTINE SCREENING

Biosensor detection techniques employing nanomaterials will be applied for quarantine screening at the point of plant importation to ensure that invasive pests and pathogens are not near. These biosensors are characterized with relatively fast and sensitive detection capabilities that make them more efficient compared to other identifying and quarantine managing systems [4].

21.10.1.4 MONITORING PATHOGEN SPREAD

Nanosensors embedded in biosensors enable convenient monitoring of pathogen evolution in plantation agricultural ecosystems as it happens. Working with generating information about the location and routes of pathogens enables getting a better understanding how diseases work and creating corresponding control measures [7].

21.10.1.5 EARLY DISEASE DIAGNOSIS

Nanomaterial-based biosensors that are highly sensitive are one of the biosensors features that have been capable of detecting diseases at a much lower concentration than any other sensor type. Fast disease diagnosis and reduced risk of crop damage can be achieved through biosensors, thus corrective actions can be done timely before the crop fails economically [4].

21.10.1.6 PRECISION DISEASE MANAGEMENT

Sensors and nanotechnology lend themselves to the precision disease management by permitting the precision and localized information about pathogen presence and localization. Such data is highly useful for the purpose of formulating site-specific interventions which includes enhanced application of site-specific pesticides and resource optimization coupled with minimum environmental impact [8]. Biosensors and nanotechnology play pivotal roles in various aspects of plant disease detection, offering versatile solutions for rapid, sensitive, and specific detection of pathogens in agricultural settings. These technologies hold promise for enhancing disease surveillance, enabling early diagnosis, and supporting precision disease management practices, ultimately contributing to sustainable agriculture and global food security (Figure 21.6).

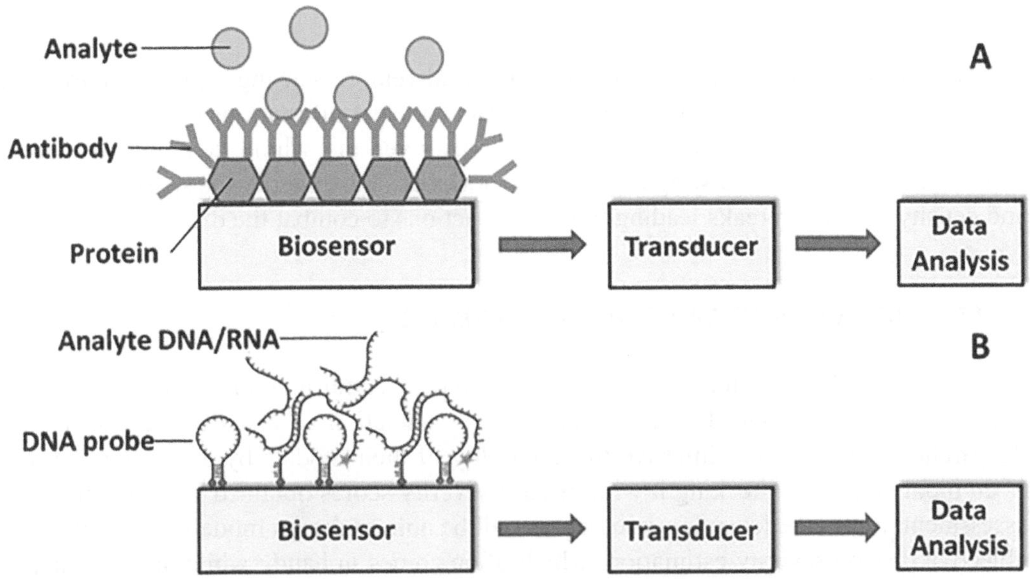

FIGURE 21.6 Illustration depicting use of biosensors for detection of plant diseases.

Source: Reprinted from Ref. [18]. https: //creativecommons.org/licenses/by/4.0/

21.10.2 REMOTE SENSING AND SPECTROSCOPY FOR DISEASE IDENTIFICATION

The use of remote sensing and spectroscopy in plant pathology has changed the science profoundly in two major ways: nondestructive and efficient methods of pathogen identification and monitoring become available to the researchers. In this particular part, principles, appliances, and contributions of remote sensing and spectroscopy in disease identification are gracefully developed bringing in different original articles in support.

Remote sensing is a technique where sensors gather data about the particular object or phenomenon in question without direct contact. The analysis of radiation signatures known as spectroscopy is the main technical solution used by remote sensing techniques to ascertain the detailed structure of the sample being investigated. The remote sensing/spectroscopic methods, involve electromagnetic radiation within varying wavelengths and have distinct spectral professionals of leaves infected with different plant diseases [5].

21.11 APPLICATIONS OF REMOTE SENSING AND SPECTROSCOPY IN DISEASE IDENTIFICATION

21.11.1 DISEASE STRESS DETECTION

The remote sensing techniques like hyperspectral imaging and multispectral imaging allow us to identify the plant response to the stressor on the basis of the phenomenon called plant-induced spectral changes. Vegetable pathogens manifest atypical spectral reflectance properties, which are applied for disease detection and monitoring [5].

21.11.2 MAPPING DISEASE DISTRIBUTION

The utilization of spectrometry and satellite-based remote sensing equipment for crop scouting makes surveillance of disease outbreaks and pest infestations possible. Through analyzing spectral reflectance collected not only over the whole area but also more precisely, researchers can distinguish "disease hotspots" and quantify the spatial expansion and density of the outbreaks leading to targeted actions to combat the disease [5].

21.11.3 QUANTIFICATION OF DISEASE SEVERITY

The specific molecular methods, for example, hyperspectral imaging accompanied with machine learning, enable, however, quantitative evaluation of disease status in plants. Researchers can gradually improve the capability of their models by using spectral data as an input and cross-checking it with disease severity scores obtained from an empirical assessment. So, in the long run, researchers will be able to design models for reliable and objective disease severity estimation, which often comes in handy while taking management decisions [5].

21.12 CONTRIBUTIONS TO DISEASE MANAGEMENT

Satellite data analysis and hyper-spectroscopy technologies enable tailored disease progress monitoring and timely support of disease management decisions. They assist in the detection of disease early on, the mapping of diseases and the evaluation of disease severity quantitatively, increasing the effectiveness of disease control interventions, which may include pest monitoring and site-specific control measures, thereby minimizing crop losses and conserving resources by using them efficiently [5].

In summary, remote sensing imagery and spectroscopy can play a major role in agricultural plant diseases which include the quick, noninvasive, efficient, and geospatial/localized monitoring of plant health conditions. These approaches aid to recreate the phenomenon of disease movement and to enter the intervention phase at the early stage of the disease course. These methods improve management of diseases with precision, which in turn boosts sustainability of crop production and food security.

21.12.1 INTEGRATION OF ADVANCED DIAGNOSTIC TOOLS IN PLANT DISEASE MANAGEMENT

1. **Precise Pathogen Identification:** Modern diagnostic apparatuses including molecular diagnostic strategies and high-throughput screening methods offer an opportunity to define the plant pathogens quite thoroughly on, and at, the molecular level [11]. Researchers and growers can now precisely diagnose the infectious pathogens in plant samples through the usage of molecular analysis

tools—the identification of genetic markers or particular DNA sequences unique to pathogens.

2. **Understanding Pathogen Diversity and Evolution:** Genetic diversity and evolution can be determined by molecular technologies including DNA sequencing techniques [3]. Through characterization of the genetic of the pathogens, scientists can comprehend their evolutionary characteristics, detect new strain emergence, and predict on shifts in pathogen's epidemiology.
3. **Real-time Disease Monitoring:** The use of the rapid diagnostics kits and point-of-care devices makes it possible for fast assessment of diseases on site, allowing the growers to monitor outbreaks cases instantly [8]. These cars have user-friendly kits which gives quick and correct output, and hence crop losses can be curbed through timely disease outbreaks management.

21.13 BENEFITS OF INTEGRATION

1. **Targeted Control Measures:** The precision in pathogen identification through advanced diagnostic devices gives growers more options in narrow strategies, such as a targeted measure that is hopefully effective [11]. Through different pathogens identification and the genetic composition knowledge, growers could choose appropriate ways to make management like resistant crop variety or precision use of pesticide.
2. **Reduced Environmental Impact:** Advanced diagnostics strategy leading to the most efficient use of pesticides contribute to the smallest environmental effects during the disease management process (according to [11]). Undoubtedly, this technology has a potential to save the environment by reducing the overall intensity of pesticide products because there won't be that much chemical runoff and environmental pollution.

Although the incorporation of such high-performance diagnostic tools into disease control enjoys many advantages, it also has some difficulties. Among these are the requirements for cheap and suitable for the people diagnostic technologies, and the incorporation of diagnostic data on disease management decision-making systems in real time [2]. Meeting those very challenges requires the longer-term commitment of researcher's creativity along with stakeholder industry and the government policymakers.

Ultimately, the assistance of advanced diagnostic systems in the process of plant disease management permits us to work out how pathogens' interactions occur, determine the dynamics of disease and eliminate the need to install inefficient control and pest management measures. The use of molecular diagnostics and other rapid methods enables crops damage to be minimized, reduces environmental impact, and promotes sustainable agriculture. Therefore, unceasing research and efforts in the field of exact disease identification and management will be paramount in overcoming these challenges and driving the area forward.

21.13.1 CHALLENGES AND FUTURE DIRECTIONS

The rapid advancement of diagnostic tools and techniques has transformed the landscape of precision disease identification in agriculture. However, along with these advances come challenges and the need for further development to fully realize their potential. This section examines the current challenges and outlines future directions for precision disease identification, supported by relevant citations from the literature.

21.14 CHALLENGES IN PRECISION DISEASE IDENTIFICATION

1. **Cost and Accessibility**: Most of the sophisticated diagnostic procedures, for example, molecular methods and high-throughput screening technologies, are very costly and imposes the need of scarce and pricey equipment as well as highly trained workforce in order to execute them [2]. The distance the transportation is greatly affected. The resources limited and small-scale growers are the most affected.
2. **Complexity of Data Interpretation**: Complex data from the use of advanced diagnostic tools had to be analyzed by sophisticated bioinformatics means in the process [2]. Interpretation of full intelligence and useful insights from the large amounts of data is one of the many obstacles which are most likely to fall within the responsibility of field workers and the practitioners who lack formal training to fully understand data analysis methods.
3. **Integration with Decision Support Systems:** Although sophisticated monitoring tools are of great importance for figuring out the pathogens' existence and genetic characteristics, the big question is how to incorporate the gathered data into the decision support systems that will provide insight during real-time disease management [2]. An important aspect of this process is the creation of user-friendly systems which can be used to collect disease information from the diagnostics, and can as well generate applicable directions to the farmers for control of the disease.

21.15 FUTURE DIRECTIONS IN PRECISION DISEASE IDENTIFICATION

21.15.1 DEVELOPMENT OF AFFORDABLE TECHNOLOGIES

There is a strong demand for sensing tools, which are affordable, portable, and easy-to-use and this can be realized through the implementation of the early adoption of the technologies by growers and practitioners [2]. This encompasses a wide range of technologies involving development of low-cost molecular diagnostics, smartphone-based platforms, POC, precision machines that deliver accurate and rapid results in the field.

21.15.2 ADVANCEMENTS IN DATA ANALYTICS

The proper development and application of new data analytic and artificial intelligence tools are therefore required to improve the efficiency of the diagnostic data analysis and interpretation [2]. Machine learning algorithms and data-driven applications can help discover patterns and trends among the most difficult datasets, and thus can lead to improved disease management.

21.15.3 INTEGRATION WITH PRECISION AGRICULTURE TECHNOLOGIES

The diagnostic technologies, which will be integrated with remote sensing and automated monitoring systems, will implement a new concept namely, precision agriculture with the purpose of better identification and repair of diseases [2]. This integration can provide the real-time monitoring of the changing dynamics of the diseases and aid in the immediate actions at the spot level.

In respective to the precision plant disease detection techniques, there is a significant improvement in our ability to operate and manage the plant diseases through rapid detection, but these techniques are far from resolution. With specific attention to economical technologies innovation, big data analytics as well as adoption of precision agriculture methodologies, the obstacles can be overcome easily and control strategies in Agri faring would be further enhanced. Throughout research and linkage of disciplines that is to come will be a logical step to innovation driving and harvesting of the full realization potent of precision diseases identification. This then becomes necessary in the guarantee of the global food security.

21.16 CONCLUSION

The evolution of diagnostic equipment and procedures has transformed our capacity to discover and make a history-altering conclusion about plant diseases with a precision and accuracy never seen before. One example is the molecular methodology and remote sensing technology shaping the modern approaches of precision disease identification. With help of these newer approaches' farmers, researchers and growers can learn to understand the disease spread, develop the specific measures of control and prevent the emerging of global food security issues.

KEYWORDS

- **precision disease identification**
- **polymerase chain reaction**
- **next-generation sequencing**
- **loop-mediated isothermal amplification**
- **high-throughput screening technique**

REFERENCES

1. Fiorani, F., & Schurr, U. (2013). Future scenarios for plant phenotyping. *Annual Review of Plant Biology*, *64*, 267–291.
2. Gupta, S. K., et al. (2022). Challenges and future directions in precision disease identification. *Annual Review of Phytopathology*, *60*, 1–22.
3. Jones, C. M., et al. (2018). Genomic-based molecular techniques for rapid and accurate detection of plant pathogens. *Phytopathology*, *108*(5), 512–518.
4. Lee, W. M., et al. (2019). Biosensors and nanotechnology: A powerful combination for plant disease detection. *Trends in Biotechnology*, *37*(9), 1013–1026.
5. Mahlein, A. K., et al. (2020). Remote sensing for crop health management. *Plant Disease*, *104*(5), 1149–1162.
6. Roberts, P. D., et al. (2017). Imaging technologies for plant phenotyping: A review. *Sensors*, *17*(9), 1–29.
7. Shroff, H., et al. (2015). Super-resolution microscopy and tracking of fluorescent molecules in cells with high spatiotemporal resolution. *Current Opinion in Neurobiology*, *34*, 1–7.
8. Singh, R., et al. (2021). Rapid diagnostic kits for on-site detection of plant pathogens: Recent developments and challenges. *Frontiers in Plant Science*, *12*, 1–14.
9. Smith, A. B., et al. (2020). Advances in molecular diagnostic techniques for plant disease management. *Plant Disease*, *104*(5), 1124–1136.
10. Sparkes, I. A., et al. (2010). Live-cell imaging of cytosolic calcium levels in Arabidopsis guard cells. *Plant Journal*, *6*(2), 421–431.
11. Thomson, S. J., et al. (2018). Integration of diagnostic tools into plant disease management: A review. *Agricultural Systems*, *159*, 123–132.
12. Wang, D., et al. (2019). High-throughput sequencing technologies for plant disease diagnostics. *Phytopathology*, *109*(4), 716–725.
13. Mehetre, G. T., et al. (2021). Current developments and challenges in plant viral diagnostics: A systematic review. *Viruses*, *13*(3), 412.
14. Nizamani, M. M., et al. (2023). High-throughput sequencing in plant disease management: A comprehensive review of benefits, challenges, and future perspectives. *Phytopathology Research*, *5*, 44.
15. Demilie, W. B. (2024). Plant disease detection and classification techniques: A comparative study of the performances. *Journal of Big Data*, *11*, 5.
16. Pineda, M., et al. (2021). Thermal imaging for plant stress detection and phenotyping. *Remote Sensing*, *13*, 68.
17. Patel, R., et al. (2022). A review of recent advances in plant-pathogen detection systems. *Heliyon*, *8*, 12.
18. Fang, Y., & Ramasamy, R. P. (2015). Current and prospective methods for plant disease detection. *Biosensors*, *5*, 3.

Index

B

H

I

J

K

L

M

T

U

V

W

Y

Z

For Product Safety Concerns and Information please contact our
EU representative GPSR@taylorandfrancis.com Taylor & Francis
Verlag GmbH, Kaufingerstraße 24, 80331 München, Germany